Jessica F. Brinkworth • Kate Pechenkina
Editors

Primates, Pathogens, and Evolution

Editors
Jessica F. Brinkworth
Department of Pediatrics
CHU Sainte-Justine Research Center
University of Montreal
Montreal, QC, Canada

Kate Pechenkina
Department of Anthropology
Queens College
City University of New York
Flushing, NY, USA

ISBN 978-1-4899-9479-0 ISBN 978-1-4614-7181-3 (eBook)
DOI 10.1007/978-1-4614-7181-3
Springer New York Heidelberg Dordrecht London

© Springer Science+Business Media New York 2013
Softcover re-print of the Hardcover 1st edition 2013
This work is subject to copyright. All rights are reserved by the Publisher, whether the whole or part of the material is concerned, specifically the rights of translation, reprinting, reuse of illustrations, recitation, broadcasting, reproduction on microfilms or in any other physical way, and transmission or information storage and retrieval, electronic adaptation, computer software, or by similar or dissimilar methodology now known or hereafter developed. Exempted from this legal reservation are brief excerpts in connection with reviews or scholarly analysis or material supplied specifically for the purpose of being entered and executed on a computer system, for exclusive use by the purchaser of the work. Duplication of this publication or parts thereof is permitted only under the provisions of the Copyright Law of the Publisher's location, in its current version, and permission for use must always be obtained from Springer. Permissions for use may be obtained through RightsLink at the Copyright Clearance Center. Violations are liable to prosecution under the respective Copyright Law.
The use of general descriptive names, registered names, trademarks, service marks, etc. in this publication does not imply, even in the absence of a specific statement, that such names are exempt from the relevant protective laws and regulations and therefore free for general use.
While the advice and information in this book are believed to be true and accurate at the date of publication, neither the authors nor the editors nor the publisher can accept any legal responsibility for any errors or omissions that may be made. The publisher makes no warranty, express or implied, with respect to the material contained herein.

Printed on acid-free paper

Springer is part of Springer Science+Business Media (www.springer.com)

Developments in Primatology: Progress and Prospects

Series Editor: Louise Barrett

For further volumes:
http://www.springer.com/series/5852

For Jordan Aria and Jeremy

Acknowledgements

First, we must thank the 30+ authors who accepted our invitation to contribute to this book. This volume unites researchers from a wide range of biological fields including Anthropology, Biochemistry, Evolutionary Biology, Genetics, Immunology, Medicine, Veterinary medicine, Virology, and Zoology. These authors provided the collection of papers that examine the molecular interactions between primates and pathogens within the context of evolution contained within, and did so while juggling many other responsibilities. The publication of an edited book is a long process. Many of the authors represented here agreed to work with us as early as 2009, when we were first recruiting speakers for a symposium at the annual meeting of the American Association of Physical Anthropologists. We thank these authors for contributing their time and effort to write interesting works and for allowing us to curate such works here. Above all, we thank them for their support and patience.

A very special thanks to our colleagues who offered anonymous review of the chapters and who must go unnamed. Each chapter in this collection was critically examined by 2–3 researchers. Many of our peers provided detailed reviews on tight deadlines. Some even sent reviews from the field!

Thanks to our editor Janet Slobodien who approached us and encouraged us to pursue this book. Melissa Higgs, editorial assistant, guided the assembly of this book, and often helped us with the fine details of assembling figures and permissions. Thanks to Lesley Poliner and Hedge Ritya, who oversaw the production and copy editing of this volume and its supporting materials. Yurii Chinenov provided critical comments on multiple chapters. Thank you Jeremy Sykes for editorial assistance and for proofing drafts.

The initial idea to develop a collection of papers that would discuss molecular host–pathogen interactions came to us while attending a number of talks and posters across scattered presentation sections at the annual meeting of the American Association of Physical Anthropologists in Columbus, Ohio, in 2008. We wanted to provide a forum for researchers interested in the evolution of primate immunity to meet, discuss findings and brainstorm. At the 2009 AAPA meetings, we began to approach potential contributors to a symposium and edited volume that would focus on the functional outcomes of evolutionary primate–pathogen interactions. We were very fortunate to be met with great enthusiasm by our future contributors, in particular George Armelagos who immediately suggested we include the research of Graham Rook, Jenny Tung, and Kristin Harper. The original participants of the "Pathogens and evolution of human and non-human primates" symposium, held in Albuquerque, New Mexico, in April of 2010, presented some truly interesting interdisciplinary works that day. We thank those researchers for sharing their ideas and enthusiasm - George Armelagos, Nels C. Elde, Harmit Malik, Cedric Feschotte, Charlie Nunn, Kristin Harper, Jayne Raper, Jenny Tung, Susan C. Alberts, Gregory A. Wray, Felicia Gomez, Wen-Ya Ko, Sarah Tishkoff, Ajit Varki, Melanie Martin, Caleb Finch, Fabian Crespo, Rafael Fernandez-Botran, Manuael Casanova, and Christopher Tilquist. Thank you to the American Association of Physical Anthropology and the Human Biology Association who hosted this symposium at their annual meetings in Albuquerque, New Mexico, in April 2010. A very special thanks to the donors, particularly members of the law firm Labaton Sucharow, who provided funds and consideration associated with the needs of this specific symposium. A special thanks to Kelly Zieman, Leslie and Sharon Brinkworth, Cheryl and Charles Brinkworth, Deirdre O'Boy, Emerson McCallum and Michael Donnelly.

Thank you to Chris Brinkworth, Mary Wong, Eva and JoAnn Brinkworth and my parents, Cheryl and Charlie, for help and support during key stages of this book. Special mention to Jenny Tung, who was an early supporter of and contributor to this project. Thanks Luis Barreiro who offered assistance at an important stage of production and who, with Jenny, has provided me the opportunity to develop a career working on these questions of primate evolution and immunity. My deepest gratitude to my husband, Jeremy, for his patience, energy and unreserved enthusiasm for all things book, career, life. Thank you for finding symposium donors, batting around ideas, reviewing many proposal drafts and working extra hard on all other tasks so that I could complete this one. Very special thanks must be given to my daughter, Jordan, who was a considerate traveling companion and made this process rather easy. My dearest dear, thank you for all days past, present and future.

Over the course of production Brinkworth and Pechenkina's studies were supported by the Wenner-Gren Foundation Grant (7845 and 8702, JFB), the National Science Foundation (0752297, KP and JFB), the Réseau de Médicine Génétique Appliquée (JFB) and the National Institutes of Health (1R01-GM102562 to Luis Barreiro, which supports JFB). Thank you also to the City University of New York, Queens College, City College of New York, Sophie Davis School of Biomedical Education, the New York Consortium in Evolutionary Primatology, Centre Hospitalier Universitaire Sainte-Justine Research Center, and the University of Montreal for their support and the opportunity to pursue our interests in the evolution of disease and immune system function.

Montreal, QC, Canada Jessica F. Brinkworth

Contents

Primates, Pathogens and Evolution: An Introduction

Jessica F. Brinkworth and Kate Pechenkina

Introduction

Primate immune systems have evolved to interact with pathogens in different ways (Mandl et al. 2008, 2011; Pandrea et al. 2007; Sawyer et al. 2004; Song et al. 2005; Soto et al. 2010). Human and nonhuman primate immune systems have diverged with some species exhibiting strong differences in immune response to certain pathogens including immunodeficiency viruses [reviewed by Pandrea and Apetrei (2010), *Toxoplasma gondii* (Epiphanio et al. 2003), herpesviruses (Estep et al. 2010; Huang et al. 1978), and trypanosomes (Thomson et al. 2009; Welburn et al. 2001)]. Why closely related primates have evolved such divergent pathogen interaction strategies is not well understood. Despite strong public interest in human/nonhuman primate evolutionary history and the importance of various primate species as biomedical models, the current picture of interspecies differences in immunity remains fairly incomplete. Our understanding of how primate immunity evolved is hindered by disconnected research on primate–pathogen molecular interaction, an uneven focus on primate coevolution with a limited number of pathogens, and the disconnect between research on primate molecular phylogeny and primate physiology. The objective of the present collection of papers is to integrate research on the evolution of primate genomes, primate immune function, primate–pathogen biochemical interaction, and infectious disease emergence to provide a knowledge base for future research on human and nonhuman primate speciation, immunity, and disease.

J.F. Brinkworth (✉)
Department of Pediatrics, CHU Sainte-Justine Research Center, University of Montreal, Montreal, QC, Canada
e-mail: jfbrinkworth@gmail.com

K. Pechenkina
Department of Anthropology, Queens College, City University of New York, Flushing, NY, USA
e-mail: ekaterina.pechenkina@qc.cuny.edu

J.F. Brinkworth and K. Pechenkina (eds.), *Primates, Pathogens, and Evolution*, Developments in Primatology: Progress and Prospects,
DOI 10.1007/978-1-4614-7181-3_1, © Springer Science+Business Media New York 2013

The Role of Long-Term Evolutionary Processes in Shaping the Primate Immune System

Evolution of the immune system is tied to the evolutionary history of a species and intertwined with the evolution of physiological functions and developmental stages of an organism. The majority of cold-blooded vertebrates appear to experience functional shifts in their immune response depending on external temperatures, resulting, in some cases, in impaired immune responses such as the inhibition of immunoglobulin class switching (Jackson and Tinsley 2002). The importance of a graft-rejection-like immune response during tadpole to adult morphogenesis in the frog genus *Xenopus* and the failure of novel proteins associated with lactation to stimulate "nonself" immune responses in mammals, for example, both suggest that the immune system has likely evolved in parallel with the evolution of species developmental stages (Izutsu 2009; Matzinger 1994, 2002).

To interpret variation in primate immune response within and between species, the evolutionary forces that shaped the underlying molecular differences need to be examined. Of these forces, pathogen-mediated natural selection has likely been the leading factor in increasing the frequency of pathogen resistance in host populations. Factors such as an organism's environment, diet, postural behavior, and sociality led to interspecies differences in exposure to specific pathogens and the frequency with which such pathogens were encountered. Complete or partial resistance of specific human genotypes to certain pathogenic strains and the patterned distribution of these genotypes among indigenous populations around the globe is fairly well documented (Hamblin and Di Rienzo 2000; Hraber et al. 2007; Leffler et al. 2013; Marmor et al. 2001; Tishkoff et al. 2001). While the genetic variability of nonhuman primate immune factors is comparatively less well studied, specific disease resistance genotypes in some of these species have been identified. Ecological differences between savanna dwelling Guinea baboons *(Papio papio)* and chimpanzees (*Pan troglodytes*) are likely responsible for resistance on the part of the former species to the savannah-based pathogen *Trypanosoma brucei gambiensis*, and high susceptibility to fatal *T. brucei*-caused sleeping sickness in the latter. Baboons are more involved in grassland ground foraging than chimpanzees, which also exploit savannah resources but are tied to forested regions and tend to retreat to the forest for sleep (Kageruka et al. 1991). In the course of their evolution in open habitats, baboons have likely been continuously exposed to the low flying tsetse fly—the vector for *T. brucei* (Lambrecht 1985; Welburn et al. 2001). The selective pressure generated by *T. brucei* is thought to have favored fixation of two nonconservative mutations in the baboon trypanosomal lytic factor *ApoLI* that have been linked to sleeping sickness resistance and are not shared with either humans or chimpanzees (Thomson et al. 2009). In humans, two unrelated mutations in *ApoLI* that increase sleeping sickness resistance are geographically restricted to Africa and absent in European populations. Interestingly, trypanolytic variants of *ApoLI* in humans contribute to an increased risk of kidney disease, providing an example of heterozygous advantage (Genovese et al. 2010) parallel to the classic case of the HbS variant of the HBB globin locus and other hemoglobinopathies that confer heterozygous advantage by means of increasing resistance to malaria(Allison 1956; Haldane 1949; Hedrick 2004; Kwiatkowski 2005; Lapoumeroulie et al. 1992; Oner et al. 1992).

The pathogen-directed evolutionary mechanisms that contribute to the divergence of immune systems remain hypothetical. In the simplest case, an epidemic of a highly virulent infectious disease eliminates a large number of susceptible organisms very quickly, thereby favoring the disproportional reproductive success of resistant individuals. In such a scenario pathogen-mediated selection is assumed to be strong. However, such a model of host–pathogen coevolution is pertinent in extreme cases only. Pathogens of moderate-to-low virulence can also affect host immune allele frequencies by, for example, contributing to lowered fertility in the form of physical inability to produce offspring after being infected or causing a lesser ability to acquire mates as a result of decreased mobility (Cheney et al. 1988; Levin et al. 1988). Pathogens that do not kill a host, therefore, can affect sexual selection and gene flow. Of course, the consequences of a deadly epidemic may *not* be limited to removal of the alleles that contribute to disease susceptibility, such as those of surface antigens recruited by a pathogen to penetrate the host. Such strong selective pressure may also affect alleles that contribute to increased susceptibility to coinfections or to overt immune activity that leads to the development of secondary conditions (e.g., sepsis). Pathogen pressure may also select for resistance traits and in doing so affect the frequency of linked alleles in a population through selective sweeps.

The most popularized perception of host–pathogen evolutionary interaction is the concept of a host–pathogen evolutionary arms race. This idea is derived from the application of Leigh Van Valen's 1973 Red Queen hypothesis to hosts and pathogens (Van Valen 1973). The Red Queen hypothesis proposes that closely associated organisms may coevolve so tightly that the likelihood of extinction for one or the other is constant over geological time. Changes in one species affect the tightly coevolved interface with another species, threatening either species with extinction. As stated by the Red Queen in Lewis Carroll's *Through the Looking Glass*: "Now, here, you see, it takes all the running you can do, to keep in the same place." Tightly coevolved hosts and pathogens have to "run"—that is, evolve quickly—just to maintain a balance and avoid extinction. Within this framework, it is tempting to characterize host–pathogen relationships as beneficial for the host to recognize and either tolerate or eliminate a pathogen, with a pathogen's main recourse being to evade a host's defensive mechanisms. These interactions are thought to result in head-to-head collisions between a host's immune system and a pathogen, leading to selective pressure on the immune system and the pathogen and culminating in coadaptation.

However, some pathogens actually co-opt normal mechanisms of primate immune recognition and the subsequent responses, such as cytokine release, cell trafficking, and tissue destruction, using them to their advantage. *Mycobacterium tuberculosis* (tuberculosis) infection is enhanced by the release of host anti-inflammatory cytokine IL-10 (Redford et al. 2011). Once consumed by a macrophage, *Yersinia pestis* (plague) migrates to its main point of dissemination, the lymph nodes, while acquiring phagocytosis resistance (Oyston et al. 2000; Pujol and Bliska 2003; Zhou et al. 2006). Similarly, HIV can disseminate to the lymph nodes and other regions through antigen-presenting cells (Koppensteiner et al. 2012). Moreover, exaggerated and uncontrolled immune cell responses may result in host death—such is the case of severe sepsis and septic shock triggered by strong innate immune cell recognition of immune system insult (e.g., blood stream infections and injury) (Brown et al. 2006; Murdoch and Finn 2003; Zemans et al. 2009). Rather than acquiring permanent and costly adaptations, some pathogens can escape host immune defenses through transient amplification of a resistance gene that creates tandem arrays aptly named "genomic accordions." Poxviruses encode two factors, E3L and K3L, which inhibit the host antiviral factor protein kinase R (PKR). In viruses lacking E3L, K3L rapidly becomes amplified 10–15-fold via serial duplication, which increases viral fitness. However, the tradeoff for increased genome size is less efficient replication. Remarkably, an expanded genomic array of identical resistance genes in a pathogen increases the probability of emergence, fixation, and spreading of additional K3L mutants with improved host avoidance; these viruses subsequently lose the K3L duplicated array but retain the novel resistance mutation (Elde et al. 2012).

Two limiting factors for pathogen-driven evolution of the immune system are disadvantages rendered to the host by hyper-responsiveness, leading to autoimmune disorders, and inadvertent pressure on symbiotic and commensal organisms. An overactive immune system is responsible for the development of chronic conditions mediated by the immune system itself, such as systemic lupus erythematosus, antiphospholipid syndrome, polycystic ovary syndrome, and diabetes, among others, that negatively affect reproductive fitness and may therefore contribute to immune system evolution (Carp et al. 2012).

Consequently, immune system divergence between species is not driven by host–pathogen interaction alone but is profoundly affected by the living environment, which includes a complex network of interspecies interactions between hosts, symbionts, commensals, and pathogens (Klimovich 2002; Lee and Mazmanian 2010). A species, with its associated microorganisms, can be considered a "holobiont," an evolutionary unit encompassing the totality of organisms involved in commensalistic, symbiotic, and parasitic relations (Zilber-Rosenberg and Rosenberg 2008). Changes in the fitness of any holobiont-involved species affect the system as a whole, rather than just the fitness of the host species. Indeed, changes in the composition of intestinal microbiota may promote outgrowth or invasion by pathogenic microorganisms or induce an exaggerated host response that results in the onset of various autoimmune disorders such as inflammatory bowel disease (Maynard et al. 2012). Alternatively, dietary and environmental changes may cause a shift in species-associated microbiota that results in the appearance of new pathogens or symbionts. Establishing whether new species-specific pathogens are effectors or consequences of speciation is a daunting task.

The Role of Pathogens in Primate Speciation

Interaction with pathogens likely played an important role in primate speciation in several ways. Pathogens have contributed to primate genome divergence through direct integration of microorganismal genomes into the genomes of primate germ-line cells. Over millions of years, viral integration into host genomes has changed genome sequences and affected multiple biological functions (Arnaud et al. 2007; Hunter 2010). For instance, Dunlap et al. (2006) proposed that effective placenta formation in mammals is impossible without a gene coding for an envelope protein that was initially introduced by a retrovirus (HERV-W) (Dunlap et al. 2006). Once incorporated, viral genomes were inherited by offspring. Past retroviral infections of human ancestors now represent approximately 8 % of the human genome (Bannert and Kurth 2006). Due to different histories of pathogen exposure, primate genomes differ from one another in terms of the types and numbers of integrated viral sequences (Horie et al. 2010; Kim et al. 2008). As such, viral pathogens have contributed to the divergence of primate genomes and the divergent functions of primate genes (Gogvadze et al. 2009; Wang et al. 2007; Yohn et al. 2005).

While portions of primate genomes have diverged because of species-specific viral integration, some loci appear to have evolved under pathogen-driven selection. Multiple pathogens have been identified as having exerted selective pressure on primate immune factors for millions of years [i.e., retroviruses and apolipoprotein B-editing catalytic polypeptide 3G (APOBEC3G) (Sawyer et al. 2004), retroviruses and Tripartite Motif 5 alpha (TRIM5α) (Sawyer et al. 2005; Song et al. 2005), *Plasmodium falciparum* and glycophorin C (Maier et al. 2003), and *Mycobacterium tuberculosis* an granulysin (Stenger et al. 1998)]. Primate immune genes, proteins, and cells have structurally and functionally diverged. Primate immune factors show evidence of selection [CC-motif receptor 5 (Wooding et al. 2005), Toll-like Receptors 1 and 4 (Nakajima et al. 2008; Wlasiuk and Nachman 2010), TRIM5α (Sawyer et al. 2005), Cluster of Differentiation-45 (Filip and Mundy 2004), and Protein kinase R (Elde et al. 2009)] or interspecies divergence in function [Major-histocompatibility Complexes, Killer cell Immunoglobulin-like Receptors (Abi-Rached et al. 2010; Moesta et al. 2009), Toll-like Receptor 7 (Mandl et al. 2008, 2011), and ApoLI (Thomson et al. 2009)]. Reconstructions of primate evolutionary relationships based on the regulatory and coding sections of immune system genes deviate significantly from generally accepted primate phylogenies [Toll-like receptor 2 (Yim et al. 2006), CXC-motif receptor 4 (Puissant et al. 2003), and Major Histocompatability-DQA1 (Loisel et al. 2006)]. Although it appears that pathogens have directly contributed to the evolution of individual loci, we primarily understand these changes in the context of the primary structures of individual genes or proteins and not in the context of immune function. Until the functional effects of these changes are considered, it is impossible to appreciate how they contributed to speciation or disease susceptibility and progression. A goal of this volume is to integrate available information on structural and functional differences in primate immunity with data on the evolutionary analysis of gene sequences, pathogen life cycles, and evolutionary history.

Clinical Implications of Primate–Pathogen Coevolution

As a consequence of primate–pathogen evolutionary interactions, primates exhibit strong interspecies and interpopulation differences in immune response to a broad range of pathogens, some of which are major agents of human disease. Many lineages of African nonhuman primates have hosted immunodeficiency viruses (IV) over millions of years and their extant descendents do not develop the overt immune activation and white blood cell loss that typifies late stage IV infection (AIDS) in comparatively new hosts such as humans or Asian monkeys [reviewed in Pandrea and Apetrei 2010; see also Greenwood et al. 2013]. Nonhuman primate Herpes simian B virus infections in their natural hosts are fairly asymptomatic or manifest mildly, in a manner similar to human herpes simplex mucosal blisters. When transmitted to naïve primate hosts, including humans, these herpes infections can progress to encephalopathy (Artenstein et al. 1991; Chellman et al. 1992; Estep et al. 2010; Vizoso 1975). Unless severely immunocompromised, humans infected with the brain and muscle parasite *Toxoplasma gondii* are asymptomatic or develop a self-limited disease characterized by fever and enlarged lymph nodes (Jones et al. 2007). By contrast, *T. gondii* infections in New World monkeys are characterized by loss of strength, respiratory difficulty, and high mortality (Epiphanio et al. 2003; Catão-Dias et al. 2013). Research on the molecular mechanisms responsible for such variation in disease manifestation among different primate species involves multiple, often disparate areas of study, which contributes to gulfs between research on immune function, research on primate–pathogen evolution, and the clinical application of the resultant findings.

Arguably, the molecular mechanisms of immunodeficiency virus infection in primates are the best understood primate–pathogen interactions. The severity of the HIV pandemic and a strong interest in developing viable therapies have encouraged the examination of disease susceptibility and progression in primates, but mainly in a limited selection of catarrhine species. Primate-IV studies often use a comparative evolutionary approach as a starting point for the analysis of primate immunity and disease progression. An important advance in HIV therapy research has been the finding that IVs have emerged multiple times in the course of primate evolution and have closely coevolved with their hosts over millions of years (Pandrea and Apetrei 2010; Santiago et al. 2002; Switzer et al. 2005; Van Heuverswyn et al. 2006; Zhu et al. 1998). Comparative studies of primate-IV interactions have led to the identification of several immune factors thought to be under selective pressure from IVs and might serve as therapeutic targets including TRIM5α (Ortiz et al. 2006), APOBEC3G (Sawyer et al. 2004, 2005), Tetherin/BST2 (Jia et al. 2009), IL-4 (Koyanagi et al. 2010; Rockman et al. 2003), TLR7 (Mandl et al. 2008), TRAIL (Kim et al. 2007), CCR5 (Wooding et al. 2005), and MHC I (de Groot et al. 2002). A limited number of broader interspecies differences in the proportion of immune cell types, activation of immune cells, expression of immune genes, and stimulation of cell death that may affect disease progression have also been noted (Kim et al. 2007; Mandl et al. 2008; Soto et al. 2010). While primates serve as models for the study of other diseases and have been examined as xenotransplantation subjects, what we currently know about interspecies differences in primate immune function is largely derived from comparative IV-primate research.

Although the genetic and functional differences identified through IV research may also contribute to a clearer picture of general immune responses to pathogens, we do not know how many of these immune factors are activated across primate species or whether this activation is stimulated in a similar way by other pathogens. This gap points to fundamental problems in our understanding of primate–pathogen interactions. First, current research is biased toward a limited number of species such as rhesus macaques (*Macaca mulatta*), humans, and sooty mangabeys (*Cercocebus atys*). Second, these studies typically use a challenge model. As such, cross-species examinations of resting/baseline primate immunity are extremely limited. How the activation and coordination of multiple immune factors coevolved with pathogens has yet, therefore, to be thoroughly investigated. To develop a better understanding of how pathogens affect primate speciation, conservation, and health, considerable additional information is needed on the differences in resting/baseline immune function and non-IV pathogen–host interactions across a greater number of primate species. Given the expense and special care considerations inherent in acquiring experimental data from primate species, it is particularly important that these efforts are comprehensive and that the resulting reports are made readily accessible.

The Effects of Increasing Human and Wild Nonhuman Primate Contact on Primate–Pathogen Interaction

In areas where human settlements and nonhuman primate habitats overlap, the potential for interspecies disease transmission increases dramatically (Chapman et al. 2005; Daszak et al. 2000; Duval and Ariey 2012; Reynolds et al. 2012; Stothard et al. 2012; Wheatley and Harya Putra 1994). Such transmission events can decimate wild primate populations, as new infectious diseases may profoundly affect animal survival, sociality, and reproduction (Berdoy et al. 2000; Nunn et al. 2008; Nunn 2012). Increased ecotourism and residential/agricultural contact has led to heightened transmission of anthroponotic pathogens to wild primates and subsequently to increased mortality in primate populations [e.g., chimpanzees and Polio virus (Wallis and Lee 1999), gorillas and respiratory disease (Palacios et al. 2011), baboons and *Schistosoma mansoni* (Farah et al. 2003; Murray et al. 2000), chimpanzees and paramyxoviruses (Kondgen et al. 2008), chimpanzees and *Schistosoma mansoni* (Stothard et al. 2012), and baboons and *Mycobacterium* (Keet et al. 2000)]. *Plasmodium falciparum* may have been introduced anthropolonotically to the neotropical primates during the colonial era, possibly through the forced migration of African slaves during the slave trade. Neotropical primate *P. simium* has been proposed to have emerged from Asian *P. vivax* during the nineteenth century, and perhaps introduced to South America by laborers from East Asia (reviewed in Cormier 2010).

The transmission of pathogens from nonhuman primates to humans has also had a profound effect on human health. The current HIV-1/AIDS pandemic likely originated with human consumption of SIV-contaminated nonhuman primate bushmeat (Gao et al. 1999). The emergence of other diseases in humans has likewise been attributed to human and nonhuman primate contact [i.e., Human T-Lymphotropic virus 1 (Vandamme et al. 1998), Monkeypox (Mutombo et al. 1983; Reynolds et al. 2012), and Malaria (Liu et al. 2010)]. As humans encroach even farther onto nonhuman primate ranges, the need for veterinary and medical intervention will increase. To be able to develop appropriate modes of intervention, it is very important to have good information on the broad differences and similarities of primate immune systems as well as the biochemical mechanics of specific primate–pathogen interactions.

Research on Primate–Pathogen Interaction Remains Scattered and Incomplete

Despite the importance of pathogen-primate interactions over the course of primate evolution, research on the functional outcomes of pathogen-mediated primate evolution is incomplete and scattered across many disciplines. A unified approach to the evolution of primate immunity will help better define the mechanisms of disease emergence, immune function, resistance, and ecology.

To better understand the differences in human and nonhuman primate immunity and help guide future research, it is important to integrate information from researchers who study the effects of pathogens on the evolution of primate genome diversity, cell function, immune response, and gene expression. This volume is one attempt at such a synthesis, incorporating contributions from a multidisciplinary group of authors who:

1. Provide a compilation of current baseline information about primate–pathogen interaction and comparative primate immunity.
2. Describe and analyze infectious disease emergence and pathogen escape of host defense mechanisms in the context of primate–pathogen coevolution.
3. Explore divergent primate immune functions and the pathogen-mediated molecular evolution of primates.
4. Discuss the human health implications of primate–pathogen evolutionary interaction.

Overview

The first section, *Immunity and Primate Evolution,* includes chapters that discuss major elements of the primate immune system (Brinkworth and Thorn; Brinkworth and Sterner), the use of primates as models of immune system evolution (Loisel and Tung) and the pathogen-mediated evolution of primates (Gomez et al. and Allen et al.). The second section, *Emergence and Divergent Disease Manifestation,* provides data on the emergence, biochemical mechanics, and interspecies differences in immune response to immunodeficiency viruses, as well as a range of clinically important, but otherwise neglected pathogens. Attention is focused on how some primate pathogens have emerged (Harper and Knauf; Greenwood et al.), coevolved with and escaped the defenses of their hosts (Wong et al.), and triggered divergent responses in different primate species (Catao-Dias et al. and Greenwood et al.). The third and final section, *Primates, Pathogens, and Health,* focuses on how primate–pathogen coevolution affects the health of modern primates. Three papers on the health and evolutionary impact of disruptions to human and microorganism association (Rook, Martin and Blackwell, and Harper et al.) review and test the hygiene hypothesis. The volume closes with an analysis of major human and nonhuman primate cross-species pathogen transmission events, the social and biological factors that contributed to those events, and what the evolutionary, public health, and conservation consequences of these events might be (Harper et al.). We thought it a wonderful chapter with which to close.

Conclusion

Primate immune systems have been formed through complex evolutionary processes, within which pathogens have played an important role. The evolution of primate immunity has likely been more nuanced than natural selection driven by coevolutionary arms races with pathogens or large pathogen-mediated selective sweeps of hosts. Rather, the evolution of the primate immune system has likely been considerably shaped by, amongst other possibilities, moderately virulent pathogens, pathogen strategies that co-opt normal immunity, viruses integrated into the primate genome, commensal microorganism maintenance, autoreactivity, and overt immune responses, as well as the evolution of primate developmental stages. As close relatives, animals within the order Primates can be comparatively studied to not only clarify how primate immune systems have functionally diverged and highlight therapeutic targets for disease, but also to help define how such divergence contributes to disease emergence and interspecies disease transmission. As such, mapping the evolution of the primate immune system can improve our understanding of primate speciation, primate conservation, and human health. This volume is one effort to unite information on the evolutionary interactions between primate immune systems and pathogens. The chapters in this volume represent research from a broad range of disciplines involved in the study of primate–pathogen molecular interaction, primate immune function, and primate–pathogen coevolution. The work presented here discusses primate interactions with both major and neglected pathogens, attempts to bridge research on molecular evolution and primate immune function, and illustrates the impact of primate–pathogen evolutionary interactions on human and nonhuman primate health. With this effort we aim to provide a sound base of knowledge for future investigation of human and nonhuman primate evolution, immunity, and disease.

References

Abi-Rached L, Moesta AK, Rajalingam R, Guethlein LA, Parham P (2010) Human-specific evolution and adaptation led to major qualitative differences in the variable receptors of human and chimpanzee natural killer cells. PLoS Genet 6(11):e1001192

Allison AC (1956) The sickle-cell and haemoglobin C genes in some African populations. Ann Hum Genet 21(1):67–89

Arnaud F, Caporale M, Varela M, Biek R, Chessa B, Alberti A, Golder M, Mura M, Zhang YP, Yu L et al (2007) A paradigm for virus–host coevolution: sequential counter-adaptations between endogenous and exogenous retroviruses. PLoS Pathog 3(11):e170

Artenstein AW, Hicks CB, Goodwin BS Jr, Hilliard JK (1991) Human infection with B virus following a needlestick injury. Rev Infect Dis 13(2):288–291

Bannert N, Kurth R (2006) The evolutionary dynamics of human endogenous retroviral families. Annu Rev Genomics Hum Genet 7:149–173

Berdoy M, Webster JP, Macdonald DW (2000) Fatal attraction in rats infected with *Toxoplasma gondii*. Proc Biol Sci 267(1452):1591–1594

Brown KA, Brain SD, Pearson JD, Edgeworth JD, Lewis SM, Treacher DF (2006) Neutrophils in development of multiple organ failure in sepsis. Lancet 368(9530):157–169

Carp HJ, Selmi C, Shoenfeld Y (2012) The autoimmune bases of infertility and pregnancy loss. J Autoimmun 38(2–3):J266–J274

Catão-Dias JL, Epiphanio S, Martins Kierulff MC (2013) Neotropical primates and their susceptibility to *Toxoplasma gondii*: new insights for an old problem. In: Brinkworth JF, Pechenkina E (eds) Primates, pathogens, and evolution. Springer, Heidelberg

Chapman CA, Gillespie TR, Goldberg TL (2005) Primates and the ecology of their infectious diseases: How will anthropogenic change affect host-parasite interactions? Evolut Anthropol 14:134–144

Chellman GJ, Lukas VS, Eugui EM, Altera KP, Almquist SJ, Hilliard JK (1992) Activation of B virus (Herpesvirus simiae) in chronically immunosuppressed cynomolgus monkeys. Lab Anim Sci 42(2):146–151

Cheney DL, Seyfarth RM, Andelman SJ, Lee PC (1988) Reproductive success in vervet monkeys. In: Clutton-Brock TH (ed) Reproductive success. Chicago University Press, Chicago, IL, pp 384–402

Cormier LA (2010) The historical ecology of human and wild primate malarias in the new world. Diversity 2(2):256–280

Daszak P, Cunningham AA, Hyatt AD (2000) Emerging infectious diseases of wildlife–threats to biodiversity and human health. Science 287(5452):443–449

de Groot NG, Otting N, Doxiadis GG, Balla-Jhagjhoorsingh SS, Heeney JL, van Rood JJ, Gagneux P, Bontrop RE (2002) Evidence for an ancient selective sweep in the MHC class I gene repertoire of chimpanzees. Proc Natl Acad Sci USA 99(18):11748–11753

Dunlap KA, Palmarini M, Varela M, Burghardt RC, Hayashi K, Farmer JL, Spencer TE (2006) Endogenous retroviruses regulate periimplantation placental growth and differentiation. Proc Natl Acad Sci USA 103(39):14390–14395

Duval L, Ariey F (2012) Ape *Plasmodium* parasites as a source of human outbreaks. Clin Microbiol Infect 18(6):528–532

Elde NC, Child SJ, Geballe AP, Malik HS (2009) Protein kinase R reveals an evolutionary model for defeating viral mimicry. Nature 457(7228):485–489

Elde NC, Child SJ, Eickbush MT, Kitzman JO, Rogers KS, Shendure J, Geballe AP, Malik HS (2012) Poxviruses deploy genomic accordions to adapt rapidly against host antiviral defenses. Cell 150(4):831–841

Epiphanio S, Sinhorini IL, Catao-Dias JL (2003) Pathology of toxoplasmosis in captive new world primates. J Comp Pathol 129(2–3):196–204

Estep RD, Messaoudi I, Wong SW (2010) Simian herpesviruses and their risk to humans. Vaccine 28(Suppl 2):B78–B84

Farah I, Borjesson A, Kariuki T, Yole D, Suleman M, Hau J, Carlsson HE (2003) Morbidity and immune response to natural schistosomiasis in baboons (Papio anubis). Parasitol Res 91(4):344–348

Filip LC, Mundy NI (2004) Rapid evolution by positive Darwinian selection in the extracellular domain of the abundant lymphocyte protein CD45 in primates. Mol Biol Evol 21(8): 1504–1511

Gao, F, Bailes E, Robertson DL, Chen Y, Rodenburg CM, Michael SF, Cummins LB, Arthur LO, Peeters M, Shaw GM, Sharp PM, Hahn, BH (1999) Origin of HIV-1 in the chimpanzee Pan troglodytes troglodytes. Nature 397:436–441

Genovese G, Friedman DJ, Ross MD, Lecordier L, Uzureau P, Freedman BI, Bowden DW, Langefeld CD, Oleksyk TK, Uscinski Knob AL et al (2010) Association of trypanolytic ApoL1 variants with kidney disease in African Americans. Science 329(5993):841–845

Gogvadze E, Stukacheva E, Buzdin A, Sverdlov E (2009) Human-specific modulation of transcriptional activity provided by endogenous retroviral insertions. J Virol 83(12):6098–6105

Greenwood EJD, Schmidt F, Heeney JL (2013) The evolution of SIV in primates and the emergence of the pathogen of AIDS. In: Brinkworth JF, Pechenkina E (eds) Primates, pathogens, and evolution. Springer, Heidelberg

Haldane JBS (1949) The rate of mutation of human genes. Heredita 35(suppl):267–273

Hamblin MT, Di Rienzo A (2000) Detection of the signature of natural selection in humans: evidence from the Duffy blood group locus. Am J Hum Genet 66(5):1669–1679

Hedrick P (2004) Estimation of relative fitnesses from relative risk data and the predicted future of haemoglobin alleles S and C. J Evol Biol 17(1):221–224

Horie M, Honda T, Suzuki Y, Kobayashi Y, Daito T, Oshida T, Ikuta K, Jern P, Gojobori T, Coffin JM et al (2010) Endogenous non-retroviral RNA virus elements in mammalian genomes. Nature 463(7277):84–87

Hraber P, Kuiken C, Yusim K (2007) Evidence for human leukocyte antigen heterozygote advantage against hepatitis C virus infection. Hepatology 46(6):1713–1721

Huang ES, Kilpatrick B, Lakeman A, Alford CA (1978) Genetic analysis of a cytomegalovirus-like agent isolated from human brain. J Virol 26(3):718–723

Hunter P (2010) The missing link. Viruses revise evolutionary theory. EMBO Rep 11(1):28–31

Izutsu Y (2009) The immune system is involved in Xenopus metamorphosis. Front Biosci 14:141–149

Jackson JA, Tinsley RC (2002) Effects of environmental temperature on the susceptibility of *Xenopus laevis* and *X. wittei* (Anura) to *Protopolystoma xenopodis*(Monogenea). Parasitol Res 88(7):632–638

Jia B, Serra-Moreno R, Neidermyer W, Rahmberg A, Mackey J, Fofana IB, Johnson WE, Westmoreland S, Evans DT (2009) Species-specific activity of SIV Nef and HIV-1 Vpu in overcoming restriction by tetherin/BST2. PLoS Pathog 5(5):e1000429

Jones JL, Kruszon-Moran D, Sanders-Lewis K, Wilson M (2007) *Toxoplasma gondii* infection in the United States, 1999 2004, decline from the prior decade. Am J Trop Med Hyg 77(3):405–410

Kageruka P, Mangus E, Bajyana Songa E, Nantulya V, Jochems M, Hamers R, Mortelmans J (1991) Infectivity of *Trypanosoma* (Trypanozoon) *brucei gambiense* for baboons (Papio hamadryas, Papio papio). Ann Soc Belg Med Trop 71(1):39–46

Keet DF, Kriek NP, Bengis RG, Grobler DG, Michel A (2000) The rise and fall of tuberculosis in a free-ranging chacma baboon troop in the Kruger National Park. Onderstepoort J Vet Res 67(2):115–122

Kim N, Dabrowska A, Jenner RG, Aldovini A (2007) Human and simian immunodeficiency virus-mediated upregulation of the apoptotic factor TRAIL occurs in antigen-presenting cells from AIDS-susceptible but not from AIDS-resistant species. J Virol 81(14):7584–7597

Kim HS, Kim DS, Huh JW, Ahn K, Yi JM, Lee JR, Hirai H (2008) Molecular characterization of the HERV-W env gene in humans and primates: expression, FISH, phylogeny, and evolution. Mol Cells 26(1):53–60

Klimovich VB (2002) Actual problems of evolutional immunology. Zh Evol Biokhim Fiziol 38(5):442–451

Kondgen S, Kuhl H, N'Goran PK, Walsh PD, Schenk S, Ernst N, Biek R, Formenty P, Matz-Rensing K, Schweiger B et al (2008) Pandemic human viruses cause decline of endangered great apes. Curr Biol 18(4):260–264

Koppensteiner H, Brack-Werner R, Schindler M (2012) Macrophages and their relevance in human immunodeficiency virus type I infection. Retrovirology 9:82

Koyanagi M, Kerns JA, Chung L, Zhang Y, Brown S, Moldoveanu T, Malik HS, Bix M (2010) Diversifying selection and functional analysis of interleukin-4 suggests antagonism-driven evolution at receptor-binding interfaces. BMC Evol Biol 10:223

Kwiatkowski DP (2005) How malaria has affected the human genome and what human genetics can teach us about malaria. Am J Hum Genet 77(2):171–192

Lambrecht FL (1985) Trypanosomes and Hominid evolution. Bioscience 35(10):640–646

Lapoumeroulie C, Dunda O, Ducrocq R, Trabuchet G, Mony-Lobe M, Bodo JM, Carnevale P, Labie D, Elion J, Krishnamoorthy R (1992) A novel sickle cell mutation of yet another origin in Africa: the Cameroon type. Hum Genet 89(3):333–337

Lee YK, Mazmanian SK (2010) Has the microbiota played a critical role in the evolution of the adaptive immune system? Science 330(6012):1768–1773

Leffler EM, Gao Z, Pfeifer S, Segurel L, Auton A, Venn O, Bowden R, Bontrop R, Wall JD, Sella G et al (2013) Multiple instances of ancient balancing selection shared between humans and chimpanzees. Science 339(6127):1578–1582

Levin JL, Hilliard JK, Lipper SL, Butler TM, Goodwin WJ (1988) A naturally occurring epizootic of simian agent 8 in the baboon. Lab Anim Sci 38(4):394–397

Liu W, Li Y, Learn GH, Rudicell RS, Robertson JD, Keele BF, Ndjango JB, Sanz CM, Morgan DB, Locatelli S et al (2010) Origin of the human malaria parasite *Plasmodium falciparum* in gorillas. Nature 467(7314):420–425

Loisel DA, Rockman MV, Wray GA, Altmann J, Alberts SC (2006) Ancient polymorphism and functional variation in the primate MHC-DQA1 5′ cis-regulatory region. Proc Natl Acad Sci USA 103(44):16331–16336

Maier AG, Duraisingh MT, Reeder JC, Patel SS, Kazura JW, Zimmerman PA, Cowman AF (2003) *Plasmodium falciparum* erythrocyte invasion through glycophorin C and selection for Gerbich negativity in human populations. Nat Med 9(1):87–92

Mandl JN, Barry AP, Vanderford TH, Kozyr N, Chavan R, Klucking S, Barrat FJ, Coffman RL, Staprans SI, Feinberg MB (2008) Divergent TLR7 and TLR9 signaling and type I interferon production distinguish pathogenic and nonpathogenic AIDS virus infections. Nat Med 14(10):1077–1087

Mandl JN, Akondy R, Lawson B, Kozyr N, Staprans SI, Ahmed R, Feinberg MB (2011) Distinctive TLR7 signaling, type I IFN production, and attenuated innate and adaptive immune responses to yellow fever virus in a primate reservoir host. J Immunol 186(11):6406–6416

Marmor M, Sheppard HW, Donnell D, Bozeman S, Celum C, Buchbinder S, Koblin B, Seage GR 3rd (2001) Homozygous and heterozygous CCR5-Delta32 genotypes are associated with resistance to HIV infection. J Acquir Immune Defic Syndr 27(5):472–481

Matzinger P (1994) Tolerance, danger, and the extended family. Annu Rev Immunol 12: 991–1045

Matzinger P (2002) The danger model: a renewed sense of self. Science 296(5566):301–305

Maynard CL, Elson CO, Hatton RD, Weaver CT (2012) Reciprocal interactions of the intestinal microbiota and immune system. Nature 489(7415):231–241

Moesta AK, Abi-Rached L, Norman PJ, Parham P (2009) Chimpanzees use more varied receptors and ligands than humans for inhibitory killer cell Ig-like receptor recognition of the MHC-C1 and MHC-C2 epitopes. J Immunol 182(6):3628–3637

Murdoch C, Finn A (2003) The role of chemokines in sepsis and septic shock. Contrib Microbiol 10:38–57

Murray S, Stem C, Boudreau B, Goodall J (2000) Intestinal parasites of baboons (Papio cynocephalus anubis) and chimpanzees (Pan troglodytes) in Gombe National Park. J Zoo Wildl Med 31(2):176–178

Mutombo M, Arita I, Jezek Z (1983) Human monkeypox transmitted by a chimpanzee in a tropical rain-forest area of Zaire. Lancet 1(8327):735–737

Nakajima T, Ohtani H, Satta Y, Uno Y, Akari H, Ishida T, Kimura A (2008) Natural selection in the TLR-related genes in the course of primate evolution. Immunogenetics 60(12):727–735

Nunn CL (2012) Primate disease ecology in comparative and theoretical perspective. Am J Primatol 74(6):497–509

Nunn C, Thrall P, Stewart K, Harcourt A (2008) Emerging infectious diseases and animal social systems. Evolut Ecol 22(4):519–543

Oner C, Dimovski AJ, Olivieri NF, Schiliro G, Codrington JF, Fattoum S, Adekile AD, Oner R, Yuregir GT, Altay C et al (1992) Beta S haplotypes in various world populations. Hum Genet 89(1):99–104

Ortiz M, Bleiber G, Martinez R, Kaessmann H, Telenti A (2006) Patterns of evolution of host proteins involved in retroviral pathogenesis. Retrovirology 3:11

Oyston PC, Dorrell N, Williams K, Li SR, Green M, Titball RW, Wren BW (2000) The response regulator PhoP is important for survival under conditions of macrophage-induced stress and virulence in Yersinia pestis. Infect Immun 68(6):3419–3425

Palacios G, Lowenstine LJ, Cranfield MR, Gilardi KV, Spelman L, Lukasik-Braum M, Kinani JF, Mudakikwa A, Nyirakaragire E, Bussetti AV et al (2011) Human metapneumovirus infection in wild mountain gorillas, Rwanda. Emerg Infect Dis 17(4):711–713

Pandrea I, Apetrei C (2010) Where the wild things are: pathogenesis of SIV infection in African nonhuman primate hosts. Curr HIV/AIDS Rep 7(1):28–36

Pandrea I, Apetrei C, Gordon S, Barbercheck J, Dufour J, Bohm R, Sumpter B, Roques P, Marx PA, Hirsch VM et al (2007) Paucity of CD4+CCR5+ T cells is a typical feature of natural SIV hosts. Blood 109(3):1069–1076

Puissant B, Abbal M, Blancher A (2003) Polymorphism of human and primate RANTES, CX3CR1, CCR2 and CXCR4 genes with regard to HIV/SIV infection. Immunogenetics 55(5):275–283

Pujol C, Bliska JB (2003) The ability to replicate in macrophages is conserved between Yersinia pestis and Yersinia pseudotuberculosis. Infect Immun 71(10):5892–5899

Redford PS, Murray PJ, O'Garra A (2011) The role of IL-10 in immune regulation during M. tuberculosis infection. Mucosal Immunol 4(3):261–270

Reynolds MG, Carroll DS, Karem KL (2012) Factors affecting the likelihood of monkeypox's emergence and spread in the post-smallpox era. Curr Opin Virol 2(3):335–343

Rockman MV, Hahn MW, Soranzo N, Goldstein DB, Wray GA (2003) Positive selection on a human-specific transcription factor binding site regulating IL4 expression. Curr Biol 13(23):2118–2123

Santiago ML, Rodenburg CM, Kamenya S, Bibollet-Ruche F, Gao F, Bailes E, Meleth S, Soong SJ, Kilby JM, Moldoveanu Z et al (2002) SIVcpz in wild chimpanzees. Science 295(5554):465

Sawyer SL, Emerman M, Malik HS (2004) Ancient adaptive evolution of the primate antiviral DNA-editing enzyme APOBEC3G. PLoS Biol 2(9):E275

Sawyer SL, Wu LI, Emerman M, Malik HS (2005) Positive selection of primate TRIM5alpha identifies a critical species-specific retroviral restriction domain. Proc Natl Acad Sci USA 102(8):2832–2837

Song B, Javanbakht H, Perron M, Park DH, Stremlau M, Sodroski J (2005) Retrovirus restriction by TRIM5alpha variants from old world and new world primates. J Virol 79(7):3930–3937

Soto PC, Stein LL, Hurtado-Ziola N, Hedrick SM, Varki A (2010) Relative over-reactivity of human versus chimpanzee lymphocytes: implications for the human diseases associated with immune activation. J Immunol 184(8):4185–4195

Stenger S, Hanson DA, Teitelbaum R, Dewan P, Niazi KR, Froelich CJ, Ganz T, Thoma-Uszynski S, Melián A, Bogdan C et al (1998) An antimicrobial activity of cytolytic T cells mediated by granulysin. Science (New York, NY) 282(5386):121–125

Stothard JR, Mugisha L, Standley CJ (2012) Stopping schistosomes from "monkeying-around" in chimpanzees. Trends Parasitol 28(8):320–326

Switzer WM, Parekh B, Shanmugam V, Bhullar V, Phillips S, Ely JJ, Heneine W (2005) The epidemiology of simian immunodeficiency virus infection in a large number of wild- and captive-born chimpanzees: evidence for a recent introduction following chimpanzee divergence. AIDS Res Hum Retroviruses 21(5):335–342

Thomson R, Molina-Portela P, Mott H, Carrington M, Raper J (2009) Hydrodynamic gene delivery of baboon trypanosome lytic factor eliminates both animal and human-infective African trypanosomes. Proc Natl Acad Sci USA 106(46):19509–19514

Tishkoff SA, Varkonyi R, Cahinhinan N, Abbes S, Argyropoulos G, Destro-Bisol G, Drousiotou A, Dangerfield B, Lefranc G, Loiselet J et al (2001) Haplotype diversity and linkage disequilibrium at human G6PD: recent origin of alleles that confer malarial resistance. Science 293(5529):455–462

Van Heuverswyn F, Li Y, Neel C, Bailes E, Keele BF, Liu W, Loul S, Butel C, Liegeois F, Bienvenue Y et al (2006) Human immunodeficiency viruses: SIV infection in wild gorillas. Nature 444(7116):164

Van Valen L (1973) A new evolutionary law. Evolut Theor 1:1–30

Vandamme AM, Salemi M, Desmyter J (1998) The simian origins of the pathogenic human T-cell lymphotropic virus type I. Trends Microbiol 6(12):477–483

Vizoso AD (1975) Recovery of herpes simiae (B virus) from both primary and latent infections in rhesus monkeys. Br J Exp Pathol 56(6):485–488

Wallis J, Lee DR (1999) Primate conservation: the prevention of disease transmission. Int J Primatol 20(6):803–826

Wang T, Zeng J, Lowe CB, Sellers RG, Salama SR, Yang M, Burgess SM, Brachmann RK, Haussler D (2007) Species-specific endogenous retroviruses shape the transcriptional network of the human tumor suppressor protein p53. Proc Natl Acad Sci USA 104(47):18613–18618

Welburn SC, Fevre EM, Coleman PG, Odiit M, Maudlin I (2001) Sleeping sickness: a tale of two diseases. Trends Parasitol 17(1):19–24

Wheatley B, Harya Putra DK (1994) Biting the hand that feeds you: Monkeys and tourists in Balinese monkey forests. Trop Biodivers 2:317–327

Wlasiuk G, Nachman MW (2010) Adaptation and constraint at Toll-like receptors in primates. Mol Biol Evol 27(9):2172–2186

Wooding S, Stone AC, Dunn DM, Mummidi S, Jorde LB, Weiss RK, Ahuja S, Bamshad MJ (2005) Contrasting effects of natural selection on human and chimpanzee CC chemokine receptor 5. Am J Hum Genet 76(2):291–301

Yim JJ, Adams AA, Kim JH, Holland SM (2006) Evolution of an intronic microsatellite polymorphism in Toll-like receptor 2 among primates. Immunogenetics 58(9):740–745

Yohn CT, Jiang Z, McGrath SD, Hayden KE, Khaitovich P, Johnson ME, Eichler MY, McPherson JD, Zhao S, Paabo S et al (2005) Lineage-specific expansions of retroviral insertions within the genomes of African great apes but not humans and orangutans. PLoS Biol 3(4):e110

Zemans RL, Colgan SP, Downey GP (2009) Transepithelial migration of neutrophils: mechanisms and implications for acute lung injury. Am J Respir Cell Mol Biol 40(5):519–535

Zhou D, Han Y, Yang R (2006) Molecular and physiological insights into plague transmission, virulence and etiology. Microbes Infect 8(1):273–284

Zhu T, Korber BT, Nahmias AJ, Hooper E, Sharp PM, Ho DD (1998) An African HIV-1 sequence from 1959 and implications for the origin of the epidemic. Nature 391(6667):594–597

Zilber-Rosenberg I, Rosenberg E (2008) Role of microorganisms in the evolution of animals and plants: the hologenome theory of evolution. FEMS Microbiol Rev 32(5):723–735

Part I
Immunity and Primate Evolution

Vertebrate Immune System Evolution and Comparative Primate Immunity

Jessica F. Brinkworth and Mitchell Thorn

Introduction

Molecular and cellular responses as diverse as RNA interference against viral infections in plants, antimicrobial peptides production in insects, and macrophage phagocytosis of *Listeria monocytogenes* bacterium in mammals are all manifestations of the immune system—an array of defense mechanisms against pathogens, cellular debris, and cancerous and dying cells that can secure the survival of host. It is an exquisitely organized and regulated system of defenses that has diversified over hundreds of millions years and, yet, is suitably conserved such that multiple organisms (e.g., insects, lamprey, dogs, rodents, and primates) can serve as immunological models of human health.

The practice of studying the immune system within the context of evolution, or "comparative immunology", emerged in the late nineteenth century. The most famous early example of a comparative immunological approach is Elie Metchnikoff's 1882 discovery of leukocyte phagocytosis. When Metchnikoff inserted rose thorns into starfish larvae to determine if the "wandering" cells (leukocytes) of starfish responded to bodily invasion by foreign matter (Metchnikoff 1893), he found the cells aggregated around the thorns. He then reiterated his experiment through the application of microorganisms to increasingly derived species (e.g., flies, rabbits) known to maintain leukocytes to find that, universally, a subset of these cells ingested microbes and offered host protection (Metchnikoff 1887).

J.F. Brinkworth (✉)
Department of Pediatrics, Faculty of Medicine, CHU Sainte-Justine Research Center, University of Montreal, Montreal, QC, Canada
e-mail: jfbrinkworth@gmail.com

M. Thorn
Roger Williams Medical Center, Providence, RI, USA
e-mail: mitchell.thorn@gmail.com

J.F. Brinkworth and K. Pechenkina (eds.), *Primates, Pathogens, and Evolution*, Developments in Primatology: Progress and Prospects,
DOI 10.1007/978-1-4614-7181-3_2, © Springer Science+Business Media New York 2013

Thus, the innate immune strategy of phagocytosis was discovered. The comparative immunological approach Metchnikoff employed goes a step beyond simply applying a representative animal model to a question of immunity. Comparative immunology involves comparing differences and similarities in organisms' immune responses to draw broader conclusions about immune system function and its evolution. This approach can be useful for the identification of immune system components responsible for particular disease phenotypes, as well as the discovery of novel immune mechanisms. The use of other vertebrate animals such as jawless vertebrates, amphibians, and rodents as biomedical models can shed light on overriding principles of immunity. Comparisons of primate immunity in the context of vertebrate immune system evolution can be useful for the understanding of biomedical model use, primate evolution, and human health.

Primates are a recent addition to the comparative exploration of animal immune responses. Direct interspecies comparisons of primate immune system function emerged in the 1920s but did not become common until decades later. The understanding of comparative primate immunity mainly developed through nonhuman primate/human xenotransplantation studies in the 1960s, as well as immunodeficiency virus research from the late 1980s onwards (Benveniste et al. 1986; Daniel et al. 1984; Hardy et al. 1964; Hitchcock et al. 1964; Murphey-Corb et al. 1986; Reemtsma et al. 1964a, b; Starzl et al. 1964). Until the adoption of catarrhine (Old World monkey, apes, and humans) species for HIV research, few immunological studies using different primate species compared interspecies differences. It is now well established that primates exhibit within-order variation and, sometimes, biologically unique manifestations of infectious diseases (Epiphanio et al. 2003; Ngampasutadol et al. 2008; Pandrea and Apetrei 2010; Thomson et al. 2009; Walker 1997). Still, monkey and ape species are commonly used in immunological research as corollaries for human disease progression due to their biochemical, physiological, and genetic similarity to humans. Because there are so many immune system similarities between primate species, there is a tendency in non-HIV literature to make the assumption that what is represented in one primate species is represented in other primate species and, possibly, other nonprimate models. By making this assumption a researcher risks overlooking aspects of primate immune systems that are unique and using primates unnecessarily to explore immunological traits that they share with many other model organisms. Comparative information on baseline primate immunity, particularly the anatomy and function of major immune organs and cell types, is rare and scattered across many fields. In certain cases it is entirely unexplored. The goal of this chapter is to illustrate the place of primates in immune system evolution by (1) putting the emergence of major primate immune system components in the context of the evolution of vertebrate immunity as a whole and (2) illustrating how baseline primate immunity has diversified by uniting and highlighting the available information on interspecies functional differences in baseline primate immune system structures and components.

Overview of the Mammalian Immune System

As jawed vertebrates, mammals maintain an immune system that can be broken into two major arms based on function. The innate immune response is the more ancient of these two arms, having invertebrate origins (Leulier et al. 2003; Yoshida et al. 1986). Innate immune defenses are inherited, germline encoded, nonspecific, and typified by barriers (e.g., mucosa, skin), antimicrobial peptides, phagocytosis (initiated by cells such as macrophages and neutrophils), and inflammation (Janeway and Medzhitov 2002; Kumar et al. 2009). This kind of immunity limits initial infections by recognizing "nonself", and damage through a variety of sophisticated but generalized mechanisms, including inherited pattern recognition receptors (e.g., Toll-like receptors, NOD-like receptors) that detect foreign material through molecular patterns associated with pathogens or cellular damage. These patterns can be shared broadly by microorganisms or may signal tissue damage. They are conventionally and somewhat imprecisely referred to as pathogen- or danger- associated molecular patterns (PAMPS or DAMPS) (Seong and Matzinger, 2004).

By contrast, adaptive immunity is highly specific, not immediate, key to immunological memory, modulated by innate immunity, and acquired over a lifetime. While phagocytosis is an important tool in innate immune defenses, the targeting of matter bearing specific epitopes by lymphocytes (e.g., T and B cells) and the retention of some of these target-specific lymphocytes is key to adaptive immunity. Lymphocytes express membrane receptors (T-cell receptors for T cells and B-cell receptors for B cells) that recognize antigens. Unlike innate immunity, which makes use of germline encoded receptors, adaptive immunity has been traditionally viewed as reliant on receptors and immunoglobulins that are made highly variable through recombination activating gene (RAG)-mediated gene rearrangement/somatic recombination that occurs during lymphocyte development. From a limited number of receptor genes is borne a broad repertoire of specific receptors. As a result, rather than recognizing pathogens through PAMPS, the lymphocytes and immunoglobulins of the adaptive system recognize and "remember" distinct epitopes [reviewed in (Hardy 2003)].

The simplified view of the vertebrate immune system function is one of immediate recognition of invading pathogens by the innate immune system and subsequent initiation of a specific adaptive immune response. In mammals, for example, when innate immune cells recognize foreign antigens, they initiate the release of reactive signaling proteins known as cytokines. Cytokines degrade pathogens nonspecifically and initiate activation of an epitope/pathogen-specific T and B cell-mediated adaptive immune response. T and B cells can become activated when receptors they bear belonging to the immunoglobulin receptor superfamily, T- and B-cell receptors (TCRs and BCRs), recognize specific epitopes that are presented to them via major histocompatibility complexes (MHC) on phagocytic cells (e.g., macrophages and dendritic cells). B cells can also become activated through direct encounters with pathogens bearing these epitopes. Activated T and B cells then clonally replicate in secondary lymphoid tissues, to be released as cytotoxic, phagocytic, or antibody producing cells that recognize and attempt to eliminate a specific epitope target. Engagement of the adaptive immune response typically occurs after the 4th hour of infection. The first clonal adaptive immune cells are released ~96 hours from the point of initial T- or B-cell activation. The first minutes and days of infection, therefore, are mainly mediated by innate immunity. As an infection is cleared over successive hours, most clonal T and B cells die off. A small percentage, however, remain in circulation as memory T and B cells. Memory cells speed the adaptive response to reencountered foreign epitopes and are the basis for immunological memory of past infections (Davis and Chien 2003; Jenkins 2003; Paul 2003).

Even this simplified description of mammalian immune system function only partially represents the immune system of other vertebrate classes as certain key components (e.g., lymph nodes, spleen components, and particular immunoglobulins/BCRs) did not appear until the emergence of recent vertebrate classes such as birds and mammals. The traditional paradigm of adaptive immunity is that it emerged in the last common ancestor of jawed vertebrates (gnathostomata) approximately 625 million years ago (mya). All lower vertebrates were thought to survive microbial assault only by initiating a generalized innate immune response. However, in the last decade it has become apparent that some components we associate with adaptive immunity emerged much earlier than previously assumed. The first evidence of lymphocyte-derived cytokines and receptors with immunoglobulin (Ig)-like domains, for example, can be found in sponges (Blumbach et al. 1999). In 2007, first evidence of somatic diversification of antigen receptors in lamprey came to light, supporting the existence of an "adaptive" immune system in jawless vertebrates (agnatha) (Guo et al. 2009; Rogozin et al. 2007). While these receptors, known as variable lymphocyte receptors (VLRs, discussed further below), are not related to the immunoglobulin receptor superfamily and are therefore not precursors to TCRs and BCRS, they appear to serve a similar function as antigen receptors (reviewed in Boehm et al. 2012b). Since the discovery of VLRs, it now seems possible that adaptive immune systems based on antigen receptor diversity mediated by gene rearrangement/somatic recombination may have evolved more than once in Metazoan history (Boehm et al. 2012b). The possibility that such sophisticated immune strategies in diverse vertebrate clades may be the outcome of convergent evolution attests to the immense evolutionary pressures that have shaped the vertebrate immune system.

The Evolution of Vertebrate Immunity: Major Lymphoid Tissues and Organs

Lymphoid tissues function as the sites of lymphoid cell development, selection of antigen–receptor repertoires, and effector cell coordination. These structures can be divided into primary or secondary lymphoid tissues based on their main function. Primary lymphoid tissues are sites of lymphocyte effector cell poiesis and

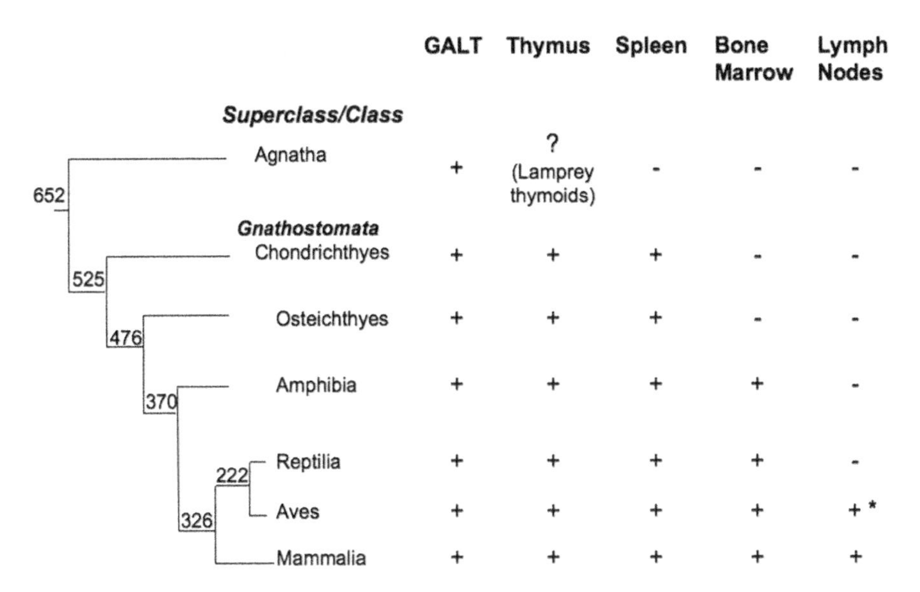

Fig. 1 Major lymphoid structures and their emergence in vertebrate clades over time. The symbols "+" and "-" indicate the presence/occurrence of and the absence of lymphoid structure, respectively. "*" while most birds do not appear to maintain lymph nodes, they are present ducks (Anatidae). Molecular divergence dates from Blair and Hedges (2005)

development, as well as antigen–receptor selection (e.g., thymus, bone marrow). The secondary lymphoid tissues are mainly coordination sites for mature immune effector cell interactions between antigen-presenting cells and adaptive lymphocytes during the immune response (e.g., spleen, lymph nodes) (reviewed in Boehm et al. 2012a; Karrer et al. 1997). The suite of major lymphoid tissues we see in primates—thymus and bone marrow (primary), gut-associated lymphoid tissue, spleen, and lymph nodes (secondary)—emerged in vertebrates in a complex pattern. Some of these structures are certainly very ancient and were already established before the emergence of jawed vertebrates (gnathostomata), while others appear sporadically in later extant clades, only becoming well established in extant mammals (Fig. 1) (e.g., lymph nodes). Here, the evolution of these tissues and organs as they emerged in geological time is described.

GALT and Peyer's Patches

Gut-associated lymphoid tissue (GALT) is a type of mucosa-associated lymphoid tissue (MALT) and secondary lymphoid structure found in the gut tube of all vertebrates. It is an ancient type of tissue dedicated to immune defense within an organ that is subject to constant antigen exposure from both food and microbial organisms. The primary role of mammalian MALT and GALT structures is the production of immunoglobulin A (IgA) in support of T helper 2 cell-dependent responses, though other T-cell activities occur in these regions (Kiyono and Fukuyama 2004). GALT can also be a site of other important immune functions such as immunoglobulin class switching and B-cell clonal expansion (Pospisil and Mage 1998a, b; Shikina et al. 2004). In primates, GALT serves as a site of Th2, Th1, and cytotoxic T-cell reactions (Cesta 2006a; Kiyono and Fukuyama 2004; Kwa et al. 2006).

Prenatally, GALT is mainly organized by lymphoid cell activities (Pearson et al. 2012; Veiga-Fernandes et al. 2007). However, much of its development appears to occur after birth, with exposure to microorganisms and development of gut flora leading to the aggregation of T cells, the development and expansion of T-cell areas, and the subsequent development of follicular germinal centers for B-cell development (Cesta 2006a). Exposure to food-sourced antigens appears to affect GALT structure as well, with intravenous feeding leading to a significant decrease in the number of lymphocytes residing in rodent GALT (King et al. 1997a; Tanaka et al. 1991). Individual environment and ecology, therefore, likely contribute to intraspecies variation in structure.

GALT organization and location is flexible and varies considerably between animal classes and species. The earliest example of GALT is found in cold-blooded vertebrates and is not an organized tissue. It appears as aggregates of lymphoid cells and tissues in the lamina propria, mesentery and, sometimes, in intraepithelial regions (Bernard et al. 2006; Mussmann et al. 1996a; Zapata and Amemiya 2000). The first organized GALT structure emerged in the submucosa of the ileum and the lamina propria of the small intestine of birds and mammals (Hofmann et al. 2010). The best characterized of these GALT structures are Peyer's patches (PP), which appear as ovular lymphoid follicles with germinal centers containing many B cells that are surrounded by T cells (Haley 2003). PPs are found in their highest concentrations in the jejunum or ileum of most mammalian species (Cornes 1965; Haley 2003; HogenEsch and Hahn 2001; Owen et al. 1991). They are initially formed during fetal development, but the maturation and development of important PP structures, such as germinal centers, is dependent on exposure to microbial organisms (Adachi et al. 1997; Cornes 1965; Hooper 2004; Kyriazis and Esterly 1971; Spencer et al. 1986). It should not be surprising, then, that the number, location, and size of PPs differ considerably between mammalian species. Dogs, for example, maintain 26–29 patches that can be categorized as two types - small patches in the jejunum and superior ileum and a large patch with dome-shaped regions in the inferior ileum (HogenEsch and Felsburg 1992; HogenEsch and Hahn 2001). Rabbits, on the other hand, have a cluster of lymphoid nodules in an ileal–caecal appendix and maintain a single, large Peyer's patch that covers the terminal end of the ileum (Haley 2003).

GALT represents the majority of the lymphoid tissue in the primate body, yet interspecies differences in GALT structure are neither well characterized nor are the functional ramifications of such differences well understood. In primates GALT appears to be an important location for immunodeficiency virus (IV) replication (Mehandru et al. 2004; Veazey et al. 1998). In AIDS-susceptible/naïve IV host primates, such as humans and macaques (*Macaca* sp.), immunodeficiency virus infection leads to the loss of CD4+ T cells in this region as well as Th17 CD4+ T-cell dysfunction, threatening the integrity of the gut barrier (Brenchley and Douek 2008; Favre et al. 2009; Raffatellu et al. 2008). This process does not appear to occur in AIDS-resistant/natural IV host species such as sooty mangabeys (*Cercocebus atys*), suggesting that the regulation of GALT components (including T cell subsets) has diverged since these species last shared a common ancestor approximately 30 mya and that these differences can affect infectious disease progression (Brenchley et al. 2008; Fabre et al. 2009; Steiper and Young 2006).

The Caecal Appendix

The caecal appendix is a controversial manifestation of GALT in mammals. A narrow, terminal extension at the end of the caecum, this tissue occurs in a handful of mammals, including primates, opossums, wombats, and rabbits, and is rich in lymphoid structures (Chivers and Hladik 1980; Dalquest et al. 1952; Fisher 2000; Mitchell 1916). It has been proposed to have evolved at least twice within mammals, appearing independently in diprotodont marsupials (e.g., the wallaby) and again in Euarchontoglires (e.g., rabbits, primates) (Smith et al. 2009). Variations on an appendix-like structure have been noted in animals lacking caecums (e.g., monotremes and some birds), suggesting that organs of similar function may have emerged multiple times over the evolution of animal life and possibly preceded the canonical caecum (Laurin et al. 2011; Smith et al. 2009). An analysis of cellulose content in lemur diet and caecal morphology suggests that, at least in the case of primates, the presence of an appendix is likely under evolutionary pressures imposed by diet as well as immune function (Campbell et al. 2000).

Amongst primates, a vermiform caecal appendix is found in all extant hominoids, and variably described appendices are noted for *Nycticebus, Loris, Perodicticus, Euoticus, Eulemur, Varecia, Daubentonia, Saguinus, Callithrix, Callicebus, Aotus, Cercopithecus, Chlorocebus, Macaca, Papio, Procolobus,* and *Piliocolobus* (reviewed in Fisher 2000). A relative reduction in caecum length and an increase in appendix length have been noted to occur over the course of primate evolution, with strepsirrhines maintaining comparatively longer caecums relative to gut length than monkeys or apes. The appendix, in turn, tends to be small and variably present in strepsirrhines, while it is larger and more consistently present in catarrhines (Fisher 2000).

Long described as a digestive tract vestige, the mammalian appendix is now thought to fulfill a subset of GALT-related immune functions including the maintenance of normal gut flora (Gorgollon 1978; Randal Bollinger et al. 2007). Not surprisingly, given that appendectomies in humans do not lead to immune dysfunction or notable shifts in gut flora, the necessity of the appendix in primates remains questionable (Laurin et al. 2011; Smith et al. 2009). In animals that maintain a caecum but lack a caecal appendix, the caecal apex tends to contain a high concentration of GALT and microbial biofilms which may compensate for the absence of an appendix (Smith et al. 2009). Frogs, which lack caecums entirely, share a very similar pattern of microbial biofilm distribution in the rest of the gut with mammals that maintain an appendix. This suggests that maintenance of flora by GALT in the primate large bowel has very ancient origins (Marjanovic and Laurin 2007; Smith et al. 2009).

Thymus

The thymus is the oldest primary lymphoid organ, the earliest example of which can be found in lampreys (*Lampetra planeri* and *Petromyzon marinus*) (Bajoghli et al. 2011). While in most jawed vertebrates both B- and T-cell poiesis occur in the bone marrow, thymocytes (immature T cells) move to the thymus to mature (reviewed in Boehm and Bleul 2007). There, they undergo T-cell receptor rearrangement, positive selection for MHC interaction, and negative selection for self-recognition. These processes lead to the development of an enormous repertoire of T cells that communicate with other immune cells and bear receptors that recognize individual and specific epitopes.

Lampreys bear a thymus precursor at the tips of their gill filaments that maintains T-like cells and expresses genes associated with the thymus in other animals (i.e., *FOXN4L*, a *FOXN1* homologue; *CDA1*) (Bajoghli et al. 2011). All extant jawed vertebrates with a canonical thymus also have T cells, TCRs, and MHCs, so it is difficult to assess how these immune components evolved together. Considerable vertebrate extinctions over the last 650 million years give the impression that this entire system of receptor repertoire expansion and immune cell development simply appeared, fully formed in cartilaginous fish (class Chrondrichthyes). This complex system did not, of course, arise in this manner, but there is very little information about the intermediate steps in its evolution. The appearance of the canonical thymus and this system of T-cell development coincides with the acquisition of two genes responsible for immunoglobulin domain receptor recombination (*RAG1* and *RAG2*) as well as with two rounds of whole genome duplication preceding the divergence of jawed and jawless vertebrates (round 1: 652 mya) and the divergence of bony and cartilaginous fish (round 2: 525 mya), respectively (Blair and Hedges 2005; Flajnik and Kasahara 2010; Ohno 1970; Rast et al. 1997). Boehm and Bleul (2007) have proposed that the acquisition of the thymus is the outcome of natural selection for a specialized site to eliminate autoreactive lymphocytes that might be generated through RAG-mediated Ig domain receptor recombination. The discovery of thymoid tissue in lampreys that predates the emergence of the TCR/MHC system, however, conflicts with their hypothesis (Bajoghli et al. 2011; Boehm and Bleul 2007).

The vertebrate thymus is derived from pharyngeal arches and pouches, the number of which differs across vertebrates. Unsurprisingly, the number, anatomical position, and structure of thymi differ between species as well. Some bony fish (superclass Osteichthyes), for example, have one thymus composed of two lobes, whereas other cold-blooded vertebrates, birds, and placental mammals have multiple thymi (reviewed in Rodewald 2008). By the time bony fish emerge, thymi derived from the 3rd and 4th pharyngeal pouches with an adult location ranging from the cervical to the thoracic region, like those in extant mammals, are established (Grevellec and Tucker 2010; Rodewald 2008). While some mammals, such as mice, have recently been described as having more than one thymus, primates have a single thoracic thymus (Dooley et al. 2006; Terszowski et al. 2006; Wong et al. 2011). Unlike fish and birds, whose thymi are attached to the pharynx for the whole of their lives, the mammalian thymus tends to migrate during development, introducing the possibility of intra- and interspecies variation in its location (Dooley et al. 2006; Grevellec and Tucker 2010; Lam et al. 2002). In humans, variation in thymus position is rare and tends to be associated with disease (Shah et al. 2001; Sturm-O'Brien et al. 2009; Tovi and Mares 1978).

Interspecies comparisons of thymus development and function in primates are not common, however, some differences in development have been suggested by the findings of Buse et al. (2006). Thymus development in cynomolgus/crab-eating macaques (*Macaca fascicularis*) appears to be accelerated in comparison to humans, with macaque lymphoid cells preceding the appearance and functional maturation of human lymphoid cells by approximately four gestational weeks (Buse et al. 2006). By comparison, the infiltration of other cells into the thymus occurs later in cynomolgus macaques than in humans, with macrophages appearing at week 10 in the former and week 8 in the latter. By parturition, however, the thymi of these species appears to be similarly developed. Some of these differences in gestational development are likely associated with a near 120-day discrepancy in gestation between these species.

Bone Marrow

The presence of bone marrow in some species of cartilaginous fish suggests that marrow emerged in some species in the absence of bone (Tavassoli 1986). In mammals, bone marrow becomes a site of lymphopoiesis and takes on the role of a primary lymphoid organ. In other animal classes, lymphocyte development can occur in other locations. In adult amphibians, for example, bone marrow expresses genes important for lymphocyte development (e.g., *RAG* genes), but B cells mainly undergo differentiation in the spleen and liver (Du Pasquier et al. 2000; Greenhalgh et al. 1993).

Amongst mammals, strong interspecies differences in haematopoiesis have been noted between rodents and hominoid primates. A simple indication of potential differences in bone marrow function between mice and hominoids is the disparity in the relative proportions of circulating neutrophils between species. Neutrophils emerge from bone marrow as terminally differentiated granulocytes and only circulate for a few hours. They represent 50–70 % of the circulating blood leukocytes in hominoids such as humans, and only 10–25 % in mice, suggesting an interspecies difference in the rate of production (Doeing et al. 2003; Mestas and Hughes 2004). Similarly, Old World monkeys have been noted to have lower percentages of circulating neutrophils (10–42 %) than humans, though these numbers vary depending on the living conditions of the animals (reviewed in Haley 2003). More complex differences between haematopoietic stem cell receptors have been noted between primates and other mammals, with humans exhibiting low CD117 and high CD135 expression on stem cell progenitors and mice exhibiting the inverse pattern (Sitnicka et al. 2003). Both of these stem cell markers are tyrosine kinase receptors. Differences in their expression suggest interspecies differences in tyrosine kinase activity and possibly signal transduction in these cells in the marrow.

Spleen

The spleen is a highly vascularized organ that filters dead/damaged cells and foreign material. Its primary immune function in mammals is to protect the host from blood-borne pathogens (Diamond 1969; Evans 1985). In the healthy primate spleen, this is achieved through either the filtering of antigens by macrophages and dendritic cells located in marginal zones surrounding arterioles, migration of activated dendritic cells to splenic T cells regions and presentation of antigen to T cells, or trapping of activated B cells in the splenic T-cell zone and their subsequent interaction with antigen-specific T helper cells and proliferation (reviewed in Cesta 2006b).

Though lampreys maintain a primordial spleen, the organ is found in cartilaginous fish and first emerged as an independent structure in the last common ancestor to extant jawed vertebrates 652–525 million years ago (reviewed in Boehm et al. 2012a)(Blair and Hedges 2005). Splenic function and structure has changed considerably over evolutionary time and varies strongly from animal class to animal class. In mammals, for example, the spleen is also a very important site of B-cell development. Mammalian B cells initially develop in the bone marrow, but typically mature in the spleen (Brendolan et al. 2007; Drayton et al. 2006). This is not the case for other animal classes. In cartilaginous fish B-cell development starts in the liver and migrates to the kidneys before the final stages occur in the spleen (Du Pasquier 1973). In bony fish, B-cell development bypasses the spleen all together and occurs mainly in the kidneys (or pronephros). In birds, B cells spend the earliest stages of their development in the spleen before migrating to the bursa of Fabricius, located near the cloaca (Boehm et al. 2012a; Pickel et al. 1993). While a perpetual blood filter, the precise role a spleen plays in the immune system of an animal varies considerably across the animal kingdom.

The complex structure of the primate spleen, with red pulp with a white pulp compartmentalized into peri-arteriolar lymphoid sheath (PALS), follicular dendritic cell clusters, and a marginal zone with germinal centers for B-cell proliferation, only emerged with the last common ancestor of mammals (Fig. 2) (Cesta 2006b; Hofmann et al. 2010). The spleen of cartilaginous fish bears a familiar structure of blood filtering and component storing red pulp containing lymphocyte organizing white pulp, with well-defined lymphocyte zones (Rumfelt et al. 2002). The white pulp has become more complex over evolutionary time, with the spleens of more recently emerged animal classes taking on progressively more complex white pulp structure and function (Jeurissen 1991; Zapata and Amemiya 2000).

In the simplest of terms, mammalian white pulp is comprised of lymphoid tissue that surrounds arterioles, with a centralized area containing resting or proliferating B cells surrounded by a peripheral area where T cells congregate. This is, however, a highly derived structure. The white pulp of cartilaginous fish is structured in the opposite fashion of more recently emerged species, with centralized zones of T cells, dendritic cells, and immunoglobulin-producing cells, encircled by B-cell clusters (Rumfelt et al. 2002). Bony fish maintain spleens with these same basic components, but have developed end capillary structures known as ellipsoids, covered in macrophages and involved in antigen capture. Because of the position of

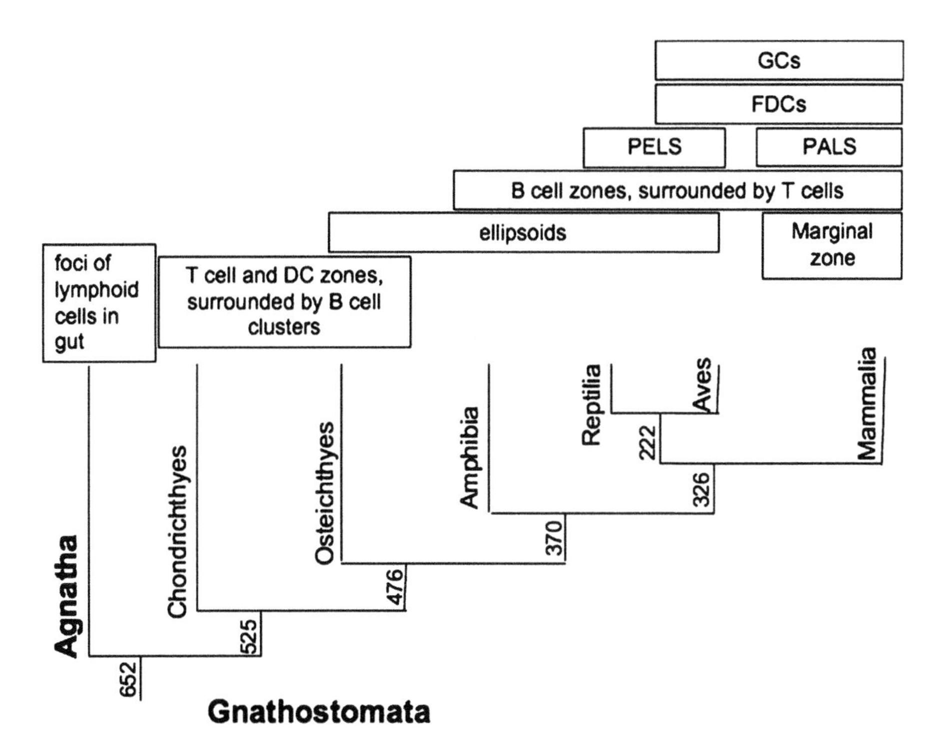

Fig. 2 Evolution of the splenic white pulp. Positioning of Reptilia, Aves, and Mammalia based on Blair and Hedges (2005) and Shedlock and Edwards (2009). *GC* germinal centers, *FDCs* follicular dendritic cells, *PELS* peri-ellipsoid lymphocyte sheath, *PALS* peri-arteriolar lymphoid sheath

macrophages that capture antigens, ellipsoids bear a resemblance to the marginal zone that serves this same function in mammals. They have, however, also been posited to function as protogerminal centers where lymphocytes might proliferate (Hofmann et al. 2010).

As the site where foreign antigens are captured and presented to lymphocytes, the marginal zone is an extremely important acquisition for the mammalian spleen. The marginal zone is found within the PALS, which encircles arterioles and functions as a site where T cells interact with passing B cells and antigen-containing dendritic cells (reviewed in Mebius and Kraal 2005). The white pulp of teleost fish (Teleostei) and anuran amphibians (Anura), though less developed than that of reptiles and birds, is known to antigen trap in ellipsoid structures physically associated with capillaries and arterioles (Sailendri and Muthukkaruppan 1975; Secombes and Manning 1980; Turner and Manning 1973). The current best first evidence of a marginal zone-like structure is found in the increasingly compartmentalized white pulp of reptiles (Murata 1959). Both reptiles and birds share a structure possibly analogous to, or a precursor of the marginal zone, known as the peri-ellipsoid lymphocyte sheath (PELS). The PELS covers ellipsoids, appears to antigen trap, and contains cells similar to those found in the mammalian marginal zone—lymphocytes and cells resembling follicular dendritic cells (reviewed in Olah and Vervelde 2008). Unlike the mammalian marginal zone, however, reptile PELS do not contain traditional germinal centers, where activated B cells undergo proliferation and differentiation (Zapata and Amemiya 2000). Birds are the earliest animals known to exhibit germinal centers in follicular sites (Jeurissen 1991).

With mammals, a macrophage-rich marginal zone specifically involved in antigen capture emerged, along with germinal centers for B-cell proliferation (Zapata and Amemiya 2000). How the development of these structures over time precisely relates to differences in splenic function between animal classes is difficult to discern. Comparative information on amphibian and mammalian B-cell development suggests that the absence of splenic structures such germinal centers can affect immune function. In frogs, an absence of such centers has been associated with lowered antibody affinity (Hsu 1998; Marr et al. 2007). Unfortunately, there is little comparative information available on the structure and function of the primate spleen, though interspecies differences in shape have been noted (von Krogh 1936).

Lymph Nodes

Lymph nodes are a relatively new immune innovation in the evolution of vertebrates. They are found only in mammals and some birds [e.g., ducks (Anatidae)], though reptiles maintain clusters of lymphoid nodules that bear some resemblance to lymph nodes (Lawn and Rose 1981; Sugimura et al. 1977; Zapata and Amemiya 2000). The canonical lymph node consists of encapsulated lymphoid lobules, vessels, and sinuses and acts as a local station of leukocyte responses to infection. Lymph nodes mainly function to filter local lymph of antigens, act as sites for activated antigen-presenting cells to present antigens to naïve lymphocytes, and to host and organize the clonal expansion of reactive lymphocytes (Kaldjian et al. 2001; Karrer et al. 1997).

Lymph node number and location is highly variable between species. Mice, for example, have 22 identified lymph nodes, while primates are prolific developers of lymph nodes and maintain hundreds of these structures (e.g., humans have ~450 lymph nodes) (Van den Broeck et al. 2006; Willard-Mack 2006). Moreover, persistent inflammation can lead to *de novo* synthesis of lymph nodes and contribute to variation between individuals (Drayton et al. 2006).

Mammals also vary in their expression of specific types of lymph nodes. Rodents, for example, do not exhibit tonsils (nasopharyngeal lymph nodes), unlike many other mammals. It has been suggested that the extensive nasal-associated lymphoid tissue (NALT) found in rodents is analogous to tonsils (Heritage et al. 1997). Many nonhuman primates, however, maintain more extensive NALT than rodents and maintain tonsils (Haley 2003). It is tempting to assume that many rodents, with their heads a few millimeters from the ground, may have lost nasopharyngeal lymph nodes due to the selection pressure of complications arising from increased lymph node infections. However, the current information on the importance of tonsils in other species is sufficiently murky and incomplete that it does not support this notion. The effect of pharyngeal tonsillectomy in both child and adult humans with persistent pharyngeal infections has not been systematically studied and reviews of case studies have argued that the procedure does not affect the occurrence of these infections or lead to immune dysfunction (Blakley and Magit 2009; Burton and Glasziou 2009; Burton et al. 2000).

The absence of tonsils in rodents does, however, suggest a significant alteration to the embryonic structure from which they originate, the 2nd pharyngeal pouch (Grevellec and Tucker 2010; Heritage et al. 1997). Similarly, second pouch development has undergone modification in different primate species. Cesta (2006) has noted that humans have four sets of tonsils (lingual, palatine, pharyngeal, and tubal), while only three sets of tonsils have been described for other primates (lingual, palatine, and pharyngeal). The acquisition of tubal tonsils in humans suggests a departure in second pouch development from that other primate species in the last 5–6 million years. The functional ramifications of the acquisition of these tonsils, however, are not known.

Major Cellular Components of Innate Immunity: The Phagocytes

Monocytes/Macrophages

Phagocytic immune cells form a critical link between the mammalian innate and adaptive immune systems by engulfing, processing, and presenting fragments of pathogens to T cells and subsequently initiating adaptive responses. The phagocytes include monocytes/macrophages, dendritic cells, and neutrophils (both a phagocyte and granulocyte). Phagocytosis is likely the oldest technique of sequestering and destroying foreign material in the interest of host defense. Indeed, phagocyte-like cells have been identified in many lifeforms, suggesting that the phagocytes may originate at the base of the Eukaryota domain and represent the earliest immune cells in multicellular life (Waddell and Duffy 1986). Social acrasid amoebae, for example, are often referenced as sharing a last common ancestor with all organisms that either participate in or maintain cellular components that identify nonself and phagocytose for nutrition or defense (Dzik 2010). Similarities in cellular organization, phagocytosis, and the migration of unicellular amoebic organisms, such as *Acanthamoeba* and human macrophages, have also been noted (Anderson et al. 2005). Several genera of amoebae and human macrophages share subsets of cell receptors used for nonself detection and phagocytosis, experience manipulation of similar cellular cascades by the same intracellular pathogens, and share a similar mode of shifting ectoplasm and hyaloplasm to "crawl" (Al-Khodor et al. 2008; Allen and Dawidowicz 1990; Yan et al. 2004). Particularly compelling evidence that the origins of primate phagocytes might be rooted in single-celled eukaryotic organisms is the tendency of *Acanthamoeba* to phagocytose a wide range of viruses, bacteria, and fungi and maintain an oxidase respiratory burst system to degrade phagocytosed material similar to that found in human macrophages (Allen and Dawidowicz 1990; Davies et al. 1991). Of course, how these amoebic activities are evolutionarily linked to primate phagocytes is not clear. As Siddiqui and Khan (2012) point out, despite the similarities between some amoebae genera and human macrophages, amoebae have evolved into organisms that live either as singletons or as colonies and, in some cases, can become cysts when resources are in short supply (Siddiqui and Khan 2012). These are important characteristics not shared with human macrophages.

By the emergence of invertebrate life, phagocytosis became an essential component of immune defense, and multiple phagocytic cell types had emerged (reviewed in Wood and Jacinto 2007). Monocytes/macrophages are likely the oldest of these cells and multiple authors have compared their morphology and function to that of amoebae (Al-Khodor et al. 2008; Chen et al. 2007; Cooper 2010; Davies et al. 1991; Siddiqui and Khan 2012). Typically derived from circulating monocytes that venture into tissues and differentiate there, vertebrate macrophages act as body "housekeepers", as well as immune system sentinels. Both cell types identify "nonself" and phagocytose foreign material and cellular debris, eliminate tumour cells, secrete a vast array of both proinflammatory and anti-inflammatory cytokines, modulate other immune cells, and act as antigen-presenting cells (APCs) (Mosser and Edwards 2008).

Macrophage-like cells have been noted in a wide variety of invertebrate species, within which they identify nonself agents and initiate phagocytosis and cytotoxic activities (Dzik 2010; Ehlers et al. 1992; Fautin and Mariscal 1991; Kurtz 2002; Moita et al. 2005; Pech and Strand 1996; Rhodes et al. 1982). By the emergence of cartilaginous fish macrophages became antigen-presenting cells expressing major histocompatability complexes (MHC) (Flajnik and Kasahara 2001). MHC molecules are polymorphic APC receptors, capable of binding a variety of both phagocytosed host and pathogen-derived antigenic peptides, presenting them to lymphocytes, and subsequently initiating adaptive responses. From the appearance of cartilaginous fish onwards, macrophages tend to distribute throughout bodily tissues, accumulate at areas of pathogen entry, phagocytose invading pathogens, and load MHC receptors bearing peptides from these pathogens onto the cell surface, initiating T-cell responses through TCR interaction (reviewed in Flajnik and Kasahara 2010).

How macrophage morphology and "behavior" has changed across vertebrate classes over time is difficult to assess, primarily because phagocytic cell and macrophage subsets have historically been difficult to accurately identify using antibody-based technologies in major classes of animals such as fish and birds and few markers have been identified (Kaspers et al. 2008). There is reason to believe that while macrophage morphology and behaviour is fairly well conserved across vertebrates, the cell type functions differently in certain classes. Macrophages appear to reside in different body regions across animal classes. For example, they are persistently present in the peritoneal cavity of mammals but appear to be completely absent from this region in healthy birds and must be recruited from circulation or tissues (Conrad 1981; Rose and Hesketh 1974; Sabet et al. 1977).

Baseline comparative studies of primate monocytes/macrophages and examinations of stimulated monocytes suggest the monocyte function of closely related primates differs considerably. An examination of baseline sialic acid-recognizing Ig-superfamily lectins (siglec) – immune cell surface molecules that can downregulate immune cell responses – expression across human, rhesus macaque (*Macaca mulatta*), and sooty mangabey (*Cercocebus atys*) monocytes found marked interspecies differences in both type and level of siglec constitutively expressed (Jaroenpool et al. 2007). The same study also found that monocytes from SIV+ rhesus macaques more strongly expressed two siglecs (1 and 7) compared to monocytes from SIV+ sooty mangabeys. The authors suggest that these differences may contribute to the strong immune activation and pathology typically seen in SIV+ macaques and typically absent in SIV+ mangabeys. They also suggest lineage-specific adaptation to IVs in sooty mangabeys since they shared a last common ancestor with rhesus macaques approximately 9 million years ago (Fabre et al. 2009). Barreiro et al (2010) found that human, chimpanzee (*Pan troglodytes*), and rhesus macaque monocytes stimulated with lipopolysaccharide (LPS) initiated species-specific transcription of genes associated with immune responses to viral infection and cancer. Specifically, human and chimpanzee responses have diverged from one another. Human responses were enriched for genes associated with apoptosis and cancer pathology, while the responses of chimpanzees, a natural immunodeficiency virus host thought to also have a low incidence of cancer, were enriched for HIV-interacting genes (Barreiro et al. 2010). Such interspecies differences in monocyte/macrophage responses suggest lineage-specific adaptation has occurred in catarrhines since cercopithecoids and hominoids last shared a common ancestor ~30 million years ago (Steiper and Young 2006).

Dendritic Cells

Like all phagocytes, dendritic cells (DCs) are ancient in origin, though likely younger than macrophages. DCs maintain a tree-like shape, with multiple dendrites as "branches." They tend to be found in tissues that have a direct interface with the external environment (e.g., mucosa) and lymphoid organs, where they participate in pathogen detection and antigen presentation. Canonical dendritic cells first emerged in teleost fish, which suggests that DCs have existed since at least the Devonian period ~420–360 million years ago (Setiamarga et al. 2009; Wolfle et al. 2009). Mixed leukocyte reactions in jawless fish such as hagfish and lampreys, however, suggest that these animals maintain similar antigen-presenting cells that recognize nonself, introducing the possibility of a much earlier emergence date for a DC precursor (Raison et al. 1987). Once emerged DC morphology and function has remained conserved in subsequent animal classes. Mammalian and teleost DCs, for example, are derived from similar precursor cells, maintain dendrites, express similar cell markers, are activated by TLR ligands, participate in phagocytosis, and are found in similar lymphoid organs (e.g., the spleen) (Bassity and Clark 2012).

In primates, DCs are a heterogeneous group of sentinel cells that arise from a haematopoietic or peripheral blood mononucleocyte progenitor, are present throughout body tissues and organs (skin, intestines, liver, lungs, etc.), survey for pathogens, and engage bacteria and viruses through a variety of receptors (reviewed in Wu and Liu 2007). Unlike other phagocytes, however, mammalian DCs are not directly involved in early clearance of pathogens. While the primary function of monocytes/macrophages is to destroy nonself, the primary function of DCs appears to be antigen presentation via MHC I and MHC II molecules. Through antigen presentation DCs are exceptional stimulators of T-cell responses (reviewed in Nussenzweig et al. 1980; Savina and Amigorena 2007). DCs mature through this process of antigen presentation, switching from a phagocytosing cell to an antigen-presenting cell that migrates to secondary lymphoid tissues, presents MHC bound antigen to T cells, and initiates T-cell responses through a two-signal system (reviewed in Gilliet et al. 2008; Kushwah and Hu 2011).

In primates DCs represent a small proportion of leukocytes (e.g., less than 1 % of circulating leukocytes), so most comparisons of primate DC phenotype and function are comparisons of DCs cultured from monocytes, bone marrow, or peripheral blood stem cells (Ashton-Chess and Blancho 2005; Gabriela et al. 2005; Ohta et al. 2008; Prasad et al. 2010; Soderlund et al. 2000). Such in vitro culture methods can differ across studies and species, making interspecies comparisons of these DCs difficult. Interspecies comparisons of DC phenotypes are also frustrated by inconsistent antibody affinity for DC markers across species. In vivo examinations of DCs have found that nonhuman catarrhine species maintain the two major subsets of DCs found in humans—the monocyte-like myeloid DCs (mDCs) and plasma cell-like plasmacytoid (pDCs) (Coates et al. 2003; Diop et al. 2008; MacDonald et al. 2002; Malleret et al. 2008; Pichyangkul et al. 2001). However, markers for these cells differ between primate species. Rhesus macaque monocyte-derived DC (moDCs) markers CD11b, CD16, and CD56 are expressed on human monocytes and natural killer cells, rather than DCs (Brown and Barratt-Boyes 2009; Carter et al. 1999). Such differences in protein expression can frustrate interspecies analyses by antibody-based flow cytometry and may have consequences for immune function. Moreover, antibody affinity for human DC markers can be depressed in other primates, making the identification of pDCs particularly difficult for more distantly related primates such as platyrrhines (specific antibodies discussed in Jesudason et al. 2012).

There are other notable interspecies differences in catarrhine DC phenotype that suggest the function of these cells have diverged multiple times since cercopithecoids and hominoids last shared a common ancestor ~30 million years ago. Human, African green monkey (*Chlorocebus* sp.), and cynomolgus macaque moDCs strongly express DC-SIGN/CD209, a receptor involved in pathogen recognition, cell migration, and T-cell activation (Geijtenbeek et al. 2000a, b, Geijtenbeek et al. 2003). The weak expression of this receptor on rhesus macaque moDCs suggests that these activities are mediated differently between catarrhine species (Wu et al. 2002). Information on primate DC baseline function is very limited, with most data on interspecies differences in function having been collected via a challenge model. An in vitro comparison of moDC responses to TLR3 ligand and common vaccine adjuvant Poly I:C, for example, found that despite similar TLR expression levels, human moDCs initiate stronger anti-inflammatory and antiviral responses (IL-12, IFNα) to the ligand than rhesus macaque moDCs (Ketloy et al. 2008). As Poly I:C is a common adjuvant, these differences may have implications for vaccine studies using rhesus macaques as a model. During in vivo infection catarrhine DCs exhibit notable differences in function. Sooty mangabeys, in particular, exhibit signs of lineage-specific adaptation of pDC function to viral infection. Unlike the pDCs of naïve hosts, such as macaques and humans, pDCs from natural immunodeficiency virus (IV) host sooty mangabey infected with IVs fail to express CCR7, an important receptor for cell homing to lymphoid tissue. Sooty mangabey pDCs stimulated with ligands that are viral mimetics do not migrate to the lymph nodes (Mandl et al. 2008). Given that pDCs are implicated in the dissemination of IVs in lymph nodes, it is possible that the absence of IV-mediated homing of pDCs to the lymph nodes in sooty mangabeys partially explains AIDS resistance in this species. Similarly, sooty mangabey pDCs have been found to produce less IFNα in response to yellow fever virus 17D vaccine than pDCs isolated from rhesus macaques and humans, suggesting that antiviral responses of pDCs differ between catarrhine species (Mandl et al. 2011).

Neutrophils

Neutrophils are the most recently emerged phagocyte, first appearing in mammals. Neutrophils are both phagocytes and granulocytes. Granulocytes are immune cells that maintain intracellular/membrane-bound vesicles containing biologically active compounds that can be quickly released in response to stimulus. While neutrophils are unique to mammals, birds, and reptiles, amphibians and fish maintain a related and considerably more granular phagocytic cell known as a heterophil. Both cell types engulf and destroy microbes, maintain granules that contain antimicrobial substances, circulate in the blood, and initiate tissue repair (Benoit et al. 2008; Robert and Ohta 2009) (reviewed in Harmon 1998; Yoder 2004). Neutrophils/heterophils differ considerably from other phagocytic cells in a number of functionally important ways. Unlike macrophages and dendritic cells, neutrophils/heterophils emerge from bone marrow myeloblasts terminally differentiated and do not differentiate further when they phagocytose material or migrate to and from tissues. While macrophages and dendritic cells tend to traverse tissues in low numbers, neutrophils/heterophils circulate in blood and cross into tissues in very large numbers. Neutrophils/heterophils are also some of the shortest lived immune cells, maintaining lifespans of hours or a few days, rather than weeks (reviewed in Amulic et al. 2012).

The primary function of neutrophils appears to be to limit bacterial infections (Borregaard and Cowland 1997). They are often the first innate immune cells at the site of an infection, migrating in large numbers to sites of inflammation in the tissues by rolling along endothelium until they are arrested and trafficked to these sites by expressed chemokines and selectins (Ley et al. 1998; Smith et al. 2004; Zhang et al. 2001). There they amplify inflammation through the release of cytokines and chemoattractants and are implicated in overt immune responses that lead to tissue injury or systemic inflammation (Ear and McDonald 2008; Kobayashi 2008; Zemans et al. 2009). It is perhaps due to the potential of these cells to cause tissue injury that neutrophils have evolved several mechanisms that limit their antimicrobial activities. They undergo apoptosis after phagocytosis and can initiate suicidal pathogen trapping. They also participate in a unique feedback mechanism that controls exuberant neutrophils. Mass migration of neutrophils to a site is followed by the recruitment of monocytes (Scapini et al. 2001; Soehnlein et al. 2008). These monocytes differentiate into macrophages, which then limit neutrophil activities through lipoxin-triggered phagocytosis (Godson et al. 2000).

Neutrophils participate in several antimicrobial activities that the other phagocytes do not. While phagocytosis in macrophages occurs through an endocytic pathway, neutrophil phagosomes become active upon granule–phagosome fusion (reviewed in Amulic et al. 2012; Lee et al. 2003; Savina and Amigorena 2007). Subsequent degranulation serves as an antimicrobial mechanism, nonspecifically destroying both pathogens and host tissue as well as enhancing phagocytosis by other phagocytes (Soehnlein et al. 2009). Neutrophils are known to swarm lymph nodes during parasitic infection, an immune strategy that other phagocytes do not appear to use (Chtanova et al. 2008). In an interesting potential adaptation to large pathogens (e.g., helminths), neutrophils participate in a kind of antimicrobial "hara-kiri" by releasing decondensed chromatin into the environment and creating extracellular traps (NETS) (Brinkmann et al. 2004; Fuchs et al. 2007). This action inevitably kills the neutrophil, but nets and kills the pathogens as well.

Neutrophils and heterophil granular contents differ, with neutrophils relying more heavily on an oxidative burst response during phagocytosis (Penniall and Spitznagel 1975; Stabler et al. 1994). Heterophils and neutrophils differ in how they contribute to lesion pathology as well. In birds and reptiles, heterophils that invade a lesion tend to form granulomas, while in mammals neutrophils may form granulomas but can also liquefy and form abscesses (Harmon 1998; Montali 1988). The number of circulating neutrophils/heterophils differ strongly between animal classes and orders as well, with neutrophils representing between 50 and 70 % of circulating blood leukocytes in primates, while they represent only 15–20 % of the circulating population in rodents and zebrafish (Haley 2003; Hawkey 1985; Martin and Renshaw 2009).

Interspecies comparisons of neutrophil phenotypes are difficult to complete for many reasons. In vitro comparisons are hard to undertake as the cells are very difficult to immortalize and primary neutrophils have a very short life span (Amulic et al. 2012). In vivo interspecies comparisons of neutrophil activities are made difficult by how widely the proportions of neutrophils to other circulating leukocytes differ between mammalian species. It is apparent that neutrophils phenotypically differ within the mammalian order. Mouse neutrophils, for example, lack defensins, while human neutrophils maintain these antimicrobial proteins (Eisenhauer and

Lehrer 1992). Human neutrophils tend to maintain large doses of histamine, a pro-inflammatory amine that increases blood vessel permeability, triggers cytokine release by neighboring leukocytes, triggers smooth muscle constriction, and is active in allergic reactions. Rabbits and guinea pigs tend to maintain lower quantities of this molecule (Haley 2003).

Even within the context of specific infections, comparative functional studies of neutrophils in primates are rare. Through an SIV challenge model Elbim et al (2008) found some evidence that neutrophil function differs between primate species. Rhesus macaques exhibit increased expression of an SIV co-receptor on neutrophils, as compared to African green monkeys. This difference in expression is correlated with increased neutrophil death in SIV+ rhesus macaques during the early stages of infection, suggesting that neutrophil function has diverged since these catarrhine lineages last shared a common ancestor ~9 million years ago (Elbim et al. 2008; Fabre et al. 2009; Steiper and Young 2006). More broadly, primates exhibit interspecies differences in the proportions of circulating neutrophils. Neutrophils represent 50–70 % of circulating leukocytes in humans and other apes and 10–42 % in cercopithecoid primates (Doeing et al. 2003; Haley 2003). An explanation for the intensely neutrophil-rich blood of hominoids is difficult to determine. A number of primate traits appear to be correlated with increased proportions of basal neutrophils, including traits that may increase exposure to bacterial pathogens such as greater terrestriality/body mass and female mating promiscuity (Nunn 2002). However, there are few comparative interspecies studies of primate neutrophil function to supplement these findings. As such, the functional ramifications of these differences in neutrophil proportions are not known.

Major Cellular Components of Innate Immunity: Other Granulocytes

Along with neutrophils (described above), vertebrate granulocytes include mast cells, basophils, and eosinophils. Granulocytes are important combatants of invading bacteria and mediators of inflammation, roles they are able to complete through degranulation and sometimes phagocytosis. While the evolutionary roots of phagocytosis have been postulated to predate multicellular eukaryotic life, the granulocyte activity of "degranulation," releasing compounds from internal vesicles known as "granules," has been postulated to have appeared with an expansion of the Metazoa during the pre-Ediacaran periods of the Neoproterozoic era, between 1,000 and 600 mya (reviewed in Crivellato et al. 2010) (Peterson and Butterfield 2005). Key functions of primate immune system granulocytes are rooted in the origins of piecemeal degranulation of specific granule components, a function known to occur in the immune cells of teleost fish that may have emerged in invertebrate life (reviewed in Crivellato et al. 2010).

Mast Cells

A mast cell progenitor, sharing qualities with basophils, appears to have emerged in tunicata during the Paleozoic (de Barros et al. 2007). Mast cells have subsequently been found in all vertebrate chordates. Mast cells are innate immune effector cells that also modulate the activities of innate and adaptive immune cells. They are a heterogeneous population of bone marrow-derived granulocyte cells that store biologically active compounds in membrane-bound granules. Mast cell precursors emerge from bone marrow and differentiate into mast subsets when they migrate to mucosal and connective tissues (Galli et al. 2008; Galli et al. 2005). At these sites, mast cells often directly interface with the environment and can live for weeks or months (Church and Levi-Schaffer 1997). Similar to other granulocytes, mast cells release various granular proteins upon activation, as well as proinflammatory molecules such as prostaglandins and leukotrienes, cytokines, and chemokines. The original function of mast cells was likely inflammation and host defense against bacteria and parasites, though they now also play a very important role in immune regulation, tissue repair, and blood vessel development (Malaviya et al. 1996) (reviewed in Crivellato and Ribatti 2010). In mammals, mast cell cytokine production can also induce a subset of dendritic cells called Langerhans cells to migrate to the local lymph nodes, where antigen presentation takes place (Jawdat et al. 2004). Intriguingly, mast cells are major mediators of detrimental hypersensitivity reactions such as anaphylaxis and yet are maintained in all vertebrate chordates (Krishnaswamy et al. 2001). Their persistence, despite their contribution to sudden and fatal immune reactions suggests that mast cells must also play a very beneficial role in immune protection.

Mast cell size, granule contents, and granule numbers vary considerably between vertebrate classes and species (Dvorak 2005). A sharp divide in histamine content in the mast cells of warm-blooded and some cold-blood vertebrates has been well noted (Reite 1965; Takaya 1969; Takaya et al. 1967). All descendants of early reptiles store large amounts of histamine in mast cells. Mammalian and avian mast cells are particularly prolific producers of histamine, while histamine is either absent or present in very low levels in amphibians and most fish. The stark contrast in histamine production between animals that emerged before and after reptiles has led to the conclusion that mast cell granule storage of histamine emerged in the last common ancestor of extant reptilia, possibly as early as the Permian period (300–250 mya) (Chieffi Baccari et al. 1998; Mulero et al. 2007; Shedlock and Edwards 2009). In fact, many fish seem unable to respond to injections of histamine, suggesting that a histamine receptor system is absent in much of this animal class (Mulero et al. 2007; Reite 1972). Mulero et al. (2007) have concluded that the presence of histamine in the mast cells of perciform fish (e.g., trout), the largest order of teleost fish, suggests mast cell storage of histamine evolved twice in vertebrate history (Mulero et al. 2007). This view has been contested, as some amphibians and earlier emerged classes such as ascidians (e.g., sea squirts) have been found to maintain low levels of histamine in mast/immune cell granules (reviewed in Crivellato and Ribatti 2010).

Mammalian mast cells have acquired at least one trait that sets them apart from the mast cells of other vertebrate classes. They uniquely become activated in response to immunoglobulin E (IgE) (reviewed in Crivellato and Ribatti 2010). This reaction appears to be very important for defense against helminth parasites. Mice bearing an IgE deletion, for example, experience increased *Schistosoma* sp. worm loads and inflammation (King et al. 1997b). It was long thought that IgE and its receptors on mast cells represented a system of mast cell activation that newly emerged in mammals, but this view has recently been challenged by detection of an IgE receptor-like protein on zebrafish mast cells (Da'as et al. 2011). While the roots of an IgE receptor-based mast cell activation system may be considerably older than previously thought, it appears mammals alone use IgE in antiparasitic strategies.

As with many leukocyte subtypes, primate mast cells have not been well investigated for baseline functional differences. Some processes, such as the cleavage of particular granule contents (e.g., chymases) appear to be conserved in cercopithecoids and hominoids (Thorpe et al. 2012). There are, however, indications that the granular contents of primate mast cells differ between species. Tryptases, for example, are serine peptidases stored in mast cell granules and implicated in pathological immune conditions such as anaphylaxis and asthma (Clark et al. 1995; Schwartz et al. 1987). δ-trypase (*TPSD1*) is functional and active in cercopithecoids but has undergone several nonsense mutations and truncations in hominoids over the last 30 million years, and the human–chimpanzee lineage over the last 6 million years. As a result δ-trypase is almost nonfunctional in humans and chimpanzees (Trivedi et al. 2008). Rather β-tryptases appear to be the more active tryptase in humans and chimpanzees, suggesting that cercopithecoid models of mast cell disorders such as allergy and anaphylaxis may not accurately reflect human tryptase activity (Trivedi et al. 2008, 2007).

Basophils

Basophils are granulocytes and the rarest of circulating leukocytes in vertebrates. Typically representing less than 0–0.3 % of circulating leukocytes, basophils share a number of functions with mast cells, including contributing to antiparasitic activities and allergic reactions (reviewed in Haley 2003; Karasuyama et al. 2011). Basophils appear to share a common cellular ancestor with mast cells that first arose in the hemolymph of tunicata (de Barros et al. 2007). Like mast cells, mammalian basophils express IgE receptors and produce considerable levels of histamine (Ohnmacht and Voehringer 2009; Stone et al. 2010). However, basophils are a distinct, short-lived cell type that emerges from bone marrow terminally differentiated, circulates in blood perennially, and fails to proliferate after maturity (reviewed in Hida et al. 2005; Karasuyama et al. 2011). Recently, it has been found that mammalian basophils promote Th2-cell differentiation, specifically participating in antigen presentation and stimulating T cells via IL-4 secretion (Sokol et al. 2009; Yoshimoto et al. 2009). This represents a considerable departure from mast cell function.

Basophils are exceptionally rare in most vertebrate species, with only turtles and rabbits reported to maintain comparatively high numbers in circulation (Canfield 1998; Haley 2003). The very low circulating numbers of basophils in vertebrates has been a significant hindrance to the comparative functional study of this cell type. Comparative studies of primate basophil function have not been completed.

Eosinophils

Eosinophils are terminally differentiated granulocytes with phagocytic capabilities that circulate in the blood, traverse tissues easily to attack parasites and other pathogens, degranulate rapidly, and modulate Th2 responses (reviewed in Hamann et al. 1991; Shamri et al. 2011; Teixeira et al. 2001). Like basophils, they are implicated in hypersensitivity reactions such as asthma and atopic dermatitis. Their granule contents are distinct from those of other granulocytes and include cationic proteins such as eosinophil-associated RNases and ribonuclease-3, which are antihelminthic and are only expressed in very small quantities by other immune cells (Hogan et al. 2008; Olsson and Venge 1974). Eosinophil degranulation releases these cytotoxic molecules in response to a variety of pathogenic challenges (viral, bacterial, and fungal) (Butterworth 1977; Rosenberg and Domachowske 2001; Yoon et al. 2008; Yousefi et al. 2008). Similar to neutrophils and basophils, eosinophils tend to be short lived, spending 8–12 hours in circulation and possibly 8–12 days in tissues if left unstimulated (reviewed in Uhm et al. 2012). In all vertebrate species they represent a very small proportion of leukocytes (1–4 %) (Uhm et al. 2012).

It is thought that eosinophils arose from a heterophil/eosinophil-like ancestor in invertebrates that maintained both amoeboid-like locomotion and phagocytic capabilities as well as granulocytic secretion and tissue remodeling capabilities (Lee and Lee 2005). By the time jawless fish emerged, eosinophils and heterophils had diverged into two different cell types (Kelenyi and Nemeth 1969; Rowley and Page 1985). Given the propensity of eosinophils to cluster at sites of tissue stress and that the emergence of current eosinophil pathogen targets (e.g., helminths) postdate the appearance of eosinophils themselves, Lee and Lee (2005) have proposed that heterophil/eosinophil divergence was driven by nonimmune system factors. Specifically, they hypothesize that eosinophils arose to fulfill a need for a cell that mediated the tissue repair and modeling needs created by an increase in body size and complexity seen in emerging vertebrates. This divergence then led to a heterophil/neutrophil cell lineage dedicated to host defense and an eosinophil lineage taking on tissue remodeling, as well as defense responsibilities.

Though eosinophil phenotypes appear evolutionarily conserved across vertebrates, there are a few differences worth noting. Avian eosinophils tend to be involved in delayed hypersensitivity reactions, while mammalian eosinophils are typically not involved in this sort of response (e.g., allograft rejection and contact dermatitis) (reviewed in Campbell 2004; Lind et al. 1990). Eosinophil degranulation can also have a species-specific phenotype. Human eosinophils, for example, are capable of very quickly and extensively degranulizing in response to both specific and nonspecific stimulation, whereas direct evidence that the eosinophils of other mammalian species, such as mice, can do the same is very limited (Lee et al. 2012). If tissue remodeling is a primary function of eosinophils, interspecies differences in degranulation would be expected, as tissue remodeling mechanisms would likely differ between species (Lee and Lee 2005).

Eosinophil granule content also differs markedly between mammalian species, which may reflect differences in cell function. While many primates appear to maintain two types of ribonucleases of varying functions in their eosinophil granules, mice maintain six, all of which are paralogous to primate ribonucleases and exhibit strong ribonuclease activity (Lee et al. 2012). Eosinophil granule content differs strongly between major primate clades, suggesting lineage-specific evolution of eosinophil function. Catarrhines, for example, store ribonucleases ribonuclease A family 3 (RNASE3) and ribonuclease A family 2 (RNASE2) at high levels in their granules, while Platyrrhines only store a ribonuclease homologous to RNASE3 (Olsson et al. 1977; Rosenberg and Dyer 1995; Rosenberg et al. 1995). RNASE3 is particularly toxic to host tissue cells. These differences in ribonuclease production and storage suggest that ribonuclease-mediated eosinophil apoptotic, cytotoxic, and antimicrobial activities have changed in these clades since they last shared a common ancestor 40 million years ago (Fabre et al. 2009; Steiper and Young 2006).

Major Cellular Components of Innate Immunity: Natural Killer Cells

Natural Killer (NK) cells are granular lymphocytes. NK cells recognize "altered" self (e.g., tumours) and nonself and have acquired their name because of their ability to destroy cells spontaneously without the need to be primed as other lymphocytes require (Herberman et al. 1975). Through direct contact mammalian NK cells can target and lyse tumour cells or virally infected cells that downregulate MHC I expression as a result of oncogenesis or infection (reviewed in Purdy and Campbell 2009). MHC I class molecules found on other immune cells are NK ligands and interactions between MHCs and NK killer cell immunoglobulin-like receptors (KIR) control NK effector functions (reviewed in Lanier 2008). Target cell killing is initiated when activated NK cells release perforin and granzyme that induce apoptosis (Lieberman 2003). For their importance, NKs are a rare cell population. In humans, NK cells represent 10–15 % of blood lymphocytes, or approximately 2–3 % of all circulating leukocytes (Purdy and Campbell 2009).

Cells exhibiting NK-like activities, such as direct killing of foreign materials via perforin-like proteins, have been noted in invertebrate life (Kauschke et al. 2001). NK-like cells are present in all tunicata, suggesting that this cell type was established by the time this clade emerged between 520 and 794 mya (Blair and Hedges 2005; Chen et al. 2000). As the earliest evidence of MHC molecules is found in cartilaginous fish, it seems that NK adoption of MHC class I molecules as ligands occurred long after the emergence of NK cells (Azumi et al. 2003; Khalturin et al. 2003). Thus, the origins of NK lymphocytes appear to predate the adaptive immune system as a whole.

The role of NK-like cells in nonmammalian classes is not well investigated but appears to be conserved. In tunicata NK-like cells appear to mainly participate in allorecognition and killing nonself materials via granule release of perforin and granzyme-like proteins (Bielek 1988; Shen et al. 2004). NK-like cells with cytotoxic, and allogeneic and tumor-killing capabilities have been found in amphibians and avian species as well (Goyos and Robert 2009; Horton et al. 1996; Jansen et al. 2010; Wainberg et al. 1983). While interclass comparisons of NK-cell distribution and characteristics have led to conclusions that mammalian but not avian NK cells can circulate in blood, it is very likely that a dearth of cross-reactive NK marker antibodies for nonmammals has frustrated these studies (Rogers et al. 2008b). While mammalian NK cells are the focus of intense study, the comprehensive set of markers that can reliably identify NK cells in all mammalian species has yet to be determined (Walzer et al. 2007). Even very closely related species such as humans and rhesus macaques exhibit sharp differences in NK marker expression. Human NKs strongly express CD56, a marker mainly expressed by monocytes in rhesus macaques (Webster and Johnson 2005). The primate species for which markers are well defined tend to be those species important to HIV research (e.g., rhesus macaques and sooty mangabeys). Interspecies comparisons of primate NK function also tend to be limited to these two species. Rhesus macaque and sooty mangabey NK-cell functional comparisons do indicate the evolution of baseline interspecies differences since these species last shared a common ancestor ~9 million years ago (Fabre et al. 2009). The NK cells of healthy sooty mangabeys appear to be twice as cytotoxic as rhesus macaque NK cells when activated, suggesting that sooty mangabeys may be able to more quickly mount a strong NK-based response than rhesus macaques when viral infection occurs (Pereira and Ansari 2009).

An interesting feature of primate NK cells is the acquisition and rapid evolution of primate killer-cell immunoglobulin-like receptors (KIRs). Primate NK cells alone express KIRs. Like KIRs, the NK-cell receptors of other mammals [e.g., killer cell lectin-like receptor, subfamily A, member 2 (KLRA2/Ly49) lectin receptors of rodents] bind with MHC class I molecules or their homologues and activate the NK cell (Abi-Rached and Parham 2005; Barten et al. 2001; Colucci et al. 2002; Lanier 2008). The KIR family of receptors appears to have expanded from a single gene, killer cell immunoglobulin-like receptor three domain long cytoplasmic tail (*KIR3DL*), over the last ~50 million years. In that time, KIR genotypes have rapidly diversified between and within haplorrhine species. Humans, for example, encode over 130 genotypes for approximately 14 KIRs, while the grey mouse lemur encodes only one functional KIR (Averdam et al. 2009; Bimber et al. 2008; Hollenbach et al. 2010). Human KIRs exhibit lineage-specific evolution of receptors that specifically recognize epitopes of MHC-B and MHC-C receptors, as well as an increased number of KIR "activating receptors" as compared to chimpanzees (Abi-Rached et al. 2010). The divergence of haplorrhine and strepsirrhine NK receptors and subsequent lineage-specific diversification of these receptors have likely affected NK interactions with similarly rapidly evolving MHC molecules. Strepsirrhines appear to have a different NK receptor/ligand system in place, with lectin receptors CD94 and KLRC1/NKG2 representing the primary NK receptors, and an absence of a functional equivalent of the MHC-E molecules that serve as lectin receptor ligands in haplorrhines (Averdam et al. 2009). Moreover, CD94 and KLRC1 receptors seem to functionally diversify by recombination, as opposed to duplication as haplorrhine NK receptors do. If true, the occurrence of combinatorial diversification in strepsirrhine NK cells introduces an interesting wrinkle in our understanding of the evolution of primate NK cells. It suggests that NK cells may share recent common ancestry with cytotoxic T cells (Averdam et al. 2009).

Major Cellular Components of Innate Immunity: Small Lymphocytes

The adaptive immune system allows vertebrate organisms to recognize specific antigens, develop a targeted clonal response, and retain an immunological memory, so that a rapid and targeted immune response can be mounted when the antigen is next detected. Key to this process in jawed vertebrates (gnathostomata) are T and B lymphocytes, which bear highly diversified immunoglobulin domain-based receptors (Ig) that recognize specific antigenic epitopes. It is through recognition of foreign molecules presented to lymphocytes by other cells via Ig receptors, such as TCRs, BCRs, and MHCs, that lymphocytes initiate antigen-specific responses (reviewed in Neefjes et al. 2011).

While the adaptive immune system we recognize in primates is found in cartilaginous fish and emerged in the last common ancestor of extant jawed vertebrates ~ 652–525 mya, some of its components appeared earlier. Lymphocyte-like cells are found in jawless vertebrates (agnatha) and notochord-bearing animals, such as lancelets (cephalochordata), suggesting that the progenitor of this cell type could have emerged as early as 715 mya, during the Cryogenian period (Blair and Hedges 2005; Huang et al. 2007; Mayer et al. 2002; Nagata et al. 2002). All vertebrate life maintains cells that are functionally similar to T and B cells, suggesting that the roots of adaptive immune system lymphocytes are actually invertebrate (Boehm 2011). It is, however, in jawless vertebrates that the first lymphocyte-like cells with antigen receptors diversified through somatic assembly are found (Alder et al. 2005; Pancer et al. 2004).

For many years the conventional wisdom was that adaptive immunity was a unique and complex arm of the immune system that emerged with jawed vertebrates. However, in 2007, Rogozin et al. found receptors that undergo somatic rearrangement on lymphocyte-like cells in lampreys (Rogozin et al. 2007). This system of adaptive immunity in jawless vertebrates has been extensively reviewed by other authors and will only be touched upon briefly here (see reviews by Boehm et al. 2012b; Dzik 2010). The discovery of these leucine-rich repeat-containing variable lymphocyte receptors (VLRs) profoundly changed our understanding of vertebrate immune system evolution. VLRS are assembled and rearranged through the activities of APOBEC-like cytidine deaminases (AID) (Rogozin et al. 2007). This system of expanding receptor repertoires through gene rearrangement/somatic recombination bears a strong resemblance to RAG-mediated TCR and immunoglobulin rearrangement. It is not, however, a homologous system and appears to be outcome of convergent evolution. VLRs are not rearranged by RAG and they are not related to the Ig receptor superfamily to which TCRs, BCRs, and MHC molecules belong (Rogozin et al. 2007). Canonical T and B cells emerged with the last common ancestor of jawed vertebrates (Flajnik and Kasahara 2001). The adaptive immune systems of jawed and jawless vertebrates bear some remarkable similarities for potentially being the products of convergent evolution. VLR expressing lymphocytes, for example, can be divided into three functional categories (VLRA+, VLRB+, and VLRC+ protolymphocytes) that maintain similar gene expression profiles and respond to antigens in a manner similar to that of vertebrate T and B cells (Guo et al. 2009).

B cells

B cells represent the arm of the adaptive immune system responsible for the production of soluble antibodies, or immunoglobulins (Ig), which recognize and bind epitopes expressed on a variety of microbial pathogens. In mammals, naïve B cells exit the bone marrow and migrate to the secondary lymphoid organs where an encounter with an antigen or antigen-specific T helper cells could lead to B-cell activation, proliferation, Ig isotype class switching, and soluble antibody secretion. Canonical B cells first emerged with the TCR/BCR/MHC adaptive immune components in jawed vertebrates, 652–525 mya (Boehm and Bleul 2007). However, it seems likely that B cells emerged earlier than T cells, as T cells require the additional step of NOTCH signaling to differentiate from the lymphocyte progenitor cell and B cells do not (Benne et al. 2009). Of the two cell types, B cells have not been as intensely studied in primates as T cells. However, baseline interspecies differences in B-cell populations have been noted between primates. For example, cynomolgus macaques (*M. fascicularis*) maintain different proportions of B cell subsets than humans, which may translate as interspecies differences in immune response. Specifically, cynomolgus macaques retain higher percentage of B cells carrying the T cell costimulatory molecule CD80 and a lower percentage of B cells carrying CD21, a molecule that interacts with the complement system, than humans (Vugmeyster et al. 2004).

Much of the comparative research on baseline B cell function has focused on immunoglobulins (antibodies). Immunoglobulins may be secreted by B cells or associate with accessory molecules to form the membrane-bound B cell receptor complex (BCR). Secreted Igs can recognize and neutralize an antigen, while the BCR complex can recognize an antigen and initiate cell signaling and/or internalization and presentation of antigen to T cells. The genes that encode Igs in B cells and TCRs in T cells are arranged in a similar way: multiple variable and constant gene segments that recombine and assemble in a largely random manner to generate a vast and almost limitless repertoire of antigenic specificities. The Ig molecule is made up of two identical Ig heavy and light chains, both of which contain variable regions. Multiple variable (V), diversity (D), and junction (J) gene segments are rearranged by V(D)J recombination to form an Ig heavy chain, while the light chains are formed from V and D gene segments. The diversity of the various gene segments is further amplified by the random nucleotide additions and subtraction at the ends of Ig gene segments before they are joined (Thai et al. 2002). This system of immunoglobulin diversification appeared by the emergence of cartilaginous fish, 652–525 million years ago (Blair and Hedges 2005; Rast et al. 1997). It is estimated that the human B cell repertoire is capable of generating about 10^{11} unique immunoglobulins based on these rearrangements (Schatz et al. 1992).

The immunoglobulins produced by B cells are divided into several classes distinguished by constant regions, which impart specific effector properties to each class. Mammals maintain immunoglobulins belonging to the IgM, IgG, IgA, IgE, and IgD classes and have lost the IgY found in other animals (Kapetanovic and Cavaillon 2007; Mussmann et al. 1996b; Vollmers and Brandlein 2006; Zhao et al. 2006). Of the immunoglobulins found in primates, IgM is oldest, having emerged with the last common ancestor to all jawed vertebrates. It is highly conserved in function and structure. All jawed vertebrates appear to produce the IgM class of secreted immunoglobulin ahead of all others in response to immune system stimulation (reviewed in Flajnik 2002). Compared to other immunoglobulins, IgM has a lower affinity for antigens, is often polyreactive, and has the ability to bind a variety of microbial and viral antigens (Boes 2000).

Like IgM, an IgD homolog (e.g., IgW) appears to have been present in the last common ancestor of jawed vertebrates (Ohta and Flajnik 2006). The evolution of IgD is enigmatic, as many different isotypes are found in jawed fish, yet IgD is variably present in bird and mammalian species (Butler et al. 1996; Zhao et al. 2002). In primates IgD is expressed on naïve and memory B cells; however, the function of this immunoglobulin is not well understood (Messaoudi et al. 2011). It seems to be involved in lymphocyte and granulocyte activities. In its secreted form in bony fish, for example, it interacts primarily with granulocytes. In humans the membrane-bound form of IgD may be important in the activation of B cells, while the secreted form activates basophils (Flajnik and Kasahara 2010).

IgA is found in extant reptiles, which suggests it emerged at least ~320 million years ago (Mussmann et al. 1996a). In mammals, IgA is commonly secreted in mucosal sites by specialized B cells, where it binds and shields bacteria to maintain gut homeostasis (Strugnell and Wijburg 2010). It is the most commonly secreted immunoglobulin isotype in mammals, frequently found in GALT structures, where its production is regulated by innate and T cell pathways in germinal centers (Suzuki et al. 2010). IgA prototypes and analogs fulfill similar functions in nonmammalian species. In teleost fish, for example, Ig-based mucosal immunity is accomplished through IgT, an IgA analog (Zhang et al. 2010). Amphibian mucosal Ig immunity is fulfilled by IgX, a possible prototype of IgA (Deza et al. 2007). IgA is also an important neutralizer of bacteria, toxins, and viruses at the mucosal barrier and may also facilitate antigen take up by the mucosal dendritic cells (Corthesy 2007; Stubbe et al. 2000). When IgA does not regulate gut microflora appropriately, inflammatory responses become dysregulated. Dysfunctional IgA is implicated in variety of inflammatory gut disorders (reviewed in Sutherland and Fagarasan 2012).

In primates, IgA function appears to be very conserved. A comparison of IgA-encoding genes (*IGHA* and *IGCA*) and IgA molecules of closely related nonhuman primates [baboons (*Papio* sp.), sooty mangabeys, rhesus macaques, and pig-tailed macaques (*M. nemestrina*)] found that despite high intraspecies variation in sequence, the IgA receptor (CD89) of macaques could readily bind with the IgA of other primates (i.e., humans). (Rogers et al. 2008a). The conservation of gut microflora "enterotypes" across distant human populations has also been interpreted as conservation of IgA activities (Arumugam et al. 2011).

IgE and IgG diverged from the ancestral IgY in the last common ancestors of amphibians and mammals, respectively (Flajnik and Kasahara 2010; Warr et al. 1995; Zhao et al. 2006). IgE is a proinflammatory immunoglobulin and a potent stimulator of granulocytes. It is associated with Th2 responses to helminthic infections and allergens, playing a key role in the atopic hypersensitivity response (i.e., allergic asthma and food allergies) (Broide et al. 2010; Capron et al. 1987; Hagel et al. 2004; Stone et al. 2010). As such, it has been the subject of intense investigation.

As IgE is only found in mammals, IgE-based mechanisms at play in allergies and helminthic infections are unique to this class and comparatively recent in the context of vertebrate evolution. The essential nature of IgE, however, has been questioned as it has been maintained by mammals for over 250 million years but is frequently suppressed in humans by modern medications without ill effect (Cooper et al. 2008; Cruz et al. 2007; Vernersson et al. 2004, 2002).

In humans, helminthic infections and IgE-mediated allergic responses have been assumed to have had a complicated selective affect on the species, with high levels of IgE considered beneficial against helminthes and yet key to atopic hypersensitivity (Hagel et al. 2004). Moreover, under the hygiene hypothesis, human IgE responses have coevolved with certain helminth species which are thought to dampen IgE-mediated atopic hypersensitivity reactions (see Rook 2013; Martin et al. 2013 for reviews of this hypothesis). Loss of these species in the modern human microbiota, due to increased hygiene/use of antihelminthic medications, is thought to contribute to overt IgE responses to allergens and subsequent hypersensitivity (Cardoso et al. 2012). However, a recent finding of cross-reactivity between allergenic extracts from the helminth *Ascaris lumbricoides* and mite allergens suggests the interplay between IgE responses to helminthes and allergens may be more complicated and sometimes involve cases of "mistaken identity" (Acevedo and Caraballo 2011; Acevedo et al. 2009).

In primates, IgE appears to be functionally diverse. Analysis of membrane-bound IgE gene-encoding regions suggests that tarsiers do not produce membrane-bound IgE (Wu et al. 2012). Wu et al. (2012) note that mouse strains unable to produce membrane-bound IgE exhibit very low IgE production during parasitic infections, which suggests a functional difference between tarsier IgE and the IgE of other primates (Achatz et al. 2008). While catarrhines produce both long and short isoforms of IgE, platyrrhines do not appear to produce a short isoform, and lemurs, lorises and nonprimate mammals do not appear to produce a long. This suggests that the ancestral condition for IgE is the short isoform and that the long isoform was acquired sometime after the divergence of haplorrhines and strepsirrhines ~77 mya, but before the divergence of platyrrhines from catarrhines 43 mya (Steiper and Young 2006 isoform; Wu et al. 2012).

IgE-mediated atopic hypersensitivity reactions appear to be increasing across the globe (Rorke and Holgate 2004). Using data collected through the International Study of Asthma and Allergies in Childhood (ISAAC), Asher et al (2010) have noted a positive correlation between ecological factors associated with developed nations, such as gross national product per capita, trans fatty acids and tylenol use, and the incidence of asthma, eczema, and rhinoconjunctivitis (Asher et al. 2010). Other studies have noted a positive association between lower socioeconomic status and asthma/allergic reactions (Hagel et al. 2004; Litonjua et al. 1999; Von Behren et al. 1999). It is worth noting, however, that an association between these factors and conditions may be an outcome of increased "hygiene" in developed nations, but may also be influenced by increased disease surveillance in wealthier countries. While potential data bias should be considered, these data propose an interesting connection between human cultural behaviors and human IgE expression.

IgG is the most abundant antibody type found in mammalian serum and has high affinity and specificity for target antigens. The key function of IgG is to bind to target antigens and activate effector cells, such as NK cells or monocytes, to destroy immunoglobulin-coated targets (Schroeder and Cavacini 2010). Complement-mediated cytotoxicity is another function modulated by IgG that depends on the C1q complement component binding to the constant portion of IgG antibody bound to a target (Schroeder and Cavacini 2010). IgG and IgG receptor interactions are known to differ between primate species, with some cynomolgus macaque IgG subclasses exhibiting stronger effector function and greater binding affinity for IgG receptors, known as Fc receptors, than human IgG (Warncke et al. 2012). Additionally, the IgG receptor CD16 binds to different IgG subclasses in sooty mangabeys than it binds in humans (Rogers et al. 2006). These results suggest that IgG function has diverged in the 25–30 million years since these species shared a last common ancestor (Fabre et al. 2009; Steiper and Young 2006).

T Cells

T cells, so named because they mature in the thymus, are lymphocytes that contribute to adaptive immune defense through cytotoxic activities and the regulation of other immune cells. In jawed vertebrates, T cells are activated indirectly through antigen presentation by MHC I and MHC II molecules of professional antigen-presenting cells, such as DCs and macrophages, to the T-cell receptors (TCRs) of T cells. This activity constitutes a bridge between innate immune responses and adaptive immunity, and initiates the adaptive immune response. Both TCRs and MHC molecules are highly variable and capable of binding a wide variety of both host and pathogen-derived antigenic peptides. In humans, for example, human lymphocyte antigen (*HLA*) encodes about 2,000 MHC alleles according to the ImMunoGeneTics HLA database (http://www.ebi.ac.uk/imgt/hal/atats.html) (MHC/*HLA* diversity in Old World primates is discussed in Loisel and Tung 2013). It was previously thought that TCRs, MHCs, and T and B lymphocytes emerged in the immunological equivalent of a "big bang", all at once, with the emergence of jawed vertebrates. With the discovery of VLR+ lymphocytes in jawless fish, it appears that the origins of T-cell lymphocytes are considerably more ancient than the Ig receptor superfamily system.

The discovery of specialized lymphocytes in jawless vertebrates significantly challenged a long-held wisdom that Ig receptors such as T-cell receptors (TCR) drove the divergence of T and B cells. Specialized VLRs (VLRA, VLRB, and VLRC) are not related to Ig receptors, but they are found in jawless fish on lymphocyte-like cells that are already proficient in different functions reminiscent of T and B cells. VLRB molecules are mainly secreted by and found in the membrane of cells whose functions and other receptors best resemble those of B cells (Alder et al. 2008; Herrin et al. 2008). Cells bearing VLRA and VLRC tend to respond to similar stimuli and produce the same cytokines and receptors as T cells (Guo et al. 2009). This suggests that lymphocyte subsets became specialized before the "big bang" of Ig receptors. Moreover, lamprey have a TCR-like gene, along with VLRs, suggesting that the foundations of both systems of somatic rearrangement/recombination may have existed in some earlier animals (Pancer et al. 2004). As Hsu (2011) has pointed out, these receptors suggest that rather than lymphocyte differentiation being based on the inherent functions of the receptor, lymphocyte receptors have been selected for based on the benefit they brought to lymphocyte function.

Within jawed vertebrates, both TCR gene organization and T-cell development are highly conserved (Flajnik and Kasahara 2010; Rast et al. 1997). TCR gene segment rearrangement takes place in the thymus of all jawed vertebrates and produces a unique antigen-specific TCR for each T cell. Task-specific TCR subtypes known in mammals (alpha/beta and gamma/delta) have been found in cartilaginous fish, which suggests that TCRs had already differentiated into subtypes by the the time that jawed vertebrates emerged (Hirano et al. 2011; Kreslavsky et al. 2010). Despite conservation, TCRs do differ across vertebrate classes and species. Two unique TCR genes arose in marsupials and sharks (*TCR mu* and *NAR-TCR*) and are expressed as atypical TCRs in these animals (Criscitiello et al. 2006; Parra et al. 2007).

In mammals, T cells can be divided into task-specific categories, the two largest subsets being CD8+ cytotoxic T cells and CD4+ helper T cells (Th). Cytotoxic T cells can directly kill infected cells, while helper T cells differentiate into Th subsets (e.g., Th1, Th2, and Th17), after interacting with antigen-presenting cells, to regulate other T cells, B-cell antibody production, and Ig class switching. It is perhaps out of a historical emphasis on adaptive immune responses that many subsets of T cells have been defined and characterized [e.g., Th1, Th2, Th17, regulatory T cells (Tregs), and additional within group subsets]. When adaptive immune features such as Th cells or memory T-cell retention emerged in vertebrate immunity is very difficult to determine. In the most catholic interpretation, these traits appear with the TCR/MHC/lymphocyte system in jawed vertebrates. Whether or not progenitors of such T-cell subtypes exist in the repertoire of lymphocyte-like cells found in earlier lifeforms is matter for further investigation.

Almost all comparative studies of primate T cells are based on a challenge model. Under these experimental circumstances, primate T cells do exhibit unique characteristics. Human T cells have been described as "overreactive" to stimulation in comparison to chimpanzee T cells, a functional difference that Soto et al. (2010) attribute to increased levels of inhibitory sialec acid-recognizing Ig-superfamily lectin 5 (Siglec 5) on chimpanzee T and B cells. Siglec 5 can suppress immune cell activation (Soto et al. 2010).

As nonhuman primates are important models of immunodeficiency virus infection and T-cell activities are key to HIV infection progression, most comparative information on primate T cells has been captured in the context of SIV/HIV research. As the primary targets of immunodeficiency virus infection, CD4+ T cells have been particularly well examined for interspecies functional differences. Available information suggests that lineage-specific adaptations have evolved in this subset of T cells. The proportions of CD4+ T cells readily infected by HIV/SIV, that is those CD4+ T cells that also express CCR5, differ between catarrhine species that are natural and naïve hosts of IVs. Healthy sooty mangabeys, for example, maintain fewer CD4+ CCR5+ T cells than humans and macaques. Moreover, these cells do not upregulate CCR5 when stimulated and appear to be less susceptible to IV infection as a result (Paiardini et al. 2011). This trait may help explain sooty mangabey resistance to IV pathogenesis. CD4+ T cells also proliferate more quickly in IV infected humans and macaques than they do in IV natural hosts such as mandrills (*Mandrillus sphinx*) and sooty mangabeys, suggesting that the haemopoietic mechanisms leading to the production T cells differ between species (Chan et al. 2010; Engram et al. 2010). Healthy macaques have been found to maintain more CD4+ T cells that produce a protease, granzyme B, that induces apoptosis in virus-infected cells (granzyme B) in the lamina propria of the gut than African green monkeys, a natural IV host (Hutchison et al. 2011). Granzyme B is a highly potent protease that may contribute to the translocation of microbes from within the gut into the peritoneal cavity during IV infection in naïve hosts. Taken together, these differences suggest pathogen-mediated selection of CD4+ traits in natural IV hosts over the 30 million years since all catarrhines last shared a common ancestor.

Conclusions

The major components of the primate immune system represent hundreds of millions of years of evolution. Here, we have outlined how the vertebrate immune system evolved and how primates have become specialized within that system, providing the first summary of functional differences in primate baseline immunity within the context of vertebrate immune system evolution. Where possible, we have offered a comparative analysis of resting primate immune component function. Reflection on vertebrate immune system evolution is important, not only for understanding how primate immune systems have diverged from the immune systems of other animals, but also for understanding aspects of primate immunity that are unique. An appreciation of the distinctive qualities of nonhuman primate immune systems will help ensure the appropriate and efficient use of animal biomedical models, so that nonhuman primates are not applied to studies for which another animal could be successfully applied. In these efforts, however, comparative immunology of primates faces challenges. Specifically, the data presented here is mainly limited to examinations of baseline immunity in animals that are common biomedical models, which in the case of primates disproportionately biases our information to several catarrhine species (i.e., rhesus macaques, humans, and sooty mangabeys). Additionally, any researcher seeking a better understanding of primate immunity faces the challenges of working with a model that is very expensive, comes with special ethical considerations and is subject to care and use guidelines not applied to other mammals used in experimentation. Several nonhuman primate species housed in National Institutes of Health facilities, for example, are now only available for access opportunistically and for ex vivo experimentation, if available at all (i.e., chimpanzees and sooty mangabeys). An understanding of how other immune systems differ and have evolved can help bridge gaps in our knowledge about primate immune system function. As such, considering how specific immune components have evolved in other vertebrates can be useful in forming testable hypotheses on immune component function in primates when experimental data has been previously absent or difficult to acquire. As a backdrop for the chapters that follow, this chapter serves as a reference for the evolution of major immune components and, we hope, as background for discovering principles governing primate immune function and its evolution.

References

Abi-Rached L, Parham P (2005) Natural selection drives recurrent formation of activating killer cell immunoglobulin-like receptor and Ly49 from inhibitory homologues. J Exp Med 201(8):1319–1332

Abi-Rached L, Moesta AK, Rajalingam R, Guethlein LA, Parham P (2010) Human-specific evolution and adaptation led to major qualitative differences in the variable receptors of human and chimpanzee natural killer cells. PLoS Genet 6(11):e1001192

Acevedo N, Caraballo L (2011) IgE cross-reactivity between Ascaris lumbricoides and mite allergens: possible influences on allergic sensitization and asthma. Parasite Immunol 33(6):309–321

Acevedo N, Sanchez J, Erler A, Mercado D, Briza P, Kennedy M, Fernandez A, Gutierrez M, Chua KY, Cheong N et al (2009) IgE cross-reactivity between Ascaris and domestic mite allergens: the role of tropomyosin and the nematode polyprotein ABA-1. Allergy 64(11):1635–1643

Achatz G, Lamers M, Crameri R (2008) Membrane bound IgE: the key receptor to restrict high IgE levels. Open Immunol J 1:25–32

Adachi S, Yoshida H, Kataoka H, Nishikawa S (1997) Three distinctive steps in Peyer's patch formation of murine embryo. Int Immunol 9(4):507–514

Alder MN, Rogozin IB, Iyer LM, Glazko GV, Cooper MD, Pancer Z (2005) Diversity and function of adaptive immune receptors in a jawless vertebrate. Science 310(5756):1970–1973
Alder MN, Herrin BR, Sadlonova A, Stockard CR, Grizzle WE, Gartland LA, Gartland GL, Boydston JA, Turnbough CL Jr, Cooper MD (2008) Antibody responses of variable lymphocyte receptors in the lamprey. Nat Immunol 9(3):319–327
Al-Khodor S, Price CT, Habyarimana F, Kalia A, Abu Kwaik Y (2008) A Dot/Icm-translocated ankyrin protein of Legionella pneumophila is required for intracellular proliferation within human macrophages and protozoa. Mol Microbiol 70(4):908–923
Allen PG, Dawidowicz EA (1990) Phagocytosis in Acanthamoeba: I. A mannose receptor is responsible for the binding and phagocytosis of yeast. J Cell Physiol 145(3):508–513
Amulic B, Cazalet C, Hayes GL, Metzler KD, Zychlinsky A (2012) Neutrophil function: from mechanisms to disease. Annu Rev Immunol 30:459–489
Anderson IJ, Watkins RF, Samuelson J, Spencer DF, Majoros WH, Gray MW, Loftus BJ (2005) Gene discovery in the Acanthamoeba castellanii genome. Protist 156(2):203–214
Arumugam M, Raes J, Pelletier E, Le Paslier D, Yamada T, Mende DR, Fernandes GR, Tap J, Bruls T, Batto JM et al (2011) Enterotypes of the human gut microbiome. Nature 473(7346):174–180
Asher MI, Stewart AW, Mallol J, Montefort S, Lai CK, Ait-Khaled N, Odhiambo J (2010) Which population level environmental factors are associated with asthma, rhinoconjunctivitis and eczema? Review of the ecological analyses of ISAAC Phase One. Respir Res 11:8
Ashton-Chess J, Blancho G (2005) An *in vitro* evaluation of the potential suitability of peripheral blood CD14(+) and bone marrow CD34(+)-derived dendritic cells for a tolerance inducing regimen in the primate. J Immunol Methods 297(1–2):237–252
Averdam A, Petersen B, Rosner C, Neff J, Roos C, Eberle M, Aujard F, Munch C, Schempp W, Carrington M et al (2009) A novel system of polymorphic and diverse NK cell receptors in primates. PLoS Genet 5(10):e1000688
Azumi K, De Santis R, De Tomaso A, Rigoutsos I, Yoshizaki F, Pinto MR, Marino R, Shida K, Ikeda M, Arai M et al (2003) Genomic analysis of immunity in a Urochordate and the emergence of the vertebrate immune system: "waiting for Godot". Immunogenetics 55(8):570–581
Bajoghli B, Guo P, Aghaallaei N, Hirano M, Strohmeier C, McCurley N, Bockman DE, Schorpp M, Cooper MD, Boehm T (2011) A thymus candidate in lampreys. Nature 470(7332):90–94
Barreiro LB, Marioni JC, Blekhman R, Stephens M, Gilad Y (2010) Functional comparison of innate immune signaling pathways in primates. PLoS Genet 6(12):e1001249
Barten R, Torkar M, Haude A, Trowsdale J, Wilson MJ (2001) Divergent and convergent evolution of NK-cell receptors. Trends Immunol 22(1):52–57
Bassity E, Clark TG (2012) Functional identification of dendritic cells in the teleost model, rainbow trout (Oncorhynchus mykiss). PLoS One 7(3):e33196
Benne C, Lelievre JD, Balbo M, Henry A, Sakano S, Levy Y (2009) Notch increases T/NK potential of human hematopoietic progenitors and inhibits B cell differentiation at a pro-B stage. Stem Cells 27(7):1676–1685
Benoit M, Desnues B, Mege JL (2008) Macrophage polarization in bacterial infections. J Immunol 181(6):3733–3739
Benveniste RE, Arthur LO, Tsai CC, Sowder R, Copeland TD, Henderson LE, Oroszlan S (1986) Isolation of a lentivirus from a macaque with lymphoma: comparison with HTLV-III/LAV and other lentiviruses. J Virol 60(2):483–490
Bernard D, Six A, Rigottier-Gois L, Messiaen S, Chilmonczyk S, Quillet E, Boudinot P, Benmansour A (2006) Phenotypic and functional similarity of gut intraepithelial and systemic T cells in a teleost fish. J Immunol 176(7):3942–3949
Bielek E (1988) Ultrastructural analysis of leucocyte interaction with tumour targets in a teleost, Cyprinus carpio L. Dev Comp Immunol 12(4):809–821
Bimber BN, Moreland AJ, Wiseman RW, Hughes AL, O'Connor DH (2008) Complete characterization of killer Ig-like receptor (KIR) haplotypes in Mauritian cynomolgus macaques: novel

insights into nonhuman primate KIR gene content and organization. J Immunol 181(9):6301–6308
Blair JE, Hedges SB (2005) Molecular phylogeny and divergence times of deuterostome animals. Mol Biol Evol 22(11):2275–2284
Blakley BW, Magit AE (2009) The role of tonsillectomy in reducing recurrent pharyngitis: a systematic review. Otolaryngol Head Neck Surg 140(3):291–297
Blumbach B, Diehl-Seifert B, Seack J, Steffen R, Müller IM, Müller WEG (1999) Cloning and expression of novel receptors belonging to the immunoglobulin superfamily from the marine sponge Geodia cydonium. Immunogenetics 49:751–763.
Boehm T (2011) Design principles of adaptive immune systems. Nat Rev Immunol 11(5):307–317
Boehm T, Bleul CC (2007) The evolutionary history of lymphoid organs. Nat Immunol 8(2):131–135
Boehm T, Hess I, Swann JB (2012a) Evolution of lymphoid tissues. Trends Immunol 33(6):315–321
Boehm T, Iwanami N, Hess I (2012b) Evolution of the immune system in the lower vertebrates. Annu Rev Genomics Hum Genet 13:127–149
Boes M (2000) Role of natural and immune IgM antibodies in immune responses. Mol Immunol 37(18):1141–1149
Borregaard N, Cowland JB (1997) Granules of the human neutrophilic polymorphonuclear leukocyte. Blood 89(10):3503–3521
Brenchley JM, Douek DC (2008) The mucosal barrier and immune activation in HIV pathogenesis. Curr Opin HIV AIDS 3(3):356–361
Brenchley JM, Paiardini M, Knox KS, Asher AI, Cervasi B, Asher TE, Scheinberg P, Price DA, Hage CA, Kholi LM et al (2008) Differential Th17 CD4 T-cell depletion in pathogenic and nonpathogenic lentiviral infections. Blood 112(7):2826–2835
Brendolan A, Rosado MM, Carsetti R, Selleri L, Dear TN (2007) Development and function of the mammalian spleen. Bioessays 29(2):166–177
Brinkmann V, Reichard U, Goosmann C, Fauler B, Uhlemann Y, Weiss DS, Weinrauch Y, Zychlinsky A (2004) Neutrophil extracellular traps kill bacteria. Science 303(5663):1532–1535
Broide DH, Finkelman F, Bochner BS, Rothenberg ME (2010) Advances in mechanisms of asthma, allergy, and immunology in 2010. J Allergy Clin Immunol 127(3):689–695
Brown KN, Barratt-Boyes SM (2009) Surface phenotype and rapid quantification of blood dendritic cell subsets in the rhesus macaque. J Med Primatol 38(4):272–278
Burton MJ, Glasziou PP (2009) Tonsillectomy or adeno-tonsillectomy versus non-surgical treatment for chronic/recurrent acute tonsillitis. Cochrane Database Syst Rev (1):CD001802
Burton MJ, Towler B, Glasziou P (2000) Tonsillectomy versus non-surgical treatment for chronic/recurrent acute tonsillitis. Cochrane Database Syst Rev (2):CD001802
Buse E, Habermann G, Vogel F (2006) Thymus development in *Macaca fascicularis* (Cynomolgus monkey): an approach for toxicology and embryology. J Mol Histol 37(3–4):161–170
Butler JE, Sun J, Navarro P (1996) The swine Ig heavy chain locus has a single JH and no identifiable IgD. Int Immunol 8(12):1897–1904
Butterworth AE (1977) The eosinophil and its role in immunity to helminth infection. Curr Top Microbiol Immunol 77:127–168
Campbell TW (2004) Hematology of birds. In: Thrall MA, Weiser G, Allison R, Campbell TW (eds) Veterinary hematology and clinical chemistry. Wiley-Blackwell, New York, pp 238–276
Campbell JL, Eisemann JH, Williams CV, Glenn KM (2000) Description of the gastrointestinal tract of five lemur species: Propithecus tattersalli, Propithecus verreauxi coquereli, Varecia variegata, Hapalemur griseus, and Lemur catta. Am J Primatol 52(3):133–142
Canfield PJ (1998) Comparative cell morphology in the peripheral blood film from exotic and native animals. Aust Vet J 76(12):793–800
Capron A, Dessaint JP, Capron M, Ouma JH, Butterworth AE (1987) Immunity to schistosomes: progress toward vaccine. Science 238(4830):1065–1072

Cardoso LS, Oliveira SC, Araujo MI (2012) Schistosoma mansoni antigens as modulators of the allergic inflammatory response in asthma. Endocr Metab Immune Disord Drug Targets 12(1):24–32
Carter DL, Shieh TM, Blosser RL, Chadwick KR, Margolick JB, Hildreth JE, Clements JE, Zink MC (1999) CD56 identifies monocytes and not natural killer cells in rhesus macaques. Cytometry 37(1):41–50
Cesta MF (2006) Normal structure, function, and histology of mucosa-associated lymphoid Tissue. Toxicol Pathol 34:599. doi:10.1080/01926230600865531
Cesta MF (2006a) Normal structure, function, and histology of mucosa-associated lymphoid tissue. Toxicol Pathol 34(5):599–608
Cesta MF (2006b) Normal structure, function, and histology of the spleen. Toxicol Pathol 34(5):455–465
Chan ML, Petravic J, Ortiz AM, Engram J, Paiardini M, Cromer D, Silvestri G, Davenport MP (2010) Limited CD4+ T cell proliferation leads to preservation of CD4+ T cell counts in SIV-infected sooty mangabeys. Proc Biol Sci 277(1701):3773–3781
Chen JY, Oliveri P, Li CW, Zhou GQ, Gao F, Hagadorn JW, Peterson KJ, Davidson EH (2000) Precambrian animal diversity: putative phosphatized embryos from the Doushantuo Formation of China. Proc Natl Acad Sci USA 97(9):4457–4462
Chen G, Zhuchenko O, Kuspa A (2007) Immune-like phagocyte activity in the social amoeba. Science 317(5838):678–681
Chieffi Baccari G, de Paulis A, Di Matteo L, Gentile M, Marone G, Minucci S (1998) In situ characterization of mast cells in the frog Rana esculenta. Cell Tissue Res 292(1):151–162
Chivers DJ, Hladik CM (1980) Morphology of the gastrointestinal tract in primates: comparisons with other mammals in relation to diet. J Morphol 166(3):337–386
Chtanova T, Schaeffer M, Han SJ, van Dooren GG, Nollmann M, Herzmark P, Chan SW, Satija H, Camfield K, Aaron H et al (2008) Dynamics of neutrophil migration in lymph nodes during infection. Immunity 29(3):487–496
Church MK, Levi-Schaffer F (1997) The human mast cell. J Allergy Clin Immunol 99(2): 155–160
Clark JM, Abraham WM, Fishman CE, Forteza R, Ahmed A, Cortes A, Warne RL, Moore WR, Tanaka RD (1995) Tryptase inhibitors block allergen-induced airway and inflammatory responses in allergic sheep. Am J Respir Crit Care Med 152(6 Pt 1):2076–2083
Coates PT, Barratt-Boyes SM, Zhang L, Donnenberg VS, O'Connell PJ, Logar AJ, Duncan FJ, Murphey-Corb M, Donnenberg AD, Morelli AE et al (2003) Dendritic cell subsets in blood and lymphoid tissue of rhesus monkeys and their mobilization with Flt3 ligand. Blood 102(7):2513–2521
Colucci F, Di Santo JP, Leibson PJ (2002) Natural killer cell activation in mice and men: different triggers for similar weapons? Nat Immunol 3(9):807–813
Conrad RE (1981) Induction and collection of peritoneal exudate macrophages. Dekker, New York
Cooper EL (2010) Evolution of immune systems from self/not self to danger to artificial immune systems (AIS). Phys Life Rev 7(1):55–78
Cooper PJ, Ayre G, Martin C, Rizzo JA, Ponte EV, Cruz AA (2008) Geohelminth infections: a review of the role of IgE and assessment of potential risks of anti-IgE treatment. Allergy 63(4):409–417
Cornes JS (1965) Number, size, and distribution of Peyer's patches in the human small intestine: Part I The development of Peyer's patches. Gut 6(3):225–229
Corthesy B (2007) Roundtrip ticket for secretory IgA: role in mucosal homeostasis? J Immunol 178(1):27–32
Criscitiello MF, Saltis M, Flajnik MF (2006) An evolutionarily mobile antigen receptor variable region gene: doubly rearranging NAR-TcR genes in sharks. Proc Natl Acad Sci USA 103(13):5036–5041
Crivellato E, Ribatti D (2010) The mast cell: an evolutionary perspective. Biol Rev Camb Philos Soc 85(2):347–360

Crivellato E, Nico B, Gallo VP, Ribatti D (2010) Cell secretion mediated by granule-associated vesicle transport: a glimpse at evolution. Anat Rec (Hoboken) 293(7):1115–1124

Cruz AA, Lima F, Sarinho E, Ayre G, Martin C, Fox H, Cooper PJ (2007) Safety of anti-immunoglobulin E therapy with omalizumab in allergic patients at risk of geohelminth infection. Clin Exp Allergy 37(2):197–207

Da'as S, Teh EM, Dobson JT, Nasrallah GK, McBride ER, Wang H, Neuberg DS, Marshall JS, Lin TJ, Berman JN (2011) Zebrafish mast cells possess an FcvarepsilonRI-like receptor and participate in innate and adaptive immune responses. Dev Comp Immunol 35(1):125–134

Dalquest WW, Werner HJ, Robert JH, Richmond ND, Roslund HR, Voge M, Bern HA, Wilber CG, Sealander JA, Conaway CH et al (1952) General notes. J Mammal 33:102–118

Daniel MD, King NW, Letvin NL, Hunt RD, Sehgal PK, Desrosiers RC (1984) A new type D retrovirus isolated from macaques with an immunodeficiency syndrome. Science 223(4636):602–605

Davies B, Chattings LS, Edwards SW (1991) Superoxide generation during phagocytosis by Acanthamoeba castellanii: similarities to the respiratory burst of immune phagocytes. Microbiology 137(3):705–710

Davis MM, Chien Y (2003) T-cell antigen receptors. In: Paul WE (ed) Fundamental immunology. Lippincott, Philadelphia, pp 227–258

de Barros CM, Andrade LR, Allodi S, Viskov C, Mourier PA, Cavalcante MC, Straus AH, Takahashi HK, Pomin VH, Carvalho VF et al (2007) The hemolymph of the ascidian Styela plicata (Chordata-Tunicata) contains heparin inside basophil-like cells and a unique sulfated galactoglucan in the plasma. J Biol Chem 282(3):1615–1626

Deza FG, Espinel CS, Beneitez JV (2007) A novel IgA-like immunoglobulin in the reptile Eublepharis macularius. Dev Comp Immunol 31(6):596–605

Diamond LK (1969) Splenectomy in childhood and the hazard of overwhelming infection. Pediatrics 43(5):886–889

Diop OM, Ploquin MJ, Mortara L, Faye A, Jacquelin B, Kunkel D, Lebon P, Butor C, Hosmalin A, Barre-Sinoussi F et al (2008) Plasmacytoid dendritic cell dynamics and alpha interferon production during Simian immunodeficiency virus infection with a nonpathogenic outcome. J Virol 82(11):5145–5152

Doeing DC, Borowicz JL, Crockett ET (2003) Gender dimorphism in differential peripheral blood leukocyte counts in mice using cardiac, tail, foot, and saphenous vein puncture methods. BMC Clin Pathol 3(1):3

Dooley J, Erickson M, Gillard GO, Farr AG (2006) Cervical thymus in the mouse. J Immunol 176(11):6484–6490

Drayton DL, Liao S, Mounzer RH, Ruddle NH (2006) Lymphoid organ development: from ontogeny to neogenesis. Nat Immunol 7(4):344–353

Du Pasquier L (1973) Ontogeny of the immune response in cold-blooded vertebrates. Curr Top Microbiol Immunol 61:37–88

Du Pasquier L, Robert J, Courtet M, Mussmann R (2000) B-cell development in the amphibian Xenopus. Immunol Rev 175:201–213

Dvorak AM (2005) Ultrastructural studies of human basophils and mast cells. J Histochem Cytochem 53(9):1043–1070

Dzik JM (2010) The ancestry and cumulative evolution of immune reactions. Acta Biochim Pol 57(4):443–466

Ear T, McDonald PP (2008) Cytokine generation, promoter activation, and oxidant-independent NF-kappaB activation in a transfectable human neutrophilic cellular model. BMC Immunol 9:14

Ehlers D, Zosel B, Mohrig W, Kauschke E, Ehlers E (1992) Comparison of an in vivo and in vitro phagocytosis in Galleria mellonella L. Parasitol Res 78:354–359

Eisenhauer PB, Lehrer RI (1992) Mouse neutrophils lack defensins. Infect Immun 60(8):3446–3447

Elbim C, Monceaux V, Mueller YM, Lewis MG, Francois S, Diop O, Akarid K, Hurtrel B, Gougerot-Pocidalo MA, Levy Y et al (2008) Early divergence in neutrophil apoptosis between pathogenic and nonpathogenic simian immunodeficiency virus infections of nonhuman primates. J Immunol 181(12):8613–8623
Engram JC, Cervasi B, Borghans JA, Klatt NR, Gordon SN, Chahroudi A, Else JG, Mittler RS, Sodora DL, de Boer RJ et al (2010) Lineage-specific T-cell reconstitution following in vivo CD4+ and CD8+ lymphocyte depletion in nonhuman primates. Blood 116(5):748–758
Epiphanio S, Sinhorini IL, Catao-Dias JL (2003) Pathology of toxoplasmosis in captive new world primates. J Comp Pathol 129(2–3):196–204
Evans DI (1985) Postsplenectomy sepsis 10 years or more after operation. J Clin Pathol 38(3):309–311
Fabre PH, Rodrigues A, Douzery EJ (2009) Patterns of macroevolution among Primates inferred from a supermatrix of mitochondrial and nuclear DNA. Mol Phylogenet Evol 53(3):808–825
Fautin DG, Mariscal RN (1991) Cnidaria: anthozoa. In: Harrison FW, Westfall JA (eds) Microscopic anatomy of invertebrates, placozoa, porfera, cnidaria and ctenophora. Wiley-Liss, Inc., New York, pp 267–358
Favre D, Lederer S, Kanwar B, Ma ZM, Proll S, Kasakow Z, Mold J, Swainson L, Barbour JD, Baskin CR et al (2009) Critical loss of the balance between Th17 and T regulatory cell populations in pathogenic SIV infection. PLoS Pathog 5(2):e1000295
Fisher RE (2000) The primate appendix: a reassessment. Anat Rec 261(6):228–236
Flajnik MF (2002) Comparative analyses of immunoglobulin genes: surprises and portents. Nat Rev Immunol 2(9):688–698
Flajnik MF, Kasahara M (2001) Comparative genomics of the MHC: glimpses into the evolution of the adaptive immune system. Immunity 15(3):351–362
Flajnik MF, Kasahara M (2010) Origin and evolution of the adaptive immune system: genetic events and selective pressures. Nat Rev Genet 11(1):47–59
Fuchs TA, Abed U, Goosmann C, Hurwitz R, Schulze I, Wahn V, Weinrauch Y, Brinkmann V, Zychlinsky A (2007) Novel cell death program leads to neutrophil extracellular traps. J Cell Biol 176(2):231–241
Gabriela D, Carlos PL, Clara S, Elkin PM (2005) Phenotypical and functional characterization of non-human primate Aotus spp. dendritic cells and their use as a tool for characterizing immune response to protein antigens. Vaccine 23(26):3386–3395
Galli SJ, Kalesnikoff J, Grimbaldeston MA, Piliponsky AM, Williams CM, Tsai M (2005) Mast cells as "tunable" effector and immunoregulatory cells: recent advances. Annu Rev Immunol 23:749–786
Galli SJ, Grimbaldeston M, Tsai M (2008) Immunomodulatory mast cells: negative, as well as positive, regulators of immunity. Nat Rev Immunol 8(6):478–486
Geijtenbeek TB, Kwon DS, Torensma R, van Vliet SJ, van Duijnhoven GC, Middel J, Cornelissen IL, Nottet HS, KewalRamani VN, Littman DR et al (2000a) DC-SIGN, a dendritic cell-specific HIV-1-binding protein that enhances trans-infection of T cells. Cell 100(5):587–597
Geijtenbeek TB, Torensma R, van Vliet SJ, van Duijnhoven GC, Adema GJ, van Kooyk Y, Figdor CG (2000b) Identification of DC-SIGN, a novel dendritic cell-specific ICAM-3 receptor that supports primary immune responses. Cell 100(5):575–585
Geijtenbeek TB, Van Vliet SJ, Koppel EA, Sanchez-Hernandez M, Vandenbroucke-Grauls CM, Appelmelk B, Van Kooyk Y (2003) Mycobacteria target DC-SIGN to suppress dendritic cell function. J Exp Med 197(1):7–17
Gilliet M, Cao W, Liu YJ (2008) Plasmacytoid dendritic cells: sensing nucleic acids in viral infection and autoimmune diseases. Nat Rev Immunol 8(8):594–606
Godson C, Mitchell S, Harvey K, Petasis NA, Hogg N, Brady HR (2000) Cutting edge: lipoxins rapidly stimulate nonphlogistic phagocytosis of apoptotic neutrophils by monocyte-derived macrophages. J Immunol 164(4):1663–1667
Gorgollon P (1978) The normal human appendix: a light and electron microscopic study. J Anat 126(Pt 1):87–101

Goyos A, Robert J (2009) Tumorigenesis and anti-tumor immune responses in Xenopus. Front Biosci 14:167–176
Greenhalgh P, Olesen CE, Steiner LA (1993) Characterization and expression of recombination activating genes (RAG-1 and RAG-2) in Xenopus laevis. J Immunol 151(6):3100–3110
Grevellec A, Tucker AS (2010) The pharyngeal pouches and clefts: development, evolution, structure and derivatives. Semin Cell Dev Biol 21(3):325–332
Guo P, Hirano M, Herrin BR, Li J, Yu C, Sadlonova A, Cooper MD (2009) Dual nature of the adaptive immune system in lampreys. Nature 459(7248):796–801
Hagel I, Di Prisco MC, Goldblatt J, Le Souef PN (2004) The role of parasites in genetic susceptibility to allergy: IgE, helminthic infection and allergy, and the evolution of the human immune system. Clin Rev Allergy Immunol 26(2):75–83
Haley PJ (2003) Species differences in the structure and function of the immune system. Toxicology 188(1):49–71
Hamann KJ, Barker RL, Ten RM, Gleich GJ (1991) The molecular biology of eosinophil granule proteins. Int Arch Allergy Appl Immunol 94(1–4):202–209
Hardy RR (2003) B-lymphocyte development and biology. In: Paul WE (ed) Fundamental immunology. Lippincott, Philadelphia, pp 159–194
Hardy JD, Kurrus FD, Chavez CM, Neely WA, Eraslan S, Turner MD, Fabian LW, Labecki TD (1964) Heart transplantation in man. Developmental studies and report of a case. JAMA 188:1132–1140
Harmon BG (1998) Avian heterophils in inflammation and disease resistance. Poult Sci 77(7):972–977
Hawkey CM (1985) Analysis of hematologic findings in healthy and sick adult chimpanzees (Pan troglodytes). J Med Primatol 14(6):327–343
Herberman RB, Nunn ME, Holden HT, Lavrin DH (1975) Natural cytotoxic reactivity of mouse lymphoid cells against syngeneic and allogeneic tumors. II. Characterization of effector cells. Int J Cancer 16(2):230–239
Heritage PL, Underdown BJ, Arsenault AL, Snider DP, McDermott MR (1997) Comparison of murine nasal-associated lymphoid tissue and Peyer's patches. Am J Respir Crit Care Med 156(4 Pt 1):1256–1262
Herrin BR, Alder MN, Roux KH, Sina C, Ehrhardt GR, Boydston JA, Turnbough CL Jr, Cooper MD (2008) Structure and specificity of lamprey monoclonal antibodies. Proc Natl Acad Sci USA 105(6):2040–2045
Hida S, Tadachi M, Saito T, Taki S (2005) Negative control of basophil expansion by IRF-2 critical for the regulation of Th1/Th2 balance. Blood 106(6):2011–2017
Hirano M, Das S, Guo P, Cooper MD, Frederick WA (2011) The evolution of adaptive immunity in vertebrates. Adv Immunol 109:125–157
Hitchcock CR, Kiser JC, Telander RL, Seljeskog EL (1964) Baboon renal grafts. JAMA 189:934–937
Hofmann J, Greter M, Du Pasquier L, Becher B (2010) B-cells need a proper house, whereas T-cells are happy in a cave: the dependence of lymphocytes on secondary lymphoid tissues during evolution. Trends Immunol 31(4):144–153
Hogan SP, Rosenberg HF, Moqbel R, Phipps S, Foster PS, Lacy P, Kay AB, Rothenberg ME (2008) Eosinophils: biological properties and role in health and disease. Clin Exp Allergy 38(5):709–750
HogenEsch H, Felsburg PJ (1992) Immunohistology of Peyer's patches in the dog. Vet Immunol Immunopathol 30(2–3):147–160
HogenEsch H, Hahn FF (2001) The lymphoid organs: anatomy, development, and age-related changes. In: Mohr U, Carlton WW, Dungworth DL, Benjamin SA, Capen CC, Hahn FF (eds) Pathobiology of the aging dog. Iowa State University Press, Ames, pp 127–135
Hollenbach JA, Meenagh A, Sleator C, Alaez C, Bengoche M, Canossi A, Contreras G, Creary L, Evseeva I, Gorodezky C et al (2010) Report from the killer immunoglobulin-like receptor (KIR) anthropology component of the 15th International Histocompatibility Workshop: world-

wide variation in the KIR loci and further evidence for the co-evolution of KIR and HLA. Tissue Antigens 76(1):9–17
Hooper LV (2004) Bacterial contributions to mammalian gut development. Trends Microbiol 12(3):129–134
Horton TL, Ritchie P, Watson MD, Horton JD (1996) NK-like activity against allogeneic tumour cells demonstrated in the spleen of control and thymectomized Xenopus. Immunol Cell Biol 74(4):365–373
Hsu E (1998) Mutation, selection, and memory in B lymphocytes of exothermic vertebrates. Immunol Rev 162:25–36
Huang G, Xie X, Han Y, Fan L, Chen J, Mou C, Guo L, Liu H, Zhang Q, Chen S et al (2007) The identification of lymphocyte-like cells and lymphoid-related genes in amphioxus indicates the twilight for the emergence of adaptive immune system. PLoS One 2(2):e206
Hutchison AT, Schmitz JE, Miller CJ, Sastry KJ, Nehete PN, Major AM, Ansari AA, Tatevian N, Lewis DE (2011) Increased inherent intestinal granzyme B expression may be associated with SIV pathogenesis in Asian non-human primates. J Med Primatol 40(6):414–426
Janeway CA Jr, Medzhitov R (2002) Innate immune recognition. Annu Rev Immunol 20:197–216
Jansen CA, van de Haar PM, van Haarlem D, van Kooten P, de Wit S, van Eden W, Viertlbock BC, Gobel TW, Vervelde L (2010) Identification of new populations of chicken natural killer (NK) cells. Dev Comp Immunol 34(7):759–767
Jaroenpool J, Rogers KA, Pattanapanyasat K, Villinger F, Onlamoon N, Crocker PR, Ansari AA (2007) Differences in the constitutive and SIV infection induced expression of Siglecs by hematopoietic cells from non-human primates. Cell Immunol 250(1–2):91–104
Jawdat DM, Albert EJ, Rowden G, Haidl ID, Marshall JS (2004) IgE-mediated mast cell activation induces Langerhans cell migration in vivo. J Immunol 173(8):5275–5282
Jenkins MK (2003) Peripheral T-lymphocyte responses and function. In: Paul WE (ed) Fundamental immunology. Lippincott, Philadelphia, pp 303–320
Jesudason S, Collins MG, Rogers NM, Kireta S, Coates PT (2012) Non-human primate dendritic cells. J Leukoc Biol 91(2):217–228
Jeurissen SH (1991) Structure and function of the chicken spleen. Res Immunol 142(4):352–355
Kaldjian EP, Gretz JE, Anderson AO, Shi Y, Shaw S (2001) Spatial and molecular organization of lymph node T cell cortex: a labyrinthine cavity bounded by an epithelium-like monolayer of fibroblastic reticular cells anchored to basement membrane-like extracellular matrix. Int Immunol 13(10):1243–1253
Kapetanovic R, Cavaillon JM (2007) Early events in innate immunity in the recognition of microbial pathogens. Expert Opin Biol Ther 7(6):907–918
Karasuyama H, Mukai K, Obata K, Tsujimura Y, Wada T (2011) Nonredundant roles of basophils in immunity. Annu Rev Immunol 29:45–69
Karrer U, Althage A, Odermatt B, Roberts CW, Korsmeyer SJ, Miyawaki S, Hengartner H, Zinkernagel RM (1997) On the key role of secondary lymphoid organs in antiviral immune responses studied in alymphoplastic (aly/aly) and spleenless (Hox11(-)/-) mutant mice. J Exp Med 185(12):2157–2170
Kaspers B, Kothlow S, Butter C (2008) Avian antigen presenting cells. In: Davidson F, Kaspers B, Schat KA (eds) Avian immunology. Academic, London, pp 183–202
Kauschke E, Komiyama K, Moro I, Eue I, Konig S, Cooper EL (2001) Evidence for perforin-like activity associated with earthworm leukocytes. Zoology (Jena) 104(1):13–24
Kelenyi G, Nemeth A (1969) Comparative histochemistry and electron microscopy of the eosinophil leucocytes of vertebrates. I. A study of avian, reptile, amphibian and fish leucocytes. Acta Biol Acad Sci Hung 20(4):405–422
Ketloy C, Engering A, Srichairatanakul U, Limsalakpetch A, Yongvanitchit K, Pichyangkul S, Ruxrungtham K (2008) Expression and function of Toll-like receptors on dendritic cells and other antigen presenting cells from non-human primates. Vet Immunol Immunopathol 125(1–2):18–30

Khalturin K, Becker M, Rinkevich B, Bosch TC (2003) Urochordates and the origin of natural killer cells: identification of a CD94/NKR-P1-related receptor in blood cells of Botryllus. Proc Natl Acad Sci USA 100(2):622–627
King BK, Li J, Kudsk KA (1997a) A temporal study of TPN-induced changes in gut-associated lymphoid tissue and mucosal immunity. Arch Surg 132(12):1303–1309
King CL, Xianli J, Malhotra I, Liu S, Mahmoud AA, Oettgen HC (1997b) Mice with a targeted deletion of the IgE gene have increased worm burdens and reduced granulomatous inflammation following primary infection with Schistosoma mansoni. J Immunol 158(1):294–300
Kiyono H, Fukuyama S (2004) NALT- versus Peyer's-patch-mediated mucosal immunity. Nat Rev Immunol 4(9):699–710
Kobayashi Y (2008) The role of chemokines in neutrophil biology. Front Biosci 13:2400–2407
Kreslavsky T, Gleimer M, Garbe AI, von Boehmer H (2010) alphabeta versus gammadelta fate choice: counting the T-cell lineages at the branch point. Immunol Rev 238(1):169–181
Krishnaswamy G, Kelley J, Johnson D, Youngberg G, Stone W, Huang SK, Bieber J, Chi DS (2001) The human mast cell: functions in physiology and disease. Front Biosci 6:D1109–D1127
Kumar H, Kawai T, Akira S (2009) Pathogen recognition in the innate immune response. Biochem J 420(1):1–16
Kurtz J (2002) Phagocytosis by invertebrate hemocytes: causes of individual variation in Panorpa vulgaris scorpionflies. Microsc Res Tech 57(6):456–468
Kushwah R, Hu J (2011) Complexity of dendritic cell subsets and their function in the host immune system. Immunology 133(4):409–419
Kwa SF, Beverley P, Smith AL (2006) Peyer's patches are required for the induction of rapid Th1 responses in the gut and mesenteric lymph nodes during an enteric infection. J Immunol 176(12):7533–7541
Kyriazis AA, Esterly JR (1971) Fetal and neonatal development of lymphoid tissues. Arch Pathol 91(5):444–451
Lam SH, Chua HL, Gong Z, Wen Z, Lam TJ, Sin YM (2002) Morphologic transformation of the thymus in developing zebrafish. Dev Dyn 225(1):87–94
Lanier LL (2008) Up on the tightrope: natural killer cell activation and inhibition. Nat Immunol 9(5):495–502
Laurin M, Everett ML, Parker W (2011) The cecal appendix: one more immune component with a function disturbed by post-industrial culture. Anat Rec (Hoboken) 294(4):567–579
Lawn AM, Rose ME (1981) Presence of a complete endothelial barrier between lymph and lymphoid tissue in the lumbar lymph nodes of the duck (Anas platyrhynchos). Res Vet Sci 30(3):335–342
Lee JJ, Lee NA (2005) Eosinophil degranulation: an evolutionary vestige or a universally destructive effector function? Clin Exp Allergy 35(8):986–994
Lee WL, Harrison RE, Grinstein S (2003) Phagocytosis by neutrophils. Microbes Infect 5(14):1299–1306
Lee JJ, Jacobsen EA, Ochkur SI, McGarry MP, Condjella RM, Doyle AD, Luo H, Zellner KR, Protheroe CA, Willetts L et al (2012) Human versus mouse eosinophils: "that which we call an eosinophil, by any other name would stain as red". J Allergy Clin Immunol 130(3):572–584
Leulier F, Parquet C, Pili-Floury S, Ryu JH, Caroff M, Lee WJ, Mengin-Lecreulx D, Lemaitre B (2003) The Drosophila immune system detects bacteria through specific peptidoglycan recognition. Nat Immunol 4(5):478–484
Ley K, Allietta M, Bullard DC, Morgan S (1998) Importance of E-selectin for firm leukocyte adhesion in vivo. Circ Res 83(3):287–294
Lieberman J (2003) The ABCs of granule-mediated cytotoxicity: new weapons in the arsenal. Nat Rev Immunol 3(5):361–370
Lind JP, Wolff PL, Petrini KR, Keyley CW, Olson DE, Redig PT (1990) Morphology of the eosinophil in raptors. J Assoc Avian Veterinarians 4:33–38
Litonjua AA, Carey VJ, Weiss ST, Gold DR (1999) Race, socioeconomic factors, and area of residence are associated with asthma prevalence. Pediatr Pulmonol 28(6):394–401

Loisel DA, Tung J (2013) Genetic variation in the immune system of Old World monkeys: functional and selective effects. In: Brinkworth JF, Pechenkina E (eds) Primates, pathogens, and evolution. Springer, Heidelberg
MacDonald KP, Munster DJ, Clark GJ, Dzionek A, Schmitz J, Hart DN (2002) Characterization of human blood dendritic cell subsets. Blood 100(13):4512–4520
Malaviya R, Ikeda T, Ross E, Abraham SN (1996) Mast cell modulation of neutrophil influx and bacterial clearance at sites of infection through TNF-alpha. Nature 381(6577):77–80
Malleret B, Karlsson I, Maneglier B, Brochard P, Delache B, Andrieu T, Muller-Trutwin M, Beaumont T, McCune JM, Banchereau J et al (2008) Effect of SIVmac infection on plasmacytoid and CD1c+ myeloid dendritic cells in cynomolgus macaques. Immunology 124(2):223–233
Mandl JN, Barry AP, Vanderford TH, Kozyr N, Chavan R, Klucking S, Barrat FJ, Coffman RL, Staprans SI, Feinberg MB (2008) Divergent TLR7 and TLR9 signaling and type I interferon production distinguish pathogenic and nonpathogenic AIDS virus infections. Nat Med 14(10):1077–1087
Mandl JN, Akondy R, Lawson B, Kozyr N, Staprans SI, Ahmed R, Feinberg MB (2011) Distinctive TLR7 signaling, type I IFN production, and attenuated innate and adaptive immune responses to yellow fever virus in a primate reservoir host. J Immunol 186(11):6406–6416
Marjanovic D, Laurin M (2007) Fossils, molecules, divergence times, and the origin of lissamphibians. Syst Biol 56(3):369–388
Marr S, Morales H, Bottaro A, Cooper M, Flajnik M, Robert J (2007) Localization and differential expression of activation-induced cytidine deaminase in the amphibian Xenopus upon antigen stimulation and during early development. J Immunol 179(10):6783–6789
Martin JS, Renshaw SA (2009) Using in vivo zebrafish models to understand the biochemical basis of neutrophilic respiratory disease. Biochem Soc Trans 37(Pt 4):830–837
Martin M, Blackwell AD, Gurven M, Kaplan H (2013) Make new friends and keep the old? Parasite coinfection and comorbidity in *Homo sapiens*. In: Brinkworth JF, Pechenkina E (eds) Primates, pathogens, and evolution. Springer, Heidelberg
Mayer WE, Uinuk-Ool T, Tichy H, Gartland LA, Klein J, Cooper MD (2002) Isolation and characterization of lymphocyte-like cells from a lamprey. Proc Natl Acad Sci USA 99(22):14350–14355
Mebius RE, Kraal G (2005) Structure and function of the spleen. Nat Rev Immunol 5(8):606–616
Mehandru S, Poles MA, Tenner-Racz K, Horowitz A, Hurley A, Hogan C, Boden D, Racz P, Markowitz M (2004) Primary HIV-1 infection is associated with preferential depletion of CD4+ T lymphocytes from effector sites in the gastrointestinal tract. J Exp Med 200(6):761–770
Messaoudi I, Estep R, Robinson B, Wong SW (2011) Nonhuman primate models of human immunology. Antioxid Redox Signal 14(2):261–273
Mestas J, Hughes CC (2004) Of mice and not men: differences between mouse and human immunology. J Immunol 172(5):2731–2738
Metchnikoff E (1887) Uber den Kampf der Zellen gegen Erysipelkokken. Archiv fur Pathologische Anatomie und Physiologie und fur Klinische Medicin 107:209–249
Metchnikoff E (1893) Lecon sur la pathologie comparee de inflammation. Ann Inst Pasteur 7:348–357
Mitchell PC (1916) Further observations on the intestinal tract of mammals. Proc Zool Soc Lond 86(1):183–252
Moita LF, Wang-Sattler R, Michel K, Zimmermann T, Blandin S, Levashina EA, Kafatos FC (2005) In vivo identification of novel regulators and conserved pathways of phagocytosis in A. gambiae. Immunity 23(1):65–73
Montali RJ (1988) Comparative pathology of inflammation in the higher vertebrates (reptiles, birds and mammals). J Comp Pathol 99(1):1–26
Mosser DM, Edwards JP (2008) Exploring the full spectrum of macrophage activation. Nat Rev Immunol 8(12):958–969
Mulero I, Sepulcre MP, Meseguer J, Garcia-Ayala A, Mulero V (2007) Histamine is stored in mast cells of most evolutionarily advanced fish and regulates the fish inflammatory response. Proc Natl Acad Sci USA 104(49):19434–19439

Murata H (1959) Comparative studies of the spleen in submammalian vertebrates. II. Minute structure of the spleen, with special reference to the periarterial lymphoid sheath. Bull Yamaguchi Med Sch 6:83–105

Murphey-Corb M, Martin LN, Rangan SR, Baskin GB, Gormus BJ, Wolf RH, Andes WA, West M, Montelaro RC (1986) Isolation of an HTLV-III-related retrovirus from macaques with simian AIDS and its possible origin in asymptomatic mangabeys. Nature 321(6068):435–437

Mussmann R, Du Pasquier L, Hsu E (1996a) Is Xenopus IgX an analog of IgA? Eur J Immunol 26(12):2823–2830

Mussmann R, Wilson M, Marcuz A, Courtet M, Du Pasquier L (1996b) Membrane exon sequences of the three Xenopus Ig classes explain the evolutionary origin of mammalian isotypes. Eur J Immunol 26(2):409–414

Nagata T, Suzuki T, Ohta Y, Flajnik MF, Kasahara M (2002) The leukocyte common antigen (CD45) of the Pacific hagfish, Eptatretus stoutii: implications for the primordial function of CD45. Immunogenetics 54(4):286–291

Neefjes J, Jongsma ML, Paul P, Bakke O (2011) Towards a systems understanding of MHC class I and MHC class II antigen presentation. Nat Rev Immunol 11(12):823–836

Ngampasutadol J, Tran C, Gulati S, Blom AM, Jerse EA, Ram S, Rice PA (2008) Species-specificity of Neisseria gonorrhoeae infection: do human complement regulators contribute? Vaccine 26(Suppl 8):I62–I66

Nunn CL (2002) A comparative study of leukocyte counts and disease risk in primates. Evolution 56(1):177–190

Nussenzweig MC, Steinman RM, Gutchinov B, Cohn ZA (1980) Dendritic cells are accessory cells for the development of anti-trinitrophenyl cytotoxic T lymphocytes. J Exp Med 152(4):1070–1084

Ohnmacht C, Voehringer D (2009) Basophil effector function and homeostasis during helminth infection. Blood 113(12):2816–2825

Ohno S (1970) Evolution by gene duplication. Springer, New York, p 160

Ohta Y, Flajnik M (2006) IgD, like IgM, is a primordial immunoglobulin class perpetuated in most jawed vertebrates. Proc Natl Acad Sci USA 103(28):10723–10728

Ohta S, Ueda Y, Yaguchi M, Matsuzaki Y, Nakamura M, Toyama Y, Tanioka Y, Tamaoki N, Nomura T, Okano H et al (2008) Isolation and characterization of dendritic cells from common marmosets for preclinical cell therapy studies. Immunology 123(4):566–574

Olah I, Vervelde L (2008) Structure of the avian lymphoid system. In: Davison F, Kaspers B, Schat KA (eds) Avian immunology. Elsevier, London, pp 13–50

Olsson I, Venge P (1974) Cationic proteins of human granulocytes. II. Separation of the cationic proteins of the granules of leukemic myeloid cells. Blood 44(2):235–246

Olsson I, Venge P, Spitznagel JK, Lehrer RI (1977) Arginine-rich cationic proteins of human eosinophil granules: comparison of the constituents of eosinophilic and neutrophilic leukocytes. Lab Invest 36(5):493–500

Owen RL, Piazza AJ, Ermak TH (1991) Ultrastructural and cytoarchitectural features of lymphoreticular organs in the colon and rectum of adult BALB/c mice. Am J Anat 190(1):10–18

Paiardini M, Cervasi B, Reyes-Aviles E, Micci L, Ortiz AM, Chahroudi A, Vinton C, Gordon SN, Bosinger SE, Francella N et al (2011) Low levels of SIV infection in sooty mangabey central memory CD4(+) T cells are associated with limited CCR5 expression. Nat Med 17(7):830–836

Pancer Z, Mayer WE, Klein J, Cooper MD (2004) Prototypic T cell receptor and CD4-like coreceptor are expressed by lymphocytes in the agnathan sea lamprey. Proc Natl Acad Sci USA 101(36):13273–13278

Pandrea I, Apetrei C (2010) Where the wild things are: pathogenesis of SIV infection in African nonhuman primate hosts. Curr HIV/AIDS Rep 7(1):28–36

Parra ZE, Baker ML, Schwarz RS, Deakin JE, Lindblad-Toh K, Miller RD (2007) A unique T cell receptor discovered in marsupials. Proc Natl Acad Sci USA 104(23):9776–9781

Paul WE (2003) The immune system: an introduction. In: Paul WE (ed) Fundamental immunology. Lippincott, Philadelphia, pp 1–22

Pearson C, Uhlig HH, Powrie F (2012) Lymphoid microenvironments and innate lymphoid cells in the gut. Trends Immunol 33(6):289–296
Pech LL, Strand MR (1996) Granular cells are required for encapsulation of foreign targets by insect haemocytes. J Cell Sci 109(Pt 8):2053–2060
Penniall R, Spitznagel JK (1975) Chicken neutrophils: oxidative metabolism in phagocytic cells devoid of myeloperoxidase. Proc Natl Acad Sci USA 72(12):5012–5015
Pereira LE, Ansari AA (2009) A case for innate immune effector mechanisms as contributors to disease resistance in SIV-infected sooty mangabeys. Curr HIV Res 7(1):12–22
Peterson KJ, Butterfield NJ (2005) Origin of the Eumetazoa: testing ecological predictions of molecular clocks against the Proterozoic fossil record. Proc Natl Acad Sci USA 102(27):9547–9552
Pichyangkul S, Saengkrai P, Yongvanitchit K, Limsomwong C, Gettayacamin M, Walsh DS, Stewart VA, Ballou WR, Heppner DG (2001) Isolation and characterization of rhesus blood dendritic cells using flow cytometry. J Immunol Methods 252(1–2):15–23
Pickel JM, McCormack WT, Chen CH, Cooper MD, Thompson CB (1993) Differential regulation of V(D)J recombination during development of avian B and T cells. Int Immunol 5(8):919–927
Pospisil R, Mage RG (1998a) B-cell superantigens may play a role in B-cell development and selection in the young rabbit appendix. Cell Immunol 185(2):93–100
Pospisil R, Mage RG (1998b) Rabbit appendix: a site of development and selection of the B cell repertoire. Curr Top Microbiol Immunol 229:59–70
Prasad S, Kireta S, Leedham E, Russ GR, Coates PT (2010) Propagation and characterisation of dendritic cells from G-CSF mobilised peripheral blood monocytes and stem cells in common marmoset monkeys. J Immunol Methods 352(1–2):59–70
Purdy AK, Campbell KS (2009) Natural killer cells and cancer: regulation by the killer cell Ig-like receptors (KIR). Cancer Biol Ther 8(23):2211–2220
Raffatellu M, Santos RL, Verhoeven DE, George MD, Wilson RP, Winter SE, Godinez I, Sankaran S, Paixao TA, Gordon MA et al (2008) Simian immunodeficiency virus-induced mucosal interleukin-17 deficiency promotes Salmonella dissemination from the gut. Nat Med 14(4):421–428
Raison RL, Gilbertson P, Wotherspoon J (1987) Cellular requirements for mixed leucocyte reactivity in the cyclostome, Eptatretus stoutii. Immunol Cell Biol 65(Pt 2):183–188
Randal Bollinger R, Barbas AS, Bush EL, Lin SS, Parker W (2007) Biofilms in the large bowel suggest an apparent function of the human vermiform appendix. J Theor Biol 249(4):826–831
Rast JP, Anderson MK, Strong SJ, Luer C, Litman RT, Litman GW (1997) Alpha, beta, gamma, and delta T cell antigen receptor genes arose early in vertebrate phylogeny. Immunity 6(1):1–11
Reemtsma K, McCracken BH, Schlegel JU, Pearl M (1964a) Heterotransplantation of the kidney: two clinical experiences. Science 143(3607):700–702
Reemtsma K, McCracken BH, Schlegel JU, Pearl MA, Pearce CW, Dewitt CW, Smith PE, Hewitt RL, Flinner RL, Creech O Jr (1964b) Renal heterotransplantation in man. Ann Surg 160:384–410
Reite OB (1965) A phylogenetical approach to the functional significance of tissue mast cell histamine. Nature 206(991):1334–1336
Reite OB (1972) Comparative physiology of histamine. Physiol Rev 52(3):778–819
Rhodes CP, Ratcliffe NA, Rowley AF (1982) Presence of coelomocytes in the primitive chordate amphioxus (Branchiostoma lanceolatum). Science 217(4556):263–265
Robert J, Ohta Y (2009) Comparative and developmental study of the immune system in Xenopus. Dev Dyn 238(6):1249–1270
Rodewald HR (2008) Thymus organogenesis. Annu Rev Immunol 26:355–388
Rogers KA, Scinicariello F, Attanasio R (2006) IgG Fc receptor III homologues in nonhuman primate species: genetic characterization and ligand interactions. J Immunol 177(6):3848–3856
Rogers KA, Jayashankar L, Scinicariello F, Attanasio R (2008a) Nonhuman primate IgA: genetic heterogeneity and interactions with CD89. J Immunol 180(7):4816–4824

Rogers SL, Viertlboeck BC, Gobel TW, Kaufman J (2008b) Avian NK activities, cells and receptors. Semin Immunol 20(6):353–360
Rogozin IB, Iyer LM, Liang L, Glazko GV, Liston VG, Pavlov YI, Aravind L, Pancer Z (2007) Evolution and diversification of lamprey antigen receptors: evidence for involvement of an AID-APOBEC family cytosine deaminase. Nat Immunol 8(6):647–656
Rook GAW (2013) Microbial exposures and other early childhood influences on the subsequent function of the immune system. In: Brinkworth JF, Pechenkina E (eds) Primates, pathogens, and evolution. Springer, Heidelberg
Rorke S, Holgate ST (2004) The atopy phenotype revisited. Revue francaise d'allergologie e d'immunologie clinique 44:436–444
Rose ME, Hesketh P (1974) Fowl peritoneal exudate cells: collection and use for the macrophage migration inhibition test. Avian Pathol 3(4):297–300
Rosenberg HF, Domachowske JB (2001) Eosinophils, eosinophil ribonucleases, and their role in host defense against respiratory virus pathogens. J Leukoc Biol 70(5):691–698
Rosenberg HF, Dyer KD (1995) Eosinophil cationic protein and eosinophil-derived neurotoxin. Evolution of novel function in a primate ribonuclease gene family. J Biol Chem 270(37): 21539–21544
Rosenberg HF, Dyer KD, Tiffany HL, Gonzalez M (1995) Rapid evolution of a unique family of primate ribonuclease genes. Nat Genet 10(2):219–223
Rowley AF, Page M (1985) Ultrastructural, cytochemical and functional studies on the eosinophilic granulocytes of larval lampreys. Cell Tissue Res 240(3):705–709
Rumfelt LL, McKinney EC, Taylor E, Flajnik MF (2002) The development of primary and secondary lymphoid tissues in the nurse shark Ginglymostoma cirratum: B-cell zones precede dendritic cell immigration and T-cell zone formation during ontogeny of the spleen. Scand J Immunol 56(2):130–148
Sabet T, Wen-Cheng H, Stanisz M, El-Domeiri A, Van Alten P (1977) A simple method for obtaining peritoneal macrophages from chickens. J Immunol Methods 14(2):103–110
Sailendri K, Muthukkaruppan V (1975) Morphology of lymphoid organs in a cichlid teleost, Tilapia mossambica (Peters). J Morphol 147(1):109–121
Savina A, Amigorena S (2007) Phagocytosis and antigen presentation in dendritic cells. Immunol Rev 219:143–156
Scapini P, Laudanna C, Pinardi C, Allavena P, Mantovani A, Sozzani S, Cassatella MA (2001) Neutrophils produce biologically active macrophage inflammatory protein-3alpha (MIP-3alpha)/CCL20 and MIP-3beta/CCL19. Eur J Immunol 31(7):1981–1988
Schatz DG, Oettinger MA, Schlissel MS (1992) V(D)J recombination: molecular biology and regulation. Annu Rev Immunol 10:359–383
Schroeder HW Jr, Cavacini L (2010) Structure and function of immunoglobulins. J Allergy Clin Immunol 125(2 Suppl 2):S41–S52
Schwartz LB, Metcalfe DD, Miller JS, Earl H, Sullivan T (1987) Tryptase levels as an indicator of mast-cell activation in systemic anaphylaxis and mastocytosis. N Engl J Med 316(26): 1622–1626
Secombes CJ, Manning MJ (1980) Comparative studies on the immune system of fishes and amphibians: antigen localisation in carp *Cyprinus carpio* L. J Fish Dis 3:399
Seong SY, Matzinger P (2004) Hydrophobicity: an ancient damage-associated molecular pattern that initiates innate immune responses. Nat Rev Immunol 4(6):469–478
Setiamarga DH, Miya M, Yamanoue Y, Azuma Y, Inoue JG, Ishiguro NB, Mabuchi K, Nishida M (2009) Divergence time of the two regional medaka populations in Japan as a new time scale for comparative genomics of vertebrates. Biol Lett 5(6):812–816
Shah SS, Lai SY, Ruchelli E, Kazahaya K, Mahboubi S (2001) Retropharyngeal aberrant thymus. Pediatrics 108(5):E94
Shamri R, Xenakis J, Spencer L (2011) Eosinophils in innate immunity: an evolving story. Cell Tissue Res 343(1):57–83
Shedlock AM, Edwards SV (2009) Amniotes (Amniota). In: Hedges SB, Kumar S (eds) The timetree of life. Oxford University Press, New York, pp 375–379

Shen L, Stuge TB, Bengten E, Wilson M, Chinchar VG, Naftel JP, Bernanke JM, Clem LW, Miller NW (2004) Identification and characterization of clonal NK-like cells from channel catfish (Ictalurus punctatus). Dev Comp Immunol 28(2):139–152
Shikina T, Hiroi T, Iwatani K, Jang MH, Fukuyama S, Tamura M, Kubo T, Ishikawa H, Kiyono H (2004) IgA class switch occurs in the organized nasopharynx- and gut-associated lymphoid tissue, but not in the diffuse lamina propria of airways and gut. J Immunol 172(10):6259–6264
Siddiqui R, Khan NA (2012) Acanthamoeba is an evolutionary ancestor of macrophages: a myth or reality? Exp Parasitol 130(2):95–97
Sitnicka E, Buza-Vidas N, Larsson S, Nygren JM, Liuba K, Jacobsen SE (2003) Human CD34+ hematopoietic stem cells capable of multilineage engrafting NOD/SCID mice express flt3: distinct flt3 and c-kit expression and response patterns on mouse and candidate human hematopoietic stem cells. Blood 102(3):881–886
Smith ML, Olson TS, Ley K (2004) CXCR2- and E-selectin-induced neutrophil arrest during inflammation in vivo. J Exp Med 200(7):935–939
Smith HF, Fisher RE, Everett ML, Thomas AD, Bollinger RR, Parker W (2009) Comparative anatomy and phylogenetic distribution of the mammalian cecal appendix. J Evol Biol 22(10):1984–1999
Soderlund J, Nilsson C, Ekman M, Walther L, Gaines H, Biberfeld G, Biberfeld P (2000) Recruitment of monocyte derived dendritic cells ex vivo from SIV infected and non-infected cynomolgus monkeys. Scand J Immunol 51(2):186–194
Soehnlein O, Kai-Larsen Y, Frithiof R, Sorensen OE, Kenne E, Scharffetter-Kochanek K, Eriksson EE, Herwald H, Agerberth B, Lindbom L (2008) Neutrophil primary granule proteins HBP and HNP1-3 boost bacterial phagocytosis by human and murine macrophages. J Clin Invest 118(10):3491–3502
Soehnlein O, Zernecke A, Weber C (2009) Neutrophils launch monocyte extravasation by release of granule proteins. Thromb Haemost 102(2):198–205
Sokol CL, Chu NQ, Yu S, Nish SA, Laufer TM, Medzhitov R (2009) Basophils function as antigen-presenting cells for an allergen-induced T helper type 2 response. Nat Immunol 10(7):713–720
Soto PC, Stein LL, Hurtado-Ziola N, Hedrick SM, Varki A (2010) Relative over-reactivity of human versus chimpanzee lymphocytes: implications for the human diseases associated with immune activation. J Immunol 184(8):4185–4195
Spencer J, MacDonald TT, Finn T, Isaacson PG (1986) The development of gut associated lymphoid tissue in the terminal ileum of fetal human intestine. Clin Exp Immunol 64(3):536–543
Stabler JG, McCormick TW, Powell KC, Kogut MH (1994) Avian heterophils and monocytes: phagocytic and bactericidal activities against Salmonella enteritidis. Vet Microbiol 38(4): 293–305
Starzl TE, Marchioro TL, Peters GN, Kirkpatrick CH, Wilson WE, Porter KA, Rifkind D, Ogden DA, Hitchcock CR, Waddell WR (1964) Renal heterotransplantation from baboon to man: experience with 6 cases. Transplantation 2:752–776
Steiper ME, Young NM (2006) Primate molecular divergence dates. Mol Phylogenet Evol 41(2):384–394
Stone KD, Prussin C, Metcalfe DD (2010) IgE, mast cells, basophils, and eosinophils. J Allergy Clin Immunol 125(2 Suppl 2):S73–S80
Strugnell RA, Wijburg OL (2010) The role of secretory antibodies in infection immunity. Nat Rev Microbiol 8(9):656–667
Stubbe H, Berdoz J, Kraehenbuhl JP, Corthesy B (2000) Polymeric IgA is superior to monomeric IgA and IgG carrying the same variable domain in preventing Clostridium difficile toxin A damaging of T84 monolayers. J Immunol 164(4):1952–1960
Sturm-O'Brien AK, Salazar JD, Byrd RH, Popek EJ, Giannoni CM, Friedman EM, Sulek M, Larrier DR (2009) Cervical thymic anomalies–the Texas Children's Hospital experience. Laryngoscope 119(10):1988–1993
Sugimura M, Hashimoto Y, Nakanishi YH (1977) Thymus- and bursa-dependent areas in duck lymph nodes. Jpn J Vet Res 25(1–2):7–16

Sutherland DB, Fagarasan S (2012) IgA synthesis: a form of functional immune adaptation extending beyond gut. Curr Opin Immunol 24(3):261–268

Suzuki K, Kawamoto S, Maruya M, Fagarasan S (2010) GALT: organization and dynamics leading to IgA synthesis. Adv Immunol 107:153–185

Takaya K (1969) The relationship between mast cells and histamine in phylogeny with special reference to reptiles and birds. Arch Histol Jpn 30(4):401–420

Takaya K, Fujita T, Endo K (1967) Mast cells free of histamine in Rana catasbiana. Nature 215(5102):776–777

Tanaka S, Miura S, Tashiro H, Serizawa H, Hamada Y, Yoshioka M, Tsuchiya M (1991) Morphological alteration of gut-associated lymphoid tissue after long-term total parenteral nutrition in rats. Cell Tissue Res 266(1):29–36

Tavassoli M (1986) Bone marrow in boneless fish: lessons of evolution. Med Hypotheses 20(1):9–15

Teixeira MM, Talvani A, Tafuri WL, Lukacs NW, Hellewell PG (2001) Eosinophil recruitment into sites of delayed-type hypersensitivity reactions in mice. J Leukoc Biol 69(3):353–360

Terszowski G, Muller SM, Bleul CC, Blum C, Schirmbeck R, Reimann J, Pasquier LD, Amagai T, Boehm T, Rodewald HR (2006) Evidence for a functional second thymus in mice. Science 312(5771):284–287

Thai TH, Purugganan MM, Roth DB, Kearney JF (2002) Distinct and opposite diversifying activities of terminal transferase splice variants. Nat Immunol 3(5):457–462

Thomson R, Molina-Portela P, Mott H, Carrington M, Raper J (2009) Hydrodynamic gene delivery of baboon trypanosome lytic factor eliminates both animal and human-infective African trypanosomes. Proc Natl Acad Sci USA 106(46):19509–19514

Thorpe M, Yu J, Boinapally V, Ahooghalandari P, Kervinen J, Garavilla LD, Hellman L (2012) Extended cleavage specificity of the mast cell chymase from the crab-eating macaque (*Macaca fascicularis*): an interesting animal model for the analysis of the function of the human mast cell chymase. Int Immunol 24:771–782

Tovi F, Mares AJ (1978) The aberrant cervical thymus. Embryology, pathology, and clinical implications. Am J Surg 136(5):631–637

Trivedi NN, Tong Q, Raman K, Bhagwandin VJ, Caughey GH (2007) Mast cell alpha and beta tryptases changed rapidly during primate speciation and evolved from gamma-like transmembrane peptidases in ancestral vertebrates. J Immunol 179(9):6072–6079

Trivedi NN, Raymond WW, Caughey GH (2008) Chimerism, point mutation, and truncation dramatically transformed mast cell delta-tryptases during primate evolution. J Allergy Clin Immunol 121(5):1262–1268

Turner RJ, Manning MJ (1973) Response of the toad, Xenopus laevis, to circulating antigens. Cellular changes in the spleen. J Exp Zool 183(1):21–34

Uhm TG, Kim BS, Chung IY (2012) Eosinophil development, regulation of eosinophil-specific genes, and role of eosinophils in the pathogenesis of asthma. Allergy Asthma Immunol Res 4(2):68–79

Van den Broeck W, Derore A, Simoens P (2006) Anatomy and nomenclature of murine lymph nodes: descriptive study and nomenclatory standardization in BALB/cAnNCrl mice. J Immunol Methods 312(1–2):12–19

Veazey RS, DeMaria M, Chalifoux LV, Shvetz DE, Pauley DR, Knight HL, Rosenzweig M, Johnson RP, Desrosiers RC, Lackner AA (1998) Gastrointestinal tract as a major site of CD4+ T cell depletion and viral replication in SIV infection. Science 280(5362):427–431

Veiga-Fernandes H, Coles MC, Foster KE, Patel A, Williams A, Natarajan D, Barlow A, Pachnis V, Kioussis D (2007) Tyrosine kinase receptor RET is a key regulator of Peyer's patch organogenesis. Nature 446(7135):547–551

Vernersson M, Aveskogh M, Munday B, Hellman L (2002) Evidence for an early appearance of modern post-switch immunoglobulin isotypes in mammalian evolution (II); cloning of IgE, IgG1 and IgG2 from a monotreme, the duck-billed platypus, Ornithorhynchus anatinus. Eur J Immunol 32(8):2145–2155

Vernersson M, Aveskogh M, Hellman L (2004) Cloning of IgE from the echidna (Tachyglossus aculeatus) and a comparative analysis of epsilon chains from all three extant mammalian lineages. Dev Comp Immunol 28(1):61–75

Vollmers HP, Brandlein S (2006) Natural IgM antibodies: the orphaned molecules in immune surveillance. Adv Drug Deliv Rev 58(5–6):755–765
Von Behren J, Kreutzer R, Smith D (1999) Asthma hospitalization trends in California, 1983–1996. J Asthma 36(7):575–582
von Krogh C (1936) The morphology of the primate spleen. Anthropol Anz 13:89–100
Vugmeyster Y, Howell K, Bakshi A, Flores C, Hwang O, McKeever K (2004) B-cell subsets in blood and lymphoid organs in *Macaca fascicularis*. Cytometry A 61(1):69–75
Waddell DR, Duffy KT (1986) Breakdown of self/nonself recognition in cannibalistic strains of the predatory slime mold, Dictyostelium caveatum. J Cell Biol 102(1):298–305
Wainberg MA, Beaupre S, Beiss B, Israel E (1983) Differential susceptibility of avian sarcoma cells derived from different periods of tumor growth to natural killer cell activity. Cancer Res 43(10):4774–4780
Walker CM (1997) Comparative features of hepatitis C virus infection in humans and chimpanzees. Springer Semin Immunopathol 19(1):85–98
Walzer T, Jaeger S, Chaix J, Vivier E (2007) Natural killer cells: from CD3(-)NKp46(+) to post-genomics meta-analyses. Curr Opin Immunol 19(3):365–372
Warncke M, Calzascia T, Coulot M, Balke N, Touil R, Kolbinger F, Heusser C (2012) Different adaptations of IgG effector function in human and nonhuman primates and implications for therapeutic antibody treatment. J Immunol 188(9):4405–4411
Warr GW, Magor KE, Higgins DA (1995) IgY: clues to the origins of modern antibodies. Immunol Today 16(8):392–398
Webster RL, Johnson RP (2005) Delineation of multiple subpopulations of natural killer cells in rhesus macaques. Immunology 115(2):206–214
Willard-Mack CL (2006) Normal structure, function, and histology of lymph nodes. Toxicol Pathol 34(5):409–424
Wolfle U, Martin S, Emde M, Schempp C (2009) Dermatology in the Darwin anniversary. Part 2: evolution of the skin-associated immune system. J Dtsch Dermatol Ges 7(10):862–869
Wong ES, Papenfuss AT, Heger A, Hsu AL, Ponting CP, Miller RD, Fenelon JC, Renfree MB, Gibbs RA, Belov K (2011) Transcriptomic analysis supports similar functional roles for the two thymuses of the tammar wallaby. BMC Genomics 12:420
Wood W, Jacinto A (2007) Drosophila melanogaster embryonic haemocytes: masters of multitasking. Nat Rev Mol Cell Biol 8(7):542–551
Wu L, Liu YJ (2007) Development of dendritic-cell lineages. Immunity 26(6):741–750
Wu L, Bashirova AA, Martin TD, Villamide L, Mehlhop E, Chertov AO, Unutmaz D, Pope M, Carrington M, KewalRamani VN (2002) Rhesus macaque dendritic cells efficiently transmit primate lentiviruses independently of DC-SIGN. Proc Natl Acad Sci USA 99(3):1568–1573
Wu PC, Chen JB, Kawamura S, Roos C, Merker S, Shih CC, Hsu BD, Lim C, Chang TW (2012) The IgE gene in primates exhibits extraordinary evolutionary diversity. Immunogenetics 64(4):279–287
Yan L, Cerny RL, Cirillo JD (2004) Evidence that hsp90 is involved in the altered interactions of Acanthamoeba castellanii variants with bacteria. Eukaryot Cell 3(3):567–578
Yoder JA (2004) Investigating the morphology, function and genetics of cytotoxic cells in bony fish. Comp Biochem Physiol C Toxicol Pharmacol 138(3):271–280
Yoon J, Ponikau JU, Lawrence CB, Kita H (2008) Innate antifungal immunity of human eosinophils mediated by a beta 2 integrin, CD11b. J Immunol 181(4):2907–2915
Yoshida H, Ochiai M, Ashida M (1986) Beta-1,3-glucan receptor and peptidoglycan receptor are present as separate entities within insect prophenoloxidase activating system. Biochem Biophys Res Commun 141(3):1177–1184
Yoshimoto T, Yasuda K, Tanaka H, Nakahira M, Imai Y, Fujimori Y, Nakanishi K (2009) Basophils contribute to T(H)2-IgE responses in vivo via IL-4 production and presentation of peptide-MHC class II complexes to CD4+ T cells. Nat Immunol 10(7):706–712
Yousefi S, Gold JA, Andina N, Lee JJ, Kelly AM, Kozlowski E, Schmid I, Straumann A, Reichenbach J, Gleich GJ et al (2008) Catapult-like release of mitochondrial DNA by eosinophils contributes to antibacterial defense. Nat Med 14(9):949–953

Zapata A, Amemiya CT (2000) Phylogeny of lower vertebrates and their immunological structures. Curr Top Microbiol Immunol 248:67–107
Zemans RL, Colgan SP, Downey GP (2009) Transepithelial migration of neutrophils: mechanisms and implications for acute lung injury. Am J Respir Cell Mol Biol 40(5):519–535
Zhang XW, Liu Q, Wang Y, Thorlacius H (2001) CXC chemokines, MIP-2 and KC, induce P-selectin-dependent neutrophil rolling and extravascular migration in vivo. Br J Pharmacol 133(3):413–421
Zhang YA, Salinas I, Li J, Parra D, Bjork S, Xu Z, LaPatra SE, Bartholomew J, Sunyer JO (2010) IgT, a primitive immunoglobulin class specialized in mucosal immunity. Nat Immunol 11(9):827–835
Zhao Y, Kacskovics I, Pan Q, Liberles DA, Geli J, Davis SK, Rabbani H, Hammarstrom L (2002) Artiodactyl IgD: the missing link. J Immunol 169(8):4408–4416
Zhao Y, Pan-Hammarstrom Q, Yu S, Wertz N, Zhang X, Li N, Butler JE, Hammarstrom L (2006) Identification of IgF, a hinge-region-containing Ig class, and IgD in Xenopus tropicalis. Proc Natl Acad Sci USA 103(32):12087–12092

Genetic Variation in the Immune System of Old World Monkeys: Functional and Selective Effects

Dagan A. Loisel and Jenny Tung

Introduction

The selective pressures exerted by pathogens and parasites have played a significant role in human and nonhuman primate evolution, especially in shaping the form and function of immune defenses (Barreiro and Quintana-Murci 2010; Nunn and Altizer 2006; Stearns and Koella 2008). While the evolutionary processes that historically influenced the primate immune system cannot be directly observed today, the extensive phenotypic diversity observed in contemporary primate populations reflects these changes. Specifically, although many aspects of basic biology are conserved among primates, both pathogen susceptibility and disease progression upon infection can greatly vary among species, subspecies, and populations. The adaptive and mechanistic origins of this variation are therefore of great interest, including as a source of insight into mechanisms of disease in humans.

Recent advances in genetic and genomic techniques, as well as steadily decreasing costs, have enabled functional and evolutionary studies of intraspecies and interspecies genetic variation in the primate immune system that would have been

D.A. Loisel
Department of Human Genetics, University of Chicago, 920 E 58th Street, Chicago, IL 60637, USA

Department of Biology, Saint Michael's College, Colchester, VT 05439

J. Tung (✉)
Department of Evolutionary Anthropology, Duke University, Box 90383, Durham, NC 27708, USA

Department of Human Genetics, University of Chicago, 920 E 58th Street, Chicago, IL 60637, USA

Duke Institute for Population Research, Duke University, Box 90420, Durham, NC 27708, USA
e-mail: jt5@duke.edu

J.F. Brinkworth and K. Pechenkina (eds.), *Primates, Pathogens, and Evolution*, Developments in Primatology: Progress and Prospects
DOI 10.1007/978-1-4614-7181-3_3, © Springer Science+Business Media New York 2013

prohibitive only a few decades ago. Thus, while interest in immune genetic variation is long-standing, research on this topic has become increasingly sophisticated. It is now possible to not only identify which immune loci may have evolved under selection (Bustamante et al. 2005; Haygood et al. 2007; Nielsen et al. 2005) but also infer the type of selection that occurred, the time scale on which it occurred, and the type of functional change (e.g., coding or regulatory) that was its target (Haygood et al. 2010; Kosiol et al. 2008). At the same time, it is becoming increasingly routine to use functional data to investigate the mechanistic and adaptive consequences of sequence-level genetic variation.

The purpose of this chapter is to highlight recent development in these areas, focusing specifically on research in Old World monkeys (OWMs). In what follows, we briefly discuss the rationale for studying genetic variation in the OWM immune system. We then review recent research on natural selection and functional genetic variation in this clade pertaining to OWM immune function.

Why Is Immune System Variation in OWMs an Interesting Topic to Study?

OWMs (family *Cercopithecidae*) represent a diverse and species-rich group of primates that radiated approximately 27 million years ago (Purvis 1995) and that currently occupy a large geographic range across much of Africa and Asia (Wilson and Reeder 2005). OWMs retain close evolutionary ties to humans (93–94 % DNA sequence identity: Gibbs et al. 2007; Silva and Kondrashov 2002) and, outside of the apes, are our closest extant relatives. In some cases, they may also serve as appropriate models for the ecological and environmental circumstances of ancestral hominins (Behrensmeyer 2006; Jolly 2001; Potts 1998). OWMs are also important in medical research. For instance, rhesus macaques are major medical models for caloric restriction and aging, alcoholism, and infectious disease resistance, while baboons have been utilized as models for steroid hormone signaling (in both captivity and in the wild) and cardiovascular disease (Comuzzie et al. 2003; Gardner and Luciw 2008; Rogers and Hixson 1997; Roth et al. 2004). In addition to their utility as models for humans, OWMs also exhibit several specific characteristics that motivate studies of phenotypic and genetic variation.

Phenotypic Diversity in Disease Response

Many OWMs exhibit an unusually large geographic range relative to other primates (macaques and baboons cover the largest geographic range of any primate genera, other than humans) and also exhibit a high level of taxonomic diversity. As a result, the OWM clade exhibits substantial ecological and phenotypic diversity. This diversity extends to variation in susceptibility to infectious disease, including malaria

(Schmidt et al. 1977) and tuberculosis (Langermans et al. 2001), and is perhaps best illustrated by the case of SIV/HIV. Surveys across OWM species indicate major differences between the response to SIV in its natural African hosts and the response to SIV in species that are not naturally exposed. Sooty mangabeys (*Cercocebus atys*), a natural host for SIV_{sm} (the strain from which HIV-2 in humans arose: Lemey et al. 2003; Sharp et al. 1995), do not develop AIDS-like symptoms, despite robust replication of the virus in the host upon infection. Comparisons between sooty mangabeys and much more susceptible hosts, such as rhesus macaques, suggest that the key to this difference may be reduced immune activation and/or reduced reliance on T-cell-mediated immunity in the mangabeys (Silvestri et al. 2007). Infected rhesus macaques also exhibit more dramatic T-cell depletion and SIV-specific antibody production in response to SIV infection than members of their sister species, cynomolgus macaques (*M. fascicularis*) (Monceaux et al. 2007; Trichel et al. 2002). Finally, variation in the response to SIV extends to the population level as well: Chinese-origin rhesus macaques exhibit a slower rate of progression and a reduced degree of immune activation compared to Indian-origin rhesus macaques (Ling et al. 2002; Reimann et al. 2005).

Diversity in the Composition of Immune System Gene Families

Many of the components of primate immunity, such as Toll-like receptors, immunoglobulin-like receptors, and major histocompatibility loci, can be grouped into related classes of genes (Box 1). While the existence of these broad classes tends to be conserved across primates, the number of genes per class and the levels of polymorphism within these genes vary substantially between populations, species, and species groups. This variation may contribute to observed species-specific differences in immunity and disease. For example, the classical MHC class I genes, *HLA-A*, *HLA-B*, and *HLA-C*, function in antigen presentation as part of the adaptive immune response to intracellular pathogens and are among the most polymorphic genes in the human genome (Shiina et al. 2009). Genetic variation at these loci has been implicated in susceptibility to a broad suite of diseases, including HIV/AIDS, hepatitis B, psoriasis, type 1 diabetes, and rheumatoid arthritis (Burton et al. 2007; Fellay et al. 2007; Fernando et al. 2008; Kamatani et al. 2009; Liu et al. 2008; Pelak et al. 2010). Compared to humans, the class I region of rhesus macaques contains only orthologs for *HLA-A* and *HLA-B* and exhibits lower levels of polymorphism (Otting et al. 2005). In rhesus macaques, it appears that repeated segmental duplications in this region (Daza-Vamenta et al. 2004; Kulski et al. 2004), together with the presence of different combinations of *HLA-A* and *HLA-B* ortholog gene copies on a given haplotype, result in the expression of a diverse repertoire of class I antigens in the absence of very high levels of polymorphism (Wiseman et al. 2009). The macaque example illustrates how structural polymorphism and sequence polymorphism can therefore serve as alternative solutions for similar evolutionary challenges.

Behavioral, Life History, and Ecological Data from Field Studies of OWMs

OWMs are important behavioral and evolutionary models for humans and other species that exhibit highly sophisticated social systems. Consequently, detailed longitudinal and comparative data about the social structure, life history, and ecology of OWMs in natural populations are available for many OWM species (for a partial summary of these field studies, see Tung et al. 2010). These data permit the comparative study of immune system evolution and its genetic effects in an ecological context, for example, by relating socioecological characteristics that may influence exposure to, and spread of, disease with molecular changes at immune loci. For instance, Nunn et al. (2000) used such an approach in OWMs and other primates to show that species with more promiscuous female mating patterns tended to have higher basal white blood cell counts. They interpreted this result as a consequence of selection on the primate immune system due to sexually transmitted disease. Indeed, in a follow-up study of 15 immune genes, Wlasiuk and Nachman (2010) identified stronger signatures of selection on branches leading to more promiscuous primate species, most notably for the subset of genes whose function involves direct interactions with pathogens. Other studies have focused on variation in the incidence of parasites and pathogens at different geographic locations (Fooden 1994; Phillips-Conroy et al. 1988; Tung et al. 2009), including how genetic effects contribute to these differences (Tung et al. 2009). In humans, immune loci exhibit high F_{st} levels (a measure of the degree of overall genetic variation explained by between-population differences) relative to the genome-wide average, suggesting a history of local adaptation and selection (Akey et al. 2002; Barreiro et al. 2008). Understanding when and whether this holds true in OWMs could provide insight into how ecological differences between populations spur the process of genetic differentiation between geographically separated groups.

Natural Selection in the Immune System of OWMs

Rapid and Repeated Evolution at Immune Loci

Parasites and pathogens have long been considered to be primary drivers of evolution. Indeed, parasite or pathogen infection can have severe consequences for the health of wild primates and can exert strong selective effects (Keele et al. 2009; Knauf et al. 2012; Sapolsky and Else 1987). Thus, genetic variation that increases an individual's ability to evade, overcome, or minimize the consequences of infection is also likely to be adaptive. Genes involved in the immune system, which often evolve rapidly (including in, but not only in, primates: Hughes et al. 2005; Schlenke and Begun 2003), are a natural place to look for such variants.

In primates, this idea has primarily been supported by studies of a small number of candidate loci, such as those in the major histocompatibility complex (MHC) (see Table 1 for examples of these genes). In the past decade, however, the availability of genome sequences for multiple primate species, including the rhesus macaque, has facilitated larger-scale genome-wide studies that provide insight into the general importance of adaptive change in the primate immune system. Typically, these studies attempt to infer the presence of lineage-specific positive selection by comparing rates of synonymous and nonsynonymous site evolution in a gene's coding sequence. When more than two species are included in such a comparison, mutations in genes that appear to have evolved under selection can be localized to specific branches within the species tree (Yang 1997). Two recent studies—a three-species analysis of selection in primates (human, chimpanzee, and macaque: Gibbs et al. 2007) and a six-species analysis of selection in these three primates and three additional mammals (mouse, rat, and dog: Kosiol et al. 2008)—demonstrated that positive selection has indeed played an unusually important role, relative to other functional groups of genes, in the evolution of immune response genes in OWMs, as well as in other mammals. Genes involved in the "immune response" were significantly enriched among positively selected genes in both studies relative to other classes of genes, whether considering selection over the entire three-species primate tree (Gibbs et al. 2007; Kosiol et al. 2008) or on the individual branches leading to each species, which included rhesus macaque (Kosiol et al. 2008). Many of these genes fell into subclasses that have been the target of previous candidate gene studies, including those whose protein products interact with MHC class I (e.g., *LILRB1*, *LAIR*) and those involved in T-cell-mediated immunity (e.g., *CD3E, TCRA*) (Gibbs et al. 2007; Kosiol et al. 2008). Interestingly, the overall rate of nonsynonymous to synonymous changes on the macaque lineage appears to be somewhat slower than the comparable rate in humans and chimpanzees (Gibbs et al. 2007), perhaps reflecting more efficient negative selection for genes evolving under constraint. The ability of selection to influence phenotypic variation is correlated with effective population size, which is larger in macaques than in humans or chimpanzees (Hernandez et al. 2007). Potentially, therefore, positive selection may also be more efficient in macaques and other OWMs with large effective population sizes. If true, such a scenario would increase power to detect selective events in OWMs once a larger number of genome sequences become available.

Interestingly, genes that have undergone selection somewhere in the primate tree often have evolved under selection across more than one lineage. Sequencing of the rhesus macaque genome, for example, identified 67 genes that carried signatures of positive selection on all three branches (rhesus macaque, human, and chimpanzee: Gibbs et al. 2007). Similarly, Kosiol et al. (2008) estimated that the majority of the genes they identified as positively selected in any lineage (including rodents and dogs) were identified as positively selected in more than one lineage. For instance, glycophorin C (*GYPC*), an erythrocytic cell surface antigen, is believed to have evolved under recent selection in humans (Wilder et al. 2009) in response to pathogen pressure from the malarial parasite *Plasmodium falciparum*. However, *GYPC* also appears to have been positively selected on all examined branches of the primate tree, suggesting that this gene has been the target of pathogen-mediated adaptive evolution for a much longer period of time (Kosiol et al. 2008). Several

Table 1 Summary of genes showing signatures of selection in Old World Monkeys

Gene	Function	Immune function	Type of selection	Timescale of selection	Reference
GYPA	Erythrocyte surface receptor	Adaptive?	Positive selection	Macaque lineage	Baum et al. (2002)
MHC Class II gene *DQA1*	Antigen presentation	Adaptive	Balancing selection	Intra-specific (baboons)	Alberts (1999); Loisel et al. (2006)
MHC Class II genes (*DQA1, DQB1, DRB1, DPB1*)	Antigen presentation	Adaptive	Balancing selection (trans-species polymorphism)	Inter-specific (baboons and rhesus macaques)	Loisel et al. (2006)
				Inter-specific (rhesus macaques & cynomolgus)	Doxiadis et al. (2006)
PTPRC	B- and T-cell maturation and activation	Adaptive	Positive selection	Across Old World Monkeys	Filip and Mundy (2004)
APOBEC3G	Antiviral DNA-editing enzyme	Innate	Positive selection	Across Old World Monkeys	Sawyer et al. (2004); Zhang and Webb (2004)
APOBEC3H	Antiviral DNA-editing enzyme	Innate	Positive selection	Across Old World Monkeys	OhAinle et al. (2006)
APOL genes	Parasite resistance	Innate	Positive selection	Across Old World Monkeys	Smith and Malik (2009)
β Defensin 2	Anti-microbial defense	Innate	Positive selection	Across Old World Monkeys	Sawyer et al. (2007); Boniotto et al. (2003)
Lysozyme	Anti-bacterial activity	Innate	Positive selection	Across Old World Monkeys	Messier and Stewart (1997)
MEFR	Inflammatory response	Innate	Positive selection	Across Old World Monkeys	Schaner et al. (2001)
PKR	Viral response	Innate	Positive selection	Across Old World Monkeys	Elde et al. (2009)

(continued)

Table 1 (continued)

Gene	Function	Immune function	Type of selection	Timescale of selection	Reference
Tetherin	Antiviral restriction factor	Innate	Positive selection	Across Old World Monkeys	Liu et al. (2010); Lim et al. (2010c)
TLR4, TLR5, and other TLR genes	Pathogen recognition	Innate	Positive selection	Across Old World Monkeys	Nakajima et al. (2008); Wlasiuk et al. (2009); Wlasiuk and Nachman (2010)
TRIM5	Antiviral restriction factor	Innate	Balancing selection	Rhesus macaques and Sooty mangbeys	Newman et al. (2006)

nonhuman primate species are also susceptible to *Plasmodium* infection (Prugnolle et al. 2010), raising the possibility that adaptive change in orthologous genes in response to hematoprotozoan parasites may be common. As in the case of the general pervasiveness of selection on the immune system, these findings reinforce the suggestion from single-gene studies that selection in the primate immune system may often be recurrent and episodic (Elde et al. 2009; Messier and Stewart 1997; Schaner et al. 2001).

Biases in Current Studies of Selection: Where and How Does Selection Operate?

Evidence from genome-wide analysis and studies of individual genes and gene families suggests that positive selection has helped shape the evolution of gene protein-coding sequences in OWMs. However, protein-coding sequence accounts for only a small percentage of the genome. Both theoretical arguments and empirical evidence suggest that other components of the genome, especially those involved in gene regulation, may also have been targets of selection (King and Wilson 1975; Wray 2007; Wray et al. 2003). In particular, a recent meta-analysis of six scans for selection (all of which included human, chimpanzee, and macaque data; some included sequence from additional mammalian species as well) revealed a significant enrichment for selection on genes involved in T-cell-mediated immunity in both the coding and noncoding regions of these genes (Haygood et al. 2010; see also Torgerson et al. 2009).

Box 1 Major Targets of Natural Selection in Old World Monkeys

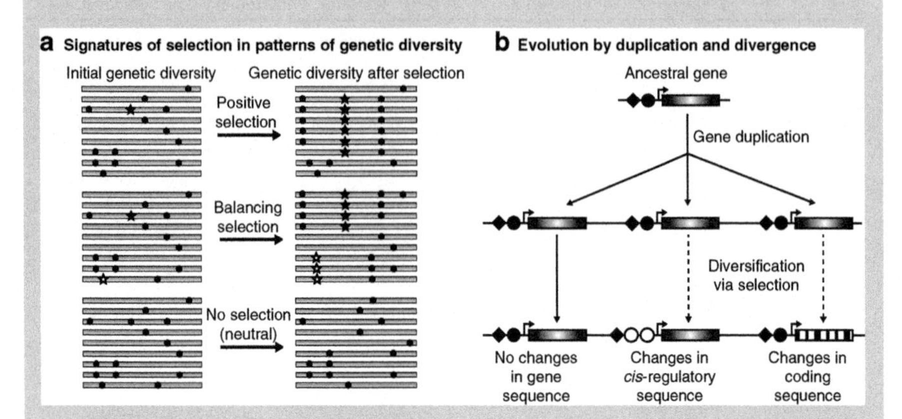

Episodes of natural selection leave characteristic signatures in the sequence of gene targets (panel A). Many of the most compelling examples of natural selection in OWMs involve genes that fall in one of a few well-studied families (note that although genes involved in immunity commonly evolve under selection across the tree of life, these families represent those that are both present and well studied in primates). In addition to MHC class I and II genes (discussed elsewhere in this chapter), these include:

APOBEC Proteins: APOBEC genes encode a family of proteins that inhibit replication of HIV and other retroviruses (Neil and Bieniasz 2009). Their antiretroviral function makes them prime candidates for coevolution with rapidly evolving viral pathogens. The evolutionary history of these genes is characterized by gene duplication (e.g., the seven functional primate genes found at the APOBEC3 locus resulted from a series of duplications from a single ancestral *APOBEC3* gene: Jarmuz et al. 2002) and by diversification via positive selection (panel B). Human–chimpanzee sequence comparisons revealed that six APOBEC genes showed evidence for positive selection, while two other evolved under purifying selection (Sawyer et al. 2004). In comparisons involving OWMs, two of these genes (*APOBEC3G* and *APOBEC3H*) also exhibited a significant excess of nonsynonymous changes, indicating a history of positive selection (OhAinle et al. 2006; Sawyer et al. 2004; Zhang and Webb 2004). The rapid expansion of this gene family in primates, combined with evidence for recurrent selection, suggests that these genes may play a particularly important role in primate antiviral immune defenses.

Defensins: Defensins are antimicrobial peptides that function in the innate immune response against a broad spectrum of bacterial, fungal, and viral invaders (Lehrer 2004). Like KIRs, the evolution of primate defensins is characterized by repeated tandem gene duplication followed by gene sequence diversification due to episodic positive selection (Das et al. 2010; Patil et al.

(continued)

Box 1 (continued)

2004; Semple et al. 2003), in contrast to many other innate immune system genes (Mukherjee et al. 2009). In primates, evidence of positive selection has been observed in genes from both the alpha and beta defensin families (Boniotto et al. 2003; Das et al. 2010; Lynn et al. 2004; Semple et al. 2003), and changes at positively selected sites have been shown to alter antimicrobial activity in vitro (Antcheva et al. 2004).

Killer Cell Immunoglobulin-Like Receptors (KIRs): KIR genes encode cell-surface receptors that interact with MHC class I genes to modulate natural killer cell and T-cell function, thus playing a central role in both innate and adaptive immunity. Radiation of the KIR family, as well as genetic diversity within specific KIR genes, may have evolved in response to either direct interactions with pathogens or as a result of coevolution with their MHC class I ligands (Hao and Nei 2005; Parham 1997). In OWMs, KIR family genes are marked by extensive intraspecific allelic diversity and substantial interspecific differences in KIR gene number and haplotype structure (Bimber et al. 2008; Hershberger et al. 2005; Kruse et al. 2010; Palacios, et al. 2011).

Toll-Like Receptors (TLRs): TLRs are innate immune system genes that are necessary for recognizing conserved patterns associated with pathogen infection (e.g., lipopolysaccharide (LPS) in gram-negative bacteria). Although TLRs share a general functional role, only some primate TLRs are associated with strong signature of selection. In OWMs, for example, interspecific comparisons of TLR sequences in 10 genes indentified evidence for positive selection at only three loci (*TLR4, TLR1,* and *TLR8*: Nakajima et al. 2008; Wlasiuk and Nachman 2010), echoing variation in the selective histories of this family in human populations (Barreiro et al. 2009).

The potential importance of immune-related selection on noncoding sequence is reinforced by the case of the OWM MHC. Population genetic diversity at the MHC *DQA1 cis*-regulatory region in wild baboons revealed a strong signature of balancing selection, with a deep split between two major and ancient allelic lineages (Loisel 2007). Comparisons of sequence diversity in this region across multiple OWM species indicated that the history of balancing selection identifiable in baboons extends even deeper into primate evolution, likely predating the split of baboons and macaques from their common ancestor (Loisel et al. 2006). This pattern is unusual and strongly suggestive of a functional role for these differences. Studies in other primates also indicate an important role for noncoding evolution in the immune system. For instance, high levels of noncoding sequence diversity have been characterized in a number of primate species at *CCR5*, the main entry point for HIV-1 and poxviruses (Lalani et al. 1999; Mummidi et al. 2000), and changes suggestive of clade-specific selection have been identified in the promoter sequence of

TNF, a cytokine involved in acute inflammation (Baena et al. 2007). Recently, studies in cell lines derived from various primate species demonstrated widespread differences in gene regulatory responses after LPS stimulation, some of which may be a consequence of selection on gene regulation (Barreiro et al. 2010). Investigating the role of regulatory sequence in the adaptive evolution of the immune system, including how regulatory and coding changes may act in concert, remains an important topic for future studies. Similarly, although maintenance of genetic diversity at immune loci by means of balancing selection or frequency-dependent selection is likely to be important (Charlesworth 2006; Dean et al. 2002; Garrigan and Hedrick 2003), we know relatively little about the extent to which this and other alternative modes of selection play a role in OWM immune system evolution (genome-wide studies have focused largely on positive selection; however, see Andres et al. 2009).

The generation of large-scale sequence data sets will be useful for addressing these gaps in our current knowledge. For instance, relatively recent selective sweeps can be localized using information about linkage disequilibrium (LD: the degree to which genetic variants segregate non-independently due to physical linkage) across the genome: recently selected loci exhibit more extensive LD with neighboring regions than is typical genome-wide (Sabeti et al. 2006, 2007; Voight et al. 2006). Similarly, local adaptation can be investigated by identifying genetic markers that show unusually high levels of genetic differentiation between populations, relative to the genomic distribution of the same metric (Akey et al. 2004; Barreiro et al. 2008; Sabeti et al. 2006, 2007; Voight et al. 2006). Both of these methods have identified a large number of candidates for selection among human populations, including many immune-related loci (Barreiro and Quintana-Murci 2010). Selective change on these time scales may also be common in OWMs, especially among geographically widespread species. Indeed, functionally important population-specific differences at MHC class I genes in rhesus macaques (Otting et al. 2007) and at MHC class I (Kita et al. 2009) and class II genes in cynomolgus macaques (Ling et al. 2011; Sano et al. 2006) have already been described. Additionally, in a broader comparison including a number of OWMs, Garamszegi and Nunn (2011) found that both levels of MHC *DRB* allelic variation and rates of nonsynonymous substitutions at this locus were correlated with parasite species richness (defined as the total number of parasite species per host species) in 41 primates.

Functional Variation at Selectively Relevant Loci

A signature of selection implies the presence, either in the past or in contemporary populations, of functional genetic variation at the same locus. Thus, data that establish the molecular- and organism-level effects of selected variants act as important corroborating evidence for the adaptive history of a given region. Perhaps more importantly, they also provide biological insight into the fitness-related effects of specific allelic variants.

Although establishing the functional relevance of specific genetic variants remains challenging, a number of methods are now well established, particularly those that focus on the molecular mechanisms that act as intermediate links between genotype and organism-level phenotypes (Box 2). The combination of selection and

Box 2 Functional Characterization of Sequence Diversity

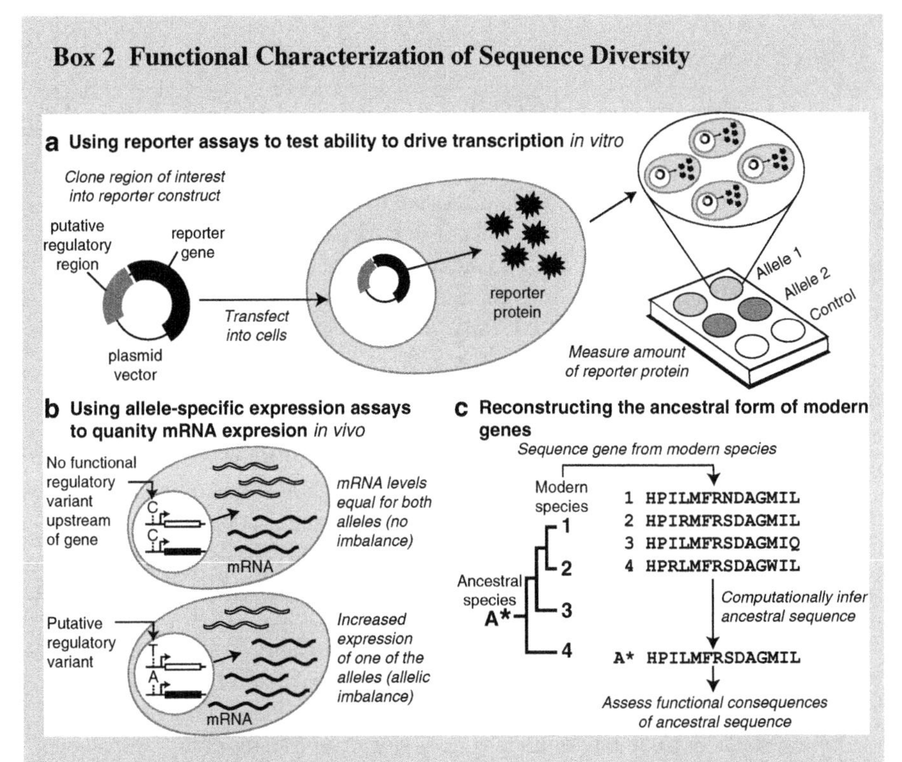

Studies of natural selection or genotype-phenotype associations identify regions of the genome harboring (or in strong linkage disequilibrium with) putatively functional genetic variation. Genetic variants in coding regions may affect protein structure, while those in regulatory regions can affect the developmental timing, magnitude, and tissue distribution of expression. To validate these effects, molecular experimental approaches are necessary. Common approaches include:

Linking Cis-Regulatory Sequence Variation and Gene Expression In Vitro: Reporter assays test the ability of regulatory sequences to drive transcription in vitro (Chorley et al. 2008; Knight 2003). Allelic variants of a putative regulatory region are inserted into plasmid vectors containing a reporter gene that lacks endogenous promoter activity (panel A). The resulting "constructs" are introduced into an appropriate cell line and protein expression of the reporter gene is measured, thus providing an indirect estimate of the ability of individual sequences to drive gene transcription. See, for example, Loisel et al. 2006; Tung et al. 2009; Vallender et al. 2008a; Vallender et al. 2008b.

Strengths: Can be used to isolate the effects of specific alleles or mutations, environmental and genetic backgrounds are controlled, and can be combined with experimental stimuli to test for differences in basal expression among alleles versus differences in expression after induction. *Limitations:* In vitro

(continued)

Box 2 (continued)

behavior may not always reflect in vivo behavior (particularly when using immortalized cells), reporter constructs are evaluated outside of the context of normal chromatin architecture, cell lines that match the tissue and/or species of interest may not be available, assays are low throughput.

Measuring Cis-Regulatory Effects via Allele-Specific Gene Expression Assays In Vivo: Allele-specific expression assays quantify the relative abundance of mRNA from the two alleles of a gene found within the same individual (panel B) (Pastinen 2010; Yan et al. 2002). Differences in the expression levels of these alleles imply the presence of variation in the *cis*-regulatory region controlling that gene, which can be localized by testing for an association between putative regulatory variants and the magnitude of allele-specific gene expression across individuals (e.g., Tung et al. 2011).

Strengths: In vivo measurements reflect the natural behavior of *cis*-regulatory functional variants, amenable to studies of natural populations, environmental and genetic background effects are controlled because comparisons are conducted between alleles within individuals, potentially feasible on a genome-wide scale (Fontanillas et al. 2010; Pastinen 2010). *Limitations:* Putative regulatory variants are identified by association and are not experimentally isolated, RNA samples may be difficult to obtain for the population or tissue of interest, the magnitude of allele-specific effects may be modified by epistatic or gene-environment interactions (de Meaux et al. 2005; Tung et al. 2011; von Korff et al. 2009).

Reconstructing the Ancestral Forms and Functional Properties of Modern Genes: Ancestral state reconstruction, which focuses on the inference and synthesis of sequences no longer found in contemporary populations (panel C), can be used to compare the functional consequences of evolutionary changes on a longer time scale (Harms and Thornton 2010; Thornton 2004). This approach permits the recreation of likely mutational pathways leading to extant genes. See, for example, Goldschmidt et al. 2008; Zhang and Rosenberg 2002.

Strengths: Permits experimental tests of the relationship between ancient sequence changes and phenotypic change, series of reconstructed proteins can indicate the probable sequence of mutations necessary to evolve novel functions. *Limitations:* Requires gene-appropriate functional tests that do not necessarily generalize well (i.e., assays for steroid hormone binding affinity work only for steroid hormone receptors: Bridgham et al. 2009), dependent on the accuracy of the inference method used to determine ancestral states, does not take into account dependence of the focal gene's function on other genes in the ancestral genome or on the ancestral environment (Harms and Thornton 2010).

functional analysis provides a far more comprehensive perspective on the evolution of immune genes than either approach alone and is therefore likely to become the standard to meet. From an applied perspective, such analysis may identify immunologically important genetic variation relevant to disease and pathogen resistance. Here, we review examples that reveal the potential for such work in OWMs.

MHC Diversity and Molecular and Disease-Related Phenotypes: Widespread and Repeated Evolution of an Immune Gene Cluster Within and Between Species

Genes of the MHC function in the presentation of self and nonself peptides to T cells and thus play a central role in the recognition of and response to pathogens and parasites. Extensive study of MHC sequence diversity suggests that selection has often contributed to shaping genetic variation at these loci (reviewed in Apanius et al. 1997; Garrigan and Hedrick 2003; Hughes and Yeager 1998; Klein 1986; Meyer and Thomson 2001). Evidence for adaptive evolution at OWM MHC loci takes several different forms. At the population level, researchers have observed allele frequency distributions that are inconsistent with those predicted by neutral evolution (cynomolgus macaques: Bonhomme et al. 2007). On a deeper time scale, comparisons of closely related species have revealed that some MHC polymorphisms seem to have been maintained since prior to divergence from the species' common ancestor. These "trans-species" polymorphisms are unusual among distinct species but are relatively common among both class I (Otting et al. 2007) and class II MHC genes (Doxiadis et al. 2006; Huchard et al. 2006; Loisel 2007), suggesting that selection has acted to maintain segregating variation over long periods of time (Klein et al. 2007). Finally, molecular evolution studies have shown that the exons encoding the antigen-binding regions in MHC proteins consistently exhibit elevated nonsynonymous to synonymous substitution ratios; this hallmark of strong positive selection has been observed at several MHC class II genes in baboons (Huchard et al. 2008; Loisel 2007) and in rhesus macaque class I genes (Urvater et al. 2000). As MHC diversity on all three of these levels may contribute to adaptively relevant phenotypic variation, these signatures of selection may harbor important clues to the location of functionally important genetic variants.

For instance, Loisel et al. (2006) demonstrated that MHC population genetic diversity is associated with functional consequences for gene regulation. The promoter region of the classical MHC class II gene *DQA1* is highly polymorphic in many primate species, including yellow baboons, rhesus macaques, and pigtail macaques, and shows evidence of trans-species polymorphism (Loisel et al. 2006). As a result, the phylogenetic relationships among alleles for this region (the "gene tree") are not congruent with the phylogenetic relationships between OWM species (the "species tree"), a pattern that is particularly unusual over such extended periods of evolutionary time (~27 million years; Fig. 1). To assess the functional effects of *DQA1* promoter region variation, Loisel et al. (2006) cloned 12 of the alleles found

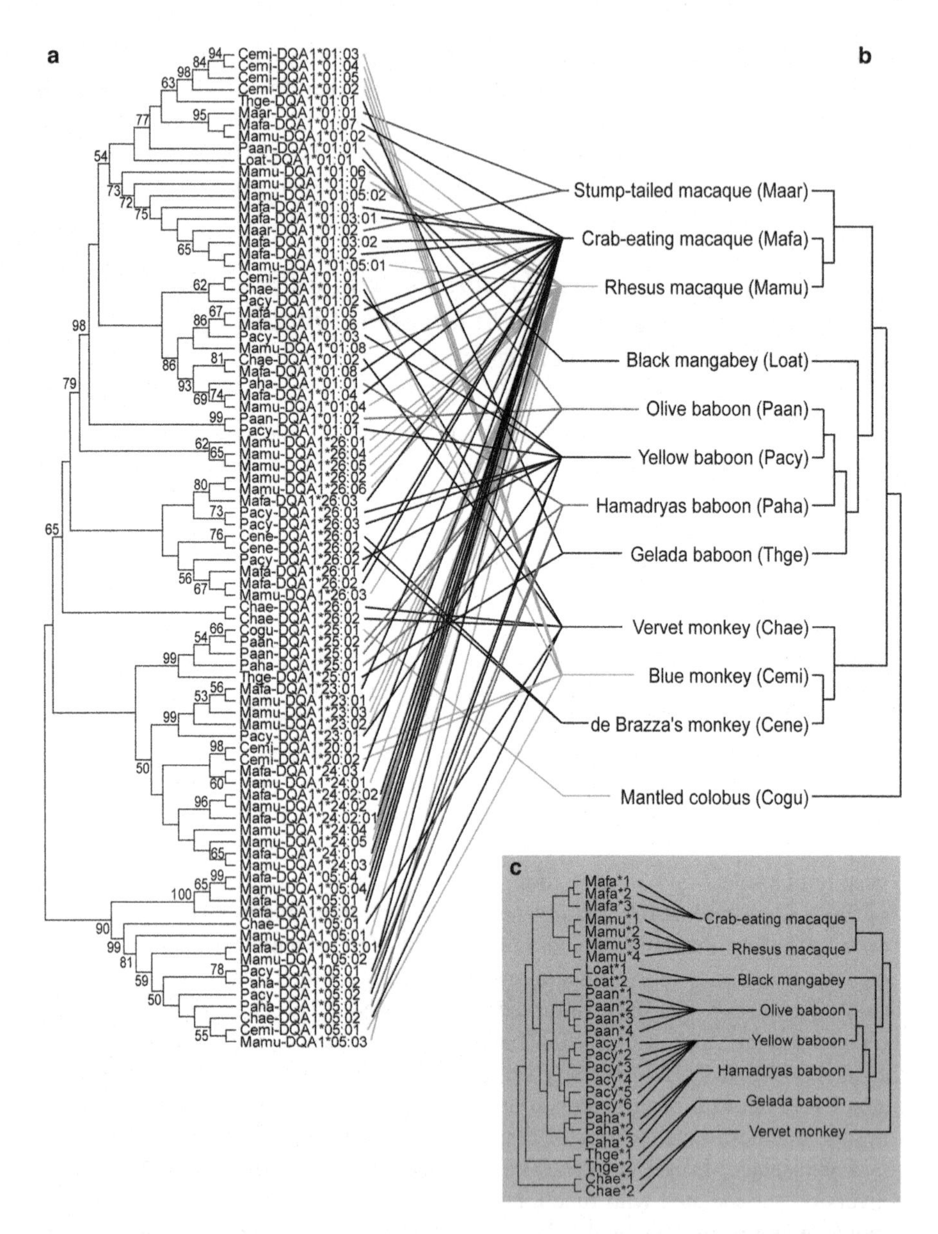

Fig. 1 Evolutionary relationships of MHC-*DQA1* exon 2 coding sequence in Old World monkeys. (**a**) The allelic genealogy for MHC-*DQA1* exon 2 sequences (containing the antigen-binding region) in 12 species of Old World monkeys (OWMs) is characterized by the long-term maintenance of ancestral allelic lineages across speciation events, a phenomenon known as trans-species polymorphism. The incongruity between the *DQA1* exon 2 gene tree and the OWM species tree (shown in **b**.) reflects the presence of this trans-species polymorphism, in contrast to the typical expected relationship between gene and species trees (shown in inset **c**.). The relationship between *DQA1* exon 2 sequences was inferred using neighbor-joining with evolutionary distances computed using the Tamura-Nei method. The percentage of replicate trees in which the associated taxa clustered together in the bootstrap test (1,000 replicates) is shown next to the branches. *DQA1* exon 2 sequences were obtained from the IPD-MHC nonhuman primate database (Robinson et al. 2003)

in a contemporary yellow baboon population into a luciferase reporter construct and introduced these constructs into a baboon lymphoblast cell line. The strongest promoter allele, as reflected by increased production of the luciferase reporter gene, drove gene expression twice as strongly as the weakest promoter allele—an effect on *DQA1* expression that is paralleled in humans (Fernandez et al. 2003; Haas et al. 2009; Morzycka-Wroblewska et al. 1997).

In addition to intracellular effects, OWM MHC variation may also influence intercellular regulation of tissue composition. For example, the MHC class II *DRB* region in cynomolgus macaques is characterized by gene duplication, high levels of interspecies divergence, trans-species polymorphism, and extensive intraspecific variation, which together strongly point to a history of natural selection (Leuchte et al. 2004). Although the phenotypic outcomes of these effects are not well characterized, at least one possible consequence includes an effect on the composition of the peripheral blood, resulting from the role MHC molecules play in thymic lymphocyte (T cell) maturation. In support of this hypothesis, a recent study in captive cynomolgus macaques indicates that genetic variation at *DRB* is significantly associated with interindividual variation in basal $CD4^+$ T-cell counts (Aarnink et al. 2011).

By acting through regulatory, developmental, and structural intermediates such as these, MHC genetic variation may contribute to variation in OWM organism-level traits, particularly disease susceptibility (although MHC polymorphism has also been suggested to influence behavior and fertility-related traits as well: Ober 1999; Ober et al. 1998; Penn and Potts 1999; Ruff et al. 2011). Associations between MHC polymorphisms and diseases are common in humans, including a robust relationship between MHC class I genetic variation and HIV infection and disease progression (Fellay et al. 2007; Goulder and Watkins 2008; Pelak et al. 2010). Interestingly, work on macaque models of HIV/SIV—a primate that, like humans, progresses to potentially fatal AIDS-like symptoms—has revealed a parallel association between MHC class I genes and SIV control and progression (Bontrop and Watkins 2005; Goulder and Watkins 2008). In particular, specific alleles of the class I loci MHC-*B* and MHC-*A* (orthologs of human *HLA-B* and *HLA-A*) were found to significantly influence the viral set point and progression to disease in experimental SIV infections involving rhesus macaques, cynomolgus macaques, and pig-tailed macaques. This parallelism is remarkable considering the contrasting evolutionary histories of this region in humans and macaques. Whereas macaque MHC class I diversity has largely been driven by gene duplication events and relatively low allelic diversity within genes, human MHC class I loci exhibit extraordinarily high levels of allelic variation. In both cases, however, this genetic diversity appears to influence HIV/SIV control.

Functional Evolution at TRIM5*: Interspecific and Intraspecific Diversity in Response to Retroviral Threat*

While the spread of HIV/SIV in humans is relatively recent, infection by retroviral pathogens in general has been ongoing for millions of years. Retroviral sequences can be identified as "fossils" in the genomes of all fully sequenced primates (Gibbs et al. 2007; Han et al. 2007), suggesting that retroviruses have exerted significant selective pressures on primate evolution for a very long time. The evolution of the antiretroviral gene, tripartite motif protein 5 (*TRIM5*), illustrates the possible effects of these pressures. *TRIM5* encodes a cytosolic protein that restricts infection by HIV-1 and other retroviruses by interacting with retrovirus capsid lattice (Stremlau et al. 2006). Through this interaction, *TRIM5* restricts retroviral infection by acting as a pattern recognition receptor for the capsid lattice and activating general innate immune signaling pathways (Pertel et al. 2011). The *TRIM5* gene is marked by numerous signatures of ancient and recurrent natural selection within primates (Liu et al. 2005; Ortiz et al. 2006; Sawyer et al. 2005, 2007), including both lineage-specific positive selection (Johnson and Sawyer 2009) and long-term balancing selection, indicated by the presence of trans-specific polymorphism between sooty mangabeys and rhesus macaques (Newman et al. 2006) and between humans and chimpanzees (Cagliani et al. 2010).

Research on natural selection at *TRIM5* has been instrumental for understanding the function of genetic diversity in the gene, which is high in both OWMs and humans (Cagliani et al. 2010; Dietrich et al. 2010; Newman et al. 2006). For instance, efforts to localize the specific nucleotide targets of positive selection in *TRIM5* indicated that two specific regions of protein, known as the B30.2 and coiled-coil domains, were likely essential to its antiviral activity (Sawyer et al. 2005; Song et al. 2005b). Indeed, the ability of the human TRIM5α protein to restrict the replication of retroviruses (including HIV-1, SIV_{agm}, and murine leukemia viruses) in vitro was significantly enhanced by mutating only a few amino acids in these regions to the version found in rhesus macaques (Sawyer et al. 2005; Yap et al. 2005). Thus, detailed analysis of selection at *TRIM5* identified at least some of the sites likely associated with species-specific differences in its ability to restrict retroviral infection (Sawyer et al. 2005; Song et al. 2005a). Indeed, long-term maintenance of *TRIM5* alleles in rhesus macaques and sooty mangabeys may have roots in retroviral resistance: allelic variation segregating within these species also explains differences in the ability to restrict in vitro retroviral infection (Newman et al. 2006; Wilson et al. 2008). This effect likely explains the finding that specific *TRIM5* alleles in rhesus macaques are associated with differences in plasma viral load and disease progression following experimental SIV infection (Lim et al. 2010a, b).

Genetic Variation at the Baboon FY *Locus: Intraspecific Functional Variation in Contemporary Populations*

Finally, immune-related genetic variation may also have selective and functional consequences observable within or between contemporary primate populations. For instance, a recent study of a natural baboon population in Kenya identified a significant association between genetic variation at the *FY* (*DARC*) locus and the probability of infection by the hematoprotozoan parasite *Hepatocystis* (a close relative of the malarial *Plasmodium* parasites) as well as unusually high levels of population genetic differentiation at the locus among three East African baboon populations (Tung et al. 2009). This association raised the possibility that the evolution of *FY* in baboons might parallel that documented for humans, in which a *cis*-regulatory genetic variant (i.e., a variant that influences the expression levels of a linked gene: Wittkopp et al. 2004; Wittkopp et al. 2008) upstream of the *FY* gene confers protection from *Plasmodium vivax* via eliminating expression of the gene on the erythrocyte surface. This variant has likely been a target of selection in some human populations (Hamblin and Di Rienzo 2000; Hamblin et al. 2002). A comparable association in baboons also fell in the *FY* *cis*-regulatory region, suggesting that a regulatory mechanism explained the relationship between *FY* genetic variation and *Hepatocystis* incidence. Measurements of allele-specific gene expression from blood-derived RNA samples from the same baboon population indicated that functional *cis*-regulatory variation did indeed segregate at this locus (see Box 2 for an explanation on inferring *cis*-regulatory effects through allele-specific gene expression measurements). Specifically, in individuals heterozygous for particular *cis*-regulatory variants, the two *FY* alleles for those individuals were expressed at different levels, suggesting that these variants (or nearby linked variants) were responsible for driving *FY* expression at different rates. Using promoter-based luciferase reporter constructs in cell culture, the authors were able to confirm that a few base pair changes within this region were sufficient to alter gene expression in the direction predicted by the allele-specific expression data. However, unlike in humans, the variants identified through these screens did not completely eliminate *FY* expression in erythrocytes nor did the association between *FY* variation and *Hepatocystis* yield complete protection against the parasite (Tung et al. 2009). Thus, the parallelism between *FY* and parasite infection in baboons and humans is incomplete: instead of different mechanisms contributing to a shared association, as in the case of MHC class I loci in human and rhesus macaque SIV/HIV, in this case similar mechanisms yielded associations of different strengths.

Conclusions

Taken together, studies thus far on the genetics of OWM immunity reinforce a general theme in evolutionary immunogenetics: adaptive change in the immune system is pervasive, is recurrent, and takes a variety of forms. Specifically, studies of individual immune-related genes reveal that *selective regimes include both directional selection and selection for maintenance of genetic variation* (sometimes at the same locus); that *functional and/or selective divergence are apparent on multiple levels of comparisons*, including within populations, between populations, between species, and between larger taxonomic groups; and that *functional and selective change can target both coding and noncoding regions*.

We anticipate that newly developed high-throughput sequencing approaches will make important contributions to extending this work, particularly functional annotation of selectively important genetic changes. These methods allow data on sequence variation and quantitative measurements of genome function, including gene expression levels, chromatin accessibility, and epigenetic patterns, to be collected in tandem. For instance, comparative high-throughput sequencing efforts in humans, chimpanzees, and rhesus macaques have already identified species-specific, genome-wide changes in gene expression and isoform usage in the brain, liver, heart, and kidney (Babbitt et al. 2010; Blekhman et al. 2010), as well as differences in histone methylation in immortalized cell lines (Cain et al. 2011). Combining these assays with experimental exposure to immune-stimulating agents could also yield insight into the basis for genetic differences in the response to particular pathogens (e.g., Barreiro et al. 2010), a subject of great interest given the diversity among OWMs in susceptibility to disease. The results of such work would allow maps of functional changes to be overlaid on top of maps of targets of selection, providing the first genome-wide maps of both functionally variable and selectively relevant loci in the same species.

Ultimately, however, understanding genetic evolution and adaptation also involves understanding the phenotypic variation that is the direct target of natural selection. Thus, we wish to conclude by again highlighting the importance of the extensive field data available for natural OWM populations. In particular, the natural ecological conditions that characterize particular OWM species and populations have an important influence on the pathogens and parasites to which those individuals are exposed and thus on the evolution of the immune response. A second level of integration—combining lab-based studies of functional variation and its consequences with field-based studies of organism-level traits—therefore offers the opportunity to understand not only the mechanistic consequences of selectively relevant sequence variation but also its implications for ecologically relevant phenotypic variation. Research already underway illustrates the opportunities afforded by combining field data with genetic approaches. These studies include research testing for fine-grained correlations between pathogen and host genetic diversity (e.g., Schad et al. 2005; Schwensow et al. 2007 in lemurs); examining the relationship between sociobehavioral characteristics, such as dominance rank, on the immune response, including how these characteristics influence genotype-environment interactions (Tung et al. 2011); and investigating the consequences of pathogen-mediated selection on immune loci for other traits, such as mate choice (Ruff et al. 2011; Sommer 2005). Given the longitudinal nature of many primate field studies, these systems also present important opportunities to test for trade-offs between different arms of the immune system, as well as trade-offs between investment in immune defense and growth, reproduction, and other forms of maintenance—two major questions in ecological immunology (McDade 2003). Together, the integration of increasingly sophisticated genetic approaches into field-based ecological studies promises to provide a compelling picture of OWM immune evolution in the past and in the present day.

Acknowledgments J.T. was supported by a Chicago Fellows postdoctoral fellowship and National Institutes of Health grants 1R01-GM102562 and P30-AG03442 (Center for Aging grant). D.A.L. was supported by NIH grants F32HL095268 and T32HL007605. We thank J. Brinkworth and two anonymous reviewers for helpful comments on an earlier version of this manuscript.

References

Aarnink A, Garchon HJ, Puissant-Lubrano B, Blancher-Sardou M, Apoil PA, Blancher A (2011) Impact of MHC class II polymorphism on blood counts of CD4+T lymphocytes in macaque. Immunogenetics 63(2):95–102

Akey J, Zhang G, Zhang K, Jin L, Shriver M (2002) Interrogating a high-density SNP map for signatures of natural selection. Genome Res 12:1805–1814

Akey JM, Eberle MA, Rieder MJ, Carlson CS, Shriver MD, Nickerson DA, Kruglyak L (2004) Population history and natural selection shape patterns of genetic variation in 132 genes. PLoS Biol 2(10):1591–1599

Alberts SC (1999) Thirteen Mhc-DQA1 alleles from two populations of baboons. Immunogenetics 49:825–827

Andres AM, Hubisz MJ, Indap A, Torgerson DG, Degenhardt JD, Boyko AR, Gutenkunst RN, White TJ, Green ED, Bustamante CD et al (2009) Targets of balancing selection in the human genome. Mol Biol Evol 26(12):2755–2764

Antcheva N, Boniotto M, Zelezetsky I, Pacor S, Falzacappa MVV, Crovella S, Tossi A (2004) Effects of positively selected sequence variations in human and *Macaca fascicularis* beta-defensins 2 on antimicrobial activity. Antimicrob Agents Chemother 48(2):685–688

Apanius V, Penn D, Slev PR, Ruff LR, Potts WK (1997) The nature of selection on the major histocompatibility complex. Crit Rev Immunol 17:179–224

Babbitt CC, Fedrigo O, Pfefferle AD, Boyle AP, Horvath JE, Furey TS, Wray GA (2010) Both noncoding and protein-coding RNAs contribute to gene expression evolution in the primate brain. Genome Biol Evol 2:67–79

Baena A, Mootnick AR, Falvo JV, Tsytsykova AV, Ligeiro F, Diop OM, Brieva C, Gagneux P, O'Brien SJ, Ryder OA et al (2007) Primate TNF promoters reveal markers of phylogeny and evolution of innate immunity. PLoS One 2(7):e621

Barreiro LB, Ben-Ali M, Quach H, Laval G, Patin E, Pickrell JK, Bouchier C, Tichit M, Neyrolles O, Gicquel B, Kidd JR, Kidd KK, Alcais A, Ragimbeau J, Pellegrini S, Abel L, Casanova J-L, Quintana-Murci L (2009) Evolutionary dynamics of human Toll-like receptors and their different contributions to host defense. PLoS Genetics 5:e1000562

Barreiro LB, Quintana-Murci L (2010) From evolutionary genetics to human immunology: how selection shapes host defence genes. Nat Rev Genet 11(1):17–30

Barreiro LB, Laval G, Quach H, Patin E, Quintana-Murci L (2008) Natural selection has driven population differentiation in modern humans. Nat Genet 40(3):340–345

Barreiro LB, Marioni JC, Blekhman R, Stephens M, Gilad Y (2010) Functional comparison of innate immune signaling pathways in primates. PLoS Genet 6(12):e1001249

Baum J, Ward RH, Conway DJ (2002) Natural selection on the erythrocyte surface. Mol Biol Evol 19:223–229

Behrensmeyer AK (2006) Climate change and human evolution. Science 311(5760):476–478

Bimber BN, Moreland AJ, Wiseman RW, Hughes AL, O'Connor DH (2008) Complete characterization of killer Ig-like receptor (KIR) haplotypes in Mauritian cynomolgus macaques: novel insights into nonhuman primate KIR gene content and organization. J Immunol 181(9): 6301–6308

Blekhman R, Marioni JC, Zumbo P, Stephens M, Gilad Y (2010) Sex-specific and lineage-specific alternative splicing in primates. Genome Res 20(2):180–189

Bonhomme M, Blancher A, Jalil MF, Crouau-Roy B (2007) Factors shaping genetic variation in the MHC of natural non-human primate populations. Tissue Antigens 70(5):398–411

Boniotto M, Tossi A, DelPero M, Sgubin S, Antcheva N, Santon D, Masters J, Crovella S (2003) Evolution of the beta defensin 2 gene in primates. Genes Immun 4(4):251–257

Bontrop RE, Watkins DI (2005) MHC polymorphism: AIDS susceptibility in non-human primates. Trends Immunol 26(4):227–233

Bridgham JT, Ortlund EA, Thornton JW (2009) An epistatic ratchet constrains the direction of glucocorticoid receptor evolution. Nature 461(7263):515–578

Burton PR, Clayton DG, Cardon LR, Craddock N, Deloukas P, Duncanson A, Kwiatkowski DP, McCarthy MI, Ouwehand WH, Samani NJ et al (2007) Genome-wide association study of 14,000 cases of seven common diseases and 3,000 shared controls. Nature 447(7145): 661–678
Bustamante CD, Fledel-Alon A, Williamson S, Nielsen R, Hubisz MT, Glanowski S, Tanenbaum DM, White TJ, Sninsky JJ, Hernandez RD et al (2005) Natural selection on protein-coding genes in the human genome. Nature 437(7062):1153–1157
Cagliani R, Fumagalli M, Biasin M, Piacentini L, Riva S, Pozzoli U, Bonaglia MC, Bresolin N, Clerici M, Sironi M (2010) Long-term balancing selection maintains trans-specific polymorphisms in the human TRIM5 gene. Hum Genet 128(6):577–588
Cain C, Bleckman R, Marioni JC, Gilad Y (2011) Gene expression differences among primates are associated with changes in a histone epigenetic modification. Genetics 187(4):1225–1234
Charlesworth D (2006) Balancing selection and its effects on sequences in nearby genomic regions. PLoS Genet 2(4):e64
Chorley BN, Wang X, Campbell MR, Pittman GS, Noureddine MA, Bell DA (2008) Discovery and verification of functional single nucleotide polymorphisms in regulatory genomic regions: current and developing technologies. Mutat Res 659(1–2):147–157
Comuzzie AG, Cole SA, Martin L, Carey KD, Mahaney MC, Blangero J, VandeBerg JL (2003) The baboon as a nonhuman primate model for the study of the genetics of obesity. Obes Res 11(1):75–80
Das S, Nikolaidis N, Goto H, McCallister C, Li JX, Hirano M, Cooper MD (2010) Comparative genomics and evolution of the alpha-defensin multigene family in primates. Mol Biol Evol 27(10):2333–2343
Daza-Vamenta R, Glusman G, Rowen L, Guthrie B, Geraghty DE (2004) Genetic divergence of the rhesus macaque major histocompatibility complex. Genome Res 14:1501–1515
de Meaux J, Goebel U, Pop A, Mitchell-Olds T (2005) Allele-specific assay reveals functional variation in the chalcone synthase promoter of Arabidopsis thaliana that is compatible with neutral evolution. Plant Cell 17(3):676–690
Dean M, Carrington M, O'Brien SJ (2002) Balanced polymorphism selected by genetic versus infectious human disease. Annu Rev Genomics Hum Genet 3:263–292
Dietrich EA, Jones-Engel L, Hu SL (2010) Evolution of the antiretroviral restriction factor TRIMCyp in Old World Primates. PLoS One 5(11):e14019
Doxiadis GGM, Rouweler AJM, de Groot NG, Louwerse A, Otting N, Verschoor EJ, Bontrop RE (2006) Extensive sharing of MHC class II alleles between rhesus and cynomolgus macaques. Immunogenetics 58(4):259–268
Elde NC, Child SJ, Geballe AP, Malik HS (2009) Protein kinase R reveals an evolutionary model for defeating viral mimicry. Nature 457(7228):485–488
Fellay J, Shianna KV, Ge DL, Colombo S, Ledergerber B, Weale M, Zhang KL, Gumbs C, Castagna A, Cossarizza A et al (2007) A whole-genome association study of major determinants for host control of HIV-1. Science 317(5840):944–947
Fernandez S, Wassmuth R, Knerr I, Frank C, Haas JP (2003) Relative quantification of HLA-DRA1 and -DQA1 expression by real-time reverse transcriptase-polymerase chain reaction (RT-PCR). Eur J Immunogenet 30:141–148
Fernando MMA, Stevens CR, Walsh EC, De Jager PL, Goyette P, Plenge RM, Vyse TJ, Rioux JD (2008) Defining the role of the MHC in autoimmunity: a review and pooled analysis. PLoS Genet 4(4):e1000024
Filip LC, Mundy NI (2004) Rapid evolution by positive Darwinian selection in the extracellular domain of the abundant lymphocyte protein CD45 in primates. Mol Biol Evol 21: 1504–1511
Fontanillas P, Landry CR, Wittkopp PJ, Russ C, Gruber JD, Nusbaum C, Hartl DL (2010) Key considerations for measuring allelic expression on a genomic scale using high-throughput sequencing. Mol Ecol 19:212–227
Fooden J (1994) Malaria in macaques. Int J Primatol 15(4):573–596
Garamszegi LZ, Nunn CL (2011) Parasite-mediated evolution of the functional part of the MHC in primates. J Evol Biol 24(1):184–195
Gardner MB, Luciw PA (2008) Macaque models of human infectious disease. ILAR J 49(2):220–255

Garrigan D, Hedrick PW (2003) Perspective: detecting adaptive molecular polymorphism: lessons from the MHC. Evolution 57(8):1707–1722

Gibbs RA, Rogers J, Katze MG, Bumgarner R, Weinstock GM, Mardis ER, Remington KA, Strausberg RL, Venter JC, Wilson RK et al (2007) Evolutionary and biomedical insights from the rhesus macaque genome. Science 316(5822):222–234

Goldschmidt V, Ciuffi A, Ortiz M, Brawand D, Munoz M, Kaessmann H, Telentil A (2008) Antiretroviral activity of ancestral TRIM5 alpha. J Virol 82(5):2089–2096

Goulder PJR, Watkins DI (2008) Impact of MHC class I diversity on immune control of immunodeficiency virus replication. Nat Rev Immunol 8(8):619–630

Haas JP, Metzler M, Frank C, Haefner R, Wassmuth R (2009) HLA-DQA1 gene expression profiling in oligoarticular JIA. Autoimmunity 42(4):389–391

Hamblin MT, Di Rienzo A (2000) Detection of the signature of natural selection in humans: evidence from the Duffy blood group locus. Am J Hum Genet 66(5):1669–1679

Hamblin MT, Thompson EE, Di Rienzo A (2002) Complex signatures of natural selection at the Duffy blood group locus. Am J Hum Genet 70(2):369–383

Han K, Konkel MK, Xing J, Wang H, Lee J, Meyer TJ, Huang CT, Sandifer E, Hebert K, Barnes EW et al (2007) Mobile DNA in old world monkeys: a glimpse through the rhesus macaque genome. Science 316(5822):238–240

Hao L, Nei M (2005) Rapid expansion of killer cell immunoglobulin-like receptor genes in primates and their coevolution with MHC Class I genes. Gene 347(2):149–159

Harms MJ, Thornton JW (2010) Analyzing protein structure and function using ancestral gene reconstruction. Curr Opin Struct Biol 20(3):360–366

Haygood R, Fedrigo O, Hanson B, Yokoyama KD, Awray G (2007) Promoter regions of many neural- and nutrition-related genes have experienced positive selection during human evolution. Nat Genet 39(9):1140–1144

Haygood R, Babbitt CC, Fedrigo O, Wray GA (2010) Contrasts between adaptive coding and noncoding changes during human evolution. Proc Natl Acad Sci USA 107(17):7853–7857

Hernandez RD, Hubisz MJ, Wheeler DA, Smith DG, Ferguson B, Rogers J, Nazareth L, Indap A, Bourquin T, McPherson J et al (2007) Demographic histories and patterns of linkage disequilibrium in Chinese and Indian rhesus macaques. Science 316(5822):240–243

Hershberger KL, Kurian J, Korber BT, Letvin NL (2005) Killer cell immunoglobulin-like receptors (KIR) of the African-origin sabaeus monkey: evidence for recombination events in the evolution of KIR. Eur J Immunol 35(3):922–935

Huchard E, Cowlishaw G, Raymond M, Weill M, Knapp LA (2006) Molecular study of Mhc-DRB in wild chacma baboons reveals high variability and evidence for trans-species inheritance. Immunogenetics 58(10):805–816

Huchard E, Weill M, Cowlishaw G, Raymond M, Knapp LA (2008) Polymorphism, haplotype composition, and selection in the Mhc-DRB of wild baboons. Immunogenetics 60(10):585–598

Hughes AL, Yeager M (1998) Natural selection at major histocompatibility complex loci of vertebrates. Annu Rev Genet 32:415–435

Hughes AL, Packer B, Welch R, Chanock SJ, Yeager M (2005) High level of functional polymorphism indicates a unique role of natural selection at human immune system loci. Immunogenetics 57:821–827

Jarmuz A, Chester A, Bayliss J, Gisbourne J, Dunham I, Scott J, Navaratnam N (2002) An anthropoid-specific locus of orphan C to U RNA-editing enzymes on chromosome 22. Genomics 79(3):285–296

Johnson WE, Sawyer SL (2009) Molecular evolution of the antiretroviral TRIM5 gene. Immunogenetics 61:163–176

Jolly CJ (2001) A proper study for mankind: analogies from the papionin monkeys and their implications for human evolution. Am J Phys Anthropol 44:177–204

Kamatani Y, Wattanapokayakit S, Ochi H, Kawaguchi T, Takahashi A, Hosono N, Kubo M, Tsunoda T, Kamatani N, Kumada H et al (2009) A genome-wide association study identifies variants in the HLA-DP locus associated with chronic hepatitis B in Asians. Nat Genet 41(5):591–595

Keele BF, Jones JH, Terio KA, Estes JD, Rudicell RS, Wilson ML, Li YY, Learn GH, Beasley TM, Schumacher-Stankey J et al (2009) Increased mortality and AIDS-like immunopathology in wild chimpanzees infected with SIVcpz. Nature 460(7254):515–519

King M-C, Wilson ACC (1975) Evolution at two levels in humans and chimpanzees. Science 188(4184):107–116

Kita YF, Hosomichi K, Kohara S, Itoh Y, Ogasawara K, Tsuchiya H, Torii R, Inoko H, Blancher A, Kulski JK et al (2009) MHC class I A loci polymorphism and diversity in three southeast Asian populations of cynomolgus macaque. Immunogenetics 61(9):635–648

Klein J (1986) Natural history of the major histocompatibility complex. Wiley, New York, NY

Klein J, Sato A, Nikolaidis N (2007) MHC, TSP, and the origin of species: from immunogenetics to evolutionary genetics. Annu Rev Genet 41:281–304

Knauf S, Batamuzi EK, Mlengeya T, Kilewo M, Lejora IA, Nordhoff M, Ehlers B, Harper KN, Fyumagwa R, Hoare R et al (2012) Treponema infection associated with genital ulceration in wild baboons. Vet Pathol 49(2):292–303

Knight JC (2003) Functional implications of genetic variation in non-coding DNA for disease susceptibility and gene regulation. Clin Sci 104(5):493–501

Kosiol C, Vinar T, da Fonseca RR, Hubisz MJ, Bustamante CD, Nielsen R, Siepel A (2008) Patterns of positive selection in six mammalian genomes. PLoS Genet 4(8):e1000144

Kruse PH, Rosner C, Walter L (2010) Characterization of rhesus macaque KIR genotypes and haplotypes. Immunogenetics 62(5):281–293

Kulski JK, Anzai T, Shiina T, Inoko H (2004) Rhesus macaque class I duplicon structures, organization, and evolution within the alpha block of the major histocompatibility complex. Mol Biol Evol 21(11):2079–2091

Lalani AS, Masters J, Zeng W, Barrett J, Pannu R, Everett H, Arendt CW, McFadden G (1999) Use of chemokine receptors by poxviruses. Science 286(5446):1968–1971

Langermans JAM, Andersen P, van Soolingen D, Vervenne RAW, Frost PA, van der Laan T, van Pinxteren LAH, van den Hombergh J, Kroon S, Peekel I et al (2001) Divergent effect of bacillus Calmette-Guerin (BCG) vaccination on Mycobacterium tuberculosis infection in highly related macaque species: implications for primate models in tuberculosis vaccine research. Proc Natl Acad Sci U S A 98(20):11497–11502

Lehrer RI (2004) Primate defensins. Nat Rev Microbiol 2(9):727–738

Lemey P, Pybus OG, Wang B, Saksena NK, Salemi M, Vandamme AM (2003) Tracing the origin and history of the HIV-2 epidemic. Proc Natl Acad Sci U S A 100(11):6588–6592

Leuchte N, Berry N, Kohler B, Almond N, LeGrand R, Thorstensson R, Titti F, Sauermann U (2004) MhcDRB-sequences from cynomolgus macaques (*Macaca fascicularis*) of different origin. Tissue Antigens 63(6):529–537

Lim SY, Chan T, Gelman RS, Whitney JB, O'Brien KL, Barouch DH, Goldstein DB, Haynes BF, Letvin NL (2010a) Contributions of Mamu-a*01 status and TRIM5 allele expression, but Not CCL3L copy number variation, to the control of SIVmac251 replication in Indian-origin rhesus monkeys. PLoS Genet 6(6):e1000997

Lim SY, Rogers T, Chan T, Whitney JB, Kim J, Sodroski J, Letvin NL (2010b) TRIM5 alpha modulates immunodeficiency virus control in rhesus monkeys. PLoS Pathog 6(1):e1000738

Lim ES, Malik HS, Emerman M (2010c) Ancient adaptive evolution of tetherin shaped the functions of Vpu and Nef in human immunodeficiency virus and primate lentiviruses. J Virol 84: 7124–7134

Ling BH, Veazey RS, Luckay A, Penedo C, Xu KY, Lifson JD, Marx PA (2002) SIVmac pathogenesis in rhesus macaques of Chinese and Indian origin compared with primary HIV infections in humans. AIDS 16(11):1489–1496

Ling F, Wei LQ, Wang T, Wang HB, Zhuo M, Du HL, Wang JF, Wang XN (2011) Characterization of the major histocompatibility complex class II DOB, DPB1, and DQB1 alleles in cynomolgus macaques of Vietnamese origin. Immunogenetics 63(3):155–166

Liu HF, Wang YQ, Liao CH, Kuang YQ, Zheng YT, Su B (2005) Adaptive evolution of primate TRIM5 alpha, a gene restricting HIV-1 infection. Gene 362:109–116

Liu Y, Helms C, Liao W, Zaba LC, Duan S, Gardner J, Wise C, Miner A, Malloy MJ, Pullinger CR et al (2008) A genome-wide association study of psoriasis and psoriatic arthritis identifies new disease loci. PLoS Genet 4(4):e1000041
Liu J, Chen K, Wang J-H, Zhang C (2010) Molecular evolution of the primate antiviral restriction factor tetherin. PLoS ONE 5:e11904
Loisel DA (2007) Evolutionary genetics of immune system genes in a wild primate population. Duke University, Durham, NC
Loisel DA, Rockman MV, Wray GA, Altmann J, Alberts SC (2006) Ancient polymorphism and functional variation in the primate MHC-DQA1 5′ cis-regulatory region. Proc Natl Acad Sci U S A 103:16331–16336
Lynn DJ, Lloyd AT, Fares MA, O'Farrelly C (2004) Evidence of positively selected sites in mammalian alpha-defensins. Mol Biol Evol 21(5):819–827
McDade TW (2003) Life history theory and the immune system: steps toward a human ecological immunology. Am J Phys Anthropol 46:100–125
Messier W, Stewart CB (1997) Episodic adaptive evolution of primate lysozymes. Nature 385(6612):151–154
Meyer D, Thomson G (2001) How selection shapes variation of the human major histocompatibility complex: a review. Ann Hum Genet 65:1–26
Monceaux V, Viollet L, Petit F, Cumont MC, Kaufmann GR, Aubertin AM, Hurtrel B, Silvestri G, Estaquier J (2007) CD4(+) CCR5(+) T-cell dynamics during simian immunodeficiency virus infection of Chinese rhesus macaques. J Virol 81(24):13865–13875
Morzycka-Wroblewska E, Munshi A, Ostermayer M, Harwood JI, Kagnoff MF (1997) Differential expression of HLA-DQA1 alleles associated with promoter polymorphism. Immunogenetics 45:163–170
Mukherjee S, Sarkar-Roy N, Wagener DK, Majumder PP (2009) Signatures of natural selection are not uniform across genes of innate immune system, but purifying selection is the dominant signature. Proc Natl Acad Sci U S A 106(17):7073–7078
Mummidi S, Bamshad M, Ahuja SS, Gonzalez E, Feuillet PM, Begum K, Galvis MC, Kostecki V, Valente AJ, Murthy KK et al (2000) Evolution of human and non-human primate CC chemokine receptor 5 gene and mRNA - potential roles for haplotype and mRNA diversity, differential haplotype-specific transcriptional activity, and altered transcription factor binding to polymorphic nucleotides in the pathogenesis of HIV-1 and simian immunodeficiency virus. J Biol Chem 275(25):18946–18961
Nakajima T, Ohtani H, Satta Y, Uno Y, Akari H, Ishida T, Kimura A (2008) Natural selection in the TLR-related genes in the course of primate evolution. Immunogenetics 60:727–735
Neil S, Bieniasz P (2009) Human immunodeficiency virus, restriction factors, and interferon. J Interferon Cytokine Res 29(9):569–580
Newman RM, Hall L, Connole M, Chen GL, Sato S, Yuste E, Diehl W, Hunter E, Kaur A, Miller GM et al (2006) Balancing selection and the evolution of functional polymorphism in Old world monkey TRIM5 alpha. Proc Natl Acad Sci U S A 103(50):19134–19139
Nielsen R, Bustamante C, Clark AG, Glanowski S, Sackton TB, Hubisz MJ, Fledel-Alon A, Tanenbaum DM, Civello D, White TJ et al (2005) A scan for positively selected genes in the genomes of humans and chimpanzees. PLoS Biol 3(6):976–985
Nunn CL, Altizer SM (2006) Infectious diseases in primates: behavior, ecology and evolution. Oxford University Press, Oxford
Nunn CL, Gittleman JL, Antonovics J (2000) Promiscuity and the primate immune system. Science 290(5494):1168–1170
Ober C (1999) Studies of HLA, fertility and mate choice in a human isolate. Hum Reprod Update 5(2):103–107
Ober C, Hyslop T, Elias S, Weitkamp LR, Hauck WW (1998) Human leukocyte antigen matching and fetal loss: results of a 10 year prospective study. Hum Reprod 13(1):33–38
OhAinle M, Kerns JA, Malik HS, Emerman M (2006) Adaptive evolution and antiviral activity of the conserved mammalian cytidine deaminase APOBEC3H. J Virol 80(8):3853–3862

Ortiz M, Bleiber G, Martinez R, Kaessmann H, Telenti A (2006) Patterns of evolution of host proteins involved in retroviral pathogenesis. Retrovirology 3:11

Otting N, Heijmans CMC, Noort RC, De Groot N, Doxiadis GGM, van Rood JJ, Watkins D, Bontrop R (2005) Unparalleled complexity of the MHC class I region in rhesus macaques. Proc Natl Acad Sci U S A 102:1626–1631

Otting N, de Vos-Rouweler AJM, Heijmans CMC, De Groot NG, Doxiadis GGM, Bontrop RE (2007) MHC class I A region diversity and polymorphism in macaque species. Immunogenetics 59(5):367–375

Palacios C, Cuervo LC, Cadavid LF (2011) Evolutionary patterns of killer cell Ig-like receptor genes in Old World monkeys. Gene 474(1–2):39–51

Parham P (1997) Events in the adaptation of natural killer cell receptors to MHC class I polymorphisms. Res Immunol 148(3):190–194

Pastinen T (2010) Genome-wide allele-specific analysis: insights into regulatory variation. Nat Rev Genet 11(8):533–538

Patil A, Hughes AL, Zhang G (2004) Rapid evolution and diversification of mammalian alpha-defensins as revealed by comparative analysis of rodent and primate genes. Physiol Genomics 20(1):1–11

Pelak K, Goldstein DB, Walley NM, Fellay J, Ge D, Shianna KV, Gumbs C, Gao X, Maia JM, Cronin KD et al (2010) Host determinants of HIV-1 control in African Americans. J Infect Dis 201(8):1141–1149

Penn DJ, Potts WK (1999) The evolution of mating preferences and major histocompatibility complex genes. Am Nat 153(2):145–164

Pertel T, Hausmann S, Morger D, Zuger S, Guerra J, Lascano J, Reinhard C, Santoni FA, Uchil PD, Chatel L et al (2011) TRIM5 Is an innate immune sensor for the retrovirus capsid lattice. Nature 472(7343):361–365

Phillips-Conroy JE, Lambrecht FL, Jolly CJ (1988) Hepatocystis in populations of baboons (Papio-hamadryas S1) of Tanzania and Ethiopia. J Med Primatol 17(3):145–152

Potts R (1998) Variability selection in hominid evolution. Evol Anthropol 7(3):81–96

Prugnolle F, Durand P, Neel C, Ollomo B, Ayala FJ, Arnathau C, Etienne L, Mpoudi-Ngole E, Nkoghe D, Leroy E et al (2010) African great apes are natural hosts of multiple related malaria species, including *Plasmodium falciparum*. Proc Natl Acad Sci U S A 107(4):1458–1463

Purvis A (1995) A composite estimate of primate phylogeny. Philos Trans R Soc Lond B Biol Sci 348(1326):405–421

Reimann KA, Parker RA, Seaman MS, Beaudry K, Beddall M, Peterson L, Williams KC, Veazey RS, Montefiori DC, Mascola JR et al (2005) Pathogenicity of simian-human immunodeficiency virus SHIV-89.6P And SIVmac is attenuated in cynomolgus macaques and associated with early T-lymphocyte responses. J Virol 79(14):8878–8885

Robinson J, Waller MJ, Parham P, de Groot N, Bontrop R, Kennedy LJ, Stoehr P, Marsh SGE (2003) IMGT/HLA and IMGT/MHC: sequence databases for the study of the major histocompatibility complex. Nucleic Acids Res 31:311–314

Rogers J, Hixson JE (1997) Baboons as an animal model for genetic studies of common human disease. Am J Hum Genet 61(3):489–493

Roth GS, Mattison JA, Ottinger MA, Chachich ME, Lane MA, Ingram DK (2004) Aging in rhesus monkeys: relevance to human health interventions. Science 305(5689):1423–1426

Ruff JS, Nelson AC, Kubinak JL, Potts WK (2011) MHC signaling during social communication. In: Lopez-Larrea C (ed) Self and non-self. Landes Bioscience, Austin, TX

Sabeti PC, Schaffner SF, Fry B, Lohmueller J, Varilly P, Shamovsky O, Palma A, Mikkelsen TS, Altshuler D, Lander ES (2006) Positive natural selection in the human lineage. Science 312(5780):1614–1620

Sabeti PC, Varilly P, Fry B, Lohmueller J, Hostetter E, Cotsapas C, Xie XH, Byrne EH, McCarroll SA, Gaudet R et al (2007) Genome-wide detection and characterization of positive selection in human populations. Nature 449(7164):913–918

Sano K, Shiina T, Kohara S, Yanagiya K, Hosomichi K, Shimizu S, Anzai T, Watanabe A, Ogasawara K, Torii R et al (2006) Novel cynomolgus macaque MHC-DPB1 polymorphisms in three South-East Asian populations. Tissue Antigens 67(4):297–306

Sapolsky RM, Else JG (1987) Bovine tuberculosis in a wild baboon population - epidemiologic aspects. J Med Primatol 16(4):229–235
Sawyer SL, Emerman M, Malik HS (2004) Ancient adaptive evolution of the primate antiviral DNA-editing enzyme APOBEC3G. PLoS Biol 2(9):1278–1285
Sawyer SL, Wu LI, Emerman M, Malik HS (2005) Positive selection of primate TRIM5 alpha identifies a critical species-specific retroviral restriction domain. Proc Natl Acad Sci U S A 102(8):2832–2837
Sawyer SL, Emerman M, Malik HS (2007) Discordant evolution of the adjacent antiretroviral genes TRIM22 and TRIM5 in mammals. PLoS Pathog 3(12):1918–1929
Schad J, Ganzhorn JU, Sommer S (2005) Parasite burden and constitution of major histocompatibility complex in the Malagasy mouse lemur, Microcebus murinus. Evolution 59:439–450
Schaner P, Richards N, Wadhwa A, Aksentijevich I, Kastner D, Tucker P, Gumucio D (2001) Episodic evolution of pyrin in primates: human mutations recapitulate ancestral amino acid states. Nat Genet 27(3):318–321
Schlenke TA, Begun DJ (2003) Natural selection drives drosophila immune system evolution. Genetics 164(4):1471–1480
Schmidt LH, Fradkin R, Harrison J, Rossan RN (1977) Differences in virulence of *Plasmodium knowlesi* for macaca-irus (Fascicularis) of Philippine and Malayan origins. Am J Trop Med Hyg 26(4):612–622
Schwensow N, Fietz J, Dausmann KH, Sommer S (2007) Neutral versus adaptive genetic variation in parasite resistance: importance of major histocompatibility complex supertypes in a free-ranging primate. Heredity 99(3):265–277
Semple CAM, Rolfe M, Dorin JR (2003) Duplication and selection in the evolution of primate beta-defensin genes. Genome Biol 4(5):R31
Sharp PM, Robertson DL, Hahn BH (1995) Cross-species transmission and recombination of aids viruses. Philos Trans R Soc Lond B Biol Sci 349(1327):41–47
Shiina T, Hosomichi K, Inoko H, Kulski JK (2009) The HLA genomic loci map: expression, interaction, diversity and disease. J Hum Genet 54(1):15–39
Silva JC, Kondrashov AS (2002) Patterns in spontaneous mutation revealed by human-baboon sequence comparison. Trends Genet 18(11):544–547
Silvestri G, Paiardini M, Pandrea I, Lederman MM, Sodora DL (2007) Understanding the benign nature of SIV infection in natural hosts. J Clin Invest 117(11):3148–3154
Sommer S (2005) The importance of immune gene variability (MHC) in evolutionary ecology and conservation. Front Zool 2:16
Song B, Javanbakht H, Perron M, Park DH, Stremlau M, Sodroski J (2005a) Retrovirus restriction by TRIM5 alpha variants from old world and new world primates. J Virol 79(7):3930–3937
Song BW, Gold B, O'hUigin C, Javanbakht H, Li X, Stremlau M, Winkler C, Dean M, Sodroski J (2005b) The B30.2(SPRY) domain of the retroviral restriction factor TRIM5 alpha exhibits line age-specific length and sequence variation in primates. J Virol 79(10):6111–6121
Stearns SC, Koella JC (eds) (2008) Evolution in health and disease, 2nd edn. Oxford University Press, New York
Stremlau M, Perron M, Lee M, Li Y, Song B, Javanbakht H, Diaz-Griffero F, Anderson DJ, Sundquist WI, Sodroski J (2006) Specific recognition and accelerated uncoating of retroviral capsids by the TRIM5 alpha restriction factor. Proc Natl Acad Sci U S A 103(14): 5514–5519
Thornton JW (2004) Resurrecting ancient genes: experimental analysis of extinct molecules. Nat Rev Genet 5(5):366–375
Torgerson DG, Boyko AR, Hernandez RD, Indap A, Hu XL, White TJ, Sninsky JJ, Cargill M, Adams MD, Bustamante CD et al (2009) Evolutionary processes acting on candidate cis-regulatory regions in humans inferred from patterns of polymorphism and divergence. PLoS Genet 5(8):e1000592
Trichel AM, Rajakumar PA, Murphey-Corb M (2002) Species-specific variation in SIV disease progression between Chinese and Indian subspecies of rhesus macaque. J Med Primatol 31(4–5):171–178

Tung J, Primus A, Bouley AJ, Severson TF, Alberts SC, Wray GA (2009) Evolution of a malaria resistance gene in wild primates. Nature 460(7253):388–391
Tung J, Alberts SC, Wray GA (2010) Evolutionary genetics in wild primates: combining genetic approaches with field studies of natural populations. Trends Genet 26:353–362
Tung J, Akinyi MY, Mutura S, Altmann J, Wray GA, Alberts SC (2011) Allele-specific gene expression in a wild nonhuman primate population. Mol Ecol 20(4):725–739
Urvater JA, Otting N, Loehrke JH, Rudersdorf R, Slukvin II, Piekarczyk MS, Golos TG, Hughes AL, Bontrop RE, Watkins DI (2000) Mamu-I: a novel primate MHC class I B-related locus with unusually low variability. J Immunol 164:1386–1398
Vallender EJ, Priddy CM, Chen GL, Miller GM (2008a) Human expression variation in the mu-opioid receptor is paralleled in rhesus macaque. Behav Genet 38(4):390–395
Vallender EJ, Priddy CM, Hakim S, Yang H, Chen GL, Miller GM (2008b) Functional variation in the 3′ untranslated region of the serotonin transporter in human and rhesus macaque. Genes Brain Behav 7(6):690–697
Voight BF, Kudaravalli S, Wen XQ, Pritchard JK (2006) A map of recent positive selection in the human genome. PLoS Biol 4(3):446–458
von Korff M, Radovic S, Choumane W, Stamati K, Udupa SM, Grando S, Ceccarelli S, Mackay I, Powell W, Baum M et al (2009) Asymmetric allele-specific expression in relation to developmental variation and drought stress in barley hybrids. Plant J 59(1):14–26
Wilder JA, Hewett EK, Gansner ME (2009) Molecular evolution of GYPC: evidence for recent structural innovation and positive selection in humans. Mol Biol Evol 26(12):2679–2687
Wilson DE, Reeder DM (eds) (2005) Mammal species of the world: a taxonomic and geographic reference, 3rd edn. The Johns Hopkins University Press, Baltimore, MD
Wilson SJ, Webb BLJ, Maplanka C, Newman RM, Verschoor EJ, Heeney JL, Towers GJ (2008) Rhesus macaque TRIM5 alleles have divergent antiretroviral specificities. J Virol 82(14):7243–7247
Wiseman RW, Karl JA, Bimber BN, O'Leary CE, Lank SM, Tuscher JJ, Detmer AM, Bouffard P, Levenkova N, Turcotte CL et al (2009) Major histocompatibility complex genotyping with massively parallel pyrosequencing. Nat Med 15(11):1322–1326
Wittkopp P, Haerum B, Clark A (2004) Evolutionary changes in cis and trans gene regulation. Nature 430:85–88
Wittkopp PJ, Haerum BK, Clark AG (2008) Independent effects of cis- and trans-regulatory variation on gene expression in drosophila melanogaster. Genetics 178(3):1831–1835
Wlasiuk G, Khan S, Switzer WM, Nachman MW (2009) A history of recurrent positive selection at the Toll-Like Receptor 5 in primates. Mol Biol Evol 26:937–949
Wlasiuk G, Nachman MW (2010) Adaptation and constraint at toll-like receptors in primates. Mol Biol Evol 27(9):2172–2186
Wray GA (2007) The evolutionary significance of cis-regulatory mutations. Nat Rev Genet 8(3):206–216
Wray GA, Hahn MW, Abouheif E, Balhoff JP, Pizer M, Rockman MV, Romano LA (2003) The evolution of transcriptional regulation in eukaryotes. Mol Biol Evol 20(9):1377–1419
Yan H, Yuan WS, Velculescu VE, Vogelstein B, Kinzler KW (2002) Allelic variation in human gene expression. Science 297(5584):1143
Yang Z (1997) PAML: a program package for phylogenetic analysis by maximum likelihood. Comput Appl Biosci 13:555–556
Yap MW, Nisole S, Stoye JP (2005) A single amino acid change in the SPRY domain of human Trim5 alpha leads to HIV-1 restriction. Curr Biol 15(1):73–78
Zhang JZ, Rosenberg HF (2002) Complementary advantageous substitutions in the evolution of an antiviral RNase of higher primates. Proc Natl Acad Sci U S A 99(8):5486–5491
Zhang JZ, Webb DM (2004) Rapid evolution of primate antiviral enzyme APOBEC3G. Hum Mol Genet 13(16):1785–1791

Toll-Like Receptor Function and Evolution in Primates

Jessica F. Brinkworth and Kirstin N. Sterner

Introduction

Despite close genetic relatedness, many primate species respond to infectious pathogens (e.g., viruses, bacteria, protozoa, and fungi) differently (Table 1). Perhaps most striking, many pathogen-induced diseases result in significant dysregulation of immune responses in humans when compared to similar diseases in closely related nonhuman primates (e.g., herpesviruses, Estep et al. 2010; *Neisseria gonorrhoea*, McGee et al. 1990; immunodeficiency viruses, Pandrea and Apetrei 2010; Gram-negative bacterial sepsis, Redl et al. 1993; von Bulow et al. 1992; Fischer et al. 1991; and hepatitis, Walker 1997). The most well-characterized example of disparate primate disease progression is the discrepancy observed between the typical human and African Old World monkey responses to immunodeficiency viruses (e.g., SIVs and HIVs). A long-term immunodeficiency virus infection in most humans is characterized by massive loss of CD4+ T cells, immune cell irregularities, and overt immune activation associated with acquired immune deficiency syndrome (AIDS). Most African Old World monkeys, on the other hand, remain comparatively symptomless despite high viral loads (Pandrea and Apetrei 2010; Vodros and Fenyo 2004). Recent evidence suggests that simian immunodeficiency virus may also be pathogenic in chimpanzees (Keele et al. 2009). The disparity in immunodeficiency virus associated disease progression observed between African cercopithecoids and hominoids is iterated in other pathogen-mediated diseases as well, suggesting lineage-specific immune system evolution in primates. Humans

J.F. Brinkworth (✉)
Department of Pediatrics, CHU Sainte-Justine Research Center, University of Montreal, Montreal, QC, Canada
e-mail: jfbrinkworth@gmail.com

K.N. Sterner (✉)
Department of Anthropology, University of Oregon, Eugene, Oregon, USA
e-mail: ksterner@uoregon.edu

J.F. Brinkworth and K. Pechenkina (eds.), *Primates, Pathogens, and Evolution*, Developments in Primatology: Progress and Prospects,
DOI 10.1007/978-1-4614-7181-3_4, © Springer Science+Business Media New York 2013

Table 1 Well-described disparities in primate responses to TLR-detected pathogens

Pathogen	Human response and symptoms	Nonhuman primate responses	TLR[a]	References
Neisseria gonorrhoeae	Gonorrhea (urethral infection with urinary pain, discharge, increased urination, fever, swollen testicles, and severe abdominal pain)	Chimpanzees: similar to human response Cercopithecoids (and other mammals): no symptoms	TLR2	Brown et al. (1972), Lucas et al. (1971), Massari et al. (2002), McGee et al. (1990)
Schistosoma mansoni	Schistosomiasis (diarrhea, abdominal pain, hematuria, urinary pain, liver dysfunction, and fibrosis)	Chimpanzees: similar to human response Baboons (*Papio anubis*): less intense pathology, longer term infections develop periportal fibrosis	TLR2	Farah et al. (2001), Farah et al. (2000), van der Kleij et al. (2002)
Toxoplasma gondii	Toxoplasmosis (usually subclinical, however immunosuppressed individuals develop fever, anorexia, myalgia, vomiting, and swollen lymph nodes)	Capuchin monkey: mild, little mortality Callitrichids: 100 % mortality, anorexia, foaming nares, lethargia, labored breathing Atelids: 20–80 % mortality, chills, vomiting, lethargia, and labored breathing	TLR2 & TLR4	Bouer et al. (2010), Debierre-Grockiego et al. (2007), Epiphanio et al. (2003), Poltorak et al. (1998), Werner et al. (1969) see Catão-Dias et al. (2013)
Gram-negative bacteria (e.g., *Escherichia coli*)	Severe sepsis/septic shock (high mortality, fever, neutropenia, organ dysfunction, hypotension, and hypoperfusion)	Chimpanzees: severe sepsis/septic shock Cercopithecoids: insensitive to LPS, blunted febrile responses; requires biologically improbable levels of LPS to initiate symptoms	TLR4	Redl et al. (1993), Shi et al. (2010), von Bulow et al. (1992), Zurovsky et al. (1987)
Plasmodium falciparum	Malaria (fever, chills, headache, sweats, fatigue, nausea, cough, enlarged spleen, and pain)	Chimpanzees and gorilla: less susceptible; susceptible to *Plasmodium reichenowi* but do not develop disease	TLR 2/1	Rayner et al. (2011), Zhu et al. (2011)

Dengue virus	Dengue (can be asymptomatic; or can cause dengue fever or the more life-threatening dengue hemorrhagic fever/ dengue shock syndrome)	Rhesus macaques: lower viral loads and no symptoms (although hemorrhage has been induced with higher doses)	TLR7	Diebold et al. (2004), Onlamoon et al. (2010)
Hepatitis C virus (HCV)	Chronic liver disease (ranges from mild to severe; can lead to progressive liver fibrosis, cirrhosis, end-stage liver disease, and cancer)	Chimpanzees: often acute and spontaneously cleared; self-limited; similar to mild human cases (no chronic active hepatitis, cirrhosis, or cancer)	TLR7	Reviewed in Bettauer (2010), De Vos et al. (2002), Diebold et al. (2004), Major et al. (2004), Thomson et al. (2003)
		Macaques: not susceptible		
Immunodeficiency viruses (SIV/HIV)	AIDS (long-term infection associated with low viral load, high HIV-specific cytotoxic T cells and antibodies; followed by increasing viral load and CD4+ T cell depletion, immune cell irregularities, inflammation, and overt immune activation)	Asian macaques: similar to humans Chimpanzees (*Pan troglodytes schweinfurthii* and *P.t. troglodytes* subspecies): increased likelihood of mortality; AIDS-like disease recently identified in wild SIV+ animals; few captive *P.t. verus* cases. Sooty mangabey and vervets: long-term infection with high viral load and without AIDS-like symptoms	TLR7	Benveniste et al. (1986), Daniel et al. (1984), Diebold et al. (2004), Etienne et al. (2011), Keele et al. (2009), reviewed in Pandrea and Apetrei (2010)

[a] associated Toll-like receptor

and chimpanzees (*Pan troglodytes*) also share a higher susceptibility to Gram-negative bacterial sepsis (Redl et al. 1993; van der Poll et al. 1994), schistosomiasis (Abe et al. 1993; Sadun et al. 1970), and gonorrhea (Brown et al. 1972; Lucas et al. 1971; McGee et al. 1990), while Old World monkeys and other primates exhibit reduced or null susceptibility. In addition, when infected with monkey B virus (Cercopithecine herpesvirus 1/McHV1) macaques sometimes manifest mild mucosal blisters, whereas infected humans can quickly progress to encephalopathy (Estep et al. 2010). Published observations of interspecies differences in the progression of these and other infections suggest lineage-specific immune responses and potential divergence in immune function along major branches of Primates. As immune system function plays a profound role in host survival and fitness, it is important to not only characterize *how* these responses differ but also consider *why* such variation exists.

Dysregulation of a host's innate immune responses to infectious pathogens can lead to increased pathogenicity. Such dysregulation can negatively influence host function by causing either a reduction in innate immune activation or by allowing for more prolonged/heightened immune responses (Bosinger et al. 2011; Hagberg et al. 1984; Hoshino et al. 1999; Qureshi et al. 1999). Toll-like receptors (TLRs) play an important role in the activation and regulation of innate immune responses and profoundly affect host survival. As such, they make compelling targets when studying the evolution of primate disease responses. Indeed, there is a growing body of research on human and nonhuman primate immunity that suggests TLR-triggered responses have not only diverged in these species but may also contribute to differing patterns of disease susceptibility and progression observed between primates (Barreiro et al. 2010; Bochud et al. 2008; Feterowski et al. 2003; Hawn et al. 2005, 2007; Johnson et al. 2007; Mandl et al. 2008, 2011; Mir et al. 2012; Schroder et al. 2005; Vasl et al. 2008; Yim et al. 2004). In addition, intraspecific differences in TLR-triggered responses have also been shown to influence disease progression and severity in several human infections (e.g., Johnson et al. 2007; Oh et al. 2009). This chapter reviews our current understanding of Toll-like receptor evolution in primates and discusses how an evolutionary perspective can help explain inter- and intraspecies disparities in disease susceptibility.

TLRs and TLR-Triggered Innate Immune Responses

The immune system of vertebrate animals can be divided into the innate and the adaptive immune responses. The innate immune system is an ancient, genetically inherited and nonspecific response that forms the host's first line of defense against immune insult. The innate immune response is immediate, functions continuously, requires no previous antigen exposure to be activated and is characterized by generalized immune tactics such as inflammation, barriers (e.g., skin and mucosa), and phagocytosis (reviewed in Janeway and Medzhitov 2002; Kumar et al. 2009). As the immediate response to invading antigens and a modulator of adaptive immune responses, innate immunity appears to exert great control over the course of infection. When excessive, the innate immune response can cause lethal pathologies in the host (e.g., septic shock) (reviewed in Murdoch and Finn 2003). When weak or disabled, initial infections grow unabated and adaptive responses exhibit dysfunction (Hagberg et al. 1984). As such, the innate immune system is an important determinant of immune function and host survival.

Toll-like receptors represent one type of innate immune receptor and play a key role in regulating innate immunity. The number of TLR types varies across animal classes. Like some mammals, primates have ten functional TLRs. TLRs are horseshoe shaped, type 1 membrane proteins that contain three major functional domains: an ecto- (extracellular) domain, a transmembrane domain, and a cytoplasmic domain (Fig. 1). The ectodomain interacts directly with TLR ligands through a series of leucine-rich repeats that facilitate protein–protein binding (Matsushima et al. 2007). As type 1 membrane proteins, TLRs are bound to plasma membranes and traverse either the cellular membrane (referred to as extracellular TLRs and include TLR1, TLR2, TLR4, TLR5, TLR6, and TLR10) or endosomal membranes (referred to as intracellular TLRs and include TLR3, TLR7, TLR8, and TLR9). The cytoplasmic region of all TLRs, however, is located in the cytoplasm and contains an ultra-conserved region referred to as the Toll/IL-1 receptor (TIR) domain that is important for intracellular signaling.

The recognition of *self* from *nonself* is a primary function of normal mammalian immunity. TLRs are immune system sentinels that enable phagocytic cells (i.e., moncytes/macrophages, neutrophils, and dendritic cells) to detect *nonself* or foreign antigens and respond accordingly. Specifically, TLRs recognize many molecular motifs (referred to as pathogen-associated molecular patterns or PAMPS) that are generally absent in vertebrate hosts but broadly shared across microbial organisms (Jin and Lee 2008; Kawai and Akira 2007). Although TLRs have been traditionally viewed as specialized sensors of broadly shared microbial molecules (Table 2), some receptors appear to be more promiscuous and use co-receptors to recognize a wider range of motifs. This is particularly true of extracellular TLRs. For example, when TLR2 and TLR4 form heterodimers with either other TLRs or other co-receptors (e.g., LY96) they can recognize motifs (e.g., lipomannan from *Mycobacterium* or lipopolysaccharide from *Escherichia coli*) that they cannot recognize as homodimers or single receptors (Gioannini et al. 2004; Shimazu et al. 1999; Zahringer et al. 2008).

TLRs become activated upon TLR–ligand binding. Once activated, the TIR domain of the cytoplasmic region recruits adaptor proteins and triggers signaling cascades that initiate downstream transcription factor binding (e.g., NF-κB and IFR7) (Kawai and Akira 2007, 2010). These transcription factors then induce the production of cytokines and costimulatory molecules (reviewed in Kawai and Akira 2007; Palm and Medzhitov 2009). Cytokines nonspecifically degrade antigens as well as initiate the activation of the epitope-specific T and B cell-mediated adaptive immune response. The engagement of the adaptive immune response generally occurs after the 4th hour of infection, with clonal expansion of T and B cells initiated more than 96 hours post-infection. TLR signaling pathways appear to help

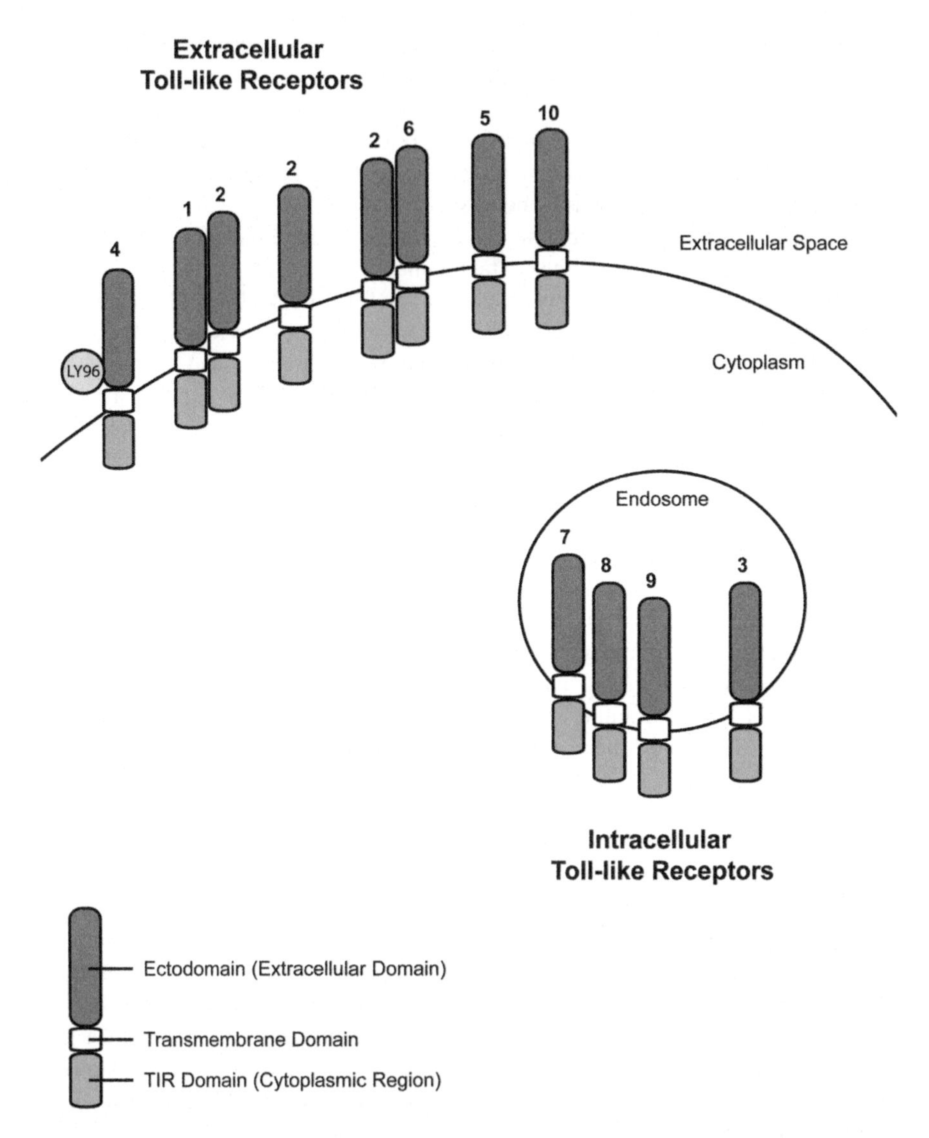

Fig. 1 Toll-like receptor and co-receptor cellular locations in primates. TLR2 is shown here heterodimerized with co-receptors TLR1 and TLR6, TLR4 with co-receptor LY96. Some (i.e., TLR2 and TLR4), possibly all TLRs, habitually exist as homodimers

regulate this arm of adaptive immunity (e.g., T-helper 1 and 2 type responses) (Dabbagh et al. 2002; Krieg 2007). TLRs, therefore, are not only key components in the early recognition of microbial invaders but also play a role in modulating the adaptive immune response. As a result, they have the potential to strongly influence and shape disease outcomes.

Table 2 Toll-like receptor location, ligands, and representative organisms

Cellular location	TLR	PAMP type	PAMP	Organisms
Cell membrane	TLR1/2 heterodimer	Lipoproteins/lipopeptides/GPI anchors/ lipoglycans	Pam3Cys lipopeptides (Aliprantis et al. 1999), Lipomannan (Quesniaux et al. 2004; Vignal et al. 2003),	bacteria, mycobacteria, Gram-positive bacteria
	TLR2		Peptidoglycan, glycosylphosphatidylinositol-anchored proteins, lipoteichoic acid, porins, zymosan (Campos et al. 2001; Massari et al. 2002; Schwandner et al. 1999; Takeuchi et al. 1999; Underhill et al. 1999)	Gram-positive bacteria, trypanosomes, yeast
	TLR2/6 heterodimer		Pam2CYs lipopeptides, MALP-2, PSM, (Bulut et al. 2001; Buwitt-Beckmann et al. 2005)	Mycobacteria, *Borrelia burgdorferi*, Gram-positive bacteria
	TLR4		Lipopolysaccharide (LPS), fungal mannans, heat-shock proteins (Ohashi et al. 2000; Poltorak et al. 1998; Tada et al. 2002)	Gram-negative bacteria
	TLR5	Protein	Flagellin (Hayashi et al. 2001)	protists
	TLR10	Unknown	Profilin-like molecule (Yarovinsky et al. 2005)	*Toxoplasma gondii*
Endosome	TLR3	Nucleic acids	Double-stranded RNA (Alexopoulou et al. 2001)	dsRNA viruses
	TLR7		Single-stranded RNA, Guanosine analogs (Hemmi et al. 2002; Lund et al. 2004)	ssRNA viruses
	TLR8		Single-stranded RNA, imidazoquinolines (Heil et al. 2004)	ssRNA viruses
	TLR9		CpG DNA motifs (Rutz et al. 2004)	Bacterial DNA

Variation in TLR-Triggered Immune Responses Between and Within Primate Species

Intra- and interspecies differences in TLR pathway function may explain within and between species differences in susceptibility to TLR-detected pathogens. Primates exhibit interspecies differences in immune response to a wide selection of TLR-detected pathogens that are major agents of human diseases (Table 1). Our current understanding of primate immune responses to such TLR-detected pathogens is biased towards inferences drawn from DNA sequence data (e.g., clinical association studies and phylogenetic analyses) and clinical/veterinarian observation, when available. Relatively few studies have directly examined primate TLR pathway function through controlled immunological challenge experiments. Due to the clinical importance of such diseases as bacterial sepsis and HIV/AIDS, most of these studies investigate primate TLR2, TLR4, and TLR7. Moreover, controlled experiments almost entirely use catarrhine species with studies biased towards the use of macaque, sooty mangabey, common chimpanzee, baboon, and human models. Despite these limitations, however, current data suggest both species- and family-level divergence in TLR pathway function. Interestingly, a combination of controlled experimentation and veterinary/clinical observations suggests a significant divergence between hominoid and cercopithecoid species in response to TLR-detected pathogens in particular (Barreiro et al. 2010; Brinkworth et al. 2012). Divergence in immune response has been observed for a subset of TLR2-, TLR4-, and TLR7-detected pathogens discussed below.

Interspecies Divergence in TLR2 and TLR4 Function

A divergence between cercopithecoid and hominoid responses to TLR2- and TLR4-detected pathogens is supported by a recent study (Brinkworth et al. 2012) that found early chemokine and cytokine transcriptional responses of human and chimpanzee total blood leukocytes to LPS (which is detected by TLR4, amongst other receptors), Lipomannan from *Mycobacterium smegmatis* (TLR2/1 ligand), and Pam3CSK4 (TLR2/1 mimetic ligand) tend to be more similar to each other than to baboon responses ex vivo. Perhaps the best-established evidence for divergence in hominoid and cercopithecoid immune responses is to LPS and Gram-negative bacteria (Fischer et al. 1991; Haudek et al. 2003; Shi et al. 2004, 2010; von Bulow et al. 1992; Zurovsky et al. 1987). Humans and common chimpanzees are very sensitive to stimulation by LPS/endotoxin and Gram-negative bacteria and are, therefore, highly susceptible to Gram-negative bacterial sepsis, severe sepsis, and septic shock (reviewed in Blackwell and Christman 1996; Redl et al. 1993). Severe sepsis is a systemic, harmful, and unregulated proinflammatory innate immune response to immune trauma (e.g., systemic bacterial infection, injury) that can involve the TLR4 pathway (reviewed in Murdoch and Finn 2003). Severe sepsis in humans can be initiated by 4–10 ng/kg of LPS and is characterized by high fever, blood clots, organ dysfunction, and, in severe cases, hypoperfusion and hypotension (septic shock). These symptoms are not observed in baboons or macaques under biologically probable conditions. These Old World monkeys are comparatively less sensitive to LPS and Gram-negative bacteria and fail to manifest fevers or broader symptoms of severe sepsis/septic shock, even when very high doses are applied intravenously or to tissues. For example, baboons require doses of LPS that are not likely to occur in nature to establish minimal symptoms (100 ng/kg) and doses that are likely impossible to establish via natural infection (6–25 mg/kg) to trigger deadly symptoms (Haudek et al. 2003; Redl et al. 1993; Shi et al. 2004, 2010; von Bulow et al. 1992; Zurovsky et al. 1987). Even when such high doses are artificially applied and baboons develop severe sepsis or septic shock, they do not typically develop febrile responses.

Similarly, opportunistic observations of primate manifestations of TLR2-detected pathogens suggest a divide between hominoid and cercopithecoid responses, though these manifestations in nonhuman primates are comparatively less well understood than sepsis. *Neisseria gonorrhoeae*, the causative agent of gonorrhoeae, is a TLR2-detected pathogen that uniquely establishes itself in the fallopian mucosa of humans and chimpanzees but not in cercopithecoids and other mammals (Arko 1989; Massari et al. 2002; McGee et al. 1983, 1990). The species specificity of *N. gonorrhoeae*-mucosa binding appears to be associated with increased human and chimpanzee complement binding affinity to *N. gonorrhoeae* porins (Ngampasutadol et al. 2008). Similarly, the TLR2-detected pathogen *Schistosoma mansoni* causes diarrhea, liver dysfunction, and fibrosis in humans but less intensively affects *Papio anubis* (Farah et al. 2000, 2001; van der Kleij et al. 2002).

TLR2 also detects *Mycobacterium* species. Some studies suggest that *Mycobacterium* sp. infections progress more rapidly in cercopithecoids than in hominoids, though these reports conflict and are affected by differences in colony disease surveillance methods and long latency periods of such infections (Flynn et al. 2003; Montali et al. 2001; Payne et al. 2011; Sapolsky and Else 1987; Tarara et al. 1985; Walsh et al. 1996, 2007). For most case reports, the date of infection is not known and is difficult to estimate within a margin of weeks. The few case studies available, in combination with a handful of deliberate *Mycobacterium* inoculations suggest complex divergence in hominoid and cercopithecoid responses to this genus of bacteria. Baboon and macaque species seem to exhibit rapid disease progression, developing pulmonary inflammation and significant lung nodules within 6 months of *M. bovis* or *M. tuberculosis* exposure (Garcia et al. 2004; Payne et al. 2011; Sapolsky and Else 1987; Tarara et al. 1985). Rhesus macaques (*Macaca mulatta*) have been described as being highly susceptible to very virulent manifestations of both *M. bovis* and *M. tuberculosis* compared to baboons and cynomolgus macaques (*Macaca fascicularis*), respectively (Garcia et al. 2004; Langermans et al. 2001; Walsh et al. 1996).

Interspecies Divergence in TLR7 Function

While some TLR2- and TLR4-detected pathogens appear to manifest very differently in hominoids vs. cercopithecoids, there is also evidence suggesting TLR7-viral interactions may vary considerably between African cercopithecoid species. The sooty mangabey (*Cercocebus atys*), in particular, exhibits very divergent TLR-initiated pathway function compared to other examined cercopithecoids. Perhaps best known in biomedical research for being a natural host of the simian precursor to HIV-2 (SIVsm), sooty mangabeys are well known to sustain lifelong infections with high viral load without experiencing overt immune activation or progressing to AIDS. Mandl et al. (2008) has provided ELISA data that suggests that when the plasmacytoid dendritic cells (pDCs) of sooty mangabeys are stimulated with SIV in vivo or TLR7 and TLR9 mimetic ligands ex vivo, they may express dampened antiviral/anti-inflammatory IFNα responses. The pDCs from SIV naïve hosts, such as rhesus macaque (*Macaca mulatta*), exhibit strong responses that are more comparable to those seen in humans (Mandl et al. 2008). In the same paper, these authors provide real-time PCR estimates of sooty mangabey IFNα expression that are similar to rhesus macaque, suggesting that either the ELISA capture antibody avidity differs between the IFNα of these species, or stark differences exist between sooty mangabey IFNα transcriptional activity and IFNα expression. Further examination of the reliability of the IFNα capture antibody used to analyze sooty mangabey IFNα in this study and more tests for sooty mangabey-specific mechanism of IFNα post-transcriptional regulation are needed.

While these results are conflicted, the observation that sooty mangabey and rhesus macaque TLR7-triggered responses may differ is recapitulated in interspecies differences in response to yellow fever virus (YFV). Yellow fever virus is a TLR7-detected pathogen for which many African cercopithecoids serve as a natural reservoir including mangabey species (*Cercocebus* sp.) (World Health Organization 2004). YFV causes hemorrhagic fevers in humans and Asian monkeys (e.g., macaques) but does not pathogenically manifest in sooty mangabey hosts (Mandl et al. 2011; Woodall 1968). When pDCs from sooty mangabeys, rhesus macaques, and humans were stimulated ex vivo with live attenuated yellow fever virus 17 day strain vaccine, the sooty mangabey pDCs expressed significantly less IFNα (Mandl et al. 2011). When both monkey species were inoculated in vivo, sooty mangabeys mounted a weaker innate immune response to YFV.

This pattern of weaker TLR-mediated overt innate immune activation in sooty mangabeys appears to hold for viral/bacterial co-infection models and other cell types as well. Monocytes from immunodeficiency virus infected mangabeys express very low levels of pro-inflammatory cytokine TNFα in response to LPS ex vivo. By contrast, monocytes from immunodeficiency virus infected humans and rhesus macaques exhibited a strong TNFα response (Mir et al. 2012). Taken together, these results suggest that TLR7-initiated signaling in the sooty mangabey can produce dampened immune responses in comparison to humans and rhesus macaques. Whether or not this trend is found in other African Old World monkeys remains to be determined.

Intraspecific Variation

Taken together, these studies suggest that differences observed in primate immune responses may be influenced by divergent TLR pathway signaling, but further controlled studies of TLR-mediated immune function are needed in this area. It is also important to consider that intraspecific variation in TLRs and TLR signaling plays an important role in mediating immune responses between individuals (Table 3). Host genetics have been shown to affect infection susceptibility, strength of inflammation and infectious disease progression (Allikmets et al. 1996; Hawn et al. 2007; Jallow et al. 2009; Kang et al. 2002; Ogus et al. 2004). Polymorphisms in both *TLR* genes (Table 3) and TLR pathway genes (e.g., Orange and Geha 2003; Puel et al. 2004) have been associated with alterations in primate infectious disease manifestation, although most association studies connecting infectious disease progression and TLR signaling focus on human *TLR* genes and their co-receptors specifically. While these data are invaluable, there are less comparable data for nonhuman primates. One recent exception examined the impact of *TLR7* polymorphisms on the degree of AIDS-like symptoms in immunized and unimmunized SIV-infected rhesus macaques and found that unimmunized rhesus macaques that were carriers for two *TLR7* polymorphisms (c.13G and c.-17C alleles) survived SIV infection longer than unimmunized noncarriers (Siddiqui et al. 2011). These findings suggest *TLR7* polymorphisms may influence not only disease progression but also vaccine efficacy in rhesus macaques and humans.

Understanding *Why* TLR-Triggered Responses Differ Between and Within Primate Species

In order to explain why TLR-dependent responses vary in primates (see examples above), it is necessary to examine species-specific patterns of pathogen exposure in the context of those genetic and environmental factors that may drive or contribute to divergent immune responses to shared pathogens. In addition, when building an evolutionary framework for understanding disease susceptibility, it is important to consider that these factors are likely interrelated and may be dynamic, varying in both space and time. In the following section we review what is known about this history of TLR-detected pathogen exposure in primates and how evolutionary processes have shaped Toll-like receptor genes.

Evidence of Both Adaptation and Constraint in Primate Toll-Like Receptors

Genes that encode proteins involved in the immune system are considered strong targets of natural selection because of their relationship to an individual's reproductive fitness and mortality (Barreiro and Quintana-Murci 2010; Nielsen et al. 2005).

Table 3 Polymorphisms in human Toll-like receptors with associated functional data

Receptor	Polymorphisms[a]	Corresponding pathogen-associated disease phenotypes
TLR1	N248S	Associated with increased tuberculosis risk (Ma et al. 2007a) and susceptibility to leprosy (Schuring et al. 2009)
	P315L	Impaired receptor function (Omueti et al. 2007)
	Y554C	Reduced NF-κB activation (Ben-Ali et al. 2011)
	S602I	Associated with increased tuberculosis risk (Ma et al. 2007a); 602S deficiencies in TLR1 trafficking to cell surface and impaired cytokine and NF-κB responses (Johnson et al. 2007); 602I increased lipopeptide-triggered and basal NF-κB signaling (Hawn et al. 2007); combination of 248S and 602S leads to impaired NF-κB signaling, 602I increased signaling (Barreiro et al. 2009)
	V651A	Reduced NF-κB activation (Ben-Ali et al. 2011)
	H720P	Reduced NF-κB activation (Ben-Ali et al. 2011)
TLR2	T411I	Reduced NF-κB activation (Ben-Ali et al. 2011)
	P631H	Reduced NF-κB activation (Ben-Ali et al. 2011)
	R753Q	Reduced NF-κB activation (Ben-Ali et al. 2011); abnormal TLR2 signaling in *Borrelia burgdorferi* infection (Schroder et al. 2005).
TLR3	P554S	Causes TLR3 deficiency and predisposition to herpes simplex encephalitis (Zhang et al. 2007)
TLR4	D299G	Endotoxin hyporesponsiveness (Arbour et al. 2000);
	T399I	Endotoxin hyporesponsiveness (Arbour et al. 2000)
TLR5	Deletion 392–858	Increased susceptibility to Legionnaires' disease and inability to mediate flagellin signaling (Hawn et al. 2003)
TLR6	L128V	Reduced NF-κB activation (Ben-Ali et al. 2011)
	L194P	Reduced NF-κB activation (Ben-Ali et al. 2011)
	249S	Associated with increased tuberculosis risk and reduced NF-κB activation (Ma et al. 2007a)
	A474V	Reduced NF-κB activation (Ben-Ali et al. 2011)
	N690T	Reduced NF-κB activation (Ben-Ali et al. 2011)
	Q708H	Reduced NF-κB activation (Ben-Ali et al. 2011)
TLR7	Q11L	Associated with increased HIV-1 susceptibility, higher viral loads and hastened progression to late stage infection (Oh et al. 2009)
	SNP c.1-120T>G	Associated with lower frequency of liver inflammation and fibrosis in individuals chronically infected with hepatitis C virus (Schott et al. 2007)
TLR8	SNP A1G (alters start; results in shorter isoform)	Protective effect on HIV disease progression and reduced NF-κB activation (Oh et al. 2008)
TLR9	SNP G1174A (located in intron 1)	Associated with rapid progression of HIV (Bochud et al. 2007)
	SNP T1237C (located in 5′ region)	Associated with increased susceptibility to malaria (Esposito et al. 2012)
	SNP A1635G (synonymous substitution)	Associated with rapid progression of HIV (Bochud et al. 2007)

[a]amino acid changes unless noted as SNP

Because TLRs play an integral part in the innate immune response, amino acid substitutions that are deleterious to receptor function and decrease a host's fitness should be selected against, resulting in strong signals of purifying selection at the DNA level (Roach et al. 2005). Using population genetics and evolutionary genomics, it is possible to identify patterns of sequence divergence within and between species, respectively, and infer what role evolutionary processes (e.g., natural selection, genetic drift, and gene flow) have played in shaping these differences. A number of recent studies have examined the protein-coding sequences of primate Toll-like receptors for evidence of adaptive evolution (Areal et al. 2011; Nakajima et al. 2008; Ortiz et al. 2008; Wlasiuk and Nachman 2010a). Briefly, statistical tests of neutrality are used to examine if patterns of genetic variation observed between species deviate from those expected given more neutral models of sequence change (Barreiro and Quintana-Murci 2010). In order to do this, a ratio of nonsynonymous (*dN*) to synonymous (*dS*) nucleotide substitution rates (*dN*/*dS*) is used as an indicator of selective pressure acting on a protein-coding gene. Using maximum likelihood-based approaches, it is possible to test alternative models that allow *dN*/*dS* to vary across sites and/or lineages (Delport et al. 2010; Kumar et al. 2009; Pond and Frost 2005; Yang 2007). Once indicated, natural selection can be classified as either purifying selection (i.e., selection for the removal of deleterious alleles), or positive selection (i.e., selection for the fixation of advantageous alleles). When *dN*/*dS* is equivalent to 1, neutral evolution is assumed. Purifying or positive selection is indicated by $dN/dS < 1$ and $dN/dS > 1$, respectively. These types of interspecific tests are useful for detecting older evolutionary events (such as those relevant to differences observed between species), whereas intraspecific tests of neutrality (reviewed in Barreiro and Quintana-Murci 2010) are better able to detect more recent selective events. Both approaches have been used to study the molecular evolution of primate Toll-like receptors.

Although primate *TLR* genes are generally conserved and show functional constraint, individual TLR domains show different degrees of underlying genetic variation (Areal et al. 2011; Nakajima et al. 2008; Ortiz et al. 2008; Wlasiuk et al. 2009; Wlasiuk and Nachman 2010a). It is important to note these studies include different primate species, which may explain some inconsistencies in their findings. That being said, patterns emerge from these and similar studies that further our understanding of TLR molecular evolution in primates. For example, the TIR domain of TLRs is highly conserved across primates and is under strong purifying selection due to its role in intracellular signaling (Areal et al. 2011; Nakajima et al. 2008; Ortiz et al. 2008; Wlasiuk et al. 2009; Wlasiuk and Nachman 2010a). The extracellular or ectodomain domain of many TLRs, on the other hand, is more divergent due to its role in ligand binding (Areal et al. 2011; Wlasiuk et al. 2009; Wlasiuk and Nachman 2010a). These findings are not necessarily primate specific and may reflect more general trends of TLR evolution observed throughout mammalian (Areal et al. 2011) and avian (Alcaide and Edwards 2011) evolution.

When more lineage-specific models of sequence evolution are applied, the distribution of positively selected sites does not vary significantly across primates except in the case of *TLR4*, where there is an increase in positively selected sites in catarrhines (Nakajima et al. 2008; Wlasiuk and Nachman 2010a). Interestingly, some of the substitutions considered targets of positive selection in *TLR4* might affect receptor function (i.e., an amino acid is replaced by another amino acid that has a different charge, polarity, size, etc.). Moreover, several of the sites identified by Wlasiuk and Nachman (2010a) as being under selection occur in the LPS and LY96 co-receptor binding region identified by Park et al. (2009) and surround a polymorphism (D299G) associated with resistance to *Legionella pneumophila* infection and, less consistently, with an increased likelihood of developing Gram-negative bacterial sepsis (Agnese et al. 2002; Barber et al. 2006; Feterowski et al. 2003; Hawn et al. 2005; Lorenz et al. 2002).

Population level genetic variation of Toll-like receptors has also been examined in humans (including Barreiro et al. 2009; Ferrer-Admetlla et al. 2008; Mukherjee et al. 2009; Wlasiuk et al. 2009; Wlasiuk and Nachman 2010a) and to a lesser extent, chimpanzees (Wlasiuk et al. 2009; Wlasiuk and Nachman 2010a). Similar to the interspecific tests described above, statistical tests of neutrality are used to examine if allele frequencies observed within or between populations deviate from those expected given more neutral models of sequence change (reviewed in Barreiro and Quintana-Murci 2010). Such intraspecific data can be used to characterize the level of genetic variation in human *TLRs* and reveal the type and degree of selective pressures acting on these important immune receptors. These data in combination with functional studies also help identify mechanisms of host defenses that may vary between human populations or individuals. In general, patterns of within species variation suggest purifying selection has had the greatest effect on the evolution of human and chimpanzee *TLRs* (Wlasiuk and Nachman 2010a). Intracellular *TLRs* (*TLR3*, *TLR7*, *TLR8*, and *TLR9*) have been subject to stronger purifying selection than *TLRs* expressed on the cell surface, suggesting these receptors in particular play essential and nonredundant roles in host survival (Barreiro et al. 2009; Wlasiuk and Nachman 2010a). Alternatively, extracellular TLRs show more variation and less constraint, which may reflect the ability of TLR2, TLR4, and their co-receptors to identify multiple ligands (Kurt-Jones et al. 2000; Termeer et al. 2002; Zahringer et al. 2008). When compared to chimpanzees, humans have a higher proportion of deleterious mutations consistent with recent relaxation of selective constraint, although greater sampling of chimpanzee population variation is needed (Wlasiuk and Nachman 2010a).

Two haplotypes in the *TLR10–TLR1–TLR6* gene cluster (i.e., these genes are located on the same chromosome and found to be in strong linkage disequilibrium) showed evidence suggesting positive selection (Barreiro et al. 2009). Interestingly, one of the three nonsynonymous variants found in a common European haplotype (H34; present in ~ half the population sampled) resulted in impaired NF-κB signaling, which may limit damaging inflammatory responses (*TLR1* 602S; see Table 3).

Understanding patterns of genetic variation in TLRs within and between primate species is of interest because variation observed today may reflect past relationships between primates and their pathogens and explain why some individuals, populations, or species are more/less vulnerable to certain pathogen-induced diseases. While some have argued that variation in TLRs is evidence of long-term coevolution between hosts and pathogens (Areal et al. 2011), Wlasiuk and Nachman (2010a) suggest that such change is more episodic in nature and does not necessarily reflect a continuous relationship between host and pathogen that persists over a long time period. As sites within the extracellular domain (or ectodomain) are variable in primates, it is tempting to speculate that selection may favor substitutions in regions important for ligand binding that increase host fitness by improving the recognition of host-specific pathogens. This variation may reflect adaptive responses of hosts to species-specific pathogens or differences in species-specific pathogen pressures. While sequence-based approaches are widely used to infer adaptive evolution, whether or not these species-specific differences result in species-specific TLR pathway function and actually increase fitness need to be tested experimentally (as discussed in Barrett and Hoekstra 2011; and see Dean and Thornton 2007). As a result, combining studies that address how TLR-dependent responses differ between primates with those that seek to understand the context in which these disparities exist will provide the most complete picture of the evolution of TLR-depended responses in primates. In addition, while it has long been assumed that variation in human TLRs is an adaptive response driven by pathogens (e.g., Fumagalli et al. 2011), genetic drift and geographic factors have also played an important role in shaping genetic diversity in TLRs (*TLR4*, Ferwerda et al. 2008; *TLR4*, Plantinga et al. 2012; and *TLR5*, Wlasiuk et al. 2009).

Which Pathogens Have Played the Largest Role in Shaping TLR Gene Evolution?

While it is possible to target potentially adaptive genetic variation in primate TLRs, it will be harder to link specific pathogens to specific genetic adaptations. Because individual TLRs often bind different pathogens and there is some redundancy in TLR signaling among cell surface receptors (Bafica et al. 2006; Kurt-Jones et al. 2000, 2004; Means et al. 2001; Ohashi et al. 2000; Ozinsky et al. 2000; Poltorak et al. 1998), it is difficult to say which pathogen in particular has played the largest role in shaping primate TLR genes and pathways. In addition, as we learn more about TLR expression in the developing brain (Ma et al. 2007b; Sterner et al. 2012) and gain a better appreciation of the interaction of TLRs and a host's microbiome (Cerf-Bensussan and Gaboriau-Routhiau 2010), it may be equally difficult to differentiate what drives selection on these important, multi-purpose receptors. Nevertheless, understanding the history of pathogen exposure in primates in the context of the genetic variation observed today in TLR genes has the potential to shed light on important aspects of human evolution (e.g., major shifts in diet, mating strategies, social group systems, and habitat) and species-specific disparities in the susceptibility to TLR-dependent diseases (e.g., HIV/AIDS).

Several TLR-detected pathogens appear to have been unevenly distributed amongst primate species over the course of primate evolution due to geographic barriers and primate behavioral ecology (e.g., diet, landscape use, and mating systems). Some authors have argued that these differences in pathogen exposure may provide the basis for pathogen or PAMP-specific immune responses in different primate species and populations (Galvani and Slatkin 2003; Martin 2003; Vitone et al. 2004; Wlasiuk and Nachman 2010b). Increased exploitation of grasslands and meat consumption, for example, have been proposed to have contributed to the emergence of the progenitor of *Mycobacterium tuberculosis* in hominins 3 million years ago (Gutierrez et al. 2005; Martin 2003). Habitual grassland use by species in the *Homo* and *Papio* lineages may have routinely exposed these species to standing watering pools contaminated with other ungulate pathogens such as the precursor to the modern TLR2-detected *Bacillus anthracis* (anthrax) (Martin 2003; Triantafilou et al. 2007).

The geographic distribution- of primate populations also influences primate–pathogen interactions, as highlighted by *Yersinia pestis.* The causative agent of bubonic plague, *Y. pestis* interacts with both TLR2 and TLR4 (Haensch et al. 2010; Matsuura et al. 2010; Schuenemann et al. 2011; Sing et al. 2002). Now pandemic and less virulent, historical *Y. pestis* potentially represents one of the strongest agents of pathogen-mediated selection in human history. It is estimated to have emerged in Asian human populations >2,600 years ago. Its spread to Europe via trade routes killed nearly 30 % of the population by the year 1357, and approximately 15–20 % of the population again in 1665–1666 (Biraben 1975; Gottfried 1983; Morelli et al. 2010). A 3–5-day infection course, broad age range of hosts, and high mortality suggests that historical plague epidemics have had a profound effect on the genetic composition of modern Eurasian populations (reviewed in Gage and Kosoy 2005). Since that time, it has re-emerged sporadically throughout Eurasia, though with less virulence than the 1357 epidemic. *Yersinia pestis* does not appear to have breeched Eurasian boundaries until very recently, which suggests that the *Y. pestis* strain of the Black Death epidemic in the 1300s bypassed African primate populations, African humans included (Morelli et al. 2010; Pollitzer 1951). While not well investigated and inconsistently supported by functional studies examining *Y. pestis* and host receptor interactions, selection by *Y. pestis* has been pointed to as possible agent that "preselected" for immune receptor anomalies that contribute to resistance to other infections in Eurasian populations (e.g., CCR5-delta32, a deletion in C-C chemokine receptor type 5 that, when homozygous in humans, confers resistance to HIV-1 infection; CCR5-delta32 allele is found in ~10 % of all Europeans) (Dean et al. 1996; Elvin et al. 2004; Galvani and Slatkin 2003; Lucotte 2001; Mecsas et al. 2004; Samson et al. 1996; Stephens et al. 1998). The virulence of the pathogen, the significant biosafety requirements for its handling, along with the an effective course of treatment for very early plague infections have meant that functional research into plague–receptor interactions has been very limited. However, it seems an interesting candidate for examining the evolutionary effects of uneven pathogen exposure across and within primate species to virulent TLR-detected pathogens.

Conclusions and Future Directions

In this chapter we review our current understanding of TLR function and evolution in primates and discuss how an evolutionary perspective can help explain why some primate species differ in susceptibility to certain pathogen-driven diseases. As we move forward, it is important to combine studies that address how TLR-dependent responses differ between primates with those that seek to understand the genetic and environmental context in which these disparities exist. Currently, there is little evidence that directly links species-specific immune responses (e.g., those described above) to fixed, species-specific *TLR* sequence differences identified as adaptive in certain primate lineages. While this may suggest that sequence-based approaches lack power to detect adaptive substitutions, it may also suggest that a more holistic, pathway-based approach is needed. This is especially important when trying to determine how and why the human innate immune response can become dysregulated following TLR activation.

References

Abe K, Kagei N, Teramura Y, Ejima H (1993) Hepatocellular carcinoma associated with chronic Schistosoma mansoni infection in a chimpanzee. J Med Primatol 22(4):237–239

Agnese DM, Calvano JE, Hahm SJ, Coyle SM, Corbett SA, Calvano SE, Lowry SF (2002) Human Toll-like receptor 4 mutations but not CD14 polymorphisms are associated with an increased risk of gram-negative infections. J Infect Dis 186(10):1522–1525

Alcaide M, Edwards SV (2011) Molecular evolution of the Toll-like receptor multigene family in birds. Mol Biol Evol 28(5):1703–1715

Alexopoulou L, Holt AC, Medzhitov R, Flavell RA (2001) Recognition of double-stranded RNA and activation of NF-kappaB by Toll-like receptor 3. Nature 413(6857):732–738

Aliprantis AO, Yang RB, Mark MR, Suggett S, Devaux B, Radolf JD, Klimpel GR, Godowski P, Zychlinsky A (1999) Cell activation and apoptosis by bacterial lipoproteins through Toll-like receptor-2. Science 285(5428):736–739

Allikmets R, Buchbinder SP, Carrington M, Dean M, Detels R, Donfield S, Goedert JJ, Gomperts E, Huttley GA, Kaslow R et al (1996) Genetic restrictions of HIV-1 infection and progression to AIDS by a deletion allele of the CKR5 structural gene. Science 273:1856–1862

Arbour NC, Lorenz E, Schutte BC, Zabner J, Kline JN, Jones M, Frees K, Watt JL, Schwartz DA (2000) TLR4 mutations are associated with endotoxin hyporesponsiveness in humans. Nat Genet 25(2):187–191

Areal H, Abrantes J, Esteves PJ (2011) Signatures of positive selection in Toll-like receptor (TLR) genes in mammals. BMC Evol Biol 11:368

Arko RJ (1989) Animal models for pathogenic *Neisseria* species. Clin Microbiol Rev 2(Suppl):S56–59

Bafica A, Santiago HC, Goldszmid R, Ropert C, Gazzinelli RT, Sher A (2006) Cutting edge: TLR9 and TLR2 signaling together account for MyD88-dependent control of parasitemia in *Trypanosoma cruzi* infection. J Immunol 177(6):3515–3519

Barber RC, Chang LY, Arnoldo BD, Purdue GF, Hunt JL, Horton JW, Aragaki CC (2006) Innate immunity SNPs are associated with risk for severe sepsis after burn injury. Clin Med Res 4(4):250–255

Barreiro LB, Quintana-Murci L (2010) From evolutionary genetics to human immunology: how selection shapes host defence genes. Nat Rev Genet 11(1):17–30

Barreiro LB, Ben-Ali M, Quach H, Laval G, Patin E, Pickrell JK, Bouchier C, Tichit M, Neyrolles O, Gicquel B et al (2009) Evolutionary dynamics of human Toll-like receptors and their different contributions to host defense. PLoS Genet 5(7):e1000562

Barreiro LB, Marioni JC, Blekhman R, Stephens M, Gilad Y (2010) Functional comparison of innate immune signaling pathways in primates. PLoS Genet 6(12):e1001249

Barrett RD, Hoekstra HE (2011) Molecular spandrels: tests of adaptation at the genetic level. Nat Rev Genet 12(11):767–780

Ben-Ali M, Corre B, Manry J, Barreiro LB, Quach H, Boniotto M, Pellegrini S, Quintana-Murci L (2011) Functional characterization of naturally occurring genetic variants in the human TLR1-2-6 gene family. Hum Mutat 32(6):643–652

Benveniste RE, Arthur LO, Tsai CC, Sowder R, Copeland TD, Henderson LE, Oroszlan S (1986) Isolation of a lentivirus from a macaque with lymphoma: comparison with HTLV-III/LAV and other lentiviruses. J Virol 60(2):483–490

Bettauer RH (2010) Chimpanzees in hepatitis C virus research: 1998–2007. J Med Primatol 39(1):9–23

Biraben JN (1975) Les hommes et la peste en France et dans les pays européens et méditerranéens. Mouton, Paris

Blackwell TS, Christman JW (1996) Sepsis and cytokines: current status. Br J Anaesth 77(1): 110–117

Bochud P-Y, Hersberger M, Taffe P, Bochud M, Stein CM, Rodrigues SD, Calandra T, Francioli P, Telenti A, Speck RF et al (2007) Polymorphisms in Toll-like receptor 9 influence the clinical course of HIV-1 infection. AIDS 21(4):441–446, 410.1097/QAD.1090b1013e328012b328018ac

Bochud PY, Hawn TR, Siddiqui MR, Saunderson P, Britton S, Abraham I, Argaw AT, Janer M, Zhao LP, Kaplan G et al (2008) Toll-like receptor 2 (TLR2) polymorphisms are associated with reversal reaction in leprosy. J Infect Dis 197(2):253–261

Bosinger SE, Sodora DL, Silvestri G (2011) Generalized immune activation and innate immune responses in simian immunodeficiency virus infection. Curr Opin HIV AIDS 6(5):411–418

Bouer A, Werther K, Machado RZ, Nakaghi AC, Epiphanio S, Catao-Dias JL (2010) Detection of anti-*Toxoplasma gondii* antibodies in experimentally and naturally infected non-human primates by Indirect Fluorescence Assay (IFA) and indirect ELISA. Rev Bras Parasitol Vet 19(1):26–31

Brinkworth J, Pechenkina E, Silver J, Goyert S (2012) Innate immune responses to TLR2 and TLR4 agonists differ between baboons, chimpanzees and humans. J Med Primatol 41:388–393

Brown WJ, Lucas CT, Kuhn US (1972) Gonorrhoea in the chimpanzee. Infection with laboratory-passed gonococci and by natural transmission. Br J Vener Dis 48(3):177–178

Bulut Y, Faure E, Thomas L, Equils O, Arditi M (2001) Cooperation of Toll-like receptor 2 and 6 for cellular activation by soluble tuberculosis factor and Borrelia burgdorferi outer surface protein A lipoprotein: role of Toll-interacting protein and IL-1 receptor signaling molecules in Toll-like receptor 2 signaling. J Immunol 167(2):987–994

Buwitt-Beckmann U, Heine H, Wiesmuller KH, Jung G, Brock R, Akira S, Ulmer AJ (2005) Toll-like receptor 6-independent signaling by diacylated lipopeptides. Eur J Immunol 35(1):282–289

Campos MA, Almeida IC, Takeuchi O, Akira S, Valente EP, Procopio DO, Travassos LR, Smith JA, Golenbock DT, Gazzinelli RT (2001) Activation of Toll-like receptor-2 by glycosylphosphatidylinositol anchors from a protozoan parasite. J Immunol 167(1):416–423

Catão-Dias JL, Epiphanio S, Martins Kierulff MC (2013) Neotropical primates and their susceptibility to *Toxoplasma gondii*: new insights for an old problem. In: Brinkworth JF, Pechenkina E (eds) Primates, pathogens, and evolution. Springer, Heidelberg

Cerf-Bensussan N, Gaboriau-Routhiau V (2010) The immune system and the gut microbiota: friends or foes? Nat Rev Immunol 10(10):735–744

Dabbagh K, Dahl ME, Stepick-Biek P, Lewis DB (2002) Toll-like receptor 4 is required for optimal development of Th2 immune responses: role of dendritic cells. J Immunol 168(9):4524–4530

Daniel MD, King NW, Letvin NL, Hunt RD, Sehgal PK, Desrosiers RC (1984) A new type D retrovirus isolated from macaques with an immunodeficiency syndrome. Science 223(4636):602–605

De Vos R, Verslype C, Depla E, Fevery J, Van Damme B, Desmet V, Roskams T (2002) Ultrastructural visualization of hepatitis C virus components in human and primate liver biopsies. J Hepatol 37(3):370–379
Dean AM, Thornton JW (2007) Mechanistic approaches to the study of evolution: the functional synthesis. Nat Rev Genet 8(9):675–688
Dean M, Carrington M, Winkler C, Huttley GA, Smith MW, Allikmets R, Goedert JJ, Buchbinder SP, Vittinghoff E, Gomperts E et al (1996) Genetic restriction of HIV-1 infection and progression to AIDS by a deletion allele of the CKR5 structural gene. Hemophilia Growth and Development Study, Multicenter AIDS Cohort Study, Multicenter Hemophilia Cohort Study, San Francisco City Cohort, ALIVE Study. Science 273(5283):1856–1862
Debierre-Grockiego F, Campos MA, Azzouz N, Schmidt J, Bieker U, Resende MG, Mansur DS, Weingart R, Schmidt RR, Golenbock DT et al (2007) Activation of TLR2 and TLR4 by glycosylphosphatidylinositols derived from *Toxoplasma gondii*. J Immunol 179(2):1129–1137
Delport W, Poon AF, Frost SD, Kosakovsky Pond SL (2010) Datamonkey 2010: a suite of phylogenetic analysis tools for evolutionary biology. Bioinformatics 26(19):2455–2457
Diebold SS, Kaisho T, Hemmi H, Akira S, Reis e Sousa C (2004) Innate antiviral responses by means of TLR7-mediated recognition of single-stranded RNA. Science 303(5663):1529–1531
Elvin SJ, Williamson ED, Scott JC, Smith JN, Perez De Lema G, Chilla S, Clapham P, Pfeffer K, Schlondorff D, Luckow B (2004) Evolutionary genetics: ambiguous role of CCR5 in *Y. pestis* infection. Nature 430(6998):417
Epiphanio S, Sinhorini IL, Catao-Dias JL (2003) Pathology of toxoplasmosis in captive new world primates. J Comp Pathol 129(2–3):196–204
Esposito S, Molteni CG, Zampiero A, Baggi E, Lavizzari A, Semino M, Daleno C, Groppo M, Scala A, Terranova L et al (2012) Role of polymorphisms of Toll-like receptor (TLR) 4, TLR9, Toll-interleukin 1 receptor domain containing adaptor protein (TIRAP) and FCGR2A genes in malaria susceptibility and severity in Burundian children. Malar J 11:196
Estep RD, Messaoudi I, Wong SW (2010) Simian herpesviruses and their risk to humans. Vaccine 28(Suppl 2):B78–84
Etienne L, Nerrienet E, LeBreton M, Bibila GT, Foupouapouognigni Y, Rousset D, Nana A, Djoko CF, Tamoufe U, Aghokeng AF et al (2011) Characterization of a new simian immunodeficiency virus strain in a naturally infected *Pan troglodytes troglodytes* chimpanzee with AIDS related symptoms. Retrovirology 8:4
Farah IO, Mola PW, Kariuki TM, Nyindo M, Blanton RE, King CL (2000) Repeated exposure induces periportal fibrosis in *Schistosoma mansoni*-infected baboons: role of TGF-beta and IL-4. J Immunol 164(10):5337–5343
Farah IO, Kariuki TM, King CL, Hau J (2001) An overview of animal models in experimental schistosomiasis and refinements in the use of non-human primates. Lab Anim 35(3):205–212
Ferrer-Admetlla A, Bosch E, Sikora M, Marques-Bonet T, Ramirez-Soriano A, Muntasell A, Navarro A, Lazarus R, Calafell F, Bertranpetit J et al (2008) Balancing selection is the main force shaping the evolution of innate immunity genes. J Immunol 181(2):1315–1322
Ferwerda B, McCall MB, Verheijen K, Kullberg BJ, van der Ven AJ, Van der Meer JW, Netea MG (2008) Functional consequences of Toll-like receptor 4 polymorphisms. Mol Med 14(5–6): 346–352
Feterowski C, Emmanuilidis K, Miethke T, Gerauer K, Rump M, Ulm K, Holzmann B, Weighardt H (2003) Effects of functional Toll-like receptor-4 mutations on the immune response to human and experimental sepsis. Immunology 109(3):426–431
Fischer E, Marano MA, Barber AE, Hudson A, Lee K, Rock CS, Hawes AS, Thompson RC, Hayes TJ, Anderson TD et al (1991) Comparison between effects of interleukin-1 alpha administration and sublethal endotoxemia in primates. Am J Physiol 261(2 Pt 2):R442–452
Flynn JL, Capuano SV, Croix D, Pawar S, Myers A, Zinovik A, Klein E (2003) Non-human primates: a model for tuberculosis research. Tuberculosis (Edinb) 83(1–3):116–118
Fumagalli M, Sironi M, Pozzoli U, Ferrer-Admetlla A, Pattini L, Nielsen R (2011) Signatures of environmental genetic adaptation pinpoint pathogens as the main selective pressure through human evolution. PLoS Genet 7(11):e1002355

Gage KL, Kosoy MY (2005) Natural history of plague: perspectives from more than a century of research. Annu Rev Entomol 50:505–528

Galvani AP, Slatkin M (2003) Evaluating plague and smallpox as historical selective pressures for the CCR5-Delta 32 HIV-resistance allele. Proc Natl Acad Sci USA 100(25):15276–15279

Garcia MA, Yee J, Bouley DM, Moorhead R, Lerche NW (2004) Diagnosis of tuberculosis in macaques, using whole-blood in vitro interferon-gamma (PRIMAGAM) testing. Comp Med 54(1):86–92

Gioannini TL, Teghanemt A, Zhang D, Coussens NP, Dockstader W, Ramaswamy S, Weiss JP (2004) Isolation of an endotoxin-MD-2 complex that produces Toll-like receptor 4-dependent cell activation at picomolar concentrations. Proc Natl Acad Sci USA 101(12):4186–4191

Gottfried R (1983) The black death: natural and human disaster in medieval Europe. Free Press, New York, p 203

Gutierrez MC, Brisse S, Brosch R, Fabre M, Omais B, Marmiesse M, Supply P, Vincent V (2005) Ancient origin and gene mosaicism of the progenitor of *Mycobacterium tuberculosis*. PLoS Pathog 1(1):e5

Haensch S, Bianucci R, Signoli M, Rajerison M, Schultz M, Kacki S, Vermunt M, Weston DA, Hurst D, Achtman M et al (2010) Distinct clones of *Yersinia pestis* caused the black death. PLoS Pathog 6(10):e1001134

Hagberg L, Hull R, Hull S, McGhee JR, Michalek SM, Svanborg EC (1984) Difference in susceptibility to gram-negative urinary tract infection between C3H/HeJ and C3H/HeN mice. Infect Immun 46(3):839–844

Haudek SB, Natmessnig BE, Furst W, Bahrami S, Schlag G, Redl H (2003) Lipopolysaccharide dose response in baboons. Shock 20(5):431–436

Hawn TR, Verbon A, Lettinga KD, Zhao LP, Li SS, Laws RJ, Skerrett SJ, Beutler B, Schroeder L, Nachman A et al (2003) A common dominant TLR5 stop codon polymorphism abolishes flagellin signaling and is associated with susceptibility to legionnaires' disease. J Exp Med 198(10):1563–1572

Hawn TR, Verbon A, Janer M, Zhao LP, Beutler B, Aderem A (2005) Toll-like receptor 4 polymorphisms are associated with resistance to Legionnaires' disease. Proc Natl Acad Sci USA 102(7):2487–2489

Hawn TR, Misch EA, Dunstan SJ, Thwaites GE, Lan NT, Quy HT, Chau TT, Rodrigues S, Nachman A, Janer M et al (2007) A common human TLR1 polymorphism regulates the innate immune response to lipopeptides. Eur J Immunol 37(8):2280–2289

Hayashi F, Smith KD, Ozinsky A, Hawn TR, Yi EC, Goodlett DR, Eng JK, Akira S, Underhill DM, Aderem A (2001) The innate immune response to bacterial flagellin is mediated by Toll-like receptor 5. Nature 410(6832):1099–1103

Heil F, Hemmi H, Hochrein H, Ampenberger F, Kirschning C, Akira S, Lipford G, Wagner H, Bauer S (2004) Species-specific recognition of single-stranded RNA via Toll-like receptor 7 and 8. Science 303(5663):1526–1529

Hemmi H, Kaisho T, Takeuchi O, Sato S, Sanjo H, Hoshino K, Horiuchi T, Tomizawa H, Takeda K, Akira S (2002) Small anti-viral compounds activate immune cells via the TLR7 MyD88-dependent signaling pathway. Nat Immunol 3(2):196–200

Hoshino K, Takeuchi O, Kawai T, Sanjo H, Ogawa T, Takeda Y, Takeda K, Akira S (1999) Cutting edge: Toll-like receptor 4 (TLR4)-deficient mice are hyporesponsive to lipopolysaccharide: evidence for TLR4 as the Lps gene product. J Immunol 162(7):3749–3752

Jallow M, Teo YY, Small KS, Rockett KA, Deloukas P, Clark TG, Kivinen K, Bojang KA, Conway DJ, Pinder M et al (2009) Genome-wide and fine-resolution association analysis of malaria in West Africa. Nat Genet 41(6):657–665

Janeway CA Jr, Medzhitov R (2002) Innate immune recognition. Annu Rev Immunol 20:197–216

Jin MS, Lee J-O (2008) Structures of the Toll-like receptor family and its ligand complexes. Immunity 29(2):182–191

Johnson CM, Lyle EA, Omueti KO, Stepensky VA, Yegin O, Alpsoy E, Hamann L, Schumann RR, Tapping RI (2007) Cutting edge: a common polymorphism impairs cell surface trafficking and functional responses of TLR1 but protects against leprosy. J Immunol 178(12):7520–7524

Kang TJ, Lee SB, Chae GT (2002) A polymorphism in the Toll-like receptor 2 is associated with IL-12 production from monocyte in lepromatous leprosy. Cytokine 20(2):56–62
Kawai T, Akira S (2007) TLR signaling. Semin Immunol 19(1):24–32
Kawai T, Akira S (2010) The role of pattern-recognition receptors in innate immunity: update on Toll-like receptors. Nat Immunol 11(5):373–384
Keele BF, Jones JH, Terio KA, Estes JD, Rudicell RS, Wilson ML, Li Y, Learn GH, Beasley TM, Schumacher-Stankey J et al (2009) Increased mortality and AIDS-like immunopathology in wild chimpanzees infected with SIVcpz. Nature 460(7254):515–519
Krieg AM (2007) Antiinfective applications of Toll-like receptor 9 agonists. Proc Am Thorac Soc 4(3):289–294
Kumar H, Kawai T, Akira S (2009) Pathogen recognition in the innate immune response. Biochem J 420(1):1–16
Kurt-Jones EA, Popova L, Kwinn L, Haynes LM, Jones LP, Tripp RA, Walsh EE, Freeman MW, Golenbock DT, Anderson LJ et al (2000) Pattern recognition receptors TLR4 and CD14 mediate response to respiratory syncytial virus. Nat Immunol 1(5):398–401
Kurt-Jones EA, Chan M, Zhou S, Wang J, Reed G, Bronson R, Arnold MM, Knipe DM, Finberg RW (2004) Herpes simplex virus 1 interaction with Toll-like receptor 2 contributes to lethal encephalitis. Proc Natl Acad Sci USA 101(5):1315–1320
Langermans JA, Andersen P, van Soolingen D, Vervenne RA, Frost PA, van der Laan T, van Pinxteren LA, van den Hombergh J, Kroon S, Peekel I et al (2001) Divergent effect of bacillus Calmette-Guerin (BCG) vaccination on *Mycobacterium tuberculosis* infection in highly related macaque species: implications for primate models in tuberculosis vaccine research. Proc Natl Acad Sci USA 98(20):11497–11502
Lorenz E, Mira JP, Frees KL, Schwartz DA (2002) Relevance of mutations in the TLR4 receptor in patients with gram-negative septic shock. Arch Intern Med 162(9):1028–1032
Lucas CT, Chandler F Jr, Martin JE Jr, Schmale JD (1971) Transfer of gonococcal urethritis from man to chimpanzee. An animal model for gonorrhea. JAMA 216(10):1612–1614
Lucotte G (2001) Distribution of the CCR5 gene 32-basepair deletion in West Europe. A hypothesis about the possible dispersion of the mutation by the Vikings in historical times. Hum Immunol 62(9):933–936
Lund JM, Alexopoulou L, Sato A, Karow M, Adams NC, Gale NW, Iwasaki A, Flavell RA (2004) Recognition of single-stranded RNA viruses by Toll-like receptor 7. Proc Natl Acad Sci USA 101(15):5598–5603
Ma X, Liu Y, Gowen BB, Graviss EA, Clark AG, Musser JM (2007a) Full-exon resequencing reveals Toll-like receptor variants contribute to human susceptibility to tuberculosis disease. PLoS One 2(12):e1318
Ma Y, Haynes RL, Sidman RL, Vartanian T (2007b) TLR8: an innate immune receptor in brain, neurons and axons. Cell Cycle 6(23):2859–2868
Major ME, Dahari H, Mihalik K, Puig M, Rice CM, Neumann AU, Feinstone SM (2004) Hepatitis C virus kinetics and host responses associated with disease and outcome of infection in chimpanzees. Hepatology 39(6):1709–1720
Mandl JN, Barry AP, Vanderford TH, Kozyr N, Chavan R, Klucking S, Barrat FJ, Coffman RL, Staprans SI, Feinberg MB (2008) Divergent TLR7 and TLR9 signaling and type I interferon production distinguish pathogenic and nonpathogenic AIDS virus infections. Nat Med 14(10):1077–1087
Mandl JN, Akondy R, Lawson B, Kozyr N, Staprans SI, Ahmed R, Feinberg MB (2011) Distinctive TLR7 signaling, type I IFN production, and attenuated innate and adaptive immune responses to yellow fever virus in a primate reservoir host. J Immunol 186(11):6406–6416
Martin R (2003) Earth history, disease, and the evolution of primates. In: Greenblatt C, Spigelmann M (eds) Emerging pathogens: archaeology, ecology and evolution of infectious disease. Oxford University Press, New York
Massari P, Henneke P, Ho Y, Latz E, Golenbock DT, Wetzler LM (2002) Cutting edge: immune stimulation by neisserial porins is Toll-like receptor 2 and MyD88 dependent. J Immunol 168(4):1533–1537

Matsushima N, Tanaka T, Enkhbayar P, Mikami T, Taga M, Yamada K, Kuroki Y (2007) Comparative sequence analysis of leucine-rich repeats (LRRs) within vertebrate Toll-like receptors. BMC Genomics 8:124

Matsuura M, Takahashi H, Watanabe H, Saito S, Kawahara K (2010) Immunomodulatory effects of *Yersinia pestis* lipopolysaccharides on human macrophages. Clin Vaccine Immunol 17(1):49–55

McGee ZA, Stephens DS, Hoffman LH, Schlech WF 3rd, Horn RG (1983) Mechanisms of mucosal invasion by pathogenic *Neisseria*. Rev Infect Dis 5(Suppl 4):S708–714

McGee ZA, Gregg CR, Johnson AP, Kalter SS, Taylor-Robinson D (1990) The evolutionary watershed of susceptibility to gonococcal infection. Microb Pathog 9(2):131–139

Means TK, Jones BW, Schromm AB, Shurtleff BA, Smith JA, Keane J, Golenbock DT, Vogel SN, Fenton MJ (2001) Differential effects of a Toll-like receptor antagonist on *Mycobacterium tuberculosis*-induced macrophage responses. J Immunol 166(6):4074–4082

Mecsas J, Franklin G, Kuziel WA, Brubaker RR, Falkow S, Mosier DE (2004) Evolutionary genetics: CCR5 mutation and plague protection. Nature 427(6975):606

Mir KD, Bosinger SE, Gasper M, Ho O, Else JG, Brenchley JM, Kelvin DJ, Silvestri G, Hu SL, Sodora DL (2012) Simian immunodeficiency virus-induced alterations in monocyte production of tumor necrosis factor alpha contribute to reduced immune activation in sooty mangabeys. J Virol 86(14):7605–7615

Montali RJ, Mikota SK, Cheng LI (2001) *Mycobacterium tuberculosis* in zoo and wildlife species. Rev Sci Tech 20(1):291–303

Morelli G, Song Y, Mazzoni CJ, Eppinger M, Roumagnac P, Wagner DM, Feldkamp M, Kusecek B, Vogler AJ, Li Y et al (2010) *Yersinia pestis* genome sequencing identifies patterns of global phylogenetic diversity. Nat Genet 42(12):1140–1143

Mukherjee S, Sarkar-Roy N, Wagener DK, Majumder PP (2009) Signatures of natural selection are not uniform across genes of innate immune system, but purifying selection is the dominant signature. Proc Natl Acad Sci USA 106(17):7073–7078

Murdoch C, Finn A (2003) The role of chemokines in sepsis and septic shock. Contrib Microbiol 10:38–57

Nakajima T, Ohtani H, Satta Y, Uno Y, Akari H, Ishida T, Kimura A (2008) Natural selection in the TLR-related genes in the course of primate evolution. Immunogenetics 60(12):727–735

Ngampasutadol J, Tran C, Gulati S, Blom AM, Jerse EA, Ram S, Rice PA (2008) Species-specificity of *Neisseria gonorrhoeae* infection: do human complement regulators contribute? Vaccine 26(Suppl 8):I62–66

Nielsen R, Bustamante C, Clark AG, Glanowski S, Sackton TB, Hubisz MJ, Fledel-Alon A, Tanenbaum DM, Civello D, White TJ et al (2005) A scan for positively selected genes in the genomes of humans and chimpanzees. PLoS Biol 3(6):e170

Ogus AC, Yoldas B, Ozdemir T, Uguz A, Olcen S, Keser I, Coskun M, Cilli A, Yegin O (2004) The Arg753GLn polymorphism of the human Toll-like receptor 2 gene in tuberculosis disease. Eur Respir J 23(2):219–223

Oh DY, Taube S, Hamouda O, Kucherer C, Poggensee G, Jessen H, Eckert JK, Neumann K, Storek A, Pouliot M et al (2008) A functional Toll-like receptor 8 variant is associated with HIV disease restriction. J Infect Dis 198(5):701–709

Oh DY, Baumann K, Hamouda O, Eckert JK, Neumann K, Kucherer C, Bartmeyer B, Poggensee G, Oh N, Pruss A et al (2009) A frequent functional Toll-like receptor 7 polymorphism is associated with accelerated HIV-1 disease progression. AIDS 23(3):297–307

Ohashi K, Burkart V, Flohe S, Kolb H (2000) Cutting edge: heat shock protein 60 is a putative endogenous ligand of the Toll-like receptor-4 complex. J Immunol 164(2):558–561

Omueti KO, Mazur DJ, Thompson KS, Lyle EA, Tapping RI (2007) The polymorphism P315L of human Toll-like receptor 1 impairs innate immune sensing of microbial cell wall components. J Immunol 178(10):6387–6394

Onlamoon N, Noisakran S, Hsiao HM, Duncan A, Villinger F, Ansari AA, Perng GC (2010) Dengue virus-induced hemorrhage in a nonhuman primate model. Blood 115(9):1823–1834

Orange JS, Geha RS (2003) Finding NEMO: genetic disorders of NF-[kappa]B activation. J Clin Invest 112(7):983–985
Ortiz M, Kaessmann H, Zhang K, Bashirova A, Carrington M, Quintana-Murci L, Telenti A (2008) The evolutionary history of the CD209 (DC-SIGN) family in humans and non-human primates. Genes Immun 9(6):483–492
Ozinsky A, Underhill DM, Fontenot JD, Hajjar AM, Smith KD, Wilson CB, Schroeder L, Aderem A (2000) The repertoire for pattern recognition of pathogens by the innate immune system is defined by cooperation between Toll-like receptors. Proc Natl Acad Sci USA 97(25): 13766–13771
Palm NW, Medzhitov R (2009) Pattern recognition receptors and control of adaptive immunity. Immunol Rev 227(1):221–233
Pandrea I, Apetrei C (2010) Where the wild things are: pathogenesis of SIV infection in African nonhuman primate hosts. Curr HIV/AIDS Rep 7(1):28–36
Park BS, Song DH, Kim HM, Choi BS, Lee H, Lee JO (2009) The structural basis of lipopolysaccharide recognition by the TLR4-MD-2 complex. Nature 458(7242):1191–1195
Payne KS, Novak JJ, Jongsakul K, Imerbsin R, Apisitsaowapa Y, Pavlin JA, Hinds SB (2011) *Mycobacterium tuberculosis* infection in a closed colony of rhesus macaques (*Macaca mulatta*). J Am Assoc Lab Anim Sci 50(1):105–108
Plantinga TS, Ioana M, Alonso S, Izagirre N, Hervella M, Joosten LA, van der Meer JW, de la Rua C, Netea MG (2012) The evolutionary history of TLR4 polymorphisms in Europe. J Innate Immun 4(2):168–175
Pollitzer R (1951) Plague studies. 1. A summary of the history and survey of the present distribution of the disease. Bull World Health Organ 4(4):475–533
Poltorak A, He X, Smirnova I, Liu MY, Van Huffel C, Du X, Birdwell D, Alejos E, Silva M, Galanos C et al (1998) Defective LPS signaling in C3H/HeJ and C57BL/10ScCr mice: mutations in Tlr4 gene. Science 282(5396):2085–2088
Pond SL, Frost SD (2005) Datamonkey: rapid detection of selective pressure on individual sites of codon alignments. Bioinformatics 21(10):2531–2533
Puel A, Picard C, Ku CL, Smahi A, Casanova JL (2004) Inherited disorders of NF-kappaB-mediated immunity in man. Curr Opin Immunol 16(1):34–41
Quesniaux VJ, Nicolle DM, Torres D, Kremer L, Guerardel Y, Nigou J, Puzo G, Erard F, Ryffel B (2004) Toll-like receptor 2 (TLR2)-dependent-positive and TLR2-independent-negative regulation of proinflammatory cytokines by mycobacterial lipomannans. J Immunol 172(7):4425–4434
Qureshi ST, Lariviere L, Leveque G, Clermont S, Moore KJ, Gros P, Malo D (1999) Endotoxin-tolerant mice have mutations in Toll-like receptor 4 (Tlr4). J Exp Med 189(4):615–625
Rayner JC, Liu W, Peeters M, Sharp PM, Hahn BH (2011) A plethora of *Plasmodium* species in wild apes: a source of human infection? Trends Parasitol 27(5):222–229
Redl H, Bahrami S, Schlag G, Traber DL (1993) Clinical detection of LPS and animal models of endotoxemia. Immunobiology 187(3–5):330–345
Roach JC, Glusman G, Rowen L, Kaur A, Purcell MK, Smith KD, Hood LE, Aderem A (2005) The evolution of vertebrate Toll-like receptors. Proc Natl Acad Sci USA 102(27):9577–9582
Rutz M, Metzger J, Gellert T, Luppa P, Lipford GB, Wagner H, Bauer S (2004) Toll-like receptor 9 binds single-stranded CpG-DNA in a sequence- and pH-dependent manner. Eur J Immunol 34(9):2541–2550
Sadun EH, von Lichtenberg F, Cheever AW, Erickson DG (1970) *Schistosomiasis mansoni* in the chimpanzee. The natural history of chronic infections after single and multiple exposures. Am J Trop Med Hyg 19(2):258–277
Samson M, Libert F, Doranz BJ, Rucker J, Liesnard C, Farber CM, Saragosti S, Lapoumeroulie C, Cognaux J, Forceille C et al (1996) Resistance to HIV-1 infection in caucasian individuals bearing mutant alleles of the CCR-5 chemokine receptor gene. Nature 382(6593):722–725
Sapolsky RM, Else JG (1987) Bovine tuberculosis in a wild baboon population: epidemiological aspects. J Med Primatol 16(4):229–235
Schott E, Witt H, Neumann K, Taube S, Oh DY, Schreier E, Vierich S, Puhl G, Bergk A, Halangk J et al (2007) A Toll-like receptor 7 single nucleotide polymorphism protects from advanced inflammation and fibrosis in male patients with chronic HCV-infection. J Hepatol 47(2):203–211

Schroder NW, Diterich I, Zinke A, Eckert J, Draing C, von Baehr V, Hassler D, Priem S, Hahn K, Michelsen KS et al (2005) Heterozygous Arg753Gln polymorphism of human TLR-2 impairs immune activation by *Borrelia burgdorferi* and protects from late stage Lyme disease. J Immunol 175(4):2534–2540

Schuenemann VJ, Bos K, Dewitte S, Schmedes S, Jamieson J, Mittnik A, Forrest S, Coombes BK, Wood JW, Earn DJ et al (2011) From the cover: targeted enrichment of ancient pathogens yielding the pPCP1 plasmid of *Yersinia pestis* from victims of the Black Death. Proc Natl Acad Sci USA 108(38):E746–752

Schuring RP, Hamann L, Faber WR, Pahan D, Richardus JH, Schumann RR, Oskam L (2009) Polymorphism N248S in the human Toll-like receptor 1 gene is related to leprosy and leprosy reactions. J Infect Dis 199(12):1816–1819

Schwandner R, Dziarski R, Wesche H, Rothe M, Kirschning CJ (1999) Peptidoglycan- and lipoteichoic acid-induced cell activation is mediated by Toll-like receptor 2. J Biol Chem 274(25):17406–17409

Shi Q, Wang J, Wang XL, VandeBerg JL (2004) Comparative analysis of vascular endothelial cell activation by TNF-alpha and LPS in humans and baboons. Cell Biochem Biophys 40(3):289–303

Shi Q, Cox LA, Glenn J, Tejero ME, Hondara V, Vandeberg JL, Wang XL (2010) Molecular pathways mediating differential responses to lipopolysaccharide between human and baboon arterial endothelial cells. Clin Exp Pharmacol Physiol 37(2):178–184

Shimazu R, Akashi S, Ogata H, Nagai Y, Fukudome K, Miyake K, Kimoto M (1999) MD-2, a molecule that confers lipopolysaccharide responsiveness on Toll-like receptor 4. J Exp Med 189(11):1777–1782

Siddiqui RA, Krawczak M, Platzer M, Sauermann U (2011) Association of TLR7 variants with AIDS-like disease and AIDS vaccine efficacy in rhesus macaques. PLoS One 6(10):e25474

Sing A, Rost D, Tvardovskaia N, Roggenkamp A, Wiedemann A, Kirschning CJ, Aepfelbacher M, Heesemann J (2002) Yersinia V-antigen exploits Toll-like receptor 2 and CD14 for interleukin 10-mediated immunosuppression. J Exp Med 196(8):1017–1024

Stephens JC, Reich DE, Goldstein DB, Shin HD, Smith MW, Carrington M, Winkler C, Huttley GA, Allikmets R, Schriml L et al (1998) Dating the origin of the CCR5-Delta32 AIDS-resistance allele by the coalescence of haplotypes. Am J Hum Genet 62(6):1507–1515

Sterner KN, Weckle A, Chugani HT, Tarca AL, Sherwood CC, Hof PR, Kuzawa CW, Boddy AM, Abbas A, Raaum RL et al (2012) Dynamic gene expression in the human cerebral cortex distinguishes children from adults. PLoS One 7(5):e37714

Tada H, Nemoto E, Shimauchi H, Watanabe T, Mikami T, Matsumoto T, Ohno N, Tamura H, Shibata K, Akashi S et al (2002) *Saccharomyces cerevisiae*- and Candida albicans-derived mannan induced production of tumor necrosis factor alpha by human monocytes in a CD14- and Toll-like receptor 4-dependent manner. Microbiol Immunol 46(7):503–512

Takeuchi O, Hoshino K, Kawai T, Sanjo H, Takada H, Ogawa T, Takeda K, Akira S (1999) Differential roles of TLR2 and TLR4 in recognition of Gram-negative and gram-positive bacterial cell wall components. Immunity 11(4):443–451

Tarara R, Suleman MA, Sapolsky R, Wabomba MJ, Else JG (1985) Tuberculosis in wild olive baboons, *Papio cynocephalus anubis* (Lesson), in Kenya. J Wildl Dis 21(2):137–140

Termeer C, Benedix F, Sleeman J, Fieber C, Voith U, Ahrens T, Miyake K, Freudenberg M, Galanos C, Simon JC (2002) Oligosaccharides of Hyaluronan activate dendritic cells via Toll-like receptor 4. J Exp Med 195(1):99–111

Thomson M, Nascimbeni M, Havert MB, Major M, Gonzales S, Alter H, Feinstone SM, Murthy KK, Rehermann B, Liang TJ (2003) The clearance of hepatitis C virus infection in chimpanzees may not necessarily correlate with the appearance of acquired immunity. J Virol 77(2):862–870

Triantafilou M, Uddin A, Maher S, Charalambous N, Hamm TS, Alsumaiti A, Triantafilou K (2007) Anthrax toxin evades Toll-like receptor recognition, whereas its cell wall components trigger activation via TLR2/6 heterodimers. Cell Microbiol 9(12):2880–2892

Underhill DM, Ozinsky A, Smith KD, Aderem A (1999) Toll-like receptor-2 mediates mycobacteria-induced proinflammatory signaling in macrophages. Proc Natl Acad Sci USA 96(25):14459–14463 van der Kleij D, Latz E, Brouwers JF, Kruize YC, Schmitz M, Kurt-Jones EA, Espevik T, de Jong EC, Kapsenberg ML, Golenbock DT et al (2002) A novel host-parasite lipid cross-talk. Schistosomal lyso-phosphatidylserine activates Toll-like receptor 2 and affects immune polarization. J Biol Chem 277(50):48122–48129

van der Poll T, Levi M, van Deventer SJ, ten Cate H, Haagmans BL, Biemond BJ, Buller HR, Hack CE, ten Cate JW (1994) Differential effects of anti-tumor necrosis factor monoclonal antibodies on systemic inflammatory responses in experimental endotoxemia in chimpanzees. Blood 83(2):446–451

Vasl J, Prohinar P, Gioannini TL, Weiss JP, Jerala R (2008) Functional activity of MD-2 polymorphic variant is significantly different in soluble and TLR4-bound forms: decreased endotoxin binding by G56R MD-2 and its rescue by TLR4 ectodomain. J Immunol 180(9):6107–6115

Vignal C, Guerardel Y, Kremer L, Masson M, Legrand D, Mazurier J, Elass E (2003) Lipomannans, but not lipoarabinomannans, purified from Mycobacterium chelonae and Mycobacterium kansasii induce TNF-alpha and IL-8 secretion by a CD14-Toll-like receptor 2-dependent mechanism. J Immunol 171(4):2014–2023

Vitone N, Altizer S, Nunn CL (2004) Body size, diet and sociality influence the species richness of parasitic worms in anthropoid primates. Evol Ecol Res 6:183–199

Vodros D, Fenyo EM (2004) Primate models for human immunodeficiency virus infection. Evolution of receptor use during pathogenesis. Acta Microbiol Immunol Hung 51(1–2):1–29

von Bulow GU, Puren AJ, Savage N (1992) Interleukin-1 from baboon peripheral blood monocytes: altered response to endotoxin (lipopolysaccharide) and *Staphylococcus aureus* stimulation compared with human monocytes. Eur J Cell Biol 59(2):458–463

Walker CM (1997) Comparative features of hepatitis C virus infection in humans and chimpanzees. Springer Semin Immunopathol 19(1):85–98

Walsh GP, Tan EV, Dela Cruz EC, Abalos RM, Villahermosa LG, Young LJ, Cellona RV, Nazareno JB, Horwitz MA (1996) The Philippine cynomolgus monkey (*Macaca fasicularis*) provides a new nonhuman primate model of tuberculosis that resembles human disease. Nat Med 2(4):430–436

Walsh DS, Dela Cruz EC, Abalos RM, Tan EV, Walsh GP, Portaels F, Meyers WM (2007) Clinical and histologic features of skin lesions in a cynomolgus monkey experimentally infected with mycobacterium ulcerans (*Buruli ulcer*) by intradermal inoculation. Am J Trop Med Hyg 76(1):132–134

Werner H, Janitschke K, Kohler H (1969) Observations on marmoset monkeys of the species *Saguinus (Oedipomidas) oedipus* following oral and intraperitoneal infection by different cyst-forming *Toxoplasma* strains of varying virulence. I. Clinical, pathological anatomical and parasitological findings. Zentralbl Bakteriol Orig 209(4):553–569

World Health Organization (2004) Manual for the monitoring of yellow fever virus infection. Immunization VaB Geneva, World Health Organization, Switzerland, p 68

Wlasiuk G, Nachman MW (2010a) Adaptation and constraint at Toll-like receptors in primates. Mol Biol Evol 27(9):2172–2186

Wlasiuk G, Nachman MW (2010b) Promiscuity and the rate of molecular evolution at primate immunity genes. Evolution 64(8):2204–2220

Wlasiuk G, Khan S, Switzer WM, Nachman MW (2009) A history of recurrent positive selection at the Toll-like receptor 5 in primates. Mol Biol Evol 26(4):937–949

Woodall JP (1968) The reaction of a mangabey monkey (*Cercocebus galeritus agilis* Milne-Edwards) to inoculation with yellow fever virus. Ann Trop Med Parasitol 62(4):522–527

Yang Z (2007) PAML 4: phylogenetic analysis by maximum likelihood. Mol Biol Evol 24(8): 1586–1591

Yarovinsky F, Zhang D, Andersen JF, Bannenberg GL, Serhan CN, Hayden MS, Hieny S, Sutterwala FS, Flavell RA, Ghosh S et al (2005) TLR11 activation of dendritic cells by a protozoan profilin-like protein. Science 308(5728):1626–1629

Yim JJ, Ding L, Schaffer AA, Park GY, Shim YS, Holland SM (2004) A microsatellite polymorphism in intron 2 of human Toll-like receptor 2 gene: functional implications and racial differences. FEMS Immunol Med Microbiol 40(2):163–169

Zahringer U, Lindner B, Inamura S, Heine H, Alexander C (2008) TLR2 - promiscuous or specific? A critical re-evaluation of a receptor expressing apparent broad specificity. Immunobiology 213(3–4):205–224
Zhang SY, Jouanguy E, Ugolini S, Smahi A, Elain G, Romero P, Segal D, Sancho-Shimizu V, Lorenzo L, Puel A et al (2007) TLR3 deficiency in patients with herpes simplex encephalitis. Science 317(5844):1522–1527
Zhu J, Krishnegowda G, Li G, Gowda DC (2011) Proinflammatory responses by glycosylphosphatidylinositols (GPIs) of *Plasmodium falciparum* are mainly mediated through the recognition of TLR2/TLR1. Exp Parasitol 128(3):205–211
Zurovsky Y, Laburn H, Mitchell D, MacPhail AP (1987) Responses of baboons to traditionally pyrogenic agents. Can J Physiol Pharmacol 65(6):1402–1407

Impact of Natural Selection Due to Malarial Disease on Human Genetic Variation

Felicia Gomez, Wen-Ya Ko, Avery Davis, and Sarah A. Tishkoff

Introduction

An examination of malaria from an evolutionary perspective is important to understand its etiology and effects on human genetic variation, including disease susceptibility. In 1949, JBS Haldane proposed that individuals who are heterozygous for

F. Gomez
Department of Genetics and Biology, School of Medicine and School of Arts and Sciences, University of Pennsylvania, Philadelphia, PA 19104, USA

Department of Anthropology, Hominid Paleobiology Doctoral Program, The George Washington University, Washington, DC 20052, USA

Department of Anthropology, Center for the Advanced Study of Hominid Paleobiology, The George Washington University, Washington, DC 20052, USA

Division of Biostatistics and Statistical Genomics, Washington University School of Medicine in St Louis, 4444 Forest Park Blvd St. Louis, MO 63108, USA

W.-Y. Ko
Department of Genetics and Biology, School of Medicine and School of Arts and Sciences, University of Pennsylvania, Philadelphia, PA 19104, USA

CIBIO, Research Center in Biodiversity and Genetic Resources, University of Porto, 4485–661 Vairão, Portugal

A. Davis
Department of Genetics and Biology, School of Medicine and School of Arts and Sciences, University of Pennsylvania, Philadelphia, PA 19104, USA

Department of Biology, Swarthmore College, Swarthmore, PA 19081, USA

Graduate Program in Biological and Biomedical Sciences, Harvard Medical School, Boston, MA 02115, USA

S.A. Tishkoff (✉)
Department of Genetics and Biology, School of Medicine and School of Arts and Sciences, University of Pennsylvania, Philadelphia, PA 19104, USA
e-mail: tishkoff@mail.med.upenn

J.F. Brinkworth and K. Pechenkina (eds.), *Primates, Pathogens, and Evolution*, Developments in Primatology: Progress and Prospects,
DOI 10.1007/978-1-4614-7181-3_5, © Springer Science+Business Media New York 2013

thalassemic traits may be more fit in malaria-endemic environments than wild-type individuals (e.g., "The Malaria Hypothesis"; Haldane 1949). Since that time, many scholars have shown that the distribution of thalassemias and other hemoglobin variants is consistent with Haldane's idea that resistance to malaria can maintain otherwise deleterious mutations in malaria-endemic environments. Haldane's foundational hypothesis has led to the widely accepted notion that malaria is one of the most important selective pressures in recent human evolution.

The purpose of this chapter is to provide an overview of genetic variants that may confer resistance to malarial disease. These include variants that affect the red blood cell (RBC), variants within endothelial receptors involved in malaria pathogenesis, as well as genetic variants that affect immune functions involved in resistance to malarial disease. For each of the genes discussed here, we review patterns of genetic variation at these loci and signatures of natural selection. We also briefly review some of the pathways and mechanisms that are thought to provide resistance or protection from disease. Finally, we argue that studying our genome as a whole—rather than looking at specific candidate loci—holds potential for the discovery of new variants that confer resistance to malarial disease.

Malaria and Its Life Cycle

Malaria is a parasitic disease caused by species in the genus *Plasmodium*, which is a member of a large phylum called Apicomplexa that includes parasites responsible for many life-threatening human and animal diseases including toxoplasmosis, cryptosporidiosis, coccidiosis, and babesiosis (Roberts and Janovy 2005). Several *Plasmodium* species—including *P. falciparum*, *P. ovale*, *P. malariae*, and *P. vivax*—infect humans. There have also been reports that *P. knowlesi*, a species that generally infects monkeys, can infect humans. However, infections from *P. knowlesi* are rare (Van den Eede et al. 2009; van Hellemond et al. 2009).

Plasmodium parasites are transmitted by female *Anopheles* mosquitoes (Fig. 1). When a female mosquito takes a blood meal from an infected human, the mosquito ingests the malaria parasites along with its blood meal. The parasites subsequently develop in the mosquito, and when the mosquito takes its next blood meal the developed parasites are injected with the mosquito's saliva into a new individual, thus transmitting the disease from one human to another. Once the parasites have entered the bloodstream, they enter hepatic (liver) cells. In hepatic cells, the parasites undergo a brief cycle of asexual reproduction and develop further into preerythrocytic parasites. The hepatic cells then rupture, and the parasites go on to infect the host's erythrocytes (red blood cells). At this point, the host has entered the erythrocytic stage of infection. This is the stage of infection when common symptoms associated with malaria, including high fevers and chills, occur. Additionally, when in the red blood cell, the parasite completes the asexual portion of its life cycle. It produces gametes, which are ultimately taken up by a mosquito to begin the infective process again (Fig. 1).

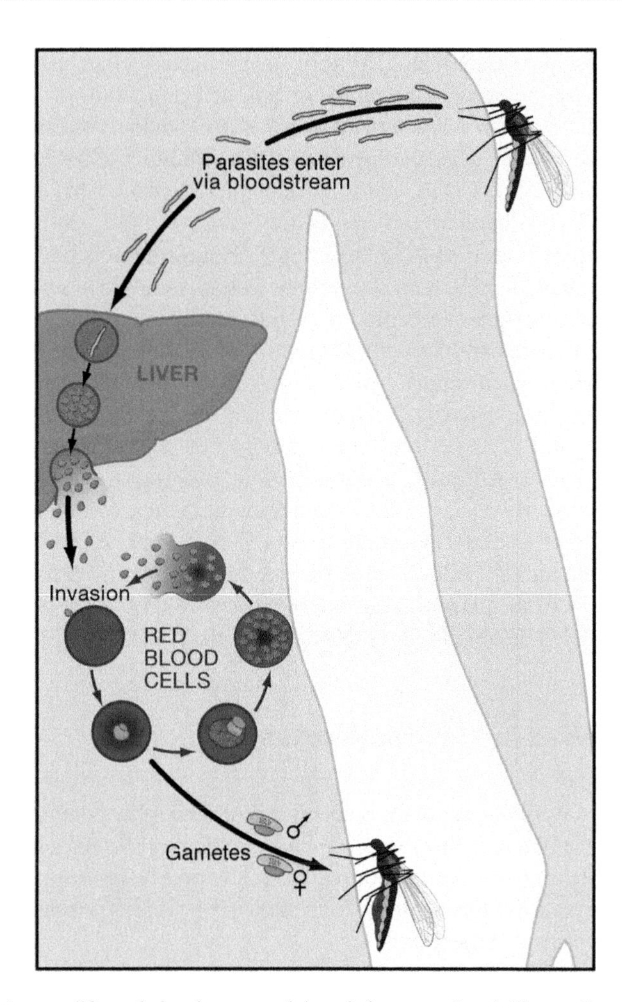

Fig. 1 *P. falciparum* life cycle in a human and *Anopheles* mosquito. A *Plasmodium*-infected mosquito injects parasites into the human host during a bite that initiates a blood meal. Following the initial bite, the parasites migrate to the liver where they invade hepatic cells. The parasites develop within hepatic cells and are then released into the bloodstream, where they invade erythrocytes. During the erythrocytic stage of infection, the parasites develop and replicate. Some parasites will go on to infect other red blood cells (merozoite form). Other intraerythrocytic parasites will develop into male and female gametes. Gametes are taken up by a mosquito during feeding and will fuse in the mosquito gut to become a zygote. The zygote develops within the mosquito and eventually migrates to the mosquito salivary gland, from which it will be injected into a human host during the next blood meal (adapted from Cowman and Crabb 2006)

Phenotypically, malarial disease symptoms include fever, headaches, and vomiting, which typically occur between a week and a half and 2 weeks after infection via mosquito bite (WHO 2010). Malaria infection is heterogeneous, manifesting itself differently among patients and at different points in the life cycle of the parasite. "Severe malaria" is characterized by high fever and anemia due to an accumulation of parasitized erythrocytes and subsequent oxygen deprivation of vital organs.

"Cerebral malaria" is a particular form of severe malaria characterized by reduced oxygen supply to the brain resulting in coma (Chen et al. 2000).

Severe malaria is also characterized by parasite cytoadherence and sequestration. Cytoadherence occurs when parasitized RBCs (PRBCs) adhere to specific host proteins (e.g., ICAM-1, CD36, CD31, E-selectin, or/and VCAM) on microvascular endothelium. Cytoadherence can also occur between PRBCs and noninfected red blood cells—this process is called "rosetting." Sequestration is the result of cytoadherence. It occurs when PRBCs adhere to the vascular endothelium and the cells are *sequestered* from peripheral circulation. Sequestration and rosetting cause accumulation of both infected and noninfected red blood cells, resulting in blocked blood flow and limited oxygen supply.

Plasmodium falciparum is the most virulent of the *Plasmodium* species that infect humans and it causes the largest number of deaths worldwide (Wahlgren 1999; WHO 2010). Although a cosmopolitan disease at one time, *falciparum* malaria is now mainly concentrated in Sub-Saharan Africa, Central, South, and East Asia, and parts of Central and South America (Hay et al. 2009). It has been estimated that there are 2.37 billion people at risk for infection by *P. falciparum* worldwide; ~47.48 % of those individuals are in Africa, ~49.58 % are in Central, South, and East Asia, and the remaining ~2.94 % are in the Americas (Guerra et al. 2008).

Genes Involved in Malaria Resistance

In the sections to follow, we discuss specific genes that play potential roles in resistance to malaria. We describe the genes, discuss the patterns of genetic variation at these loci, and describe evidence for natural selection at each locus. We also review some of the suggested mechanisms for protection or resistance that are associated with genetic variants at these genes.

Erythrocyte Genetic Variation and Malaria

Hemoglobin Variants

Adult hemoglobin is comprised of four globin chains (two α and two β chains). These chains are encoded by the α-globin genes, *HBA1* and *HBA2*, and by the β-globin gene, *HBB*. Hundreds of polymorphisms that cause hemoglobinopathies (disorders of hemoglobin) have been identified worldwide. There are two broad categories of hemoglobinopathies. These categories include pathologies that affect the structure of hemoglobin, namely HbS, HbC, and HbE, and pathologies that are related to the production of hemoglobin or the regulation of hemoglobin production (i.e., thalassemias, which will not be discussed here). Hemoglobinopathies were among the first genetic disorders to be associated with malarial disease because of the strong correlation between the global distribution of many of these diseases (and the genetic variants that cause them) and the current or historic distribution of malarial disease.

HbS

HbS is the mutation that causes sickle-cell anemia, which is one of the most common hemoglobinopathies associated with protection from malaria. HbS is caused by a point mutation in the *HBB* gene that results in a glutamic acid to valine substitution at codon 6. Erythrocytes with this form of hemoglobin are prone to deform into a characteristic sickle shape. When this mutation is in its homozygous form (HbSS), the formation of sickled red blood cells is quite serious and can often be fatal or extremely debilitating (Ashley-Koch et al. 2000). However, in the heterozygous form, HbAS, there are very limited clinical symptoms. These individuals are relatively healthy and benefit from protection from malaria. Indeed, studies have confirmed that HbAS is protective against severe and lethal malaria (Aidoo et al. 2002; Williams et al. 2005; Williams 2006).

The HbS allele is widespread over areas of the world where malaria is, or was, historically endemic. It is especially common in Sub-Saharan Africa, where it reaches frequencies of 15–20 % (Weatherall 2001; Allison 2009; Verra et al. 2009). The mechanisms by which HbS confers protection against *P. falciparum* malaria are not well understood. Some common hypotheses postulate that HbS hemoglobin may interfere with parasite growth in RBCs (Nagel and Roth 1989) or that HbS could enhance splenic clearance of PRBCs (Shear et al. 1993; Roberts and Williams 2003; Kwiatkowski 2005). It has also been suggested that HbS is protective against malaria because the mutation causes an increase in the concentration of heme and consequently heme oxygenase-1 (HO-1). HO-1 is an enzyme that catabolizes free heme and has been shown to be protective against a number of diseases including cerebral malaria (Pamplona et al. 2007; Ferreira et al. 2011). Recently, using a mouse model, Ferreira et al. (2011) confirmed the protective effects of HO-1 against malaria and showed that the protective effect of HbS occurs irrespective of parasite load. In another recent study, Glushakova et al. (2010) showed that sickled RBCs may be protective against malaria because they do not support parasite replication. In their study, they showed that infected HbS RBCs have inefficient parasite egress or an aborted parasite life cycle. It has also been hypothesized that the protective mechanism of the HbS allele may involve the disruption of cytoadherence and sequestration. Compared to HbAA-infected erythrocytes, HbAS-infected erythrocytes show a 54 % reduction in adherence to vascular endothelium (Cholera et al. 2008).

The HbS mutation is associated with five "classical" haplotypes that are defined by different RFLP (restriction fragment length polymorphism) patterns across a 70-kb region surrounding *HBB*. These haplotypes have different geographic distributions and are named accordingly—Benin, Bantu (central Africa), Cameroon, Senegal, and Arab (Hanchard et al. 2007). Because the HbS mutation is known to occur on several haplotype backgrounds, some have suggested that this mutation may have arisen multiple times (Pagnier et al. 1984; Nagel et al. 1985; Chebloune et al. 1988). However, recombination and gene conversion have also been suggested as alternative explanations for the occurrence of HbS on multiple haplotype backgrounds (Orkin et al. 1982; Webster et al. 2003; Hanchard et al. 2007).

Aside from the well-documented correlation between the distribution of HbS and *P. falciparum* malaria (Fig. 2), there is additional evidence of recent positive

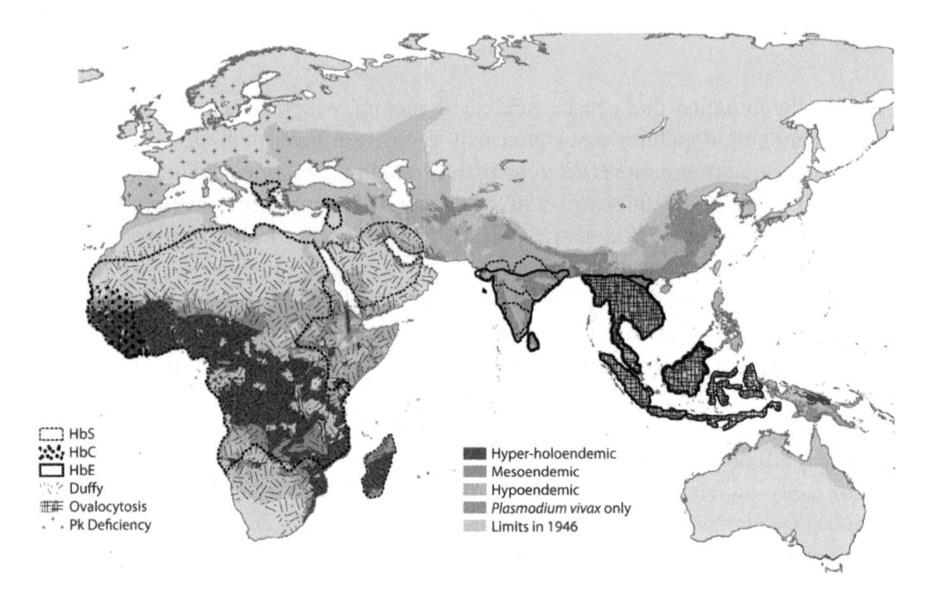

Fig. 2 A global map of the spatial limits and endemic levels of *P. falciparum* malaria and the geographic distribution of malaria susceptibility alleles. Hyper-holoendemic=areas where childhood infection prevalence is 50 % or more of population; mesoendemic = areas where childhood infection prevalence ranges between 11 and 50 %; hypoendemic = areas where childhood infection prevalence is ≤10 %. Areas where only *P. vivax* is prevalent and the spatial limits of malaria transmission are also shown. (Adapted from Snow et al. 2005.) Allelic distribution of HbS, HbC, HbE, Fy*O allele, Southeast Asian Ovalocytosis (SAO), and Pyruvate Kinase (PK) deficiency overlay the map of malaria endemicity

selection at this locus. Using 20 high-frequency SNP (single nucleotide polymorphism) markers across a 414-kb region of the *HBB* locus in Gambian, Jamaican, and Yoruban individuals, Hanchard et al. (2007) demonstrated a high degree of haplotype similarity among HbS haplotypes. This extended haplotype homozygosity was primarily observed for the Benin haplotype but was also observed for other HbS haplotypes. The same pattern of extended haplotype homozygosity on HbS haplotypes was observed by Liu et al. (2009). In addition to evidence of recent selection, there is evidence for a recent origin of the HbS allele (Table 2), which suggests that this allele could have arisen ~10,000 years ago when selection due to malaria is expected to have increased (Livingstone 1958; Wiesenfeld 1967; Livingstone 1971). Indeed, the selection coefficient is inferred to be ~0.15, which is amongst the highest in humans (Currat et al. 2002; Hedrick 2011).

HbC

HbC is another mutation that causes a structural change in hemoglobin. The HbC allele causes a glutamic acid to lysine substitution in codon 6 of *HBB*, the same codon at which the HbS substitution occurs (Table 1). This mutation is primarily found in the malaria-endemic regions of West Africa and is much less common than

Table 1 Genetic polymorphisms, adaptive consequences, and associated pathologies

	Alleles	Sequence alteration/ haplotype	Adaptive consequences	Associated pathology and severity	References
Erythrocyte					
Hemoglobin structural variants					
HBB					
	HbS	Glutamic acid/valine substitution at codon 6	HbS heterozygotes have >90 % protection lethal malaria (study conducted in Kenya)	HbS homozygotes have sickle-cell disease, which can be fatal and debilitating; HbS heterozygotes have mild to minimal pathologies	Kwiatkowski (2005), Williams et al. (2005), Williams (2006)
	HbC	Glutamic acid/lysine substitution at codon 6	HbC heterozygotes have 29 % reduction in risk for clinical malaria; HbC homozygotes have 93 % reduction in risk (study conducted in Burkina Faso)	HbC homozygotes have mild clinical symptoms; HbC heterozygotes are relatively healthy	Modiano et al. (2001b), Diallo et al. (2004)
	HbE	Glutamic acid/lysine substitution at codon 26	HbE heterozygotes have less parasitic invasion than HbE homozygotes and other red blood cells (study conducted in Thailand)	HbE homozygotes usually have symptomless anemia; HbE heterozygotes are generally asymptomatic	Rees et al. (1998), Chotivanich et al. (2002)

(continued)

Table 1 (continued)

	Alleles	Sequence alteration/ haplotype	Adaptive consequences	Associated pathology and severity	References
Erythrocyte surface variants					
Spectrin (*SPTA1*)	Many alleles	Missense mutations; insertion/deletion polymorphisms; frameshift mutations	In vitro studies show mutation *SPTA1* in that cause elliptocytosis resulted in 42 % reduction in parasite invasion	Elliptocytosis—elliptic RBCs that can result in hemolysis	Palek (1987), Gallagher and Forget (1996), Dhermy et al. (2007)
SLC4A1	SAO (Southeast Asian ovalocytosis)	27-base pair deletion (SLC4A1D27) causes a nine amino acid deletion	In vitro studies show reduced parasite invasion; also protection from cerebral malaria	Ovalocytosis—uncommon form of elliptocytosis; thought to be lethal	Cortes et al. (2004), Cortes et al. (2005)
DARC (Duffy antigen receptor for chemokines)	FY*A/FY*B; Promoter SNP	Glycine/aspartic acid substitution at codon 42; T/C SNP at –33 disrupts erythroid expression (Fy^{nul})	Fy*O prevents *P. vivax* erythrocyte invasion	Occasional clinical significance but no significant pathologies	Miller et al. (1976), Mercereau-Puijalon and Menard (2010)
ABO	O (O01/O02)	1-Base deletion in exon 6 (D261) shared in O01 a O02;O01 and O02 differ by 10 mutations in exons 6 and 7 and intron 6	Rosetting is significantly reduced in type O individuals	No significant pathologies	Carlson and Wahlgren (1992), Olsson and Chester (1996), Rowe et al. (2007), Calafell et al. (2008)

Erythrocyte enzyme variants					
G6PD	A-	Asparagine/aspartic acid at codon 156; valine/methionine at codon98	In vitro studies show that parasite growth is slower in G6PD deficient RBCs	A– variant cause a 10–15 % reduction in G6PD enzyme activity, which reduces RBC ability to counteract oxidative stress	Cappellini and Fiorelli (2008), Johnson et al. (2009)
PKLR	Many alleles	SNPs, insertion/deletion polymorphisms; frame shift; splice site mutations	In vitro studies show reduced invasion of PKLR deficient RBCs	RBC PKLR deficiency can cause hemolytic anemia; there is a wide variety in clinical severity of PKLR deficiency	Zanella et al. (2005), Ayi et al. (2008)
Endothelial receptors for cytoadhesion and sequestration					
ICAM-1	*ICAM*^{Kilifi}/rs5491	Lysine/methionine at codon 29	In vitro studies show that *ICAM-1*^{Kilifi} can alter the binding ability of some strains of *P. falciparum*; association studies are inconsistent (studies in Africa and Asia)	No significant pathologies	Fernandez-Reyes et al. (1997), Adams et al. (2000), Tse et al. (2004)

(continued)

Table 1 (continued)

	Alleles	Sequence alteration/ haplotype	Adaptive consequences	Associated pathology and severity	References
	rs5498	Lysine/glutamic acid at codon 469	Association studies are inconsistent	No significant pathologies; possible associations with Type 1 diabetes and asthma	Amodu et al. (2005), Ma et al. (2006), Puthothu et al. (2006)
CD36	T1264G/ rs3211938	Tyrosine/stop	Association studies are inconsistent	CD36 deficiency	Aitman et al. (2000), Fry et al. (2009)
Immune system variants					
HLA					
	HLA-B (MHC-class I)	Bw53 haplotype	Associated with reduced risk for severe malaria (study conducted in The Gambia)	No significant pathologies	Hill et al. (1991), Hill et al. (1992)
	HLA-DR (MHC class II)	DRB1*1302 haplotype	Associated with reduced risk for severe malaria (study conducted in The Gambia)	No significant pathologies	Hill et al. (1991)

TNF					
	−238	G/A SNP at −238	Multiple association studies with conflicting results; some suggest protection some suggest increased susceptibility	Associations with many autoimmune (e.g., rheumatoid arthritis, asthma, and multiple sclerosis) and infectious diseases but no consistency across studies	Knight et al. (1999), Ubalee et al. (2001), Bayley et al. (2004), Flori et al. (2005)
	−308	G/A SNP at −308			McGuire et al. (1994), Wattavidanage et al. (1999), Aidoo et al. (2001), Ubalee et al. (2001), Meyer et al. (2002), Bayley et al. (2004), Flori et al. (2005)
	−376	G/A SNP at −376			Knight et al. (1999), Bayley et al. (2004)

HbS. As described by Diallo et al. (2004), HbC is associated with phenotypes that are much less severe than HbS. HbCC homozygotes are described as having mild anemia and heterozygotes (HbAC) are relatively healthy. Like HbS, this structural variant of hemoglobin is thought to provide protection from malaria. In a large case–control study conducted in the Mossi of Burkina Faso, ~90 % of HbCC individuals and ~29 % of HbAC individuals showed a reduction in clinical malaria (Modiano et al. 2001b; Min-Oo and Gros 2005). Other studies conducted in Ghana (Mockenhaupt et al. 2004) and Mali (Agarwal et al. 2000) support the protective effect of the HbC allele against malarial infection (Min-Oo and Gros 2005). It has been suggested that HbC may inhibit the invasion or growth of the parasite in HbC RBCs (Fairhurst et al. 2003). Fairhurst et al. (2005) also reported that HbC-positive PRBCs showed 25 % reduced adhesion to vascular endothelium. They also showed rosetting occurred 33 % less often in HbAC PRBCs and they showed rosetting rarely occurred in HbCC PRBCs.

There have been a limited number of studies that focus on evidence for natural selection on HbC haplotypes. Wood et al. (2005) estimated the age and strength of selection associated with the HbC allele by sequencing a ~5.2 kb region of the *HBB* gene and analyzing the extent of LD (linkage disequilibrium) (Table 2). Based on allele frequencies, they estimated that the HbC allele is <5,000 years old and that the selection coefficient at HbC is 0.04–0.09. However, despite this relatively strong selection, they observed very little LD upstream from the HbC allele. They suggested that either recombination or gene conversion may have weakened signatures of recent selection. Similar results were also reported by Modiano et al. (2008). They demonstrated that HbC haplotypes displayed more evidence of recombination than did HbS Benin haplotypes and suggested that HbC is likely to be older than the HbS Benin haplotype.

HbE

An additional variant that affects the structure of hemoglobin is HbE, which is prevalent in Southeast Asia. This variant is caused by a glutamic acid to lysine substitution in codon 26 of *HBB* (Table 1). Like HbC, HbE is relatively benign. HbEE individuals usually have symptomless anemia, and HbAE individuals are generally asymptomatic (Rees et al. 1998). A retrospective study of malaria patients conducted in Thailand found that severe malaria complications were less acute and less prevalent in individuals with the HbE allele (Hutagalung et al. 1999). In vitro experiments suggest that HbAE RBCs showed reduced parasitic invasion compared to HbEE RBCs, normal RBCs, and RBCs with other hemoglobinopathies (Chotivanich et al. 2002).

To examine whether natural selection has affected the pattern of LD on haplotypes that contain the HbE variant, Ohashi et al. (2004) sequenced a portion of the *HBB* locus in Thai patients with mild malaria. They reported that there is extended LD associated with the HbE variant extending >100 kb. Additionally, their haplotype analysis showed that the HbE variant was found on one haplotype background in the Thai patients they surveyed. They suggested that the age of the HbE variant

Table 2 Age of selected polymorphisms described here, selection coefficients, and corresponding evidence of selection

Gene and protective allele	Age: Generations	Age: Years	Selection coefficient	Evidence for natural selection	References
HBB					
S	45–70	1,125–1,750	0.152*	Extended LD (>60 kB); long range haplotype similarity (>400 kb); evidence of positive directional selection	Currat et al. (2002), Hanchard et al. (2007), Liu et al. (2009), Hedrick (2011)
	10–28	250–700	–		Modiano et al. (2008), Hedrick (2011)
C	75–150	1,875–3,750	0.04–0.09	Evidence for selection is not	Wood et al. (2005)
	38–120	950–3,000	–	strong; very little LD is associated with HbC perhaps due to age, recombination or gene conversion	Modiano et al. (2008), Hedrick (2011)
E	100.3 (62–222)	2,006(1,240–4,440)	0.079 (0.035–0.099)	Extended LD (>100 kb); positive directional selection	Ohashi et al. (2004), Hedrick (2011)
DARC					
ES (null)	1323	33,075 (6,500–97,200)	–	No clear cut pattern; fixation in Africa implies a selective	Hamblin and Di Rienzo (2000)
ES (null)	490/310	12,250(4,250–26,500)/7,750 (3,625–13,125)	–	advantage; reduced haplotype variability in Africa also suggests positive selection in Africa	Seixas et al. (2002), Hedrick (2011)
ABO					
O01		1.15 million	–	Strong population differentiation; significantly positive Tajima's D and Fu and Li's F suggests balancing selection	Calafell et al. (2008), Fry et al. (2008b)
O02		2.5 million	–		

(continued)

Table 2 (continued)

Gene and protective allele	Age: Generations	Age: Years	Selection coefficient	Evidence for natural selection	References
G6PD					
A-	254.3	6,357 (3,840–11,760)	0.044	Constrained variability and extended LD; evidence of positive directional selection	Tishkoff et al. (2001)
	93	2,325 (1,200–3,862)		Extended LD at least 413 kb	Sabeti et al. (2002)
	1800	4,500 (25,000–65,000)	–	Observed an excess of nonsynonymous substitutions and date of A- allele suggest balancing selection	Verrelli et al. (2002)
	100	2,500–3,750	0.1	Extended LD >1.6 Mb; positive directional selection	Saunders et al. (2002)
HLA-B					
B53	86	2,150	0.041*	Extraordinary nucleotide polymorphism; evidence of balancing selection	Garrigan and Hedrick (2003), Hedrick (2011)

The age of the allele generally corresponds to the time when strong selection for malaria resistance (due to high mortality) resulted in an increase of the frequency of a given variant; "—" indicates that no selection (neutrality) is assumed in the estimation. Selection coefficients with an * are assumed, not estimated, values (adapted from Hedrick 2011) is between 1,240 and 4,440 years old (Table 2). The results from Ohashi et al. (2004) support the conclusion that the HbE mutation occurred recently and rose to high frequency because of selection due to malarial disease.

As a whole, the population genetic data described here suggest that variants involved in changing the structure of hemoglobin are relatively new mutations (Table 2). All of the variants discussed here are all likely to have risen to high frequency within the past 10,000 years due to strong selective pressures from *falciparum* malaria. Because these are new mutations and the selection pressures on these mutations were strong, most loci exhibit long range LD on the haplotypes that contain the functional alleles (i.e., a classic selective sweep). One exception is the *HBC* haplotypes with shorter tracts of LD due to a recombination hotspot. These haplotypes, however, show other signatures of recent strong selection.

Erythrocyte Membrane Proteins and Surface Antigens

Alleles that cause erythrocyte membrane abnormalities are another class of RBC polymorphisms that confer resistance to *P. falciparum* malaria. These mutations likely perturb the complex process of parasite invasion into erythrocytes, and therefore play an important role in providing protection from disease.

Variants Associated with Elliptocytosis and Ovalocytosis

One type of RBC membrane abnormality that can affect malaria pathogenesis is elliptocytosis. Just as it sounds, elliptocytosis is a condition in which RBCs are elliptical, rather than round, in shape. This condition is considered pathologic because individuals affected by this condition suffer from anemia and hemolysis (Palek 1987). The most common form of elliptocytosis is hereditary elliptocytosis. Hereditary elliptocytosis (HE) includes a heterogeneous group of disorders and a number of variants in different genes are associated with these disorders (Roux et al. 1989).

Well-known mutations that cause HE are polymorphisms in the α and β chains of spectrin, a structural protein that supports the cell membrane and regulates movements of membrane proteins in erythrocytes. A number of these mutations have been reviewed in Gallagher and Forget (1996). Within African populations there are two common spectrin mutations, Sp$\alpha^{I/65}$ and Sp$\alpha^{I/46}$. These mutations result in abnormal 46- or 65-kDa amino acid variants in Spectrin αI (*SPTA1*). Studies show that in vitro invasion and growth of *P. falciparum* is reduced in both the Sp$\alpha^{I/65}$ and Sp$\alpha^{I/46}$ RBCs (Dhermy et al. 2007).

In addition to spectrin mutations, deletions in the gene *SLC4A1* have been studied extensively and may confer resistance to *P. falciparum* malaria. The primary mutation that confers protection from malaria is a 27-bp deletion that causes what is commonly known as Southeast Asian Ovalocytosis, or SAO (Table 1). As with the spectrin mutations, this deletion causes the formation of oval-shaped RBCs. The oval shape of the RBC is thought to be the result of modifications to band 3, a major anion transporter of erythrocytes (Liu et al. 1990). This deletion is generally not found in the homozygous state, and it is therefore thought to be lethal when an individual inherits two copies (Genton et al. 1995; Allen et al. 1999; Patel et al. 2004; Williams 2006). SAO is found at high frequencies (up to 30 %) in many parts of Southeast Asia and in Papua New Guinea (Verra et al. 2009). Like the spectrin mutations, it is thought that SAO confers protection from malaria because the structural change in the RBC membrane reduces parasite invasion. It has also been shown that SAO is associated with reduced in vitro invasion by some strains of *P. falciparum* (Cortes et al. 2004). Upon further investigation, Cortes et al. (2005) showed a marked increase in SAO RBCs infected with *P. falciparum* that adhere to CD36, an adhesion receptor that is generally not associated with cerebral malaria. They suggested that this increased adhesion to CD36 may prevent the parasite from adhering to neurovascular adhesion molecules, such as ICAM-1, thereby protecting the carriers of the SAO-causing deletion from cerebral malaria.

The population genetics of the *SLC4A1* deletion in Japan, Taiwan, and Indonesia were studied by Wilder et al. (2009), who addressed whether SAO chromosomes differ from wild-type *SLC4A1* chromosomes and whether or not these chromosome show a pattern of variation that is consistent with natural selection. They resequenced ~5 kb of the *SLC4A1* gene and typed all individuals for the SAO deletion. They found significantly negative values of Fay and Wu's H in all populations, which implies a larger than expected number of high-frequency derived alleles in these populations. They also showed that the SAO deletion is on one haplotype background and the non-SAO haplotype that is closest to the SAO haplotype is rare and found only in Japan and Indonesia. The Indonesian sample showed the highest level of diversity, and this was the only population in which the SAO deletion was polymorphic. From these data, they concluded that the SAO deletion is relatively recent and that it occurred on a rare haplotype that segregates in Asian populations.

Glycophorin A and B Antigen Receptors

P. falciparum has multiple pathways to invade erythrocytes through interacting with different erythrocyte surface receptors (Cowman and Crabb 2006). Among them are glycophorin A and B, which are common glycoproteins on the erythrocyte surface. The carbohydrate groups on the extracellular domains of these glycoproteins are recognized by *P. falciparum* and are essential for parasite invasion of the erythrocyte (Pasvol et al. 1982; Sim et al. 1994; Mayer et al. 2009; Crosnier et al. 2011).

Although recent studies have revealed some evidence of adaptive evolution at the glycophorin A and B genes (*GYPA*/*GYPB*), there are conflicting interpretations of the patterns of genetic diversity at glycophorin A. Baum et al. (2002) observed an excess of intermediate-frequency alleles (signatures of balancing selection) and rapid rates of protein evolution at *GYPA* in humans. They suggested that the observed pattern of selection is consistent with the decoy hypothesis, which argues that surface proteins like glycophorin A can act as decoy receptors to attract various pathogens to nonnucleated cells (Gagneux and Varki 1999). Wang et al. (2003) observed only accelerated rates of protein evolution at *GYPA* and suggested that these patterns might have evolved adaptively to escape pathogen recognition during erythrocyte invasion (evasion hypothesis).

Ko et al. (2011) studied sequence evolution of *GYPA*, *GYPB*, and another homologue (*GYPE*) across diverse African ethnic groups residing within a broad geographic range. They observed signatures of balancing and positive selection on *different* extracellular domains of *GYPA*. Ko et al. (2011) observed accelerated rates of adaptive protein evolution at exons 3 and 4, which encode O-sialoglycan-poor protein domains that are likely to be selected for maintaining the stability of the protein structure. They also observed a skewed frequency spectrum toward an excess of intermediate-frequency alleles (evidence of balancing selection) at exon 2, which encodes an O-sialoglycan-rich domain that can modify the binding affinity of *P. falciparum* ligands. Ko et al. (2011) further showed that the magnitude of skewness in the frequency spectrum at exon 2 is significantly correlated with malaria exposure. Ko et al. (2011) suggested that the observed selection patterns at *GYPA* can be better explained by the evasion hypothesis rather than the decoy hypothesis because the decoy hypothesis usually predicts both fast evolutionary rates and high genetic diversity at a genetic region, which Ko et al. (2011) did not observe.

Additionally, because the glycophorin genes are highly homologous to each other at the nucleotide level (>95 %), high levels of gene conversion have been observed among these homologues (Blumenfeld and Huang 1995; Wang et al. 2003; Ko et al. 2011). In particular, Ko et al. (2011) showed that two *GYPA* nonsynonymous SNPs that code for the N allele of the MN blood group polymorphism were introduced from *GYPB* through gene conversion. These two SNPs were also identified as candidate variants targeted by balancing selection. Thus, gene conversion may be a mechanism for introducing high levels of variation upon which natural selection may act (Wang et al. 2003; Ko et al. 2011).

Duffy Antigen Receptor for Chemokines

Another example of an erythrocyte surface antigen showing variability that may be related to malaria resistance is the Duffy surface antigen (Duffy Antigen Receptor for Chemokines, or DARC). The Duffy system consists of two principle antigens, Fy^a and Fy^b. These antigens are encoded by two co-dominant alleles called FY*A and FY*B, which differ by a single nonsynonymous SNP that causes an amino acid change from glycine to aspartic acid at codon 42. Another functionally important polymorphic site at this locus, a T/C substitution at position −33 in the FY promoter (the FY*O allele), disrupts a GATA box and silences expression of the FY antigen in erythroid lineages without affecting transcription or expression in other cells (Mercereau-Puijalon and Menard 2010). Most commonly, the FY*O alleles occur on a FY*B haplotype background. These chromosomes are referred to as FY*B^{ES}

(erythroid silent) or FY*B^{null}. An FY*A^{null} allele called FY*A^{ES} has been identified in Papua New Guinea (Zimmerman et al. 1999); however, these haplotypes are generally rare (Mercereau-Puijalon and Menard 2010).

There are strong differences in the global geographic distribution patterns of the Duffy alleles. For example, the FY*O allele is almost fixed in West and Central Africa; thus, most Africans do not express the Duffy antigen (Hamblin and Di Rienzo 2000; Fig. 2). The FY*B^{ES} allele is rare among Europeans, American Indians, and Asian populations. Recently, Howes et al. (2011) used an extensive collection of literature-based surveys of FY allele frequencies and a Bayesian geostatistical model to generate a continuous surface map of the FY alleles across the globe. Their model showed that the FY*B^{ES} allele is fixed in West, Central, and East Africa and at high frequencies in Madagascar and the Arabian Peninsula. They also demonstrate that the most prevalent allele outside of Africa is FY*A.

The driving force behind fixation of the Duffy null allele in Africa is a long-standing question within the fields of anthropology and human genetics. In a landmark study, Miller et al. (1976) reported that African Americans who do not express the Duffy antigen on their RBCs are resistant to malarial infections caused by *Plasmodium vivax*. Miller et al. (1976) suggested that because Duffy negativity offers a selective advantage to people of African descent, the alleles that are responsible for Duffy negativity have reached extremely high frequencies in African populations. It is now well understood that Duffy-negative individuals are resistant to malaria caused by *P. vivax* because the Duffy antigen is the receptor that *P. vivax* uses to enter the RBC and is necessary to begin the blood stage of a *P. vivax* infection.

Studies of the FY (*DARC*) locus in a population genetics context loosely support the idea that directional selection has influenced the patterns of genetic variation and LD at this locus in African populations (Hamblin and Di Rienzo 2000; Hamblin et al. 2002). In 2000, Hamblin and Di Rienzo showed that the *DARC* locus is two- to threefold more diverse in Europeans than in Africans, which is expected if positive directional selection has occurred in Africa. However, they also observed two major FY*O haplotypes, which is not expected if Duffy negativity recently evolved and quickly gained a selective advantage. Additionally, Hamblin et al. (2002) observed patterns of genetic variation that departed from neutral evolution on chromosomes with FY*A alleles in Italian and Chinese populations, suggesting that selection has acted at this locus in regions other than Africa.

Using sequence data from the *FY* locus, Hamblin and DiRienzo proposed that the time of fixation of the FY* O allele is 33,000 years ago, with a confidence interval of 6,500–97,200 years (Table 2), which is inconsistent with recent selection (Hamblin and Di Rienzo 2000; Hamblin et al. 2002). However, when Seixas et al. (2002) used microsatellites to date the fixation of the FY*O allele, they suggested a date of either 7,750 or 12,250 years ago, depending on the haplotype used for estimation (see Hedrick 2011), which is more consistent with recent selection at *DARC*.

Because *P. vivax* infections are generally considered to be benign, or at a minimum less serious than infection caused by *P. falciparum*, some have wondered whether the strength of selection from *P. vivax* infection is sufficient to cause the

FY*O allele to fix across the African continent (Livingstone 1984). When global variation within the genome of *P. vivax* has been examined, an Asian origin for this parasite has been suggested (Garnham 1966; Carter 2003; Escalante et al. 2005; Cornejo and Escalante 2006). Additionally, using samples obtained from South and Central America, Africa, Southeast Asia, and Melanesia, Mu et al. (2005) estimated that the time to the most recent common ancestor (TMRCA) of *P. vivax* is 53,000–265,000 years ago. Although these observations do not necessarily remove *P. vivax* as a possible selective agent underlying the FY*O allele fixation in Africa, an old TMRCA combined with an Asian origin of *P. vivax* does suggest that this parasite was not present in Africa with enough time to cause a single SNP to sweep across West and Central Africa. Thus, it has been suggested that some other selective pressure besides *P. vivax* may have caused the FY*O allele to fix across most of Sub-Saharan Africa (Mu et al. 2005).

ABO

Human blood group antigens have long been considered one of the many human polymorphisms that are associated with resistance to malaria (Athreya and Coriell 1967; Martin et al. 1979; Kassim and Ejezie 1982; Cavalli-Sforza and Feldman 2003). The ABO blood antigens are coded for by one gene (*ABO*) on chromosome 9 (Table 1). There are four amino acid changes—Arg176Gly, Gly235Ser, Leu266-Met, and Gly268Ala—that determine whether or not the A or B antigens are expressed on the red blood cell surface (Type A—Arg176, Gly235, Leu266, and Gly268; Type B—Gly176, Ser235, Met266, and Ala268; Seto et al. 1997). Type O is the result of a single nucleotide deletion that causes a reading frame shift, which creates a premature stop codon, resulting in the expression of a shortened protein that lacks a necessary catalytic site for determining a blood cell surface antigen (Yamamoto et al. 1990; Daniels 2005). There are two common O haplotypes, O01 and O02, which differ at exons 6 and 7 (Yamamoto et al. 1990; Olsson and Chester 1996; Roubinet et al. 2004; Calafell et al. 2008).

ABO blood types are differently distributed in global human populations—an intriguing pattern that has made this a classic human genetic polymorphism of interest to geneticists and anthropologists (reviewed in Uneke 2007). Over the years, a number of studies have shown that there is a high prevalence of the O allele in tropical areas of the world and in places where malaria was historically prevalent—and that blood group A is typically found in colder parts of the world, where malaria is not historically prevalent (Cserti and Dzik 2007). In addition, numerous case–control studies have investigated the relationship between *P. falciparum* malaria and ABO blood groups. These studies are comprehensively reviewed in Cserti and Dzik (2007) and Cserti-Gazdewich et al. (2010). For example, Fry et al. (2008b) conducted a case–control study with patients from The Gambia, Kenya, and Malawi and showed that A, B, and AB individuals appear to be at significantly greater risk for severe malaria than are individuals with blood group O. They also suggest that individuals with type B have a slightly lower risk than blood group A, and blood group AB has the greatest risk of the non-O blood types. Therefore, it has been hypothesized that type O confers protection from *P. falciparum* malaria, the A group confers a disadvantage for protection from malarial disease, and the B antigen has an intermediate effect (Cserti and Dzik 2007).

The mechanism by which the O blood group confers protection from malaria is not well understood. It has been shown that red blood cells with A, B, and O phenotypes show differing amounts of cytoadherence to endothelial receptors and uninfected RBCs (Cserti and Dzik 2007; Cserti-Gazdewich et al. 2010). Chen et al. (2000) reviewed the data that describe differences in rosette formation among the blood groups and suggested that because A and B antigens are found in abundance on the surface of non-O RBCs, they are important rosetting receptors. Carlson and Wahlgren (1992) showed that rosettes can be formed with O RBCs, but they found the greatest rosetting among A, B, and AB RBCs. Recently, Rowe et al. (2007) conducted a matched case–control study of 567 Malian children and showed that type O was least frequent among the patients with severe malaria. They were also able to show that rosetting was significantly reduced in patients with type O RBCs, suggesting that the protective effect of type O may be related to a reduced level of rosetting.

The *ABO* locus has long been considered a target of natural selection (Akey et al. 2004; Sabeti et al. 2006), and polymorphisms at ABO are considered to be maintained due to balancing selection that preceded the divergence of modern humans (Saitou and Yamamoto 1997; Calafell et al. 2008; Fry et al. 2008b). Thus, there may be ancient selection maintaining genetic diversity at the *ABO* locus. In fact, in addition to malaria, ABO blood groups have also been implicated in resistance to other infections, including *Escherichia coli* (Blackwell et al. 2002), *Helicobacter pylori* (Boren et al. 1993), *Campylobacter jejuni* (Ruiz-Palacios et al. 2003), and Norwalk virus (Lindesmith et al. 2003). Infections from these organisms could also have acted as selective forces influencing variation at *ABO* (Calafell et al. 2008). Fry et al. (2008b) analyzed SNP data from the International HapMap Project and observed significantly low levels of population differentiation, measured by F_{ST}, when the YRI (Yoruban population), ASN (combined Chinese and Japanese population), and CEU (northwestern European population) were compared across the *ABO* locus. Their observations are consistent with a model of long-standing balancing selection at this locus across all HapMap populations. In another study of patterns of genetic diversity at *ABO*, Calafell et al. (2008) also found evidence for balancing selection at this locus. They estimated the age of the two primary O lineages (O01 and O02) to be >1 million years old, consistent with a signature of long-standing balancing selection (Table 2). Thus, it is likely that genetic variation at *ABO* has been maintained due to multiple selective forces acting over different time periods.

Studies of genetic variation at loci that code for erythrocyte membrane proteins and surface antigens show differing patterns of variation. The *SLC4A1* locus shows evidence of recent positive selection, while *GYPA* and *DARC* show mixed signatures of balancing and positive selection depending on the gene region that is examined. By contrast, the *ABO* locus shows evidence of long-standing balancing selection. Although most of these loci are likely to participate in the process of *Plasmodium* entry into the RBC, they play fundamentally different roles in normal host biology, which affects the patterns of natural selection and evolution observed at each locus.

Erythrocyte Enzymes and Enzyme Deficiency

G6PD

The gene that encodes the enzyme glucose-6-phosphate dehydrogenase (G6PD) is an important human housekeeping gene. G6PD is the first important and rate-limiting enzyme in the pentose phosphate metabolic pathway. It plays a critical role in the metabolism of glucose and maintaining the balance of reduced, or oxidized, states of glutathione, which is important for handling oxidative stress. The *G6PD* locus, which is located on the X chromosome, is known to harbor many mutations (>140) that are associated with enzyme deficiency (Cappellini and Fiorelli 2008). Indeed, G6PD-enzyme deficiency is one of humanity's most common enzymopathies, affecting >400 million people worldwide (Ruwende and Hill 1998; Nkhoma et al. 2009). Interestingly, some of the polymorphisms that cause G6PD deficiency are highly correlated with the distribution of *P. falciparum* malaria. This observation has led to the well-established hypothesis that G6PD deficiency confers reduced risk of infection by *Plasmodium falciparum* (Beutler 1994; Ruwende and Hill 1998).

The A- variant, which is defined by two nonsynonymous point mutations, A376G and G202A, is the most common deficiency mutation (10–50 % of normal enzyme activity) in Sub-Saharan Africa and is believed to be associated with resistance to malaria (Ruwende et al. 1995). A study conducted in Mali suggests that hemizygous males and homozygous females with the A- variant have protection against severe malaria. However, no protection was observed in heterozygous females (Guindo et al. 2007).

Evidence for natural selection at *G6PD*-deficiency alleles has been addressed by a number of studies. Tishkoff et al. (2001) examined RFLP and microsatellite variation in African and non-African individuals at *G6PD*. They identified low microsatellite haplotype variation and high LD on haplotypes with the A- variant, which is consistent with natural selection affecting variation at this locus. In addition, through the use of coalescent analyses Tishkoff et al. (2001) inferred the age of the A- allele to be approximately 6,357 years old (3,840–11,760 range), consistent with malaria being the selective force that caused the A- variants to increase in frequency in Africa. Sabeti et al. (2002) also found evidence of long-range haplotype homozygosity at *G6PD* associated with the A- allele and estimated the age of the allele to be 2,325 (1,200–3,862 range). These results were also confirmed by Saunders et al. (2002), who used resequencing analyses to show that African chromosomes with the A- allele have overall low level of variation and high levels of LD. They also inferred the A- variant to be between 2,500 and 3,750 years old. To examine SNP variation at *G6PD* in African populations, Verrelli et al. (2002) sequenced a total of 5.2 kb of the *G6PD* locus in eight African and five non-African populations. Similar to what Tishkoff et al. (2001) reported, Verrelli et al.'s (2002) results show that there is less variation associated with A- haplotypes than is expected by chance alone. However, Verrelli et al. (2002) also suggested that there is an excess of amino acid polymorphism and that the A- variant is much older than 2,000 years, which may be consistent with a model of balancing selection maintaining diversity (Table 2).

Pyruvate Kinase

Pyruvate kinase (PK) is an enzyme that is important for the reactions involved in glucose metabolism. It catalyzes the conversion of phosphoenolpyruvate to pyruvate, which results in the generation of one molecule of ATP. This reaction is considered to be the rate-limiting step of glycolysis, which is critically important for energy production in red blood cells because they lack mitochondria and rely exclusively on glycolysis for the production of ATP (Ayi et al. 2008; Berghout et al. 2012). First identified in 1960 (Valentine et al. 1961; Zanella et al. 2005), RBC PK deficiency is known to be the most frequently inherited enzymatic disorder of the glycolytic pathway and is one of the most common causes of nonspherocytic hemolytic anemia (Ayi et al. 2008; Durand and Coetzer 2008; Tekeste and Petros 2010). Similar to *G6PD*, many mutations in the human *PKLR* gene cause PK deficiency. For example, Zanella et al. (2005) reported that there are at least 158 known mutations in *PKLR* that cause enzyme deficiency. However, unlike *G6PD*, mutations that cause PK deficiency are generally not in areas where malaria is prevalent (Fig. 1).

Although, PK-deficiency mutations do not show a general pattern of global distribution characteristic of genetic variation that is correlated with malarial disease (Fig. 2), some data suggest that variation at this gene could confer protection against malaria. The first line of evidence in this regard has come from mouse models. Min-Oo et al. (2003) identified *PKL* (mouse *PLKR* gene) using QTL mapping in mouse strains resistant to malaria. At *PKL*, Min-Oo et al. (2003) showed that a nonsynonymous SNP that is known to cause PK deficiency in humans, is significantly associated with decreased parasitemia, and is associated with malarial infection survival in mice. (Min-Oo et al. 2003; Min-Oo et al. 2004; Min-Oo and Gros 2005). Since this discovery in the mouse, further studies in humans have suggested that PK deficiency is related to protection from malaria (Ayi et al. 2008; Durand and Coetzer 2008). Ayi et al. (2008) showed that invasion of RBCs by *P. falciparum* is significantly lower in subjects with homozygous mutations for PK deficiency and phagocytosis of PRBCs in patients who are homozygous for PK-deficiency mutations is markedly higher than phagocytosis in control PRBCs.

Patterns of genetic diversity at *PKLR* in humans are also suggestive of selection acting at this locus. In a study of SNPs at the *PKLR* and neighboring *GBA* loci, Mateu et al. (2002) showed that there is strong LD over about 90 kb in this region. Additionally, Machado et al. (2010) examined SNP and STR variation in individuals from Angola, Mozambique, and Portugal. In this study they showed, using F_{ST}, that there is strong population differentiation between Africans and Portuguese for SNPs at *PKLR*. These data suggest selection could have resulted in significantly different alleles frequencies among Africans and Europeans.

In a recent resequencing study of *PKLR* in 387 individuals from the Human Genome Diversity Project (HGDP; Berghout et al. 2012), seven nonsynonymous SNPs were described, three of which have been reported in PK-deficient individuals. Using the algorithms in the computer program SIFT (Henn et al. 2012), Berghout et al. (2012) showed that these sites are likely to affect the function of PK. The authors also identified potential signatures of recent positive selection at the *PKLR* locus in a pooled population from Pakistan and Sub-Saharan Africa; they observed significantly negative value of Tajima's D in the Pakistani population and significantly negative values of Fu and Li's D* and F* in the combined Sub-Saharan population and the population from Pakistan. It should be noted that there is likely to be some substructure among the pooled populations examined by Berghout et al. (2012), which can also result in negative values of Tajima's D and Fu and Li's D* and F* (Simonsen et al. 1995; Ptak and Przeworski 2002).

Endothelial Receptors, Cytoadhesion, and Sequestration

As discussed above, one of the unique attributes of *P. falciparum* malaria pathogenesis is parasite sequestration caused by cytoadherence (Chen et al. 2000). There are several human receptors known to mediate cytoadherence phenotypes (Rowe et al. 2009). ICAM-1 and CD36 are two important and well-studied receptors. ICAM-1 is thought to be a key neurovascular receptor that mediates the pathogenesis of cerebral malaria, a syndrome that causes a number of serious complications that can be fatal (Turner et al. 1994; Rowe et al. 2009). CD36 is a receptor known to adhere to surface antigens from many different parasite lines but is generally not associated with cerebral malaria (Newbold et al. 1997; Rowe et al. 2009). It is thought that cytoadherence and sequestration are adaptive strategies that the parasite has evolved to escape immune clearance and to maintain a chronic infection (Pasloske and Howard 1994; Craig and Scherf 2001; Sherman et al. 2003).

ICAM-1

ICAM-1 (CD54) is a transmembrane glycoprotein that is expressed at low levels on the surface of endothelial cells and leukocytes (Chakraborty and Craig 2004). Its expression is markedly upregulated in response to many cytokines including TNF-α and IL-1β (Chakraborty and Craig 2004; Amodu et al. 2005). ICAM-1 on the endothelium of vascular tissue plays an important role in adhering to and facilitating the migration of activated leukocytes to sites of infection. Because ICAM-1 is known to be involved in severe malaria pathogenesis, a number of studies have tested whether genetic variation at *ICAM-1* provides protection from severe malaria (Fernandez-Reyes et al. 1997; Bellamy et al. 1998; Kun et al. 1999; Tse et al. 2004; Fry et al. 2008a). In 1997, Fernandez-Reyes et al. described a nonsynonymous allele that interacts with PfEMP-1 (*Plasmodium falciparum* Erythrocyte Membrane Protein-1), which is the *Plasmodium* surface protein associated with cytoadherence and sequestration. In this study it was shown that this allele is associated with *increased* susceptibility to malaria *and* occurs at a relatively high frequency (minor allele frequency >30 %) within the study population. Fernandez-Reyes et al. (1997) named this allele *ICAM-1*^{Kilifi}, after the area in Kenya where they were working.

Since this initial paper, several other scholars (e.g., Bellamy et al. 1998; Kun et al. 1999; Ndiaye et al. 2005) have studied association of *ICAM-1*^{Kilifi} with malaria susceptibility and obtained conflicting results. Bellamy et al. (1998) were unable to show a significant association between *ICAM-1*^{Kilifi} and severe malaria in The Gambia, whereas in a study of children from Gabon, Kun et al. (1999) showed that the *ICAM-1*^{Kilifi} homozygous and heterozygous genotypes confer protection against severe malaria. Fry et al. (2008a) examined the association of common SNPs in *ICAM-1* with severe malaria phenotypes in families and unrelated individuals from Kenya, The Gambia, and Malawi. In their family-based analyses, Fry et al. (2008a) were unable to show a significant association between *ICAM-1* SNPs and severe malaria phenotypes. In their population-based analysis, they showed a potential association between cerebral malaria susceptibility and *ICAM-1*^{Kilifi} homozygotes in The Gambia, but the authors interpreted this result cautiously because this SNP was at relatively low frequency in The Gambia and because previous association studies conducted in The Gambia have failed to show a significant correlation between *ICAM-1* and severe malaria. These studies and others (Ohashi et al. 2001; Amodu et al. 2005; Ndiaye et al. 2005) exemplify the uncertainty behind the role that nucleotide diversity at *ICAM-1* plays in susceptibility to malaria.

The global pattern of allele frequency at *ICAM-1*^{Kilifi} is consistent with the possibility that it plays a role in malaria resistance. The *ICAM-1*^{Kilifi} allele is at moderate frequency in African and Asian populations with high prevalence of malaria (Ohashi et al. 2001; Fry et al. 2008a) and is absent or rare in European and European American populations (Zimmerman et al. 1997; Vijgen et al. 2003; Register et al. 2004; Ma et al. 2006). Ryan et al. (2006) showed, using publicly available resequencing data of African and European Americans, that the *ICAM-1*^{Kilifi} allele has an unusually high F_{ST} value when compared to SNPs in genes that encode cytokines, adhesion molecules, cytokine receptors, and Toll-like receptors. These data suggest that the allele frequency differences at *ICAM-1*^{Kilifi} between African and European Americans are quite large. In a recent study of genetic diversity at *ICAM-1* in Africa, Gomez et al. (2013) showed that the *ICAM-1*^{Kilifi} allele is significantly correlated with malaria endemicity in Africa. They also demonstrate through haplotype analyses that the *ICAM-1*^{Kilifi} allele potentially arose independently in African and Asian populations. These data suggest that the *ICAM-1*^{Kilifi} allele may confer a selective advantage in malaria-endemic environments.

Although association studies have provided conflicting results, functional studies suggest that the *ICAM-1*^{Kilifi} allele could play a role in malaria resistance.

The *ICAM-1^{Kilifi}* allele is located in the exon that codes for the Ig-like domain that interacts with PfEMP-1. Experiments have shown that proteins containing the *ICAM-1^{Kilifi}* variant amino acid (29M) are less adherent to certain laboratory strains of *P. falciparum*. Adams et al. (2000) showed, under static and flow conditions, that the *P. falciparum* strain A4 is dramatically less adherent to proteins with the 29M mutation than the reference protein. However, they also showed that adherence to the ItG *P. falciparum* strain was less affected by the *ICAM-1^{Kilifi}* mutation. These data suggest that the changes in adhesion efficacy due to the *ICAM-1^{Kilifi}* variant may be strain specific.

CD36

CD36 is a transmembrane glycoprotein within the class B scavenger receptor family (Armesilla and Vega 1994). Also known as platelet glycoprotein IV, or GP IIIb, CD36 is found on the surface of many different cell types including platelets, adipocytes, endothelial cells, macrophages, dendritic cells, striated muscle cells, and hematopoietic cells (Rac et al. 2007). CD36 has a variety of functions depending on the cell type and tissue in which the protein is expressed. When expressed on macrophages, CD36 binds to oxidized low-density lipoproteins (ox-LDL) that can no longer interact with their usual LDL receptors (Abbas and Lichtman 2005). In the literature, the removal of ox-LDL is described as one of the primary functions of CD36.

As with *ICAM-1*, several groups (Aitman et al. 2000; Gelhaus et al. 2001; Pain et al. 2001; Fry et al. 2009) have examined genetic variation at *CD36* to test whether alleles at this locus confer protection from malaria. Aitman et al. (2000) showed that a T/G nonsense mutation in exon 10 (T1264G) and a compound mutation in exon 12 (G/C SNP with a frame-shift deletion) are associated with increased susceptibility to severe malaria. However, Pain et al. (2001) also examined the T1264G variant and suggested that the G allele may confer protection against, rather than increased susceptibility to, severe malaria.

The T1264G SNP is of particular interest because it is the most common SNP in Africa responsible for CD36 deficiency. CD36 deficiency is a condition in which CD36 is not expressed on platelets and monocytes/macrophages (type I), or when CD36 is not expressed on platelets but shows a normal expression profile on monocytes/macrophages (type II) (Imai et al. 2002). CD36 deficiency has been reported in East Asians, Africans, and African Americans. The pathologies associated with CD36 deficiency (types I and II) include hereditary hypertrophic cardiomyopathy, increased serum LDL and triglycerides, high blood pressure, and insulin resistance (Yamamoto et al. 1994; Kashiwagi et al. 2001; Tanaka et al. 2001; Ma et al. 2004; Corpeleijn et al. 2006). Because CD36 deficiency can be deleterious and is also common in some African and Asian populations, it has been suggested that this is another selected trait because it confers protection from malaria (Sabeti et al. 2006; Fry et al. 2009). Sabeti et al. (2006) used HapMap phase I data to show that the 1264G allele is associated with extended haplotype homozygosity in the Yoruba HapMap population. Fry et al. (2009) genotyped the T1264G polymorphism in 3,420 individuals from 66 ethnic groups including several African populations. Their results showed that the G allele is at relatively high frequency in the Yoruba population and is at lower frequencies in other African groups. They also observed unusually long extended haplotypes for the Yoruba but not for the populations in the HGDP panel and the Gambia. Overall, Fry et al. (2009) did not find strong evidence for an association between SNPs at CD36 and malarial phenotypes but suggested that there was a nonsignificant trend toward heterozygous advantage for the 1264 G allele. In addition, Bhatia et al. (2011) examined levels of population differentiation across the genomes of African Americans, Nigerians, and Gambians and identified strong signal of excessive population differentiation at *CD36*, which could be interpreted as a signal of natural selection at this locus.

These population genetics results are supported by functional studies that suggest CD36 plays a role in malaria susceptibility. McGilvray et al. (2000) examined whether CD36 may be involved in monocyte/macrophage-mediated malaria clearance. Through several in vitro experiments, they showed that CD36 plays a significant role in nonopsonic (nonmediated) phagocytosis of PRBCs and therefore is an important component leading to infection clearance. Using a mouse model, Patel et al. (2007) showed that CD36 contributes significantly to the success of an innate inflammatory response to *Plasmodium* infections and that CD36 also influences the duration and severity of malarial disease. These studies imply that CD36 has a complex involvement in *Plasmodium* infections; it can facilitate infection and pathogenesis through adherence to PfEMP-1, but it may also play a critical role in mounting an efficient immune response to disease.

The patterns of genetic variation at *ICAM-1* and *CD36* do not show a consistent signature of strong recent natural selection. Although cytoadherence is an important and fatal component of malarial disease, there are multiple host receptors that the parasite exploits. This is an important advantage for the parasite, because it creates flexibility in the ways in which cytoadherence can be achieved and limits the host's ability to protect itself. Thus, interpreting patterns of genetic variation and identifying a classic "selective sweep" at these loci can be difficult. However, at *ICAM-1* Gomez et al. (2013) showed that the frequency of the $ICAM\text{-}1^{Kilifi}$ allele is correlated with malaria endemicity. These data suggest correlation analyses can be used to detect potentially functional alleles in the absence of other signatures of natural selection. With regard to *CD36*, it is difficult to explain the inconsistent signatures of selection at this locus across Africa. Further investigation of sequence variation at this locus (including regulatory regions) in diverse African populations may help to reveal the evolutionary history of this locus and its potential role in malaria susceptibility (Gomez 2012).

Genetic Variation at Genes of the Immune System

The immune response to *P. falciparum* infection is an interesting and complex part of the coevolution of the human and *P. falciparum* genomes (Langhorne et al. 2008). The immune response to *P. falciparum* is complex because it is often considered to be both beneficial (i.e., in the elimination and clearance of parasites) and detrimental (i.e., in the excessive release of proinflammatory cytokines) to the human host. Doolan et al. (2009) comprehensively reviewed the literature that discusses immunity to malaria and suggested that there are various types of acquired and adaptive immunity to malaria. They described "antidisease immunity," which is immunity that affects the level of morbidity and mortality associated with an individual's infection; "antiparasite immunity," which describes protection against overall infection, parasitemia, or both; and "premunition," which describes protection against new infections through persistent low levels of parasitemia. Unfortunately, despite a sizable amount of literature describing the immune response to *P. falciparum* and experimental model systems to explore the functional basis of genetic variants that confer an immune response, we still have an inadequate understanding of mechanisms that provide immunity to malaria, especially the level of immunity that Doolan et al. (2009) described as "premunition."

However, it has been shown that there is an important heritable component to malaria susceptibility (Mackinnon et al. 2005) and specifically to the immune response that is mounted against this parasite (Sjoberg et al. 1992; Jepson et al. 1997; Stirnadel et al. 1999; Stirnadel et al. 2000; Phimpraphi et al. 2008; Duah et al. 2009). Therefore, it is informative to examine nucleotide diversity at genes implicated in malarial immune response. These data will help to uncover the heritable and potentially functional variants involved in disease protection.

Human Leukocyte Antigen

The major histocompatibility complex (MHC) system (HLA, or human leukocyte antigen in humans) has been studied over a number of years in relation to malarial disease susceptibility (reviewed in Kwiatkowski 2005; Verra et al. 2009; Hedrick 2011). In humans the HLA locus is located on chromosome six and encodes cell-surface proteins that bind to and present intracellular and extracellular peptide fragments, which are often pathogen derived, to immune effector cells. The presentation of pathogen-derived proteins to the immune system is a fundamental component of a successful immune response to most human pathogens. In humans, this process is driven by *HLA* genes, which are divided into two classes, class I and class II. The HLA region is very polymorphic (reviewed in Traherne 2008). It has been hypothesized that the high degree of variability at the *HLA* locus has evolved through natural selection in response to infectious disease (Potts and Wakeland 1993; Prugnolle et al. 2005). A number of polymorphisms within the *HLA* region have been shown to be associated with malarial disease susceptibility (reviewed in Ghosh 2008).

One of the first studies to show an association between genetic diversity at *HLA* and malarial disease was conducted in Sardinia by Piazza et al. (1985). In this study, they found large allele frequency differences at *HLA-B* (one of several *HLA* class I genes) between highland regions of Sardinia, which is relatively malaria free, and lowland areas where malaria was highly prevalent, analogous to patterns of variation observed at well-known malaria resistance loci (i.e., *G6PD*).

Hill et al. (1991) conducted a very influential study of *HLA* variation and malaria in The Gambia. In this study of over 2,000 malaria cases and controls, they showed that the HLA-Bw53 allele and the DRB1*1302-DQBB1*0501 haplotype (an HLA class II gene) are associated with reduced susceptibility to severe malaria. Hill et al. (1991) also highlight the fact that these alleles are common in West Africans and are less common in other global populations, suggesting that natural selection may have played a role in shaping the allele frequency distribution at these genes.

Although these studies and those reviewed by Ghosh et al. (2008) show that variation at *HLA* is potentially important for protection against malaria, the extent and relevance of this protection is unclear. Once the parasite has entered a human host, it spends some time in liver cells, but it causes the most serious symptoms and spends the remainder of its life cycle in the red blood cell. This is an important point because erythrocytes do not express HLA molecules after enucleation, which severely limits the ability of the erythrocyte to present antigens to the immune system (de Villartay et al. 1985; Cserti-Gazdewich et al. 2010). Furthermore, when Jepson et al. (1997) compared the humoral and cellular immune response to malaria antigens in dizygotic and haploidentical dizygotic twins, they found that genes in the *HLA* class II region did not significantly contribute to the heritable component of an immune response to malaria. This result suggests that variance in *HLA* genes does not greatly contribute to variance in immune response to malarial antigens.

Inflammatory Cytokines and Antibody Response

Many other genes involved in immune function, especially those that encode cytokines, have been suggested to harbor variants that influence susceptibility to malaria. Unfortunately, many of the associations uncovered at cytokine genes have inconsistent results depending on study location and the malarial phenotypes investigated. One locus that is consistently shown to have polymorphisms that influence a number of different malarial phenotypes is the gene that encodes tumor necrosis factor (TNFα), an important proinflammatory cytokine. TNFα is among the initial inflammatory cytokines that are released in response to infection. The gene, *TNF,* which encodes this cytokine has several promoter polymorphisms (*TNF*-308, *TNF*-238, and *TNF*-376) that have been shown in several ethnic groups to be related to susceptibility to severe malaria, symptomatic reinfections with *P. falciparum*, and parasite density (McGuire et al. 1994; Meyer et al. 2002; Flori et al. 2005). However, despite some evidence that genetic variation at *TNF* may confer protection against some malarial phenotypes, the exact function of these promoter SNPs and how those functions relate to malaria susceptibility remains unclear (reviewed in Bayley et al. 2004; Kwiatkowski 2005; Smith and Humphries 2009).

Additionally, a number of other cytokine genes have been shown to contain mutations that may affect malaria susceptibility. These include *IL1A* (Walley et al.

2004), *IL1B* (Walley et al. 2004), *IL4* (Gyan et al. 2004), *IL12B* (Morahan et al. 2002; Marquet et al. 2008), and *IL10* (Wilson et al. 2005; Ouma et al. 2008). However, as discussed above, many of these associations require replication and further study to understand the functional mechanisms of these genes that are related to malaria.

Fulani Immunity to Malaria

When compared to neighboring populations, the Fulani people, who inhabit North Sudan, Central Africa, and much of West Africa, demonstrate markedly lower susceptibility to malarial infection despite facing similar exposure levels compared to neighboring populations (Modiano et al. 1995; Modiano et al. 1996; Luoni et al. 2001; Dolo et al. 2005). Thus, the Fulani are speculated to have distinct genetic traits that confer increased protection against *P. falciparum* infection. However, many of the known genetic traits that influence infection rate (e.g., HbS, HbC, and G6PD deficiency) are not overly represented in the Fulani (Modiano et al. 2001a). It has been proposed that the functional basis of the increased protection is the result of a more effective immune response (Luoni et al. 2001; Dolo et al. 2005). When Torcia et al. (2008) studied the expression profiles of a panel of genes involved in immune response in the Fulani and the Mossi (a neighboring ethnic group), it was shown that peripheral blood mononuclear cells (PBMCs) from the Fulani have higher expression of genes related to Th1 and Th2 immune response compared to the Mossi. Additionally, they also showed a decrease in the expression of genes related to T-cell regulatory activity in both PBMCs and T regulatory cells. The authors suggest these data point to decreased activity of T regulatory cells as the underlying factor mediating the Fulani's immunity to malaria.

More recent studies continue to demonstrate that, when compared to other neighboring ethnic groups, the Fulani show differences in immune response to malaria infection and different patterns of genetic variation at genes involved in the immune response to malaria (Cserti and Dzik 2007; Senga et al. 2007; Carvalho et al. 2010; Arama et al. 2011; Henn et al. 2011). For example, Israelsson et al. (2011) showed that the Fulani have a higher frequency of SNPs associated with a proinflammatory immune response to malaria compared to neighboring groups. Specifically, their study showed Fulani have a higher frequency of SNPs that increase the expression of *IL1B* (a proinflammatory cytokine) and SNPs that decrease the expression of *IL10* (an inhibitory cytokine). Additionally, Israelsson et al. (2011) showed that when compared to other ethnic groups, the Fulani have a higher frequency of SNPs in *TNF* that are associated with decreased severity of malarial infection (Israelsson et al. 2011). It should be noted that cultural practices (i.e., pastoralism) and lifestyle have also been suggested to contribute to the observed difference in malarial disease in Fulani communities (Wallace and Wallace 2002).

Genome-Wide Association Studies

Several genome-wide linkage and association studies have been conducted to identify the genes that contribute to protection against malaria. For example, Timmann et al. (2007) used 10,000 SNPs from all the autosomes to conduct a linkage study in families from Ghana. The results of the linkage analysis revealed three regions associated with several malarial phenotypes ($p < 10^{-4}$). These include loci mapped to10p15.3–10p14, which was linked to malaria fever episode; 13q, which was linked to parasite density; and 1p36, which was linked to both parasite prevalence and the level of anemia. Sakuntabhai et al. (2008) also conducted a genome-wide linkage analysis in two Senegalese villages using 400 genome-wide microsatellite markers and 66 SNPs in regions known to influence malaria susceptibility (i.e., markers near *TNF*, *ICAM-1*, and *CD36*). This study showed notable linkage results at markers associated with 5q31 ($p < 10^{-4}$), a region that was previously shown to be associated with parasitemia and asymptomatic parasite infection (Garcia et al. 1998; Rihet et al. 1998; Flori et al. 2003). One of the interesting components of this study is that it was conducted in two Senegalese villages (Dielmo and Ndiop) where exposure to malaria differs, and villagers' ethnicities differ. This study design allowed the authors to look for associations between genotype and phenotype in different genetic backgrounds as well as in different environmental contexts. The significant linkage that was identified at 5q31was found in Dielmo, the village with intense and perennial (potentially lasting year-round) malaria transmission, but they did not find the same association in Ndiop, the village with seasonal malaria transmission. The most recent genome-wide linkage study was conducted in Senegalese families (626 individuals—249 parents and 377 children) using 250k SNPs (Milet et al. 2010). The results of their linkage analyses showed an association between mild malaria attack and SNPs located at 6p25.1 and 12q22, between parasite prevalence in asymptomatic infection and SNPs located at 20p11–q11, and between the intensity of parasite infection and SNPs located at 9q34.

To date, the two largest genome-wide association studies involving malaria phenotypes were conducted by Jallow et al. (2009) and Timmann et al. (2012). Jallow et al. examined >400k SNPs in 2,500 children from The Gambia. After correcting for population structure, the strongest signal of association they achieved was close to the *HBB* gene ($p < 10^{-6}$). Additionally, after genotyping the HbS allele in a greater number of samples they achieved a much stronger association signal ($p = 1.3 \times 10^{-28}$). However, this study failed to identify any of the other commonly accepted genes that affect malaria susceptibility and also did not find significant overlap with any of the genome-wide studies mentioned above.

Timmann et al. (2012) examined 1,325 severe malaria cases and 828 unaffected controls from Ghana using the Affymetrix Genome-Wide Human SNP Array 6.0 (>900k SNPs) and SNPs imputed from the 1,000 Genomes Project (http://www.1000genomes.org) to create a panel of >5 million SNPs for analysis. After analyses of the initial sample and replication experiments, four loci showed genome-wide significance level of association ($p < 5 \times 10^{-8}$). These loci included *HBB* and

ABO, as well as SNPS at two novel loci—the *ATP2B4* (ATPase, Ca^{2+}-transporting, and plasma membrane 4) gene and an intergenic region of chromosome 16 between the *TAT* (Tyrosine Aminotransferase) gene and *MARVELD3* (MARVEL domain-containing protein 3 gene). The authors suggested that both novel loci are interesting genome-wide hits because *ATP2B4* encodes the main calcium pump for RBCs and *MARVELD3* encodes part of the tight junction structures of epithelial and vascular endothelial cells. Besides the large sample size and number of SNPs used by Timmann et al. (2012), part of their success can also be attributed to their focus on well-defined severe malaria phenotypes. The Timmann et al. (2012) results suggest that with enough samples and enough power (i.e., genomic variation), new discoveries can be made about the underlying genetic architecture of susceptibility and resistance to malaria.

Taken as a whole, these studies demonstrate that genome-wide investigations are still in their infancy in terms of identifying causal variants that are associated with malarial phenotypes. These studies generally do not overlap with each other, aside from the consistent identification of *HBB*, and they do not replicate most associations found at previously identified candidate genes. There are a number of reasons for the inconsistencies. These include the possibility that different genes or genetic variants confer protection to malaria in different populations. Additionally, the current commercially available chips that assay genome-wide variation do a very poor job at assaying genetic variability in Africa (Albrechtsen et al. 2010). Recent whole genome sequencing studies in Africa (Abecasis et al. 2012; Lachance et al. 2012) indicate that common SNP arrays are missing much of the underlying genomic variation in Africa, much of which may be population or region specific. Therefore, it is quite likely that current genome-wide studies simply have not assayed enough of the genetic variability in Africa to understand what portions of the genome are involved in susceptibility to malaria. Additionally, studies have shown that there are low levels of LD in African populations (Tishkoff and Williams 2002), which may hamper conventional genotype-phenotype association tests. Lastly, even with large sample sizes and large number of genetic markers, it is likely that there will still be a large amount of "missing heritability" to be discovered, some of which may be the result of complex gene–gene and gene–environment interactions as well as complex regulatory networks.

Conclusions and Future Directions

Human genetic studies have identified a number of genetic variants that play a role in malaria resistance and susceptibility. Because malaria is a strong selection pressure, mutations that are potentially deleterious can be maintained in populations and will evolve adaptively if they confer protection from malarial infection. Genes that carry these adaptive mutations will exhibit signatures of natural selection that vary depending on the age and functional consequences of a particular variant.

Several of the examples discussed above include malaria-protective variants that have arisen relatively recently and have strong deleterious effects or result in obvious clinical abnormalities (Table 2). The HbS mutation at *HBB* and the A- allele at *G6PD* are examples of mutations that fall into this category. These genes are generally characterized by distinct haplotypes on which the protective mutation occurred and the adaptive haplotypes tend to have low haplotype variability and high LD. Other loci, such as *ABO*, *GYPA*, *ICAM-1*, and *CD36* have pleiotropic effects and have complex signatures of selection resulting from different selective forces acting over different time periods. These loci may not exhibit classic signatures of recent positive selection. However, studies that examine the correlation of patterns of variation at these loci with malaria endemicity (e.g., *GYPA* and *ICAM-1*) may be informative for identifying signatures of selection that are associated with malaria susceptibility.

Looking forward, as the cost of genomic sequencing becomes more affordable and population genomic studies become a reality, we will be able to examine many diverse African populations with varying risk for malaria. We can use these data to identify new candidate loci that influence susceptibility to malaria and test whether genetic variation is correlated with malaria endemicity. These data may help to explain the prevalence of deleterious genetic variants in specific ethnic groups and populations of recent African descent. Low-cost whole-genome sequencing will also create the opportunity to study genomic variation in *Plasmodium* genomes. These data combined with human genomic sequences will provide the means to explore host–pathogen coevolution and will help us better characterize malaria as a foundational selective pressure in human evolution.

In summary, malaria has been and continues to be an important selective pressure in modern human evolution. When we are able to combine our improved technology with access to all human populations at risk for malaria, we will better understand how our genes play a role in susceptibility to malaria and the role malaria has played in shaping the human genome.

Acknowledgments We would like to thank the two anonymous reviewers for their critiques and helpful suggestions. We also thank Dr. Katrina Van Heest for her editorial assistance. S.A. Tishkoff is supported by a National Institutes of Health grants R01GM076637 and DP1-OD-006445-01, and NSF Hominid grant (BCS0827436). A Doctoral Dissertation Improvement Grant from the US National Science Foundation (NSF) (BCS0925802) was given to F. Gomez An NSF IGERT grant (9987590) to F. Gomez and S.A. Tishkoff supported this research. F. Gomez was also supported by a Ford Foundation Pre-doctoral fellowship, a Cosmos Club research award, a Sigma Xi (GWU) Grant-in-Aid of Research (GIAR), and an American Anthropological Association Minority Dissertation Writing Fellowship.

References

Abbas A, Lichtman A (2005) Cellular and molecular immunology. Elsevier Saunders, Philadelphia

Abecasis GR, Auton A, Brooks LD, DePristo MA, Durbin RM, Handsaker RE, Kang HM, Marth GT, McVean GA (2012) An integrated map of genetic variation from 1,092 human genomes. Nature 491(7422):56–65

Adams S, Turner GD, Nash GB, Micklem K, Newbold CI, Craig AG (2000) Differential binding of clonal variants of *Plasmodium falciparum* to allelic forms of intracellular adhesion molecule 1 determined by flow adhesion assay. Infect Immun 68(1):264–269

Agarwal A, Guindo A, Cissoko Y, Taylor JG, Coulibaly D, Kone A, Kayentao K, Djimde A, Plowe CV, Doumbo O, Wellems TE, Diallo D (2000) Hemoglobin C associated with protection from severe malaria in the Dogon of Mali, a West African population with a low prevalence of hemoglobin S. Blood 96(7):2358–2363

Aidoo M, McElroy PD, Kolczak MS, Terlouw DJ, ter Kuile FO, Nahlen B, Lal AA, Udhayakumar V (2001) Tumor necrosis factor-alpha promoter variant 2 (Tnf2) is associated with pre-term delivery, infant mortality, and malaria morbidity in western Kenya: Asembo Bay Cohort Project Ix. Genet Epidemiol 21(3):201–211

Aidoo M, Terlouw DJ, Kolczak MS, McElroy PD, ter Kuile FO, Kariuki S, Nahlen BL, Lal AA, Udhayakumar V (2002) Protective effects of the sickle cell gene against malaria morbidity and mortality. Lancet 359(9314):1311–1312

Aitman TJ, Cooper LD, Norsworthy PJ, Wahid FN, Gray JK, Curtis BR, McKeigue PM, Kwiatkowski D, Greenwood BM, Snow RW, Hill AV, Scott J (2000) Malaria susceptibility and Cd36 mutation. Nature 405(6790):1015–1016

Akey JM, Eberle MA, Rieder MJ, Carlson CS, Shriver MD, Nickerson DA, Kruglyak L (2004) Population history and natural selection shape patterns of genetic variation in 132 genes. PLoS Biol 2(10):e286

Albrechtsen A, Nielsen FC, Nielsen R (2010) Ascertainment biases in Snp chips affect measures of population divergence. Mol Biol Evol 27(11):2534–2547

Allen SJ, O'Donnell A, Alexander ND, Mgone CS, Peto TE, Clegg JB, Alpers MP, Weatherall DJ (1999) Prevention of cerebral malaria in children in Papua New Guinea by Southeast Asian ovalocytosis band 3. Am J Trop Med Hyg 60(6):1056–1060

Allison AC (2009) Genetic control of resistance to human malaria. Curr Opin Immunol 21(5): 499–505

Amodu OK, Gbadegesin RA, Ralph SA, Adeyemo AA, Brenchley PE, Ayoola OO, Orimadegun AE, Akinsola AK, Olumese PE, Omotade OO (2005) *Plasmodium falciparum* malaria in South-West Nigerian children: is the polymorphism of Icam-1 and E-selectin genes contributing to the clinical severity of malaria? Acta Trop 95(3):248–255

Arama C, Giusti P, Bostrom S, Dara V, Traore B, Dolo A, Doumbo O, Varani S, Troye-Blomberg M (2011) Interethnic differences in antigen-presenting cell activation and Tlr responses in Malian children during *Plasmodium falciparum* malaria. PLoS One 6(3):e18319

Armesilla AL, Vega MA (1994) Structural organization of the gene for human Cd36 glycoprotein. J Biol Chem 269(29):18985–18991

Ashley-Koch A, Yang Q, Olney RS (2000) Sickle hemoglobin (Hbs) allele and sickle cell disease: a huge review. Am J Epidemiol 151(9):839–845

Athreya BH, Coriell LL (1967) Relation of blood groups to infection. I. A survey and review of data suggesting possible relationship between malaria and blood groups. Am J Epidemiol 86(2):292–304

Ayi K, Min-Oo G, Serghides L, Crockett M, Kirby-Allen M, Quirt I, Gros P, Kain KC (2008) Pyruvate kinase deficiency and malaria. N Engl J Med 358(17):1805–1810

Baum J, Ward RH, Conway DJ (2002) Natural selection on the erythrocyte surface. Mol Biol Evol 19(3):223–229

Bayley JP, Ottenhoff TH, Verweij CL (2004) Is there a future for Tnf promoter polymorphisms? Genes Immun 5(5):315–329

Bellamy R, Kwiatkowski D, Hill AV (1998) Absence of an association between intercellular adhesion molecule 1, complement receptor 1 and interleukin 1 receptor antagonist gene polymorphisms and severe malaria in a West African population. Trans R Soc Trop Med Hyg 92(3):312–316

Berghout J, Higgins S, Loucoubar C, Sakuntabhai A, Kain KC, Gros P (2012) Genetic diversity in human erythrocyte pyruvate kinase. Genes Immun 13(1):98–102

Beutler E (1994) G6pd deficiency. Blood 84(11):3613–3636

Bhatia G, Patterson N, Pasaniuc B, Zaitlen N, Genovese G, Pollack S, Mallick S, Myers S, Tandon A, Spencer C, Palmer CD, Adeyemo AA, Akylbekova EL, Cupples LA, Divers J, Fornage M, Kao WH, Lange L, Li M, Musani S, Mychaleckyj JC, Ogunniyi A, Papanicolaou G, Rotimi CN, Rotter JI, Ruczinski I, Salako B, Siscovick DS, Tayo BO, Yang Q, McCarroll S, Sabeti P,

Lettre G, De Jager P, Hirschhorn J, Zhu X, Cooper R, Reich D, Wilson JG, Price AL (2011) Genome-wide comparison of African-ancestry populations from care and other cohorts reveals signals of natural selection. Am J Hum Genet 89(3):368–381

Blackwell CC, Dundas S, James VS, Mackenzie DA, Braun JM, Alkout AM, Todd WT, Elton RA, Weir DM (2002) Blood group and susceptibility to disease caused by *Escherichia Coli* O157. J Infect Dis 185(3):393–396

Blumenfeld OO, Huang CH (1995) Molecular genetics of the glycophorin gene family, the antigens for Mnss blood groups: multiple gene rearrangements and modulation of splice site usage result in extensive diversification. Hum Mutat 6(3):199–209

Boren T, Falk P, Roth KA, Larson G, Normark S (1993) Attachment of Helicobacter pylori to human gastric epithelium mediated by blood group antigens. Science 262(5141):1892–1895

Calafell F, Roubinet F, Ramirez-Soriano A, Saitou N, Bertranpetit J, Blancher A (2008) Evolutionary dynamics of the human Abo gene. Hum Genet 124(2):123–135

Cappellini MD, Fiorelli G (2008) Glucose-6-phosphate dehydrogenase deficiency. Lancet 371(9606):64–74

Carlson J, Wahlgren M (1992) *Plasmodium falciparum* erythrocyte rosetting is mediated by promiscuous lectin-like interactions. J Exp Med 176(5):1311–1317

Carter R (2003) Speculations on the origins of *Plasmodium vivax* malaria. Trends Parasitol 19(5):214–219

Carvalho DB, de Mattos LC, Souza-Neiras WC, Bonini-Domingos CR, Cosimo AB, Storti-Melo LM, Cassiano GC, Couto AA, Cordeiro AJ, Rossit AR, Machado RL (2010) Frequency of Abo blood group system polymorphisms in *Plasmodium falciparum* malaria patients and blood donors from the Brazilian Amazon region. Genet Mol Res 9(3):1443–1449

Cavalli-Sforza LL, Feldman MW (2003) The application of molecular genetic approaches to the study of human evolution. Nat Genet 33(Suppl):266–275

Chakraborty S, Craig AG (2004) The role of Icam-1 in *Plasmodium falciparum* cytoadherence. Eur J Cell Biol 84:15–27

Chebloune Y, Pagnier J, Trabuchet G, Faure C, Verdier G, Labie D, Nigon V (1988) Structural analysis of the 5′ flanking region of the beta-globin gene in African sickle cell anemia patients: further evidence for three origins of the sickle cell mutation in Africa. Proc Natl Acad Sci USA 85(12):4431–4435

Chen Q, Schlichtherle M, Wahlgren M (2000) Molecular aspects of severe Malaria. Clin Microbiol Rev 13(3):439–450

Cholera R, Brittain NJ, Gillrie MR, Lopera-Mesa TM, Diakite SA, Arie T, Krause MA, Guindo A, Tubman A, Fujioka H, Diallo DA, Doumbo OK, Ho M, Wellems TE, Fairhurst RM (2008) Impaired cytoadherence of *Plasmodium falciparum* infected erythrocytes containing sickle hemoglobin. Proc Natl Acad Sci USA 105(3):991–996

Chotivanich K, Udomsangpetch R, Pattanapanyasat K, Chierakul W, Simpson J, Looareesuwan S, White N (2002) Hemoglobin E: a balanced polymorphism protective against high parasitemias and thus severe P falciparum malaria. Blood 100(4):1172–1176

Cornejo OE, Escalante AA (2006) The origin and age of *Plasmodium vivax*. Trends Parasitol 22(12):558–563

Corpeleijn E, van der Kallen CJ, Kruijshoop M, Magagnin MG, de Bruin TW, Feskens EJ, Saris WH, Blaak EE (2006) Direct association of a promoter polymorphism in the Cd36/fat fatty acid transporter gene with type 2 diabetes mellitus and insulin resistance. Diabet Med 23(8):907–911

Cortes A, Benet A, Cooke BM, Barnwell JW, Reeder JC (2004) Ability of *Plasmodium falciparum* to invade Southeast Asian ovalocytes varies between parasite lines. Blood 104(9):2961–2966

Cortes A, Mellombo M, Mgone CS, Beck HP, Reeder JC, Cooke BM (2005) Adhesion of *Plasmodium falciparum* infected red blood cells to Cd36 under flow is enhanced by the cerebral malaria-protective trait South-East Asian ovalocytosis. Mol Biochem Parasitol 142(2): 252–257

Cowman AF, Crabb BS (2006) Invasion of red blood cells by malaria parasites. Cell 124(4):755–766

Craig A, Scherf A (2001) Molecules on the surface of the *Plasmodium falciparum* infected erythrocyte and their role in malaria pathogenesis and immune evasion. Mol Biochem Parasitol 115(2):129–143

Crosnier C, Bustamante LY, Bartholdson SJ, Bei AK, Theron M, Uchikawa M, Mboup S, Ndir O, Kwiatkowski DP, Duraisingh MT, Rayner JC, Wright GJ (2011) Basigin is a receptor essential for erythrocyte invasion by *Plasmodium falciparum*. Nature 480(7378):534–537

Cserti CM, Dzik WH (2007) The Abo blood group system and Plasmodium falciparum malaria. Blood 110(7):2250–2258

Cserti-Gazdewich CM, Mayr WR, Dzik WH (2010) *Plasmodium falciparum* malaria and the immunogenetics of Abo, Hla, and Cd36 (Platelet Glycoprotein Iv). Vox Sang 100(1):99–111

Currat M, Trabuchet G, Rees D, Perrin P, Harding RM, Clegg JB, Langaney A, Excoffier L (2002) Molecular analysis of the beta-globin gene cluster in the Niokholo Mandenka population reveals a recent origin of the beta(S) senegal mutation. Am J Hum Genet 70(1):207–223

Daniels G (2005) The molecular genetics of blood group polymorphism. Transpl Immunol 14(3–4):143–153

de Villartay JP, Rouger P, Muller JY, Salmon C (1985) Hla antigens on peripheral red blood cells: analysis by flow cytofluorometry using monoclonal antibodies. Tissue Antigens 26(1):12–19

Dhermy D, Schrevel J, Lecomte MC (2007) Spectrin-based skeleton in red blood cells and malaria. Curr Opin Hematol 14(3):198–202

Diallo DA, Doumbo OK, Dicko A, Guindo A, Coulibaly D, Kayentao K, Djimde AA, Thera MA, Fairhurst RM, Plowe CV, Wellems TE (2004) A comparison of anemia in hemoglobin C and normal hemoglobin a children with *Plasmodium falciparum* malaria. Acta Trop 90(3):295–299

Dolo A, Modiano D, Maiga B, Daou M, Dolo G, Guindo H, Ba M, Maiga H, Coulibaly D, Perlman H, Blomberg MT, Toure YT, Coluzzi M, Doumbo O (2005) Difference in susceptibility to malaria between two sympatric ethnic groups in Mali. Am J Trop Med Hyg 72(3):243–248

Doolan DL, Dobano C, Baird JK (2009) Acquired immunity to malaria. Clin Microbiol Rev 22(1):13–36, Table of Contents

Duah NO, Weiss HA, Jepson A, Tetteh KK, Whittle HC, Conway DJ (2009) Heritability of antibody isotype and subclass responses to *Plasmodium falciparum* antigens. PLoS One 4(10):e7381

Durand PM, Coetzer TL (2008) Pyruvate kinase deficiency protects against malaria in humans. Haematologica 93(6):939–940

Escalante AA, Cornejo OE, Freeland DE, Poe AC, Durrego E, Collins WE, Lal AA (2005) A Monkey's tale: the origin of *Plasmodium vivax* as a human malaria parasite. Proc Natl Acad Sci USA 102(6):1980–1985

Fairhurst RM, Fujioka H, Hayton K, Collins KF, Wellems TE (2003) Aberrant development of *Plasmodium falciparum* in hemoglobin Cc red cells: implications for the malaria protective effect of the homozygous state. Blood 101(8):3309–3315

Fairhurst RM, Baruch DI, Brittain NJ, Ostera GR, Wallach JS, Hoang HL, Hayton K, Guindo A, Makobongo MO, Schwartz OM, Tounkara A, Doumbo OK, Diallo DA, Fujioka H, Ho M, Wellems TE (2005) Abnormal display of Pfemp-1 on erythrocytes carrying haemoglobin C may protect against malaria. Nature 435(7045):1117–1121

Fernandez-Reyes D, Craig AG, Kyes SA, Peshu N, Snow RW, Berendt AR, Marsh K, Newbold CI (1997) A high frequency African coding polymorphism in the N-terminal domain of Icam-1 predisposing to cerebral malaria in Kenya. Hum Mol Genet 6(8):1357–1360

Ferreira A, Marguti I, Bechmann I, Jeney V, Chora A, Palha NR, Rebelo S, Henri A, Beuzard Y, Soares MP (2011) Sickle hemoglobin confers tolerance to *Plasmodium* infection. Cell 145(3):398–409

Flori L, Kumulungui B, Aucan C, Esnault C, Traore AS, Fumoux F, Rihet P (2003) Linkage and association between *Plasmodium falciparum* blood infection levels and chromosome 5q31-Q33. Genes Immun 4(4):265–268

Flori L, Delahaye NF, Iraqi FA, Hernandez-Valladares M, Fumoux F, Rihet P (2005) Tnf as a malaria candidate gene: polymorphism-screening and family-based association analysis of mild malaria attack and parasitemia in Burkina Faso. Genes Immun 6(6):472–480

Fry AE, Auburn S, Diakite M, Green A, Richardson A, Wilson J, Jallow M, Sisay-Joof F, Pinder M, Griffiths MJ, Peshu N, Williams TN, Marsh K, Molyneux ME, Taylor TE, Rockett KA, Kwiatkowski DP (2008a) Variation in the Icam1 gene is not associated with severe malaria phenotypes. Genes Immun 9(5):462–469
Fry AE, Griffiths MJ, Auburn S, Diakite M, Forton JT, Green A, Richardson A, Wilson J, Jallow M, Sisay-Joof F, Pinder M, Peshu N, Williams TN, Marsh K, Molyneux ME, Taylor TE, Rockett KA, Kwiatkowski DP (2008b) Common variation in the Abo glycosyltransferase is associated with susceptibility to severe *Plasmodium falciparum* malaria. Hum Mol Genet 17(4):567–576
Fry AE, Ghansa A, Small KS, Palma A, Auburn S, Diakite M, Green A, Campino S, Teo YY, Clark TG, Jeffreys AE, Wilson J, Jallow M, Sisay-Joof F, Pinder M, Griffiths MJ, Peshu N, Williams TN, Newton CR, Marsh K, Molyneux ME, Taylor TE, Koram KA, Oduro AR, Rogers WO, Rockett KA, Sabeti PC, Kwiatkowski DP (2009) Positive selection of a Cd36 nonsense variant in sub-Saharan Africa, but no association with severe malaria phenotypes. Hum Mol Genet 18(14):2683–2692
Gagneux P, Varki A (1999) Evolutionary considerations in relating oligosaccharide diversity to biological function. Glycobiology 9(8):747–755
Gallagher PG, Forget BG (1996) Hematologically important mutations: spectrin variants in hereditary elliptocytosis and hereditary pyropoikilocytosis. Blood Cells Mol Dis 22(3):254–258
Garcia A, Marquet S, Bucheton B, Hillaire D, Cot M, Fievet N, Dessein AJ, Abel L (1998) Linkage analysis of blood *Plasmodium falciparum* levels: interest of the 5q31-Q33 chromosome region. Am J Trop Med Hyg 58(6):705–709
Garnham PC (1966) Malaria parasites and other haemosporidia. Blackwell Scientific, Oxford
Garrigan D, Hedrick PW (2003) Perspective: detecting adaptive molecular polymorphism: lessons from the Mhc. Evolution 57(8):1707–1722
Gelhaus A, Scheding A, Browne E, Burchard GD, Horstmann RD (2001) Variability of the Cd36 gene in West Africa. Hum Mutat 18(5):444–450
Genton B, al-Yaman F, Mgone CS, Alexander N, Paniu MM, Alpers MP, Mokela D (1995) Ovalocytosis and cerebral malaria. Nature 378(6557):564–565
Ghosh K (2008) Evolution and selection of human leukocyte antigen alleles by *Plasmodium falciparum* infection. Hum Immunol 69(12):856–860
Glushakova S, Humphrey G, Leikina E, Balaban A, Miller J, Zimmerberg J (2010) New stages in the program of malaria parasite egress imaged in normal and sickle erythrocytes. Curr Biol 20(12):1117–1121
Gomez F (2012) Genetic variation at Icam-1 and Cd36: a study of malaria resistance candidate loci in diverse global human populations. PhD Thesis The George Washington University
Gomez F, Tomas G, Ko WY, Ranciaro A, Froment A, Ibrahim M, Lema G, Nyambo TB, Omar SA, Wambebe C, Hirbo JB, Rocha J, Tishkoff SA (2013) Patterns of nucleotide and haplotype diversity at Icam-1 across global human populations with varying levels of malaria exposure. Hum Genet. Epub date: 2013/04/24
Guerra CA, Gikandi PW, Tatem AJ, Noor AM, Smith DL, Hay SI, Snow RW (2008) The limits and intensity of *Plasmodium falciparum* transmission: implications for malaria control and elimination worldwide. PLoS Med 5(2):e38
Guindo A, Fairhurst RM, Doumbo OK, Wellems TE, Diallo DA (2007) X-linked G6pd deficiency protects hemizygous males but not heterozygous females against severe malaria. PLoS Med 4(3):e66
Gyan BA, Goka B, Cvetkovic JT, Kurtzhals JL, Adabayeri V, Perlmann H, Lefvert AK, Akanmori BD, Troye-Blomberg M (2004) Allelic polymorphisms in the repeat and promoter regions of the interleukin-4 gene and malaria severity in Ghanaian children. Clin Exp Immunol 138(1):145–150
Haldane JBS (1949) The rate of mutation in human genes. Proc VIII Int Cong Genet Hereditas 35:267–273
Hamblin MT, Di Rienzo A (2000) Detection of the signature of natural selection in humans: evidence from the Duffy blood group locus. Am J Hum Genet 66(5):1669–1679

Hamblin MT, Thompson EE, Di Rienzo A (2002) Complex signatures of natural selection at the Duffy blood group locus. Am J Hum Genet 70(2):369–383
Hanchard N, Elzein A, Trafford C, Rockett K, Pinder M, Jallow M, Harding R, Kwiatkowski D, McKenzie C (2007) Classical sickle beta-globin haplotypes exhibit a high degree of long-range haplotype similarity in African and Afro-caribbean populations. BMC Genet 8:52
Hay SI, Guerra CA, Gething PW, Patil AP, Tatem AJ, Noor AM, Kabaria CW, Manh BH, Elyazar IR, Brooker S, Smith DL, Moyeed RA, Snow RW (2009) A world malaria map: *Plasmodium falciparum* endemicity in 2007. PLoS Med 6(3):e1000048
Hedrick PW (2011) Population genetics of malaria resistance in humans. Heredity (Edinb) 107(4):283–304
Henn BM, Gignoux CR, Jobin M, Granka JM, Macpherson JM, Kidd JM, Rodriguez-Botigue L, Ramachandran S, Hon L, Brisbin A, Lin AA, Underhill PA, Comas D, Kidd KK, Norman PJ, Parham P, Bustamante CD, Mountain JL, Feldman MW (2011) Hunter-gatherer genomic diversity suggests a Southern African origin for modern humans. Proc Natl Acad Sci USA 108(13): 5154–5162
Henn BM, Botigue LR, Gravel S, Wang W, Brisbin A, Byrnes JK, Fadhlaoui-Zid K, Zalloua PA, Moreno-Estrada A, Bertranpetit J, Bustamante CD, Comas D (2012) Genomic ancestry of North Africans supports back-to-Africa migrations. PLoS Genet 8(1):e1002397
Hill AV, Allsopp CE, Kwiatkowski D, Anstey NM, Twumasi P, Rowe PA, Bennett S, Brewster D, McMichael AJ, Greenwood BM (1991) Common West African Hla antigens are associated with protection from severe malaria. Nature 352(6336):595–600
Hill AV, Elvin J, Willis AC, Aidoo M, Allsopp CE, Gotch FM, Gao XM, Takiguchi M, Greenwood BM, Townsend AR et al (1992) Molecular analysis of the association of Hla-B53 and resistance to severe malaria. Nature 360(6403):434–439
Howes RE, Patil AP, Piel FB, Nyangiri OA, Kabaria CW, Gething PW, Zimmerman PA, Barnadas C, Beall CM, Gebremedhin A, Menard D, Williams TN, Weatherall DJ, Hay SI (2011) The global distribution of the Duffy blood group. Nat Commun 2:266
Hutagalung R, Wilairatana P, Looareesuwan S, Brittenham GM, Aikawa M, Gordeuk VR (1999) Influence of hemoglobin E trait on the severity of falciparum malaria. J Infect Dis 179(1): 283–286
Imai M, Tanaka T, Kintaka T, Ikemoto T, Shimizu A, Kitaura Y (2002) Genomic heterogeneity of type Ii Cd36 deficiency. Clin Chim Acta 321(1–2):97–106
Israelsson E, Maiga B, Kearsley S, Dolo A, Homann MV, Doumbo OK, Troye-Blomberg M, Tornvall P, Berzins K (2011) Cytokine gene haplotypes with a potential effect on susceptibility to malaria in sympatric ethnic groups in Mali. Infect Genet Evol 11(7):1608–1615
Jallow M, Teo YY, Small KS, Rockett KA, Deloukas P, Clark TG, Kivinen K, Bojang KA, Conway DJ, Pinder M, Sirugo G, Sisay-Joof F, Usen S, Auburn S, Bumpstead SJ, Campino S, Coffey A, Dunham A, Fry AE, Green A, Gwilliam R, Hunt SE, Inouye M, Jeffreys AE, Mendy A, Palotie A, Potter S, Ragoussis J, Rogers J, Rowlands K, Somaskantharajah E, Whittaker P, Widden C, Donnelly P, Howie B, Marchini J, Morris A, SanJoaquin M, Achidi EA, Agbenyega T, Allen A, Amodu O, Corran P, Djimde A, Dolo A, Doumbo OK, Drakeley C, Dunstan S, Evans J, Farrar J, Fernando D, Hien TT, Horstmann RD, Ibrahim M, Karunaweera N, Kokwaro G, Koram KA, Lemnge M, Makani J, Marsh K, Michon P, Modiano D, Molyneux ME, Mueller I, Parker M, Peshu N, Plowe CV, Puijalon O, Reeder J, Reyburn H, Riley EM, Sakuntabhai A, Singhasivanon P, Sirima S, Tall A, Taylor TE, Thera M, Troye-Blomberg M, Williams TN, Wilson M, Kwiatkowski DP (2009) Genome-wide and fine-resolution association analysis of malaria in West Africa. Nat Genet 41(6):657–665
Jepson A, Sisay-Joof F, Banya W, Hassan-King M, Frodsham A, Bennett S, Hill AV, Whittle H (1997) Genetic linkage of mild malaria to the major histocompatibility complex in Gambian children: study of affected sibling pairs. BMJ 315(7100):96–97
Johnson MK, Clark TD, Njama-Meya D, Rosenthal PJ, Parikh S (2009) Impact of the method of G6pd deficiency assessment on genetic association studies of malaria susceptibility. PLoS One 4(9):e7246

Kashiwagi H, Tomiyama Y, Nozaki S, Kiyoi T, Tadokoro S, Matsumoto K, Honda S, Kosugi S, Kurata Y, Matsuzawa Y (2001) Analyses of genetic abnormalities in type I Cd36 deficiency in Japan: identification and cell biological characterization of two novel mutations that cause Cd36 deficiency in man. Hum Genet 108(6):459–466
Kassim OO, Ejezie GC (1982) Abo blood groups in malaria and Schistosomiasis haematobium. Acta Trop 39(2):179–184
Knight JC, Udalova I, Hill AV, Greenwood BM, Peshu N, Marsh K, Kwiatkowski D (1999) A polymorphism that affects Oct-1 binding to the Tnf promoter region is associated with severe malaria. Nat Genet 22(2):145–150
Ko WY, Kaercher KA, Giombini E, Marcatili P, Froment A, Ibrahim M, Lema G, Nyambo TB, Omar SA, Wambebe C, Ranciaro A, Hirbo JB, Tishkoff SA (2011) Effects of natural selection and gene conversion on the evolution of human glycophorins coding for Mns blood polymorphisms in malaria-endemic African populations. Am J Hum Genet 88(6):741–754
Kun JF, Klabunde J, Lell B, Luckner D, Alpers M, May J, Meyer C, Kremsner PG (1999) Association of the Icam-1kilifi mutation with protection against severe malaria in Lambarene, Gabon. Am J Trop Med Hyg 61(5):776–779
Kwiatkowski DP (2005) How malaria has affected the human genome and what human genetics can teach us about malaria. Am J Hum Genet 77(2):171–192
Lachance J, Vernot B, Elbers CC, Ferwerda B, Froment A, Bodo JM, Lema G, Fu W, Nyambo TB, Rebbeck TR, Zhang K, Akey JM, Tishkoff SA (2012) Evolutionary history and adaptation from high-coverage whole-genome sequences of diverse African hunter-gatherers. Cell 150(3):457–469
Langhorne J, Ndungu FM, Sponaas AM, Marsh K (2008) Immunity to malaria: more questions than answers. Nat Immunol 9(7):725–732
Lindesmith L, Moe C, Marionneau S, Ruvoen N, Jiang X, Lindblad L, Stewart P, LePendu J, Baric R (2003) Human susceptibility and resistance to Norwalk virus infection. Nat Med 9(5):548–553
Liu SC, Zhai S, Palek J, Golan DE, Amato D, Hassan K, Nurse GT, Babona D, Coetzer T, Jarolim P et al (1990) Molecular defect of the Band 3 protein in Southeast Asian ovalocytosis. N Engl J Med 323(22):1530–1538
Liu L, Muralidhar S, Singh M, Sylvan C, Kalra IS, Quinn CT, Onyekwere OC, Pace BS (2009) High-density Snp genotyping to define beta-globin locus haplotypes. Blood Cells Mol Dis 42(1):16–24
Livingstone F (1958) Anthropological implications of sickle cell gene distribution in West Africa. Am Anthropol 60(3):533–562
Livingstone F (1971) Malaria and human polymorphisms. Annu Rev Genet 5:33–64
Livingstone FB (1984) The Duffy blood groups, vivax malaria, and malaria selection in human populations: a review. Hum Biol 56(3):413–425
Luoni G, Verra F, Arca B, Sirima BS, Troye-Blomberg M, Coluzzi M, Kwiatkowski D, Modiano D (2001) Antimalarial antibody levels and Il4 polymorphism in the Fulani of West Africa. Genes Immun 2(7):411–414
Ma X, Bacci S, Mlynarski W, Gottardo L, Soccio T, Menzaghi C, Iori E, Lager RA, Shroff AR, Gervino EV, Nesto RW, Johnstone MT, Abumrad NA, Avogaro A, Trischitta V, Doria A (2004) A common haplotype at the Cd36 locus is associated with high free fatty acid levels and increased cardiovascular risk in caucasians. Hum Mol Genet 13(19):2197–2205
Ma J, Mollsten A, Prazny M, Falhammar H, Brismar K, Dahlquist G, Efendic S, Gu HF (2006) Genetic influences of the intercellular adhesion molecule 1 (Icam-1) gene polymorphisms in development of type 1 diabetes and diabetic nephropathy. Diabet Med 23(10):1093–1099
Machado P, Pereira R, Rocha AM, Manco L, Fernandes N, Miranda J, Ribeiro L, Do Rosario VE, Amorim A, Gusmao L, Arez AP (2010) Malaria: looking for selection signatures in the human Pklr gene region. Br J Haematol 149(5):775–784
Mackinnon MJ, Mwangi TW, Snow RW, Marsh K, Williams TN (2005) Heritability of malaria in Africa. PLoS Med 2(12):e340

Marquet S, Doumbo O, Cabantous S, Poudiougou B, Argiro L, Safeukui I, Konate S, Sissoko S, Chevereau E, Traore A, Keita MM, Chevillard C, Abel L, Dessein AJ (2008) A functional promoter variant in Il12b predisposes to cerebral malaria. Hum Mol Genet 17(14):2190–2195

Martin SK, Miller LH, Hicks CU, David-West A, Ugbode C, Deane M (1979) Frequency of blood group antigens in Nigerian children with falciparum malaria. Trans R Soc Trop Med Hyg 73(2):216–218

Mateu E, Perez-Lezaun A, Martinez-Arias R, Andres A, Valles M, Bertranpetit J, Calafell F (2002) Pklr- Gba region shows almost complete linkage disequilibrium over 70 kb in a set of worldwide populations. Hum Genet 110(6):532–544

Mayer DC, Cofie J, Jiang L, Hartl DL, Tracy E, Kabat J, Mendoza LH, Miller LH (2009) Glycophorin B Is the erythrocyte receptor of *Plasmodium falciparum* erythrocyte-binding ligand, Ebl-1. Proc Natl Acad Sci USA 106(13):5348–5352

McGilvray ID, Serghides L, Kapus A, Rotstein OD, Kain KC (2000) Nonopsonic monocyte/macrophage phagocytosis of *Plasmodium falciparum* parasitized erythrocytes: a role for Cd36 in malarial clearance. Blood 96(9):3231–3240

McGuire W, Hill AV, Allsopp CE, Greenwood BM, Kwiatkowski D (1994) Variation in the Tnf-alpha promoter region associated with susceptibility to cerebral malaria. Nature 371(6497): 508–510

Mercereau-Puijalon O, Menard D (2010) *Plasmodium vivax* and the Duffy antigen: a paradigm revisited. Transfus Clin Biol 17(3):176–183

Meyer CG, May J, Luty AJ, Lell B, Kremsner PG (2002) Tnfalpha-308a associated with shorter intervals of *Plasmodium falciparum* reinfections. Tissue Antigens 59(4):287–292

Milet J, Nuel G, Watier L, Courtin D, Slaoui Y, Senghor P, Migot-Nabias F, Gaye O, Garcia A (2010) Genome wide linkage study, using a 250k Snp map, of *Plasmodium falciparum* infection and mild malaria attack in a senegalese population. PLoS One 5(7):e11616

Miller LH, Mason SJ, Clyde DF, McGinniss MH (1976) The resistance factor to *Plasmodium vivax* in blacks. The Duffy-blood-group genotype, Fyfy. N Engl J Med 295(6):302–304

Min-Oo G, Gros P (2005) Erythrocyte variants and the nature of their malaria protective effect. Cell Microbiol 7(6):753–763

Min-Oo G, Fortin A, Tam MF, Nantel A, Stevenson MM, Gros P (2003) Pyruvate kinase deficiency in mice protects against malaria. Nat Genet 35(4):357–362

Min-Oo G, Fortin A, Tam MF, Gros P, Stevenson MM (2004) Phenotypic expression of pyruvate kinase deficiency and protection against malaria in a mouse model. Genes Immun 5(3):168–175

Mockenhaupt FP, Ehrhardt S, Cramer JP, Otchwemah RN, Anemana SD, Goltz K, Mylius F, Dietz E, Eggelte TA, Bienzle U (2004) Hemoglobin C and resistance to severe malaria in Ghanaian children. J Infect Dis 190(5):1006–1009

Modiano D, Petrarca V, Sirima BS, Bosman A, Nebie I, Diallo D, Lamizana L, Esposito F, Coluzzi M (1995) *Plasmodium falciparum* malaria in sympatric ethnic groups of Burkina Faso, West Africa. Parassitologia 37(2–3):255–259

Modiano D, Petrarca V, Sirima BS, Nebie I, Diallo D, Esposito F, Coluzzi M (1996) Different response to *Plasmodium falciparum* malaria in West African sympatric ethnic groups. Proc Natl Acad Sci USA 93(23):13206–13211

Modiano D, Luoni G, Sirima BS, Lanfrancotti A, Petrarca V, Cruciani F, Simpore J, Ciminelli BM, Foglietta E, Grisanti P, Bianco I, Modiano G, Coluzzi M (2001a) The lower susceptibility to *Plasmodium falciparum* malaria of Fulani of Burkina Faso (West Africa) is associated with low frequencies of classic malaria-resistance genes. Trans R Soc Trop Med Hyg 95(2):149–152

Modiano D, Luoni G, Sirima BS, Simpore J, Verra F, Konate A, Rastrelli E, Olivieri A, Calissano C, Paganotti GM, D'Urbano L, Sanou I, Sawadogo A, Modiano G, Coluzzi M (2001b) Haemoglobin C protects against clinical *Plasmodium falciparum* malaria. Nature 414(6861): 305–308

Modiano D, Bancone G, Ciminelli BM, Pompei F, Blot I, Simpore J, Modiano G (2008) Haemoglobin S and Haemoglobin C: 'Quick but Costly' Versus 'Slow but Gratis' genetic adaptations to *Plasmodium falciparum* malaria. Hum Mol Genet 17(6):789–799

Morahan G, Boutlis CS, Huang D, Pain A, Saunders JR, Hobbs MR, Granger DL, Weinberg JB, Peshu N, Mwaikambo ED, Marsh K, Roberts DJ, Anstey NM (2002) A promoter polymorphism in the gene encoding interleukin-12 P40 (Il12b) is associated with mortality from cerebral malaria and with reduced nitric oxide production. Genes Immun 3(7):414–418

Mu J, Joy DA, Duan J, Huang Y, Carlton J, Walker J, Barnwell J, Beerli P, Charleston MA, Pybus OG, Su XZ (2005) Host switch leads to emergence of *Plasmodium vivax* malaria in humans. Mol Biol Evol 22(8):1686–1693

Nagel RL, Roth EF Jr (1989) Malaria and red cell genetic defects. Blood 74(4):1213–1221

Nagel RL, Fabry ME, Pagnier J, Zohoun I, Wajcman H, Baudin V, Labie D (1985) Hematologically and genetically distinct forms of sickle cell anemia in Africa. The Senegal type and the Benin type. N Engl J Med 312(14):880–884

Ndiaye R, Sakuntabhai A, Casademont I, Rogier C, Tall A, Trape JF, Spiegel A, Dieye A, Julier C (2005) Genetic study of Icam1 in clinical malaria in Senegal. Tissue Antigens 65(5):474–480

Newbold C, Warn P, Black G, Berendt A, Craig A, Snow B, Msobo M, Peshu N, Marsh K (1997) Receptor-specific adhesion and clinical disease in *Plasmodium falciparum.* Am J Trop Med Hyg 57(4):389–398

Nkhoma S, Nair S, Mukaka M, Molyneux ME, Ward SA, Anderson TJ (2009) Parasites bearing a single copy of the multi-drug resistance gene (Pfmdr-1) with wild-type Snps predominate amongst *Plasmodium falciparum* isolates from Malawi. Acta Trop 111(1):78–81

Ohashi J, Naka I, Patarapotikul J, Hananantachai H, Looareesuwan S, Tokunaga K (2001) Absence of association between the allele coding methionine at position 29 in the N-terminal domain of Icam-1 (Icam-1(Kilifi)) and severe malaria in the Northwest of Thailand. Jpn J Infect Dis 54(3):114–116

Ohashi J, Naka I, Patarapotikul J, Hananantachai H, Brittenham G, Looareesuwan S, Clark AG, Tokunaga K (2004) Extended linkage disequilibrium surrounding the hemoglobin E variant due to malarial selection. Am J Hum Genet 74(6):1198–1208

Olsson ML, Chester MA (1996) Frequent occurrence of a variant O1 gene at the blood group Abo locus. Vox Sang 70(1):26–30

Orkin SH, Kazazian HH Jr, Antonarakis SE, Goff SC, Boehm CD, Sexton JP, Waber PG, Giardina PJ (1982) Linkage of beta-thalassaemia mutations and beta-globin gene polymorphisms with DNA polymorphisms in human beta-globin gene cluster. Nature 296(5858):627–631

Ouma C, Davenport GC, Were T, Otieno MF, Hittner JB, Vulule JM, Martinson J, Ong'echa JM, Ferrell RE, Perkins DJ (2008) Haplotypes of Il-10 promoter variants are associated with susceptibility to severe malarial anemia and functional changes in Il-10 production. Hum Genet 124(5):515–524

Pagnier J, Mears JG, Dunda-Belkhodja O, Schaefer-Rego KE, Beldjord C, Nagel RL, Labie D (1984) Evidence for the multicentric origin of the sickle cell hemoglobin gene in Africa. Proc Natl Acad Sci USA 81(6):1771–1773

Pain A, Urban BC, Kai O, Casals-Pascual C, Shafi J, Marsh K, Roberts DJ (2001) A non-sense mutation in Cd36 gene is associated with protection from severe malaria. Lancet 357(9267): 1502–1503

Palek J (1987) Hereditary elliptocytosis, spherocytosis and related disorders: consequences of a deficiency or a mutation of membrane skeletal proteins. Blood Rev 1(3):147–168

Pamplona A, Ferreira A, Balla J, Jeney V, Balla G, Epiphanio S, Chora A, Rodrigues CD, Gregoire IP, Cunha-Rodrigues M, Portugal S, Soares MP, Mota MM (2007) Heme oxygenase-1 and carbon monoxide suppress the pathogenesis of experimental cerebral malaria. Nat Med 13(6): 703–710

Pasloske BL, Howard RJ (1994) Malaria, the red cell, and the endothelium. Annu Rev Med 45:283–295

Pasvol G, Wainscoat JS, Weatherall DJ (1982) Erythrocytes deficiency in glycophorin resist invasion by the malarial parasite *Plasmodium falciparum.* Nature 297(5861):64–66

Patel SS, King CL, Mgone CS, Kazura JW, Zimmerman PA (2004) Glycophorin C (Gerbich antigen blood group) and band 3 polymorphisms in two malaria holoendemic regions of Papua New Guinea. Am J Hematol 75(1):1–5

Patel SN, Lu Z, Ayi K, Serghides L, Gowda DC, Kain KC (2007) Disruption of Cd36 impairs cytokine response to *Plasmodium falciparum* glycosylphosphatidylinositol and confers susceptibility to severe and fatal malaria in vivo. J Immunol 178(6):3954–3961

Phimpraphi W, Paul R, Witoonpanich B, Turbpaiboon C, Peerapittayamongkol C, Louicharoen C, Casademont I, Tungpradabkul S, Krudsood S, Kaewkunwal J, Sura T, Looareesuwan S, Singhasivanon P, Sakuntabhai A (2008) Heritability of *P. falciparum* and *P. vivax* malaria in a karen population in Thailand. PLoS One 3(12):e3887

Piazza A, Mayr WR, Contu L, Amoroso A, Borelli I, Curtoni ES, Marcello C, Moroni A, Olivetti E, Richiardi P et al (1985) Genetic and population structure of four Sardinian villages. Ann Hum Genet 49(Pt 1):47–63

Potts WK, Wakeland EK (1993) Evolution of Mhc genetic diversity: a tale of incest, pestilence and sexual preference. Trends Genet 9(12):408–412

Prugnolle F, Manica A, Charpentier M, Guegan JF, Guernier V, Balloux F (2005) Pathogen-driven selection and worldwide Hla class I diversity. Curr Biol 15(11):1022–1027

Ptak SE, Przeworski M (2002) Evidence for population growth in humans is confounded by fine-scale population structure. Trends Genet 18(11):559–563

Puthothu B, Krueger M, Bernhardt M, Heinzmann A (2006) Icam1 amino-acid variant K469e is associated with paediatric bronchial asthma and elevated Sicam1 levels. Genes Immun 7(4):322–326

Rac ME, Safranow K, Poncyljusz W (2007) Molecular basis of human Cd36 gene mutations. Mol Med 13(5–6):288–296

Rees DC, Styles L, Vichinsky EP, Clegg JB, Weatherall DJ (1998) The hemoglobin E syndromes. Ann NY Acad Sci 850:334–343

Register TC, Burdon KP, Lenchik L, Bowden DW, Hawkins GA, Nicklas BJ, Lohman K, Hsu FC, Langefeld CD, Carr JJ (2004) Variability of serum soluble intercellular adhesion molecule-1 measurements attributable to a common polymorphism. Clin Chem 50(11):2185–2187

Rihet P, Traore Y, Abel L, Aucan C, Traore-Leroux T, Fumoux F (1998) Malaria in humans: *Plasmodium falciparum* blood infection levels are linked to chromosome 5q31-Q33. Am J Hum Genet 63(2):498–505

Roberts LS, Janovy J (2005) Foundations of parasitology. McGraw Hill Higher Education, Boston

Roberts DJ, Williams TN (2003) Haemoglobinopathies and resistance to malaria. Redox Rep 8(5):304–310

Roubinet F, Despiau S, Calafell F, Jin F, Bertranpetit J, Saitou N, Blancher A (2004) Evolution of the O alleles of the human Abo blood group gene. Transfusion 44(5):707–715

Roux AF, Morle F, Guetarni D, Colonna P, Sahr K, Forget BG, Delaunay J, Godet J (1989) Molecular basis of Sp Alpha I/65 hereditary elliptocytosis in North Africa: insertion of a Ttg triplet between codons 147 and 149 in the alpha-spectrin gene from five unrelated families. Blood 73(8):2196–2201

Rowe JA, Handel IG, Thera MA, Deans AM, Lyke KE, Kone A, Diallo DA, Raza A, Kai O, Marsh K, Plowe CV, Doumbo OK, Moulds JM (2007) Blood group O protects against severe *Plasmodium falciparum* malaria through the mechanism of reduced rosetting. Proc Natl Acad Sci USA 104(44):17471–17476

Rowe JA, Claessens A, Corrigan RA, Arman M (2009) Adhesion of *Plasmodium falciparum*-infected erythrocytes to human cells: molecular mechanisms and therapeutic implications. Expert Rev Mol Med 11:e16

Ruiz-Palacios GM, Cervantes LE, Ramos P, Chavez-Munguia B, Newburg DS (2003) *Campylobacter jejuni* binds intestinal H(O) antigen (Fuc Alpha 1, 2gal Beta 1, 4glcnac), and fucosyloligosaccharides of human milk inhibit its binding and infection. J Biol Chem 278(16):14112–14120

Ruwende C, Hill A (1998) Glucose-6-phosphate dehydrogenase deficiency and malaria. J Mol Med (Berl) 76(8):581–588

Ruwende C, Khoo SC, Snow RW, Yates SN, Kwiatkowski D, Gupta S, Warn P, Allsopp CE, Gilbert SC, Peschu N et al (1995) Natural selection of hemi- and heterozygotes for G6pd deficiency in Africa by resistance to severe malaria. Nature 376(6537):246–249

Ryan AW, Mapp J, Moyna S, Mattiangeli V, Kelleher D, Bradley DG, McManus R (2006) Levels of interpopulation differentiation among different functional classes of immunologically important genes. Genes Immun 7(2):179–183

Sabeti PC, Reich DE, Higgins JM, Levine HZ, Richter DJ, Schaffner SF, Gabriel SB, Platko JV, Patterson NJ, McDonald GJ, Ackerman HC, Campbell SJ, Altshuler D, Cooper R, Kwiatkowski D, Ward R, Lander ES (2002) Detecting recent positive selection in the human genome from haplotype structure. Nature 419(6909):832–837

Sabeti PC, Schaffner SF, Fry B, Lohmueller J, Varilly P, Shamovsky O, Palma A, Mikkelsen TS, Altshuler D, Lander ES (2006) Positive natural selection in the human lineage. Science 312(5780):1614–1620

Saitou N, Yamamoto F (1997) Evolution of primate Abo blood group genes and their homologous genes. Mol Biol Evol 14(4):399–411

Sakuntabhai A, Ndiaye R, Casademont I, Peerapittayamongkol C, Rogier C, Tortevoye P, Tall A, Paul R, Turbpaiboon C, Phimpraphi W, Trape JF, Spiegel A, Heath S, Mercereau-Puijalon O, Dieye A, Julier C (2008) Genetic determination and linkage mapping of *Plasmodium falciparum* malaria related traits in Senegal. PLoS One 3(4):e2000

Saunders MA, Hammer MF, Nachman MW (2002) Nucleotide variability at G6pd and the signature of malarial selection in humans. Genetics 162(4):1849–1861

Seixas S, Ferrand N, Rocha J (2002) Microsatellite variation and evolution of the human Duffy blood group polymorphism. Mol Biol Evol 19(10):1802–1806

Senga E, Loscertales MP, Makwakwa KE, Liomba GN, Dzamalala C, Kazembe PN, Brabin BJ (2007) Abo blood group phenotypes influence parity specific immunity to *Plasmodium falciparum* malaria in Malawian women. Malar J 6:102

Seto NO, Palcic MM, Compston CA, Li H, Bundle DR, Narang SA (1997) Sequential interchange of four amino acids from blood group B to blood group a glycosyltransferase boosts catalytic activity and progressively modifies substrate recognition in human recombinant enzymes. J Biol Chem 272(22):14133–14138

Shear HL, Roth EF Jr, Fabry ME, Costantini FD, Pachnis A, Hood A, Nagel RL (1993) Transgenic mice expressing human sickle hemoglobin are partially resistant to rodent malaria. Blood 81(1):222–226

Sherman IW, Eda S, Winograd E (2003) Cytoadherence and sequestration in *Plasmodium falciparum*: defining the ties that bind. Microbes Infect 5(10):897–909

Sim BK, Chitnis CE, Wasniowska K, Hadley TJ, Miller LH (1994) Receptor and ligand domains for invasion of erythrocytes by *Plasmodium falciparum*. Science 264(5167):1941–1944

Simonsen KL, Churchill GA, Aquadro CF (1995) Properties of statistical tests of neutrality for DNA polymorphism data. Genetics 141(1):413–429

Sjoberg K, Lepers JP, Raharimalala L, Larsson A, Olerup O, Marbiah NT, Troye-Blomberg M, Perlmann P (1992) Genetic regulation of human anti-malarial antibodies in twins. Proc Natl Acad Sci USA 89(6):2101–2104

Smith AJ, Humphries SE (2009) Cytokine and cytokine receptor gene polymorphisms and their functionality. Cytokine Growth Factor Rev 20(1):43–59

Stirnadel HA, Beck HP, Alpers MP, Smith TA (1999) Heritability and segregation analysis of immune responses to specific malaria antigens in Papua New Guinea. Genet Epidemiol 17(1):16–34

Stirnadel HA, Al-Yaman F, Genton B, Alpers MP, Smith TA (2000) Assessment of different sources of variation in the antibody responses to specific malaria antigens in children in Papua New Guinea. Int J Epidemiol 29(3):579–586

Tanaka T, Nakata T, Oka T, Ogawa T, Okamoto F, Kusaka Y, Sohmiya K, Shimamoto K, Itakura K (2001) Defect in human myocardial long-chain fatty acid uptake is caused by Fat/Cd36 mutations. J Lipid Res 42(5):751–759

Tekeste Z, Petros B (2010) The Abo blood group and *Plasmodium falciparum* malaria in Awash, Metehara and Ziway Areas, Ethiopia. Malar J 9:280

Timmann C, Evans JA, Konig IR, Kleensang A, Ruschendorf F, Lenzen J, Sievertsen J, Becker C, Enuameh Y, Kwakye KO, Opoku E, Browne EN, Ziegler A, Nurnberg P, Horstmann RD (2007) Genome-wide linkage analysis of malaria infection intensity and mild disease. PLoS Genet 3(3):e48

Timmann C, Thye T, Vens M, Evans J, May J, Ehmen C, Sievertsen J, Muntau B, Ruge G, Loag W, Ansong D, Antwi S, Asafo-Adjei E, Nguah SB, Kwakye KO, Akoto AO, Sylverken J, Brendel M, Schuldt K, Loley C, Franke A, Meyer CG, Agbenyega T, Ziegler A, Horstmann RD (2012) Genome-wide association study indicates two novel resistance loci for severe malaria. Nature 489(7416):443–446

Tishkoff SA, Williams SM (2002) Genetic analysis of African populations: human evolution and complex disease. Nat Rev Genet 3(8):611–621

Tishkoff SA, Varkonyi R, Cahinhinan N, Abbes S, Argyropoulos G, Destro-Bisol G, Drousiotou A, Dangerfield B, Lefranc G, Loiselet J, Piro A, Stoneking M, Tagarelli A, Tagarelli G, Touma EH, Williams SM, Clark AG (2001) Haplotype diversity and linkage disequilibrium at human G6pd: recent origin of alleles that confer malarial resistance. Science 293(5529):455–462

Torcia MG, Santarlasci V, Cosmi L, Clemente A, Maggi L, Mangano VD, Verra F, Bancone G, Nebie I, Sirima BS, Liotta F, Frosali F, Angeli R, Severini C, Sannella AR, Bonini P, Lucibello M, Maggi E, Garaci E, Coluzzi M, Cozzolino F, Annunziato F, Romagnani S, Modiano D (2008) Functional deficit of T regulatory cells in Fulani, an ethnic group with low susceptibility to *Plasmodium falciparum* malaria. Proc Natl Acad Sci USA 105(2):646–651

Traherne JA (2008) Human Mhc architecture and evolution: implications for disease association studies. Int J Immunogenet 35(3):179–192

Tse MT, Chakrabarti K, Gray C, Chitnis CE, Craig A (2004) Divergent binding sites on intercellular adhesion molecule-1 (Icam-1) for variant *Plasmodium falciparum* isolates. Mol Microbiol 51(4):1039–1049

Turner GD, Morrison H, Jones M, Davis TM, Looareesuwan S, Buley ID, Gatter KC, Newbold CI, Pukritayakamee S, Nagachinta B et al (1994) An immunohistochemical study of the pathology of fatal malaria. Evidence for widespread endothelial activation and a potential role for intercellular adhesion molecule-1 in cerebral sequestration. Am J Pathol 145(5):1057–1069

Ubalee R, Suzuki F, Kikuchi M, Tasanor O, Wattanagoon Y, Ruangweerayut R, Na-Bangchang K, Karbwang J, Kimura A, Itoh K, Kanda T, Hirayama K (2001) Strong association of a tumor necrosis factor-alpha promoter allele with cerebral malaria in Myanmar. Tissue Antigens 58(6):407–410

Uneke CJ (2007) *Plasmodium falciparum* malaria and Abo blood group: is there any relationship? Parasitol Res 100(4):759–765

Valentine WN, Tanaka KR, Miwa S (1961) A specific erythrocyte glycolytic enzyme defect (Pyruvate Kinase) in three subjects with congenital non-spherocytic hemolytic anemia. Trans Assoc Am Physicians 74:100–110

Van den Eede P, Van HN, Van Overmeir C, Vythilingam I, Duc TN, Hungle X, Manh HN, Anne J, D'Alessandro U, Erhart A (2009) Human *Plasmodium knowlesi* infections in young children in central Vietnam. Malar J 8:249

van Hellemond JJ, Rutten M, Koelewijn R, Zeeman AM, Verweij JJ, Wismans PJ, Kocken CH, van Genderen PJ (2009) Human *Plasmodium knowlesi* infection detected by rapid diagnostic tests for malaria. Emerg Infect Dis 15(9):1478–1480

Verra F, Mangano VD, Modiano D (2009) Genetics of susceptibility to *Plasmodium falciparum*: from classical malaria resistance genes towards genome-wide association studies. Parasite Immunol 31(5):234–253

Verrelli BC, McDonald JH, Argyropoulos G, Destro-Bisol G, Froment A, Drousiotou A, Lefranc G, Helal AN, Loiselet J, Tishkoff SA (2002) Evidence for balancing selection from nucleotide sequence analyses of human G6pd. Am J Hum Genet 71(5):1112–1128

Vijgen L, Van Essche M, Van Ranst M (2003) Absence of the Kilifi mutation in the rhinovirus-binding domain of Icam-1 in a Caucasian population. Genet Test 7(2):159–161

Wahlgren M (1999) Creating deaths from malaria. Nat Genet 22(2):120–121
Wallace R, Wallace RG (2002) Immune cognition and vaccine strategy: beyond genomics. Microbes Infect 4(4):521–527
Walley AJ, Aucan C, Kwiatkowski D, Hill AV (2004) Interleukin-1 gene cluster polymorphisms and susceptibility to clinical malaria in a Gambian case–control study. Eur J Hum Genet 12(2):132–138
Wang HY, Tang H, Shen CK, Wu CI (2003) Rapidly evolving genes in human. I. The glycophorins and their possible role in evading malaria parasites. Mol Biol Evol 20(11):1795–1804
Wattavidanage J, Carter R, Perera KL, Munasingha A, Bandara S, McGuinness D, Wickramasinghe AR, Alles HK, Mendis KN, Premawansa S (1999) Tnfalpha*2 marks high risk of severe disease during *Plasmodium falciparum* malaria and other infections in Sri Lankans. Clin Exp Immunol 115(2):350–355
Weatherall DJ (2001) Phenotype-genotype relationships in monogenic disease: lessons from the Thalassaemias. Nat Rev Genet 2(4):245–255
Webster MT, Clegg JB, Harding RM (2003) Common 5′ beta-globin Rflp haplotypes harbour a surprising level of ancestral sequence mosaicism. Hum Genet 113(2):123–139
WHO (2010) World malaria report 2010. WHO Press, Geneva, Switzerland
Wiesenfeld SL (1967) Sickle-cell trait in human biological and cultural evolution. Science 157:1134–1140
Wilder JA, Stone JA, Preston EG, Finn LE, Ratcliffe HL, Sudoyo H (2009) Molecular population genetics of Slc4a1 and Southeast Asian ovalocytosis. J Hum Genet 54(3):182–187
Williams TN (2006) Red blood cell defects and malaria. Mol Biochem Parasitol 149(2):121–127
Williams TN, Mwangi TW, Wambua S, Alexander ND, Kortok M, Snow RW, Marsh K (2005) Sickle cell trait and the risk of *Plasmodium falciparum* malaria and other childhood diseases. J Infect Dis 192(1):178–186
Wilson JN, Rockett K, Jallow M, Pinder M, Sisay-Joof F, Newport M, Newton J, Kwiatkowski D (2005) Analysis of Il10 haplotypic associations with severe malaria. Genes Immun 6(6): 462–466
Wood ET, Stover DA, Slatkin M, Nachman MW, Hammer MF (2005) The beta -globin recombinational hotspot reduces the effects of strong selection around Hbc, a recently arisen mutation providing resistance to malaria. Am J Hum Genet 77(4):637–642
Yamamoto F, Clausen H, White T, Marken J, Hakomori S (1990) Molecular genetic basis of the histo-blood group Abo system. Nature 345(6272):229–233
Yamamoto N, Akamatsu N, Sakuraba H, Yamazaki H, Tanoue K (1994) Platelet glycoprotein Iv (Cd36) deficiency is associated with the absence (Type I) or the presence (Type Ii) of glycoprotein Iv on monocytes. Blood 83(2):392–397
Zanella A, Fermo E, Bianchi P, Valentini G (2005) Red cell pyruvate kinase deficiency: molecular and clinical aspects. Br J Haematol 130(1):11–25
Zimmerman PA, Wieseman M, Spalding T, Boatin BA, Nutman TB (1997) A new intercellular adhesion molecule-1 allele identified in West Africans is prevalent in African-Americans in contrast to other North American racial groups. Tissue Antigens 50(6):654–656
Zimmerman PA, Woolley I, Masinde GL, Miller SM, McNamara DT, Hazlett F, Mgone CS, Alpers MP, Genton B, Boatin BA, Kazura JW (1999) Emergence of Fy*a(Null) in a *Plasmodium vivax*-endemic region of Papua New Guinea. Proc Natl Acad Sci USA 96(24):13973–13977

Parasitic Lice Help to Fill in the Gaps of Early Hominid History

Julie M. Allen, Cedric O. Worman, Jessica E. Light, and David L. Reed

Introduction

Hearing the word "lice" will immediately terrify parents and cause school nurses to spring into action. Pediculosis (a louse infestation) is not a new problem—lice have coevolved with humans over millions of years, and at this moment in human history, head lice are a worldwide epidemic. Although they can infect anyone, they are most common among children aged 3–12 and are widely spread throughout our school systems. According to the World Health Organization, it is thought that around 10–20 % of children are infested worldwide. In the USA alone, approximately 6–12 million infestations occur every year (Frankowski and Weiner 2002). Parents have attacked this problem using every method from shaving their child's head to covering the entire scalp with petroleum jelly, vinegar, and even toxic chemicals like kerosene (Meinking 1999; Frankowski and Weiner 2002). Even though we have been evolving with lice for millions of years, we are still struggling to understand and eradicate these parasites.

J.M. Allen (✉)
Florida Museum of Natural History, University of Florida, Museum Rd. and Newell Dr., Gainesville, FL 32611, USA
e-mail: juliema@illinois.edu

C.O. Worman
Biology Department, University of Florida, 223 Bartram Hall, Gainesville, FL 32611, USA

J.E. Light
Department of Wildlife and Fisheries Sciences, Texas A&M University, 210 Nagle Hall, College Station, TX 77843, USA

D.L. Reed
Florida Museum of Natural History, University of Florida, Museum Rd. and Newell Dr., Gainesville, FL 32611, USA

J.F. Brinkworth and K. Pechenkina (eds.), *Primates, Pathogens, and Evolution*, Developments in Primatology: Progress and Prospects,
DOI 10.1007/978-1-4614-7181-3_6, © Springer Science+Business Media New York 2013

Fig. 1 Male human head louse (*Pediculus humanus*) *left* and female chimpanzee louse (*Pediculus schaeffi*) *right*. The *circle* outlines the modified tibia with extended claw for hanging onto and climbing up and down hair shafts

Although it is the most common, the head louse (*Pediculus humanus capitis*) is just one of three types of lice that infest humans. Most similar to head lice are body lice (*Pediculus humanus humanus*), perhaps more aptly called "clothing lice" because they live principally in the clothing. The third and more distantly related louse is the pubic louse (*Pthirus pubis*), which lives primarily in the pubic region. Taxonomically, pubic lice belong to a different louse family, Phthiridae, whereas head and clothing lice belong to the family Pediculidae. These insects are host-specific obligate parasites that do not spend any part of their life cycle off their host. They live around 30 days and females attach eggs to the base of a hair shaft, laying around 3–5 eggs per day. Eggs hatch in 5–9 days, and in about 10 days nymphs become reproductively active. These insects feed on the blood of their hosts several times a day (Buxton 1947). Because lice have secondarily lost their wings, they cannot fly. Instead, lice move by climbing up and down hair shafts. Their tibiae are modified with claws that are adapted specifically for holding onto hair (Fig. 1). If a louse is removed from its host for a long period of time, it does not survive. In fact, most lice become so dehydrated that they cannot move after 21 h off the host (Burgess 2004).

Although we have been studying lice for hundreds of years (Darwin 1871; Hooke 1665), research slowed significantly in the 1940s when DDT was introduced and seemed likely to eradicate lice. In the 1990s, however, prevalence of lice increased worldwide due to pesticide resistance, and in 1997 there was a massive outbreak of typhus, which is transmitted by clothing lice (Raoult and Roux 1999), and louse research found a new beginning (Burgess 2004). This research reached a new peak in 2010 when the first louse genome (*Pediculus humanus humanus*) was sequenced (Kirkness et al. 2010). The genome work revealed a number of fascinating characteristics about lice. For example, clothing lice have the smallest insect genome sequenced to date (only 108 megabases), and they do not have many of the genes related to environmental sensing, which may be a result of their highly specialized lifestyle as an obligate parasite. Here we review the biology and latest research on head, clothing, and pubic lice including what these parasites can teach us about our own evolutionary history.

Head Lice: Pediculus humanus capitis

Head lice have long been considered an economical and societal problem rather than a dangerous infectious disease (Hansen and O'Havier 2004; Stafford 2008). The estimated cost of head louse infestation ranges from $367 million to $1 billion per year. Spending on over-the-counter chemical treatments, loss of school days, and loss of workdays for parents contribute to this high cost (Hansen and O'Havier 2004; Frankowski and Bocchini 2010). There are a number of myths and stigmas associated with head lice that cause undue stress to parents and family members of children infested with lice (Frankowski and Weiner 2002; Gordon 2007; Frankowski and Bocchini 2010). For example, head lice can be found in all social classes and are not correlated with cleanliness (Frankowski and Bocchini 2010). Head lice do not appear to spread disease in natural populations. Although, some studies have detected infectious bacteria in head lice (Sasaki et al. 2006a,b; Bonilla et al. 2009; Angelakis et al. 2011 but see Parola et al. 2006), and other research has suggested that head lice even have the capacity to transmit the infectious bacteria (Goldberger and Anderson 1912; Murray and Torrey 1975) to date, there has not been a known outbreak of disease caused by head louse transmission. The most common symptom from a head louse infestation is pruritis (itching) caused by a bite from the louse. In extreme cases, scratching the bite can cause an infection from common skin bacteria (Meinking 1999), but for the most part head lice are a mere annoyance rather than a dangerous parasite.

Manual removal of lice is a common means of treating head louse infestations, an activity which has shown up in artwork for centuries (the term "nit-picking" actually refers to the physical removal of lice). Recently, the most popular treatment of head lice has been chemical. Unfortunately, pesticide resistance has increased tremendously over the last few decades, (Burgess 2004) and in some places lice are even resistant to more than one chemical. Currently no pesticide is 100 % effective (Frankowski and Weiner 2002; Tebruegge et al. 2011). Other nonchemical methods (such as manual removal) have been suggested; however, these methods have been met with mixed success likely due to the effort required to remove all the lice (Frankowski and Bocchini 2010). One new method of louse eradication focuses on the temperature sensitivity of the lice. This sensitivity to temperature has been known for some time (Buxton 1947); in fact studies have shown that lice are likely to leave a person with a fever for a healthy person (Lloyd 1919 *cited in* Buxton 1947), and it has been suggested for some time that hot temperatures may be a way to kill lice. Following this literature, a modified prototype of the LouseBuster™ has recently been released. This appliance uses hot air to kill the lice by dehydrating them at high temperatures. In clinical trials, this method had a 94 % success rate in killing lice and 100 % success killing eggs on infected individuals. Thus, the LouseBuster™ may prove to be a faster, more effective, nonchemical method for treating head lice (Bush et al. 2011).

Head lice are most commonly transmitted from human to human by direct contact (Canyon and Speare 2010). It has been long suggested that lice can be transmitted via fomites, objects such as combs, and pillows (Burkhart and Burkhart 2007). However, this idea has been strongly challenged, and little to no evidence of fomite transmission has been found (Canyon and Speare 2010). Furthermore, head lice removed from the head die quickly due to dehydration (Burgess 2004), making fomite transmission unlikely. Unfortunately, methods to eradicate lice from schools have been difficult due to these types of misinformed ideas about how lice move from child to child.

An extremely controversial societal issue with lice is the "no nit" policies adopted by some schools. These policies require children to stay out of school until they are free of detectable nits (louse eggs). One issue with this policy is that empty nit casings (those from which the louse has already hatched) can remain on the hair long after a louse outbreak has ended. Furthermore, eggs are incubated by body heat so nits more than 1 cm from the scalp are unlikely viable (Frankowski and Weiner 2002). This means that nits that are farther from the scalp are likely empty and left over from a cured infestation and may remain in the hair until it grows out. Even more problematic, there is evidence that cases of head lice are frequently misdiagnosed (Pollack et al. 2000). Because of this, many children, particularly those with longer hair, miss school unnecessarily (Gordon 2007), which inflates the total cost of pediculosis per year. Not only are these children missing valuable class time, they are likely to face bullying by their classmates upon return due to the stigma associated with head lice.

Clothing Lice: Pediculus humanus humanus

Clothing lice (*Pediculus humanus humanus*), which are also called body lice, live in clothing fibers where they attach eggs to cloth fibers rather than hair (Buxton 1947). They look very similar to head lice, although there are some important differences between them. Clothing lice are generally larger than head lice and can consume a larger blood meal (Busvine 1978; Meinking 1999; Reed et al. 2004). Clothing lice only go to the skin to feed a few times a day, much less than head lice and likely why they consume a larger blood meal. Additionally, unlike head lice, clothing lice are associated with conditions of poor hygiene and are commonly found on those forced to live in crowded situations where they lack the ability to change or wash clothes regularly such as refugees, the homeless, soldiers, and victims of war or natural disasters (Meinking 1999).

It has been unclear for quite some time whether head lice and clothing lice are one species or two, and many studies have found conflicting results (Busvine 1978; Amevigbe et al. 2000; Burgess 2004; Leo and Barker 2005; Leo et al. 2005; Light et al. 2008). However, recent studies with more comprehensive sampling are finding that clothing lice and head lice are in fact the same species and that clothing lice evolve from head lice in certain conditions (such as in situations where individuals have a considerable head louse infestation and poor hygiene; Li et al. 2010). It is now clear that throughout our history clothing lice have opportunistically evolved repeatedly from head lice to fill a different ecological niche from that of their head louse counterparts (Reed et al. 2004; Light et al. 2008; Li et al. 2010).

Finally, and perhaps most importantly, clothing lice are the known vectors of three human pathogens: *Rickettsia prowazekii* (agent of epidemic typhus; Andersson and Andersson 2000), *Borrelia recurrentis*, and *Bartonella quintana* (the agents of relapsing fever and trench fever, respectively; Buxton 1947). These transmissions occur when louse feces are unintentionally rubbed into an open wound caused by the louse bite, most generally occurring when scratching the area of the bite (Buxton 1947). These diseases have had a devastating impact throughout human history. For example, epidemic typhus may have been largely responsible for the demise of Napoleon's Grand Army around 1812 (Raoult et al. 2006). These diseases have been a problem not only historically but also recently. In 1999, a large outbreak of endemic typhus broke out in Burundi, infecting more than 100,000 people. With the growing worldwide problem of lice due to pesticide resistance, these diseases will need to be carefully monitored.

Pubic Lice: Pthirus pubis

The third type of louse that parasitizes humans is *Pthirus pubis*, commonly known as the "crab" or pubic louse. Pubic lice are a sexually transmitted disease (STD), and their presence has often been found in combination with other STD infections (Anderson and Chaney 2009). Pubic lice are also a worldwide phenomenon. Although it is much more difficult to calculate the level of prevalence because infections are not often reported, recent estimates suggest that 2 % of the world's adult population are infected with pubic lice (Anderson and Chaney 2009). Pubic lice are found in all levels of society and among all ethnic groups (Meinking 1999).

Pubic lice live primarily in the androgenic hairs (hair that begins to grow at sexual maturity) around the groin; however, these lice have been found on children in the eyelashes, eyebrows, and edges of the hairline. In rare cases the presence of *Pthirus* on children has alerted authorities to incidents of child abuse; however, this is not common (Chosidow 2000). It is thought that *Pthirus* prefer hair that is more widely spaced due to the wider spacing of their claws (Waldeyer 1900; Nuttall 1918; Fisher and Morton 1970). Interestingly, it is thought that fomite transmission is more important in pubic lice, which may explain how children get an infestation in their eyebrows and eyelashes without sexual contact (Meinking 1999). Similar to human head lice, pesticides and manual removal are considered to be the primary treatment for pubic lice (Orion et al. 2004).

The genus *Pthirus* has two species of lice: the human pubic louse and the gorilla louse, *Pthirus gorillae*. The host associations of this genus of louse have puzzled researchers for some time: why are humans and gorillas, but not chimpanzees, parasitized by *Pthirus*? Furthermore, although lice are very host specific, why do humans have two genera of lice (*Pediculus* and *Pthirus*), whereas chimpanzees and gorillas each have one (*Pediculus* and *Pthirus*, respectively; Fig. 2)? In 2007, Reed et al. conducted molecular dating analyses on gorilla and human pubic lice and found that

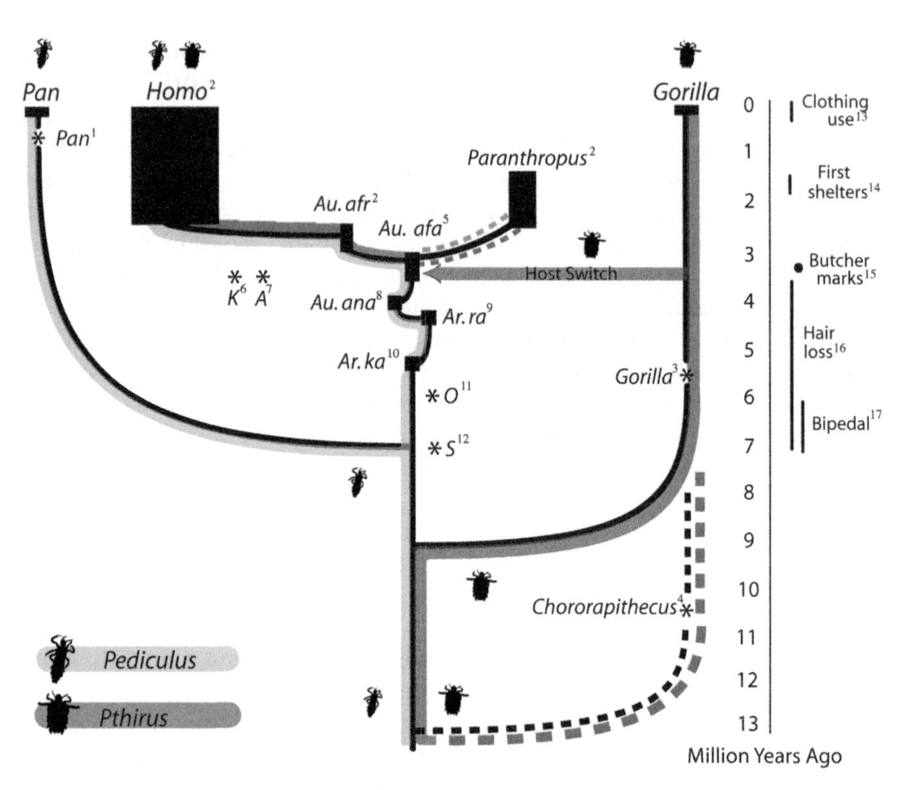

Fig. 2 The coevolutionary history of humans (*Homo*), chimpanzees (*Pan*), and gorillas (*Gorilla*) and their lice. The primate lineages are indicated by *thin black lines* and *black boxes* that depict the longevity (*box height*) and the species richness (*box width*) of the primate genera known from physical evidence (either the fossil record or extant species). Parasite lineages are indicated with *thick gray bars* with the *lighter gray* representing *Pediculus* and the *darker gray Pthirus*. *Dotted lines* indicate possible coevolutionary scenarios that remain unclear due to lack of data. The most recent estimated divergence between gorillas and the lineage leading to humans and chimpanzees is shown at 9.2 mya; however, we also show the possible divergence (and consequently the cospeciation event) between these lineages at 13 mya incorporating the gorilla-like fossil *Chororapithecus abyssinicus* (*dashed lines* at 13 mya). We further show the host switch event 3–4 mya by lice in the genus *Pthirus* from the gorilla lineage onto the hominin lineage. The extinct hominin genus *Paranthopus* and its possible association with both *Pediculus* and *Pthirus* are shown at approximately 2 mya. Events on the *right* represent our current knowledge of hominin history including time ranges for the origin of clothing use by *Homo sapiens*, the first putative shelters, the earliest known butcher marks, and the development of bipedality (as indicated from the louse data). *Asterisks* mark isolated fossil finds. Species abbreviations are as follows: *Ar. ka*=*Ardipithecus kaddaba*, *Ar. ra*=*Ardipithecus ramidus*, *Au. ana*=*Australopithecus anamensis*, *Au. afa*=*Australopithecus afarensis*, and *Au. afr*=*Australopithecus africanus*. O=*Orrorin tugenensis*, S=*Sahelanthropus tchadensis*, K=*Kenyanthropus platyops*, and A=*Australopithecus bahrelghazali*. (1) McBrearty and Jablonski (2005). (2) Foley (2002). (3) Pickford et al. (1988). (4) Suwa et al. (2007). (5) Strait et al. (1997). (6) Leakey et al. (2001). (7) Brunet et al. (1995). (8) Kimbel et al. (2006). (9) WoldeGabriel et al. (2009). (10) Haile-Selassie (2001). (11) Pickford et al. (2002). (12) Brunet et al. (2002). (13) Toups et al. (2011). (14) Rantala (1999). (15) McPherron et al. (2010). (16) Reed et al. (2007). (17) Pickford et al. (2002)

these two species were sister taxa. Additionally, these two taxa diverged only 3–4 million years ago (mya), much more recently than the gorilla/human-chimp split, which is estimated at around 9 mya (Wilkinson et al. 2011). This finding was extremely interesting as it shed light onto two details of human evolutionary history that were previously unknown. Here we go into more detail about the biology of *Pthirus* to discuss what this host switch tells us about our ancestors 3–4 mya.

Lice Tell Us About Our Past

Sucking lice have been coevolving with their hosts for at least the last 65 million years and likely much longer (Light et al. 2010). Due to their obligate nature and the fact that they mostly move between hosts via direct contact, these blood-feeding lice have been used to give us clues about their hosts' evolutionary history not easily gleaned from the fossil record (e.g., behavior). Interestingly, this idea dates back to Darwin (1871) who in *On the Decent of Man* wrote:

> "*...and the fact of races of man being infested with parasites which appear to be specifically distinct might fairly be urged as an argument that the races themselves ought to be classed as distinct species.*"

Although the idea that races of humans represent different species is no longer entertained and even the concept of race has dramatically changed since Darwin's time, the idea that lice can tell us about our evolutionary history is now well accepted and gaining momentum (Whiteman and Parker 2005; Hypsa 2006; Nieberding and Olivieri 2007; Reed et al. 2009).

In 2003, it became apparent upon genetic examination of recently collected human head and clothing lice that there were several distinct lineages of lice (Kittler et al. 2003). Reed et al. (2004) used fossil calibrations for the split between humans and chimpanzees (5.6 mya) and the split between great apes and Old World monkeys (22.5 mya) to estimate the divergence time between these human louse lineages. They found that the youngest of these lineages splits around 1.18 mya, which is similar in age to the ancestor of *Homo sapiens* and *H. neanderthalensis*. The other louse lineage is even older, suggesting its origin may date back to a *H. erectus*-like host. Because head lice are primarily transmitted through direct contact, this finding suggests that modern humans came into contact with archaic hominin species and picked up distinct lineages of head lice. Although we do not know which archaic humans our ancestors came into contact with, the timing of the divergence of the ancient head louse lineages is consistent with contact with *H. neanderthalensis* and *H. erectus*. Furthermore, while the type of contact between different hominin species is unknown, recent work sequencing the *H. neanderthalensis* genome found evidence of interbreeding between *H. neanderthalensis* and *H. sapiens* (Green et al. 2010; Yotova et al. 2011). Similar types of contact between *H. sapiens* and *H. erectus* would have been sufficient for the transfer of lice and may explain the existence of ancient lineages of head lice on modern humans.

Because clothing lice live exclusively in the clothing, they are thought to have evolved only after humans began to wear clothes, and it has long been proposed that dating the origin of clothing lice could give us a date by which *H. sapiens* must have been wearing clothing (Kittler et al. 2003, 2004). Previous estimates of when humans started wearing clothing were based on the emergence of eyed needles (which suggests complex clothing had already been developed) around 40,000 mya (Delson et al. 2000) and sometime after the loss of body hair as late as 1.2 mya (Rogers et al. 2004; based on molecular evidence) and as early as 3 mya (Reed et al. 2007; detailed below). Recent molecular evidence from clothing lice suggests that clothing use originated between 83,000 and 170,000 years ago, which is earlier than previously proposed, and suggests that clothing use by *H. sapiens* likely originated before they moved out of Africa (Toups et al. 2011). This clothing use may have enabled modern humans to more readily move into colder climates as they migrated out of Africa and eventually throughout the world.

The coevolutionary relationships between great apes and their lice have been worked out morphologically and molecularly and are illustrated in Fig. 2. Humans and chimpanzees share lice in the genus *Pediculus* as sister taxa, and humans and gorillas share lice in the genus *Pthirus* (Fig. 2). The split between human and chimpanzee lice was estimated to be 5–7 mya (Reed et al. 2004, using mitochondrial genes; Light et al. 2008, using a multigenic approach with mitochondrial and nuclear genes), strongly suggesting cospeciation between these two lice and their primate hosts (humans and chimpanzees also are believed to have diverged at this time; Wilkinson et al. 2011). On the other hand, Reed et al. (2007) found that the *Pthirus* and *Pediculus* are sister taxa and they diverged ~13 mya, long before the presumed 7 mya split between gorillas and the other African apes (Fig. 2). Reed et al. (2007) hypothesized an evolutionary scenario in which there was a louse duplication (or speciation) event on the African ape common ancestor to humans, chimpanzees, and gorillas. In other words, one louse species would have diverged into two on the ape common ancestor. More recent refinement of the somewhat troublesome great ape molecular clock has pushed the gorilla divergence back to 9.2 mya (Wilkinson et al. 2011). Additionally, a recent fossil find of the gorilla-like ape *Chororapithecus abyssinicus* dating to 10–11 mya (Suwa et al. 2007) suggests that the great ape molecular clock may still be underestimating the divergence of the gorillas by millions of years. If true, the presumed louse duplication event ~13 mya suggested by Reed et al. (2007) could have actually been a cospeciation event, suggesting that the divergence time between gorillas and the other African apes occurred ~13 mya, far earlier than currently thought (Fig. 2).

Reed et al. (2007) also examined the history of the two species of *Pthirus*, one on gorillas and the other on humans, which diverged 3–4 mya. Reed et al. (2007) hypothesized that after *Pthirus* and *Pediculus* diverged ~13 mya, the *Pthirus* lineage remained on ancestral gorillas but went extinct on the common ancestor of humans and chimpanzees, and the *Pediculus* lineage remained on the common ancestor of humans and chimpanzees but went extinct on ancestral gorillas. Then, approximately 3–4 mya, the *Pthirus* lineage from ancestral gorillas switched to a human ancestor (Fig. 2). This type of host switch is not uncommon; there have been a number of zoonotic transmissions of diseases from primates, as well as domesticated animals, to humans (Wolfe et al. 2007). For example, HIV-1 is now known to have come from a chimpanzee (Gao et al. 1999) possibly as a result of humans hunting chimpanzees for food. This particular host-switching event of *Pthirus* on ancestral gorillas to our human ancestors gives us clues about human evolutionary history that we outline and detail below.

Human Hair Loss

Reed et al. (2007) postulated that a *Pthirus*-type louse switched hosts from archaic gorillas to hominins approximately 3–4 mya. It is interesting to consider what was necessary for this host switch to have been successful: there had to have been a niche for *Pthirus* to occupy. Studies of chewing lice have found that their mouthparts (which grip the hair) are highly adapted and specialized to the hair of their hosts (Reed et al. 2000), so it is likely that hair type is similarly important to sucking lice. Rather than using their mouthparts to grip hair, sucking lice use their claws, which are also highly specialized. *Pthirus*, in particular, is highly adapted to hairs in the pubic regions as these hairs are more widely spaced, which match the wider spacing of their claws (see below). As stated previously, pubic hair is a type of androgenic hair—hair that grows in response to increased levels of androgens circulating in the human body at sexual maturity (Randall 2008). Proposed functions of pubic hair (pheromonal and visual signaling; Randall 2008) could have only come into play after the loss of typical ape body hair. Additionally, the invasion of a new host would have been far more likely if *Pediculus* had already been confined to the head by the loss of functional body hair, leaving competitor-free regions available to *Pthirus*. We hypothesize that the loss or reduction of body hair as well as the development of androgenic hair would have facilitated the success of this host switch and therefore suggest that human hair loss and the gain of androgenic hair had occurred by 3–4 mya, a date that is much older than other predictions.

Among primates, humans are unique in their apparent nakedness. Humans, however, are not actually hairless. They have a similar number and density of hair follicles as other great apes, but the hairs are much finer (i.e., smaller in diameter) and shorter and offer little protection or insulation (Kushlan 1985; Amaral 1996; Rantala 1999). There is a great variety of hypotheses ranging from the bizarre to the pedestrian as to why humans had such a drastic reduction in body hair (reviewed in Rantala 2007). Many of these hypotheses are directly related to the *Pthirus* host switch because they either incorporate habitat (as discussed below) and thermodynamics (which is closely tied to habitat) or attempt to establish the timing of hair reduction. Habitat and thermodynamics are important to the cooling device, bipedality, hunting, vestiary, allometry, and other hypotheses of why humans lost their body hair. The timing of hair loss is incorporated to some extent in any hair loss hypothesis, but it is particularly important in the clothing, vestiary, and ectoparasite hypotheses.

The reduction in body hair has obvious thermodynamic consequences. This loss of insulation increases heat exchange with the environment. Several body hair loss hypotheses (see Rantala 2007) are based on the need to shed increased heat loads resulting from either a move from forest into hotter savanna habitats (cooling device, bipedality, hunting, vestiary hypotheses), an increased activity (hunting and vestiary hypotheses), or a large body size (allometry hypothesis). The cooling device hypothesis states that the increased heat load was simply caused by the move into open savannas from the forest (Rantala 2007). The bipedality hypothesis adds to this by examining how an upright stance decreases the solar heat load experienced by an individual in a bipedal stance compared to a quadrupedal stance (Wheeler 1992). Active hunting in the savanna and the excess heat that must be shed from high levels of activity are incorporated into the hunting hypothesis (Brace and Montagu 1977). The additional need to retain heat during the cool savanna nights (fulfilled by the use of clothing) while being able to shed heat during the day, presumably through hairlessness and sweating, is the basis of the vestiary hypothesis (Kushlan 1985). The allometry hypothesis is based on the observation that larger primates have increasingly more widely spaced hair (Schwartz and Rosenblum 1981).

As intuitive as it may seem to people from cooler climes that shedding insulation increases heat loss, hot open savanna environments make body hair extremely valuable for decreasing heat gain from both solar radiation and the air (Newman 1970). The upright stance central to the bipedality hypothesis reduces solar heat gain compared to a quadrupedal stance, making it less detrimental to be hairless, but it does not make hair loss beneficial in savanna environments (Amaral 1996). In addition, it is now clear that bipedalism evolved in basal hominins by the time of *Orrorin tugenensis* (Pickford et al. 2002; Galik et al. 2004) around 5.7–6 mya (Richmond and Jungers 2008), long before the shift to dry open habitats by *Homo* (Elton 2008), and that evaporative cooling is not prevented by body hair; the patas monkey (*Erythrocebus patas*), a cursorial savanna monkey, has both thick fur and effective sweating (Mahoney 1980). Another characteristic of savannas compared to forests (addressed by the vestiary hypothesis) is colder nights unmitigated by heat-retaining forest tree cover and humidity (Amaral 1996). This makes the insulation provided by body hair even more valuable and makes loss of body hair in the savanna environment doubly detrimental.

The allometry hypothesis is not a complete explanation of human hairlessness by itself. Schwartz and Rosenblum (1981) reanalyzed Schultz's (1931, 1969) measurements of primate hair density and found that there are fewer hairs per unit of body surface in larger primates compared to smaller primates. They reasoned that this reduction in body hair was likely due to thermoregulatory constraints associated with decreasing ratios of surface area to volume, which make shedding metabolic heat difficult for larger animals. Because fossil data indicated that early australopithecine hominins weighed between 45 and 70 kg (Pilbeam and Gould 1974), Schwartz and Rosenblum (1981) hypothesized that substantial decreases in hominin hair density likely occurred prior to human shifts from forest to grassland habitats at the end of the Pliocene.

However, the effective hairlessness of humans is not just a result of hair density but also hair size. In contrast to humans, other similar-sized and larger apes (orangutans and gorillas) have substantial body hair. Thus, hominins appear to have exaggerated the typical primate strategy of shedding metabolic heat via reduced insulative effectiveness of body hair (the heavy sweating of humans and patas monkeys does not appear to be typical of primates; Amaral 1996) by reducing hair size. It seems likely that this was in answer to an additional metabolic heat load beyond that experienced by typical apes. Although it is impossible to say with any degree of certainty what this additional heat load was, the development of bipedalism, a more energetically efficient mode of locomotion than knuckle walking (Sockol et al. 2007), hints that increased daily travel may have been important to the basal hominin niche and that hair loss may have occurred very early on. The extra metabolic heat produced by travel through forests could have been shed by decreasing body hair insulation without the costs of nakedness associated with savanna environments. Based on the timing of the *Pthirus* host switch, the habitat in which this switch likely occurred (see below), and the problems of hairlessness in open habitats, hair loss in the hominin line almost certainly occurred in a forested habitat and was complete and effectively irreversible by the time savanna habitats were fully utilized. The human dependence on sweating as a cooling mechanism likely occurred long after hair loss to deal with the additional heat loads in open habitats as sweating is less effective in the humid still air of forests than in drier more open habitats (Newman 1970; Montagna 1972).

Other than lice, the only line of evidence that helps establish the timing of hair loss in the hominin line is genetic. The human melanocortin 1 receptor (MC1R) gene is involved in human skin coloration. By looking at the neutral variation in this gene, Rogers et al. (2004) estimated that human skin has been exposed to strong sunlight for at least 1.2 my. Therefore, based on the MC1R data, human ancestors became both hairless and began living in the savanna between 1.2 mya and 6–7 mya (the chimpanzee/hominin split). The lice data are consistent with the MC1R data but give a narrower range of 3–4 mya from the *Pthirus* switch to the 6–7 million year split between the human and chimpanzee lineages.

The timing of hair loss is particularly central to the clothing, vestiary, and ectoparasite hypotheses. The clothing hypothesis (Glass 1966) is similar to the vestiary hypothesis (Kushlan 1985) in that they both posit that clothing superseded the insulative value of body hair and hair loss occurred with or after the invention of clothing. However, while the vestiary hypothesis holds (erroneously, as discussed above) that the loss of body hair was advantageous during the hot days and that clothing replaced the need for body hair during the cool nights, the clothing hypothesis argues that after clothing was invented, body hair disappeared as it was no longer needed (Glass 1966; however, Glass does not propose a reason for the invention and use of clothing by hominins with functional coats of body hair). The louse and MC1R data estimates for both hairlessness (*Pthirus*, >3–4 mya in Reed et al. 2007; MC1R, >1.2 mya in Rogers et al. 2004) and the invention of clothing (*Pediculus*, 0.08–0.17 mya in Toups et al. 2011) indicate that clothing had nothing to do with the evolution of hairlessness in hominins.

The ectoparasite hypothesis states that when hominins first established long-term habitations, they were beset with new types of ectoparasites, such as fleas, that completed their life cycles in the living space but off the body of the host (Rantala 1999). Thus, the loss of body hair was a defense against increased parasite loads encouraged by the establishment of a home base. In apparent conflict with the adaptation-against-ectoparasites hypothesis is the presence of pubic hair (Pagel and Bodmer 2003). Pubic hair, however, may play an important role in sexual selection and thus may have been selected for in spite of its ability to shelter ectoparasites. The moist and humid environment of the pubic region (due to an increased density of sweat glands; Stoddart 1990) is favorable to pheromonal signaling (Guthrie 1976), and pubic hair could have initially functioned in pheromonal signaling (Randall 1994), visual signaling (Randall 2008), or both. Rantala (1999) associates the beginnings of long-term settlements with an increase in cooperative hunting and places both developments at ~1.8 mya based on excavations of *Homo habilis* artifacts at Olduvai Gorge. While this estimate is consistent with the hairlessness range provided by the MC1R gene (>1.2 mya), it is far later than the hairlessness estimate provided by the *Pthirus* host switch.

Pediculus and *Pthirus*

The two genera of human lice (*Pediculus* and *Pthirus*) occupy distinct niches on the body (head/clothing and pubic region, respectively). Many researchers have wondered why *Pediculus* and *Pthirus* do not co-occur and are apparently isolated to these different regions especially given their similar biology (Howlett 1917). Hypotheses have included *Pthirus* having a preference for darkness and moist areas (Nuttall 1918; however, this idea is not accepted as *Pthirus* survives on eyelashes and eyebrows) and that differences in hair spacing have geographically restricted these lice because *Pediculus* cannot adequately grasp the hairs of the pubic region, and *Pthirus* cannot grasp the hairs on the head. Both genera of lice have been found occasionally occupying and surviving in other regions of the body (see below), but they do not seem to be successful in these areas. Of these hypotheses, hair spacing seems to be the most likely explanation for restricting these two types of lice to their respective habitats.

Schwartz and Rosenblum (1981) found that there are fewer hairs per unit of body surface in larger primates compared to smaller primates. If the Reed et al. (2007) hypothesis that the human pubic louse (*Pthirus pubis*) is a descendent of gorilla lice is true, then based on Schwartz and Rosenblum's (1981) findings, we can postulate that *Pthirus* was adapted to living among widely spaced hairs because gorillas are the largest extant primate. Early studies of *Pthirus* support this idea. *Pthirus* uses its second and third pair of legs to cling to host hair, and these legs, when stretched apart, span a distance of 2 mm (Waldeyer 1900; Nuttall 1918; Fisher and Morton 1970). It just so happens that hairs in the pubic region are also distributed 2 mm apart (Waldeyer 1900; Nuttall 1918). Furthermore, the number of hairs present in the pubic region (34 hairs/cm^2) is significantly less than the head (220 hairs/cm^2; Waldeyer 1900; Payot 1920). All in all, *Pthirus* appears to prefer body regions with widely spaced hairs for better grasping as well as for ease of flattening itself against the skin (Burgess et al. 1983; Burgess 1995; Nuttall 1918; Buxton 1947; Fisher and Morton 1970). According to measurements made by Schultz (1931), chimpanzee hair density is more similar to humans than to gorillas. That, in addition to the lack of pubic-type hair on chimpanzees, may help explain why there are no *Pthirus* species currently parasitizing chimpanzees.

The *Pthirus* preference for widely spaced hairs is likely why this genus can be occasionally found in other sparsely haired areas on the human body, such as the margins of the scalp, eyebrows, eyelashes, and areas of the trunk such as the chest, stomach, and thighs (if sufficient body hair is present; Burgess 1995, and references therein; Buxton 1941, 1947; Elgart and Higdon 1973). *Pediculus*, in comparison, is rarely found in the pubic region of humans (Busvine 1944). *Pediculus schaeffi*, the louse on chimpanzees louse, is more catholic in habitat choice than *Pediculus humanus* and can be found almost anywhere on a chimpanzee host but favors the groin, underarms, and head (D. Cox, pers. comm.). It is likely that the *Pediculus* found on hominins before the loss of body hair was similarly widely spread but was restricted to the head region during the hominin denudation and subsequently prevented from spreading to androgenic hair because the larger spacing of pubic hair made movement from one hair to another difficult.

Differences in mobility also may prevent *Pediculus* from traveling to the pubic region as often as *Pthirus* appears to move to other parts of the body. Although several studies have found that *Pediculus* moves faster than *Pthirus* when displaced from the body (Nuttall 1918; Busvine 1944), *Pthirus* does tend to wander more (Burgess et al. 1983). Furthermore, head lice are recognized as being rather picky in how they move from hair to hair, suggesting that they are unlikely to move readily to foreign objects or fomites (Canyon et al. 2002). Closer examination of the first tarsal claws (which may facilitate movement when lice are not in contact with hair) of both *Pediculus* and *Pthirus* reveals why there may be differences in mobility between these two genera (Ubelaker et al. 1973). The inner surface of the first tarsal claw in *Pthirus* is serrated, allowing for traction even on smooth surfaces, whereas in *Pediculus* the inner surface of the claw is smooth and the lice are unable to move without hair follicles or roughened surfaces (Nuttall 1918; Ubelaker et al. 1973; Burkhart and Burkhart 2000). This simple difference, along with preferential movement patterns, may restrict *Pediculus* from moving easily on smooth, non-haired substrates. *Pthirus*, on the other hand, with their serrated first tarsal claws, may be able to move much more easily on non-haired substrates, thus allowing them to reach other parts of the body such as the perimeter of the scalp. The biology of these two parasites supports the idea that human ancestors had not only lost their body hair by the host switch 3–4 mya (isolating *Pediculus* in the head region) but that early hominins had also developed androgenic hair, providing a suitable environment for *Pthirus*.

Habitats of Early Hominins

Given the fossil species currently known, the most parsimonious scenario explaining the appearance of *Pthirus* in the human lineage is a host switch directly from gorilla ancestors to human ancestors. This scenario strongly implies that the two ancestral host species came into repeated and close contact, which further implies significant overlap in habitat. By combining data from the fossil record, paleoclimate, extant species, and the *Pthirus* host switch, we can augment the current thinking of the habitat and habits of human ancestors.

The fossil record of nonhuman African apes is abysmal to say the least. There is currently only one known chimpanzee fossil, which lived 0.5 mya (McBrearty and Jablonski 2005); one gorilla fossil from 5 to 6 mya (Pickford et al. 1988); and the very gorilla-like *Chororapithecus abyssinicus* from 10 to 11 mya (Suwa et al. 2007; Fig. 2). The reasons for the dearth of these fossils are likely due to several factors (Cote 2004). For one, apes were an uncommon component of the fauna in any region, and African fossil sites commonly produce fewer specimens than Eurasian fossil sites. Therefore, a site must produce a large number of fossils if any apes are expected to be represented in the first place. Second, with the exceptions of *Samburupithecus kiptalami* and *Nakalipithecus nakayami*, which might have been adapted to drier forests (Kunimatsu et al. 2007), nonhuman African apes are, and appear to always have been, tightly associated with moist tropical forests. The wet acidic soil in these types of habitats is much more conducive to quick bone decomposition than fossilization, so it is likely that very few specimens were fossilized to begin with (Kingston 2007). Added to those problems is the fact that sites currently under moist tropical forests are seldom found or excavated, partly due to a lack of exposed strata and partly due to the political insecurity that often inflames those regions, reducing safety for international teams and handicapping intranational capacity for, and interest in, research.

It is generally thought that the nonhuman African apes are conservative in body form and habits contrasting with the hominins that stumbled upon a new behavior/body form (bipedalism) that led to their subsequent radiation into a speciose and relatively diverse group. That assumption is supported by the dearth of non-hominin ape fossils in Africa, which indicates they were restricted to the wet forests that are particularly hostile to fossil formation (Kingston 2007), and by the nature of the few fossils that have been found. Additionally, the morphology and wear of the *Chororapithecus abyssinicus* fossil from 11 to 10 mya (Suwa et al. 2007) and the gorilla tooth from 5 to 6 mya (Pickford et al. 1988), as well as their respective faunal assemblage contexts, suggest that gorillas have been conservative in diet and habitat over the period of time during which the hominin group was rapidly developing novel traits and habits.

Molecular work has also indicated the conservatism of the gorilla lineage. Thalmann et al. (2007) have estimated that eastern and western gorillas (*Gorilla beringei* and *G. gorilla*, respectively) diverged 0.9–1.6 mya with very little subsequent gene flow between those two species. Thalmann et al. (2007) also suggest that the genetic divergence between the eastern and western forms is small enough to unite the two into a single species. The gene flow between the two groups is low enough that it is likely not the cause of their genetic similarity but a result of it. With the apparent conservatism of the gorilla lineage in mind, we can cautiously use the natural history of extant gorillas to inform us of the probable habits of their ancestors and examine how that information fits into and expands the understanding of our ancestors.

Gorillas today range across forested tropical Africa (with a large interruption between the eastern and western species) from lowland rainforest to high-altitude montane forest. In spite of these wide longitudinal and altitudinal ranges, both species share similar diets based on succulent herbaceous vegetation (Kingdon 1974). The diet often incorporates more fruit in areas where fruit is available, but herbaceous vegetation still forms a large portion of the diet, retaining its primary importance especially as a fallback food in times of fruit scarcity (Yamagiwa and Basabose 2006).

The importance of fibrous herbaceous foods makes gorillas more independent of often unreliably fruiting trees than the two chimpanzee species (*Pan troglodytes* and *P. paniscus*); however, it also limits their available habitat to moist forests with enough sunlight penetrating the canopy to support a rank herbaceous understory (Schaller 1963). While swidden agriculture produces ample areas of lush secondary growth, prior to agriculture, suitable gorilla foraging areas would have been limited to montane forests, river edges, treefall gaps, elephant tramples, and the like (Schaller 1965a), with feeding and use by gorillas likely slowing succession and extending the usable life and possibly the size of temporary clearings (Plumptre 1994). Even if the gorilla lineage had significantly different dietary preferences than extant gorillas during the host switch of *Pthirus* 3–4 mya, it is unlikely that the differences would have a meaningful impact on our analysis as the conservatism of body and tooth form and lack of fossils indicate a folivorous/frugivorous diet in a moist forest.

During the 3–4 mya range given for the *Pthirus* host switch, there were 1–4 hominin species present (Fig. 2) depending on the validities of species identifications: *Australopithecus anamensis*, *A. afarensis*, *A. bahrelghazali*, and *Kenyanthropus platyops* (Fig. 2). *K. platyops* (3.5 mya) is known from only one locality and the skull upon which the identification is based is severely fragmented and distorted (Leakey et al. 2001). Therefore, the identity of the specimen as a new genus (*Kenyanthropus*), a new species within *Australopithecus*, or another *A. afarensis* specimen is controversial (White 2003; Spoor et al. 2010). However, if *K. platyops* is a valid species, it is potentially ancestral to both *Homo* and *Paranthropus* (robust australopithecines) and lived in a well-watered forest or woodland (Leakey et al. 2001; Strait and Grine 2004).

Australopithecus bahrelghazali is another species known only from a single fossil from 3.5 mya (Brunet et al. 1995; Brunet 2010) and, like *K. platyops*, is controversial as to whether it is a separate species or an unusual *A. afarensis* (Kimbel et al. 2006; Guy et al. 2008). This is a unique find because it is the only australopithecine found in Chad rather than East or Southern Africa. Unfortunately, the state of the fossils makes establishing phylogenetic relationships difficult (Strait and Grine 2004), but it is likely that *A. bahrelghazali* lived in a gallery forest/wooded savanna/grassland mosaic context (Brunet et al. 1995).

Australopithecus anamensis (3.9–4.2 mya) was probably the anagenetic ancestor of *A. afarensis* (Kimbel et al. 2006) and therefore also an ancestor of *Homo* (Strait et al. 1997). Because *A. anamensis* and *A. afarensis* are chronospecies, they appear to be similar in habitat and diet. The habitat of *A. anamensis* was likely mosaic forest, woodland, grassland, bush, and riverine forest (Bonnefille 2010). In spite of the variety of habitats postulated, *A. anamensis* seems to have been tied to the presence of at least some trees and lived at a time of increasing tree cover in Africa (Bonnefille 2010). The diet has been postulated with many methods. Tooth morphology suggests hard brittle items (Grine et al. 2006) but with the ability to exploit fleshy fruits (Teaford and Ungar 2000). Microwear patterns suggest tough and fibrous foods (Ungar et al. 2010). Finally, enamel microstructure suggests tough, hard, and abrasive foods with limited brittle and acidic foods (Macho and Shimizu 2010). It seems likely that the majority of the *A. anamensis* diet was fibrous vegetation that required grinding with brittle foods forming an important fallback food (Ungar et al. 2010).

The fourth species, *Australopithecus* (*Praeanthropus*) *afarensis*, is by far the best understood of the four candidates and a presumed ancestor of *Homo* (Strait et al. 1997). Sites containing *A. afarensis* fossils have been found all over East Africa, which dated from 3.0 to 3.6 mya (Kimbel and Delezene 2009). The paleoenvironments of these sites, temporally and spatially, are extremely variable, ranging from steppe to woodland to forest (Bonnefille et al. 2004), but not wet dense evergreen forest (Bonnefille 2010). Because *A. afarensis* showed no apparent association with any particular habitat in a single site that fluctuated between being dominated by steppe and being dominated by forest, it has been described as a generalist (Bonnefille et al. 2004).

The teeth of *A. afarensis* are thought to be adapted for crushing hard, brittle foods such as seeds, hard fruits, and/or tubers (Luca et al. 2010); however, microwear analysis of the teeth paints a different picture entirely. Compared to a variety of other primates, including those that specialize in eating hard seeds and those that consume substantial numbers of tubers, the microwear patterns on *A. afarensis* teeth from a diversity of habitats actually most closely resemble those of the mountain gorilla (Grine et al. 2006), the least frugivorous gorilla subspecies (Yamagiwa and Basabose 2006). As Grine et al. (2006) make clear, this resemblance does not mean that *A. afarensis* had the same diet as a gorilla, and the gross tooth morphology makes it unlikely that they could eat the same foods in the same way. Rather, the similarity means that they both ate fibrous foods with fine abrasiveness but none of the hard brittle foods for which the *A. afarensis* teeth appear to be adapted, at least not in the period before each of the individuals died. However, in areas of range overlap, it seems likely that they may have often been attracted to some of the same types of food, increasing the chances of interaction.

The tooth morphology seemingly at odds with wear patterns may indicate a difference between commonly eaten preferred foods and the ability to efficiently process seasonally important fallback foods (Ungar 2004). Other authors suggest that wetland vegetation was an important *A. afarensis* food (also exploited by gorillas in certain areas) that could explain the tooth wear patterns and the insensitivity of *A. afarensis* to changes in upland habitat (Verhaegen et al. 2002). Significant for this discussion is an additional *A. afarensis* food: meat scavenged from large animals with the help of stone butchering tools (McPherron et al. 2010).

It is difficult to choose the hominin most likely to have first acquired *Pthirus* given the similarities between the hominins alive 3–4 mya and the limited, vague, and contradictory information available. However, the species with the most information available, *A. afarensis*, appears to be a good candidate because it was a habitat generalist that foraged on fibrous foods and scavenged large mammals (and therefore could have come into repeated contact with gorilla ancestors, as discussed below) and is a likely ancestor to *Homo*. However, nothing rules out other species in hominin lineage, especially as use of wet forests would not be recorded in the fossil record.

Whichever hominin species was actually first infested with *Pthirus*, the successful transfer of a disease or parasite to another species is most likely when there is relatively frequent contact between the two host species, which indicates that moist forests were a far more important hominin habitat than previously realized. The savanna/grassland model of the origin of bipedalism and hominins has been largely rejected by careful examination of the context of hominin fossils (WoldeGabriel et al. 1994; Reed 1997; WoldeGabriel et al. 2009; Luca et al. 2010; Brunet 2010). These fossils tend to indicate the importance of wooded habitats including woodlands and dry forests or at the very least forests lining bodies of water or forest/woodland/savanna/grassland mosaics (although mosaics can be an illusion created by the coarse resolution of the paleontological record and time averaging of more homogenous habitats changing over a period of time; Elton 2008). Though the emphasis has shifted from the savanna to the woodland and dry forest, moist forests have largely been ignored as potentially important habitats for hominins; however, there is no reason to think hominins would be less flexible in habitat use than those extant primates that use both wet forests and drier open habitats (Elton 2008). This oversight has certainly been reasonable based on the context of fossil finds; in fact, there has never been a hominin fossil found associated with rain forest habitat, more likely due to the difficulty of fossilization rain forests (Kingston 2007 and see above) than habitat specificity. However, the evidence of habitual contact between hominins and gorilla ancestors given by the *Pthirus* host switch provides strong support for wet forests playing a more important role in hominin evolution than normally thought.

Additionally, the loss of body hair that is likely to have occurred before the *Pthirus* switch from gorilla ancestors to human ancestors is much more likely to have happened under a closed canopy forest than in a less wooded ecosystem because the forest reduces the usefulness of body hair by reducing the solar heat load and mitigating diurnal temperature changes as discussed above (Newman 1970). Thus, the loss of body hair likely occurred when hominins were restricted exclusively to dense forests long before the evolution of the australopithecines as habitat generalists that incorporated more open areas into their ranges. While australopithecines could presumably still utilize warm moist forests, use of montane forests of australopithecines is not likely given the loss of body hair by this point and the cold temperatures experienced in these forests. Unfortunately, fossil evidence of wet forest use by hominins will likely be as difficult to come by as fossil evidence of the other great apes that are restricted to wet forests. As mentioned before, the fossil record is almost silent even on the subject of common chimpanzees, which venture into drier habitats (such as woodlands and scrublands) than gorillas but remain largely tied to closed canopy forests. Thus, if restricted to fossil evidence, our picture of hominin evolution and habits is severely limited by the taphonomic processes in the moist forests that form the origin of African ape diversity.

Our suggestion that australopithecines expanded their habitat from drier wooded areas into wet tropical forests introduces several more possibilities. Habitat use could have been seasonal with australopithecines predictably moving from drier habitats to wet forest areas and back. Alternately, a widely spread generalist species, as *A. afarensis* has been proposed to be, might have populations permanently inhabiting entirely different habitats. In this case, the question becomes which habitat was preferred, i.e., which, if any, was able to support higher densities of hominins. A third possibility is that one of the habitats was primary with the other being utilized only in times of drought, etc. The importance of moist forests with poor fossilization conditions to hominins raises the possibility that it was not the dry woodlands that were the center of hominin radiation; rather the radiation occurred in the rain forests with a minority of the species expanding out into more xeric areas to be fossilized and finally found. If this is true, we may still be missing a large portion of hominin diversity.

Of course, for *Pthirus* to have switched hosts successfully, sharing habitats would have been insufficient for transfer—far more intimate contact would have been required. However, the contact would not have to be as intimate as most people seem to gleefully assume. Although hybrids between different guenon monkey species are known (Struhsaker et al. 1988; de Jong and Butynski 2010), there are far more likely scenarios than two species as divergent as archaic gorillas and australopithecines having sexual contact.

Pthirus are known to be transmitted via fomites (Meinking 1999) and therefore could have potentially switched hosts if a hominin used an abandoned louse-infested archaic gorilla nest (Reed et al. 2007). All extant great apes including orangutans fashion nests (Schaller 1965b), so it is reasonable to assume that all apes 3–4 mya also made nests. While this scenario is possible, it seems unlikely. Great ape nests (aside from those of humans) are constructed swiftly for a single use and are simple rudimentary structures. This is especially true for gorilla nests made on the ground, which have better rims than bottoms, provide little if any padding, and are typically constructed in 1 min (though the minority that are made in the trees are more substantial; Schaller 1965a). Unfortunately (but not unexpectedly), there is no surviving evidence of australopithecine nests or nest use. However, if they practiced the single-use pattern typical of great apes, there would have been little reason for individuals who did not reuse their own nests to reuse the nests of another species.

On the other hand, if australopithecines reused nests like at least some of the later *Homo spp.*, the reuse would probably have been motivated by increased effort required to build more complex and functional structures. Again, the reuse of old slipshod gorilla-type nests would have been unlikely.

A more probable scenario for contact between human and gorilla ancestors is the existence of mixed foraging groups. Though rare compared to multiple species associations in monkeys (Yamagiwa and Basabose 2006), mixed foraging groups have been observed containing gorillas and chimpanzees. Interactions from avoidance (Yamagiwa et al. 1996) to obliviousness (Kuroda et al. 1996) have been seen taking place in mixed groups of apes (even from a distance of 3 m; Stanford 2006). The nature of the interaction likely depends on food availability and level of competition as well as the individual personalities of those involved. Habituated gorilla and chimpanzee troops tend to ignore human observers, but some physical interactions (e.g., playing, bluffing, and testing) have occurred, normally with young animals. In primate mixed foraging groups, interactions are most frequent between young animals and can involve play.

Mixed foraging groups of gorilla and human ancestors would have given opportunities for the *Pthirus* switch through play or grooming; however, primate interspecific interactions are rare even where multispecies associations are common. Aggressive interactions are the most numerous interactions, with play being relatively unusual and grooming being extremely rare and of short duration (Ihobe 1990; Heymann and Buchanan-Smith 2000). Because juveniles are typically the age group involved in interspecific interactions and young hominins would have not yet developed the pubic hair that forms the current habitat of *Pthirus* on humans, the ancestral lice would have to be passed to an adult host before they could become established on the new host species. The possibility of a host switch through mixed foraging groups does exist, but because aggression and play rarely involve physical contact (Rose 1977), which occurs quickly and generally between juveniles who would then have to pass *Pthirus* to an adult host, social interaction is a less likely route of host switching than the last possibility: the consumption of archaic gorilla meat by human ancestors.

Other than the gorillas, all the African great apes hunt and consume meat, although only humans have managed to prey on animals of similar and larger body sizes through the use of relatively sophisticated tools. Given the body size of *A. afarensis* (♀, ~29 kg; ♂, ~45 kg; McHenry 1994) compared to that of gorillas (♀, ~80 kg; ♂, ~169 kg; Smith and Jungers 1997) as well as the dangerous nature of enraged gorillas and relatively simple tools used by the australopithecines, it seems unlikely that hunting archaic gorillas by early hominins would be a particularly effective or common food acquisition strategy. It is far more likely that scavenging on gorilla carcasses led to the kind of contact most conducive to a louse host switch: repeated close contact over a substantial period of time. Additionally, lice are extremely sensitive to environmental conditions and readily abandon dead hosts. The desperate situation of lice on a dead or dying host makes the switch to any available host, even one of the incorrect species, much more likely than casual contact between a living native host and a potential novel host.

The presence of likely butcher marks on the bones of large mammals contemporary with *A. afarensis* indicates that scavenging was likely an important and effective component in their feeding repertoire (McPherron et al. 2010, 2011; but see Domínguez-Rodrigo et al. 2010, 2011). Although the validity of these butcher marks does not determine the level of carnivory by *A. afarensis* (Domínguez-Rodrigo et al. 2010), simple tools would have enabled both more efficient processing of meat than allowed by primate dentition alone and the transportation of meat away from the main carcass to a location with less predation danger. Thus, the most likely scenario is that a hominin habitually used moist tropical forests far more than previously realized or shown by the fossil record and opportunistically scavenged meat from gorilla carcasses. This feeding resulting in contact with *Pthirus* that was frequent enough to establish a population of *Pthirus* on hominins 3–4 mya.

Conclusion

There has been much research into the biology, epidemiology, and the evolutionary history of lice. These parasites have bedeviled human and nonhuman primates alike for millions of years, and the increase in louse prevalence over the last 20 years suggests they will continue to parasitize humans for some time yet. Due to their obligate host-specific nature, we can use these parasites to inform us about human evolutionary history and gather information that is not available in the host fossil record, providing an unexpected benefit to an otherwise bothersome parasite.

Although most great ape lice have strictly cospeciated with their great ape hosts, *Pthirus pubis* (the human pubic louse) has a different evolutionary history. *Pthirus* switched to the human lineage 3–4 mya from an archaic gorilla. The biology of *Pthirus* suggests that for this host switch to have occurred, suitable habitat had to be available, which indicates that hominins had not only lost their body hair but also developed androgenic hair by 3–4 mya.

Because gorillas are conservative in their habitats and diet, we postulate that this likely means that these archaic hominins were using similar habitat as gorillas (moist forest habitat). The best candidate for this host switch was *Australopithecus afarensis* (based on our current knowledge). *A. afarensis* possibly butchered large mammal carcasses during this time, presenting a scenario of *A. afarensis* scavenging archaic gorilla meat and *Pthirus* likely switching to *A. afarensis* from a dead gorilla host. *Pthirus* then continued to evolve with the hominin lineage as a sexually transmitted disease due to their placement on the body, explaining the presence of two genera of lice (*Pthirus* and *Pediculus*) on extant humans today.

References

Amaral LQ (1996) Loss of body hair, bipedality and thermoregulation: comments on recent papers in the journal of human evolution. J Hum Evol 30:357–366
Amevigbe MDD, Ferrer A, Champorie S, Monteny N, Deunff J, Richard-Lenoble D (2000) Isoenzymes of human lice: Pediculus humanus and P. capitis. Med Vet Entomol 14:419–425
Anderson AL, Chaney E (2009) Pubic lice (*Pthirus pubis*): history, biology and treatment vs. knowledge and beliefs of US college students. Int J Environ Res Public Health 6:592–600
Andersson JO, Andersson SG (2000) A century of typhus, lice and Rickettsia. Res Microbiol 151:143–150
Angelakis E, Rolain JM, Raoult D, Brouqui P (2011) Bartonella Quintana in head louse nits. FEMS Immunol Med Mic 62(2):244–246
Bonilla DL, Kabeya H, Henn J, Kramer VL, Kosoy MY (2009) *Bartonella quintana* in body lice and head lice from homeless persons, San Francisco, California, USA. Emerg Infect Dis 15(6):912–915
Bonnefille R (2010) Cenozoic vegetation, climate changes and hominid evolution in tropical Africa. Glob Planet Change 72:390–411
Bonnefille R, Potts R, Chalié F, Jolly D, Peyron O (2004) High-resolution vegetation and climate change associated with Pliocene *Australopithecus afarensis*. Proc Natl Acad Sci 101:12125–12129
Brace CL, Montagu A (1977) Human evolution, 2nd edn. Macmillan, New York
Brunet M (2010) Two new Mio-Pliocene Chadian hominids enlighten Charles Darwin's 1871 prediction. Phil Trans Roy Soc B 365:3315–3321
Brunet M, Beauvilain A, Coppens Y, Heintz E, Moutaye AHE, Pilbeam D (1995) The first australopithecine 2,500 kilometers west of the Rift Valley (Chad). Nature 378:273–275
Brunet M, Guy F, Pilbeam D, Mackaye HT, Likius A, Ahounta D, Beauvilain A, Blondel C, Bocherensk H, Boisserie J-R, De Bonis L, Coppens Y, Dejax J, Denys C, Duringer P, Eisenmann V, Fanone G, Fronty P, Geraads D, Lehmann T, Lihoreau F, Louchart A, Mahamat A, Merceron G, Mouchelin G, Otero O, Campomanes PP, De Leon MP, Rage J-C, Sapanetkk M, Schusterq M, Sudrek J, Tassy P, Valentin X, Vignaud P, Viriot L, Zazzo A, Zollikofer C (2002) A new hominid from the Upper Miocene of Chad, Central Africa. Nature 418:145–151
Burgess IF (1995) Human lice and their management. Adv Parasitol 36:271–342
Burgess IF (2004) Human lice and their control. Annu Rev Entomol 49:457–481
Burgess I, Maunder JW, Myint TT (1983) Maintenance of the crab louse, *Pthirus pubis*, in the laboratory and behavioral studies using volunteers. Community Med 5:238–241
Burkhart CN, Burkhart CG (2000) The route of head lice transmission needs enlightenment for proper epidemiologic evaluations. Int J Dermatol 39(11):878–879
Burkhart CN, Burkhart CG (2007) Fomite transmission in head lice. J Am Acad Dermatol 56:1044–1047
Bush SE, Rock AN, Jones SL, Malenke JR, Clayton DH (2011) Efficacy of the LouseBuster, a new medical device for treating head lice (Anoplura: Pediculidae). J Med Entomol 48(1):67–72
Busvine JR (1944) Simple experiments on the behaviour of body lice (Siphunculata). Proc Roy Ent Soc Lond 19:22–26
Busvine JR (1978) Evidence from double infestations for the specific status of human head lice and body lice (Anoplura). Syst Entomol 3:1–8
Buxton PA (1941) On the occurrence of the crab-louse (*Phthirus pubis:* Anoplura) in the hair of the head. Parasitology 33:117–118
Buxton P (1947) The louse an account of the lice which infest man, their medical importance and control. Edward-Arnold, London
Canyon DV, Spearce R, Muller R (2002) Spatial and kinetic factors for the transfer of head lice (*Pediculus capitis*) between hairs. J Invest Dermatol 119:629–631
Canyon DV, Speare R (2010) Indirect transmission of head lice via inanimate objects. Open Dermatol J 4:72–76

Chosidow O (2000) Scabies and pediculosis. Lancet 355:819–826
Cote SM (2004) Origins of the African hominoids: an assessment of the palaeobiogeographical evidence. CR Palevol 3:323–340
Darwin C (1871) The descent of man and selection in relation to sex, 2nd edn. John Murray, London
de Jong YA, Butynski TM (2010) Thre Sykes's Monkey *Cercopithecus mitis* × Vervet Monkey *Chlorocebus pygerythrus* hybrids in Kenya. Primat Cons 25:43–56
Delson E, Tattersall I, Van Couvering J, Brooks A (2000) Encyclopedia of human evolution and prehistory. Garland, New York
Domínguez-Rodrigo M, Pickering TR, Bunn HT (2010) Configurational approach to identifying the earliest hominin butchers. Proc Natl Acad Sci 107(49):20929–20934
Domínguez-Rodrigo M, Pickering TR, Bunn HT (2011) Reply to McPherron et al.: Doubting Dikika is about data, not paradigms. Proc Natl Acad Sci 108(21):E117
Elgart ML, Higdon RS (1973) Pediculosis pubis of the scalp. Arch Dermatol 107:916–917
Elton S (2008) The environmental context of human evolutionary history in Eurasia and Africa. J Anat 212:377–393.
Fisher I, Morton RS (1970) *Phthirus pubis* infestation. Br J Vener Dis 46:326–329
Foley R (2002) Adaptive radiations and dispersals in hominin evolutionary ecology. Evol Anthropol Suppl 1:132–137
Frankowski BL, Bocchini JA (2010) Head lice. Pediatrics 126:392–403
Frankowski BL, Weiner LB (2002) Head lice. Pediatrics 110(3):638–643
Galik K, Senut B, Pickford M, Gommery D, Treil J, Kuperavage AJ, Eckhardt RB (2004) External and internal morphology of the BAR 1002'00 *Orrorin tugenensis* femur. Science 305(5689):1450–1453
Gao, F, Bailes E, Robertson DL, Chen Y, Rodenburg CM, Michael SF, Cummins LB, Arthur LO, Peeters M, Shaw GM, Sharp PM, Hahn, BH (1999) Origin of HIV-1 in the chimpanzee Pan troglodytes troglodytes. Nature 397:436–441
Glass B (1966) The evolution of hairlessness in man. Science 152:294
Goldberger J, Anderson JF (1912) The transmission of typhus fever, with especial reference to transmission by the head louse (*Pediculus capitis*). Public Health Rep 27(9):297–307
Gordon SC (2007) Shared vulnerability: a theory of caring for children with persistent head lice. J School Nurs 5:283–292
Green RE, Krause J, Briggs AW, Maricic T, Stenzel U, Kricher M, Patterson N, Li H, Zhai W, Fritz MH, Hansen NF, Durand EY, Malaspinas A, Jensen JD, Marques-Bonet T, Alkan C, Prüfer K, Meyer M, Burbano HA, Good JM, Schultz R, Aximu-Petri A, Butthof A, Höber B, Höffner B, Siegemund M, Weihmann A, Nusbaum C, Lander ES, Russ C, Novod N, Affourtit J, Egholm M, Verna C, Rudan P, Brajkovic D, Kucan Z, Gušic I, Doronichev VB, Golovanova LV, Lalueza-Fox C, Rasilla M, Fortea J, Rosas A, Schmitz RW, Johnson PLF, Eichler EE, Falush D, Birney E, Mullikin JC, Slatkin M, Nielsen R, Kelso J, Lachmann M, Reich D, Pääbo S (2010) A draft sequence of the Neanderthal genome. Science 328:710–722
Grine FE, Ungar P, Teaford MF, El-Zaatari S (2006) Molar microwear in *Praeanthropus afarensis*: evidence for dietary stasis through time and under diverse paleoecological conditions. J Hum Evol 51
Guthrie RD (1976) Body hot spots: the anatomy of human social organs and behavior. Van Nostrand Reinhold, New York
Guy F, Mackaye H-T, Likius A, Vignaud P, Schmittbuhl M, Brunet M (2008) Symphyseal shap variation in extant and fossil hominoids, and the symphysis of *Australopithecus bahrelghazali*. J Hum Evol 55:37–47
Haile-Selassie Y (2001) Late Miocene hominids from the Middle Awash, Ethiopia. Nature 412:178–181
Hansen RC, O'Havier JO (2004) Economic considerations associated with *Pediculus humanus capitis* infestation. Clin Ped 43:523–527
Heymann EW, Buchanan-Smith HM (2000) The behavioral ecology of mixed-species troops of callitrichine primates. Biol Rev 75:169–190

Hooke R (1665) Micrographia: or some physiological description of minute bodies made by magnifying glasses with observations and inquiries thereupon. Council Royal Society of London for Improving of Natural Knowledge, London
Howlett FM (1917) Notes on head- and body-lice and upon temperature reactions of lice and mosquitoes. Parasitology 10:186–188
Hypsa V (2006) Parasite histories and novel phylogenetic tools: alternative approaches to inferred parasite evolution from molecular markers. Int J Parasitol 36:141–155
Ihobe H (1990) Interspecific interactions between wild pygmy chimpanzees (*Pan paniscus*) and red colobus (*Colobus badius*). Primates 31:109–112
Kimbel WH, Delezene LK (2009) "Lucy" redux: a review of research on *Australopithecus afarensis*. Yearbk Phys Anthropol 52:2–48
Kimbel W, Lockwood CA, Ward CV, Leakey MG, Rak Y, Johanson DC (2006) Was *Australopithecus anamensis* ancestoral to *A. afarensis*? A case of anagenesis in the hominin fossil record. J Hum Evol 51:134–152
Kingdon J (1974) East African mammals: an altas of evolution in Africa. University of Chicago Press, Chicago
Kingston JD (2007) Shifting adaptive landscapes: progress and challenges in reconstructing early hominid environments. Yearbk Phys Anthropol 50:20–58
Kirkness EF, Haas BJ, Sun W, Braig HR, Perotti MA, Clark JM et al (2010) Genome sequences of the human body louse and its primary endosymbiont provide insights into the permanent parasitic lifestyle. Proc Natl Acad Sci 107(27):12168–12173
Kittler R, Kayser M, Stoneking M (2003) Molecular evolution of Pediculus humanus and the origin of clothing. Curr Biol 13:1414–1417
Kittler R, Kayser M, Stoneking M (2004) Erratum molecular evolution of *Pediculus humanus* and the origin of clothing. Curr Biol 14:2309
Kunimatsu Y, Nakatsukasa M, Sawada Y, Sakai T, Hyodo M, Hyodo H, Itaya T, Nakaya H, Saegusa H, Mazurier A, Saneyoshi M, Tsujikawa H, Yamamoto A, Mbua E (2007) A new late Miocene great ape from Kenya and its implications for the origins of African great apes and humans. Proc Natl Acad Sci 104:19220–19225
Kuroda S, Nishihara T, Suzuki S, Oko RA (1996) Sympatric chimpanzees and gorillas in the Ndoki Forest, Congo. In: McGrew WC, Marchant LF, Nishida T (eds) Great ape societies. Cambridge University Press, Cambridge, pp 71–81
Kushlan JA (1985) The vestiary hypothesis of human hair reduction. J Hum Evol 14:29–32
Leakey MG, Spoor F, Brown FH, Gathogo PN, Kairie C, Leakey LN, McDougall I (2001) New hominin genus from eastern Africa shows diverse middle Pliocene lineages. Nature 410:433–440
Leo NP, Barker SC (2005) Unravelling the origins of the head lice and body lice of humans. Parasitol Res 98:44–47
Leo NP, Hughes JM, Yang X, Poudel SKS, Brogdon WG, Barker SC (2005) The head and body lice of humans are genetically distinct (Insecta: Phthiraptera: Pediculidae): evidence from double infestations. Heredity 95:34–40
Li W, Ortiz G, Fournier P-E, Gimenez G, Reed DL, Pittendrigh B et al (2010) Genotyping of human lice suggests multiple emergences of body lice from local head louse populations. PLoS Neglect Trop Dis 4(3):e641
Light JE, Toups MA, Reed DL (2008) What's in a name: the taxonomic status of human head and body lice. Mol Phylogenet Evol 47(3):1203–1216
Light JE, Smith VS, Allen JM, Durden LA, Reed DL (2010) Evolutionary history of mammalian sucking lice (Phthiraptera: Anoplura). BMC Evol Biol 10:292
Lloyd L (1919) Lice and their menace to man. Frowde, Oxford
Luca F, Perry GH, Di Rienzo A (2010) Evolutionary adaptations to dietary changes. Annu Rev Nutr 30:291–314
Macho GA, Shimizu D (2010) Kinematic parameters inferred from enamal microstructure: new insights into the diet of *Australopithecus anamensis*. J Hum Evol 58:23–32
Mahoney SA (1980) Cost of locomotion and heat balance during rest and running from 0 to 55°C in a patas monkey. J Appl Physiol: Respirat Environ Exercise Physiol 49:789–800

McBrearty S, Jablonski NG (2005) First fossil chimpanzee. Nature 437:105–108
McHenry H (1994) Behavioral ecological implications of early hominid body size. J Hum Evol 27:77–87
McPherron SP, Alemseged Z, Marean CW, Wynn JG, Reed D, Geraads D, Bobe R, Béarat HA (2010) Evidence for stone-tool-assisted consumption of animal tissues before 3.39 million years ago at Dikika, Ethiopia. Nature 466:857–860
McPherron SP, Alemseged Z, Marean C, Wynn JG, Reed D, Geraads D, Bobe R, Bearat H (2011) Tool-marked bones from before the Oldowan change the paradigm. Proc Natl Acad Sci 108(21):E116
Meinking TL (1999) Infestations. Curr Probl Dermatol 11(3):75–118
Montagna W (1972) The skin of nonhuman primates. Am Zool 12:109–121
Murray ES, Torrey SB (1975) Virulence of Rickettsia prowazeki for head lice. Ann N Y Acad Sci 266:25–34
Newman RW (1970) Why man is such a sweaty and thirsty naked mammal: a speculative review. Hum Biol 42:12–27
Nieberding CM, Olivieri I (2007) Parasites: proxies for host genealogy and ecology? Trends Ecol Evol 22:156–165
Nuttall GHF (1918) The Biology of *Phthirus pubis*. Parasitology 10:383–405
Orion E, Matz H, Wolf R (2004) Ectoparasitic sexually transmitted diseases: scabies and pediculosis. Clin Dermatol 22:513–519
Pagel M, Bodmer W (2003) A naked ape would have few parasites. Proc Roy Soc Lond B Biol 270:S117–S119
Parola P, Fournier PE, Raoult D (2006) *Bartonella quintana*, lice, and molecular tools. J Med Entomol 43(5):787
Payot F (1920) Contribution a l'etude du Phthirus pubis. Bull Soc Vaud Sci Nat 53:127–161
Pickford M, Senut B, Ssemmanda I, Elepu D, Obwona P (1988) Premiers resultats de la mission de l'Uganda palaeontology expedition a Nkondo (Pliocene du bassin du lac Albert, Ouganda). CR Acad Sci II 306:315–320
Pickford M, Senut B, Gommery D, Treil J (2002) Bipedalism in *Orrorin tugenensis* revealed by its femora. Comptes Rendus Palevol 1(4):191–203
Pilbeam D, Gould SJ (1974) Size and scaling in human evolution. Science 186:892–901
Plumptre AJ (1994) The effects of trampling damage by herbivores on the vegetation of the Parc National des Volcans, Rwanda. Afr J Ecol 32:115–129
Pollack RJ, Kiszewski AE, Spielman A (2000) Overdiagnosis and consequent mismanagement of head louse infestations in North America. Pediatr Infect Dis J 19(8):689–693
Randall VA (1994) Androgens and human hair growth. Clin Endocrinol 40:439–457
Randall VA (2008) Androgens and hair growth. Dermatol Ther 21:314–328
Rantala MJ (1999) Human nakedness: adaptation against ectoparasites? Int J Parasitol 29: 1987–1989
Rantala MJ (2007) Evolution of nakedness in *Homo sapiens*. J Zool 271:1–7
Raoult D, Roux V (1999) The body louse as a vector of reemerging human diseases. Clin Infect Dis 29:888–911
Raoult D, Dutour O, Houhamdi L, Jankauskas R, Fournier P-E, Ardagna Y et al (2006) Evidence for louse-transmitted diseases in soldiers of Napoleon's Grand Army in Vilnius. J Infect Dis 193:112–120
Reed KE (1997) Early hominid evolution and ecological change through the African Plio-Pleistocene. J Hum Evol 32:289–322
Reed DL, Hafner MS, Allen SK (2000) Mammalian hair diameter as a possible mechanism for host specialization in chewing lice. J Mammal 81:999–1007
Reed DL, Smith VS, Hammond SL, Rogers AR, Clayton DH (2004) Genetic analysis of lice supports direct contact between modern and archaic humans. PLoS Biol 2(11):e340
Reed DL, Light JE, Allen JM, Kirchman JJ (2007) Pair of lice lost or parasites regained: the evolutionary history of anthropoid primate lice. BMC Biol 5:7

Reed DL, Toups MA, Light JE, Allen JM, Flannigan S (2009) Lice and other parasites as markers of primate evolutionary history. In: Huffman M, Chapman C (eds) Primate parasite ecology: the dynamics and study of host-parasite relationships. Cambridge University Press, Cambridge, pp 231–250

Richmond BG, Jungers WL (2008) *Orrorin tugenensis* femoral morphology and the evolution of hominin bipedalism. Science 319:1662–1665

Rogers AR, Iltis D, Wooding S (2004) Genetic variation at the MC1R locus and the time since loss of body hair. Curr Anthropol 45:105–108

Rose MD (1977) Interspecific play between free ranging guerezas (*Colobus guereza*) and vervet monkeys (*Cercopithecus aethiops*). Primates 18:957–964

Sasaki T, Poudel SKS, Isawa H, Hayashi T, Seki N, Tomita T, Sawabe K, Kobayashi M (2006a) First molecular evidence of *Bartonella quintana* in *Pediculus humanus capitis* (Phthiraptera: Pediculidae, collected from Nepalese children. J Med Entomol 43(1):110–112

Sasaki T, Poudel SKS, Isawa H, Hayashi T, Seki N, Tomita T, Sawabe K, Kobayashi M (2006b) First molecular evidence of Bartonella quintana in Pediculus humanus capitis (Phthiraptera: Pediculidae), collected from Nepalese children. J Med Entomol 43(5):788

Schaller GB (1963) The Mountain Gorilla: ecology and behavior. University of Chicago Press, Chicago

Schaller GB (1965a) The behavior of the Mountain Gorilla. In: DeVore I (ed) Primate behavior: field studies of monkeys and apes. Holt, Rinehart, and Winston, New York, pp 324–367

Schaller GB (1965b) Behavioral comparisons of the apes. In: DeVore I (ed) Primate behavior: field studies of monkeys and apes. Holt, Rinehart, and Winston, New York, pp 474–481

Schultz AH (1931) The density of hair in primates. Hum Biol 3:303–321

Schultz AH (1969) The life of primates. Universe Books, New York

Schwartz GG, Rosenblum LA (1981) Allometry of hair density and the evolution of human hairlessness. Am J Phys Anthropol 55:9–12

Smith RJ, Jungers WL (1997) Body mass in comparative primatology. J Hum Evol 32:523–559

Sockol MD, Raichlen DA, Pontzer H (2007) Chimpanzee locomotor energetics and the origin of human bipedalism. Proc Natl Acad Sci 104:12265–12269

Spoor F, Leakey MG, Leakey LN (2010) Hominin diversity in the Middle Pliocene of eastern Africa: the maxilla of KNM-WT 40000. Proc Roy Soc Lond B 365:3377–3388

Stanford CB (2006) The behavioral ecology of sympatric African apes: implications for understanding fossil hominoid ecology. Primates 47:91–101

Stafford (2008) Head lice: evidence-based guidelines based on the Stafford Report 2008 Update. Public Health Med Environ Group

Stoddart DM (1990) The scented ape: the biology and culture of human odour. Cambridge University Press, Cambridge

Strait DS, Grine FE (2004) Inferring hominoid and early hominid phylogeny using craniodental characters: the role of fossil taxa. J Hum Evol 47:399–452

Strait DS, Grine FE, Moniz MA (1997) A reappraisal of early hominid phylogeny. J Hum Evol 32:17–82

Struhsaker TT, Butynski TM, Lwanga JS (1988) Hybridization between redtail (*Cercopithecus ascanius schmidti*) and blue (*C. mitis stuhlmanni*) monkeys in the Kibale Forest, Uganda. In: Gautier-Hion A, Bourlière F, Gautier JP, Kingdon J (eds) A primate radiation: evolutionary biology of the African Guenons. Cambridge University Press, Cambridge, pp 477–497

Suwa G, Kono RT, Katoh S, Asfaw B, Beyene Y (2007) A new species of great ape from the late Miocene epoch in Ethiopia. Nature 448:921–924

Teaford MF, Ungar PS (2000) Diet and the evolution of the earliest human ancestors. Proc Natl Acad Sci 97:13506–13511

Tebruegge M, Pantazidou A, Curtis N (2011) What's bugging you? An update on the treatment of head lice infestation. Arch Dis Child Educ Pract Ed 96:2–8

Thalmann O, Fischer A, Lankester F, Pääbo S, Vigilant L (2007) The complex evolutionary history of gorillas: Insights from genomic data. Mol Biol Evol 24:146–158

Toups MA, Kitchen A, Light JE, Reed DL (2011) Origin of clothing lice indicates early clothing use by anatomically modern humans in Africa. Mol Biol Evol 28(1):29–32
Ubelaker JE, Payne E, Allison VF, Moore DV (1973) Scanning electron microscopy of the human pubic louse, *Pthirus pubis* (Linnaeus, 1758). J Parasitol 59(5):913–919
Ungar P (2004) Dental topography and diets of *Australopithecus afarensis* and early *Homo*. J Hum Evol 46:605–622
Ungar PS, Scott RS, Grine FE, Teaford MF (2010) Molar microwear textures and the diets of *Australopithecus anamensis* and *Australopithecus afarensis*. Proc Roy Soc Lond B Biol 365:3345–3354
Verhaegen M, Puech P-F, Munro S (2002) Aquarboreal ancestors? Trends Ecol Evol 17:212–217
Waldeyer L (1900) Ein Fall von *Phthirius pubis* im Bereiche des behaarten Kopfes. Charite-Annalen, Berlin, XXV:494–499
Wheeler PE (1992) The influence of the loss of functional body hair on the water budgets of early hominids. J Hum Evol 23:379–388
White T (2003) Early hominids–diversity or distortion? Science 299:1994–1997
Whiteman NK, Parker PG (2005) Using parasites to infer host population history: a new rationale for parasite conservation. Anim Cons 8:175–181
Wilkinson RD, Steiper ME, Soligo C, Martin RD, Yang ZH, Tavare S (2011) Dating primate divergences through an integrated analysis of palaeontological and molecular data. Syst Biol 60(1):16–32
WoldeGabriel G, White TD, Suwa G, Renne P, de Heinzelin J, Hart WK, Heiken G (1994) Ecological and temporal placement of early Pliocene hominids at Aramis, Ethiopia. Nature 371:330–333
WoldeGabriel G, Ambrose SH, Barboni D, Bonnefille R, Bremond L, Currie B, DeGusta D, Hart WK, Murray AM, Renne PR, Jolly-Saad MC, Stewart KM, White TD (2009) The geological, isotopical, botanical, invertebrate and lower vertebrate surroundings of *Ardipithecus ramidus*. Science 326(5949):65e1–65e5
Wolfe ND, Dunavan CP, Diamond J (2007) Origins of major human infectious diseases. Nature 447(17):279–283
Yamagiwa J, Basabose AK (2006) Effects of fruit scarcity on foraging strategies of sympatric gorillas and chimpanzees. In: Hohmann G, Robbins M, Boesch C (eds) Feeding ecology in apes and other primates. Cambridge University Press, Cambridge, pp 73–96
Yamagiwa J, Maruhashi T, Yumoto T, Mwanza N (1996) Dietary and ranging overlap in sympatric gorillas and chimpanzees in Kahuzi-Biega National Park, Zaire. In: McGrew WC, Marchant LF, Nishida T (eds) Great ape societies. Cambridge University Press, Cambridge, pp 82–98
Yotova V, Lefebvre JF, Moreau C, Gbeha E, Hovhannesyan K, Bourgeois S, Bédarida S, Azevedo L, Amorim A, Sarkisian T, Avogbe P, Chabi N, Dicko MH, Amouzou ESKS, Sanni A, Roberts-Thomson J, Boettcher B, Scott RJ, Labuda D (2011) An X-linked haplotype of Neanderthal origin is present among all non-African populations. Mol Biol Evol 28(7):1957–1962

Part II
Emergence and Divergent Disease Manifestation

Treponema pallidum Infection in Primates: Clinical Manifestations, Epidemiology, and Evolution of a Stealthy Pathogen

Kristin N. Harper and Sascha Knauf

Introduction

Treponema pallidum is a pathogenic bacterium that causes a variety of debilitating diseases in humans and nonhuman primates (NHPs). In humans, *T. pallidum* causes the sexually transmitted disease syphilis and the nonsexually transmitted diseases yaws and bejel. Its history in humans is mysterious and has been a source of great controversy over the years. Did yaws affect our earliest ancestors, perhaps present even in *Homo erectus* (Rothschild et al. 1995)? Where and when did syphilis arise—did Columbus bring it from the New World to the Old (Crosby 1969; Diaz de Isla 1539; Harper et al. 2008, 2011)? Its history and distribution in NHPs has been much less studied, but novel data are accumulating. Although *Treponema* infection has been reported in a Pleistocene bear (Rothschild and Turnbull 1987), reports of naturally occurring *T. pallidum* infection in species outside of primates have never been replicated. Here, we summarize what is known about *T. pallidum* infection in both human and NHPs, considering diverse sources of evidence, including serology of wild animals, clinical manifestations, and genetic characterization of strains. We focus on how our understanding of the history of the disease in NHPs can inform our understanding of its transmission and evolutionary history in humans and vice versa.

K.N. Harper (✉)
Environmental Health Sciences, Columbia University, New York, NY 10032, USA
e-mail: kh2383@columbia.edu

S. Knauf
Pathology Unit, German Primate Center, Leibniz-Institute for Primate Research, Goettingen, Germany

J.F. Brinkworth and K. Pechenkina (eds.), *Primates, Pathogens, and Evolution*, Developments in Primatology: Progress and Prospects,
DOI 10.1007/978-1-4614-7181-3_7, © Springer Science+Business Media New York 2013

Treponema pallidum in Humans

T. pallidum subsp. *pallidum*, the causative agent of syphilis, is the only treponemal subspecies that is primarily transmitted via sex. An estimated 12 million new cases of syphilis occur every year and can be found throughout the world (WHO 2001). Certain socio-ecological risk factors (e.g., multiple sex partners and immunosuppression) are associated with a greater risk of contracting and transmitting the disease (Karp et al. 2009; Rolfs et al. 1990; Yahya-Malima et al. 2008). Yaws, caused by subsp. *pertenue*, is a nonvenereal disease that is usually acquired during early childhood. Transmission appears to occur primarily via skin-to-skin contact (Perine et al. 1984), although flies may also serve as vectors (Cousins 1972; Kumm and Turner 1936; Satchell and Harrison 1953). Yaws was once common in tropical regions throughout the world. Although it has never been eradicated, it grew increasingly rare after a WHO-sponsored eradication campaign mid-century (Arya and Bennett 1976; Guthe et al. 1953, 1972). Surveillance is typically poor in the areas most likely to be affected; accurate figures on the infection's current prevalence are not available. However, recent foci of infection in the Republic of Congo and the Central African Republic (Salomone 1999), the Democratic Republic of Congo (Gersti et al. 2009), Papua New Guinea (Mitjà et al. 2011), East Timor (Satter and Tokarz 2010), and Vanuatu (Fegan et al. 2010) have been documented. The disease appears to have been eradicated in India (Lahariya and Pradhan 2007) and has not been reported recently in South America either. Finally, bejel, caused by subsp. *endemicum*, was once common in arid regions such as the Middle East and the Balkans, but it has not been reported in the literature since the 1990s (Yakinci et al. 1995). Like yaws, it is typically acquired during childhood and is thought to be transmitted primarily via fomites such as utensils and drinking vessels (Perine et al. 1984).

All three diseases are chronic, multistage infections in humans that are easily treated by antibiotics in most cases, but they can be deadly if neglected (Table 1). It should be noted that increasing macrolide resistance has begun to complicate treatment worldwide, and antibiotic-resistant *T. pallidum* could one day represent a substantial health challenge (Stamm 2010). Syphilis typically begins with a hard, painless chancre at the site where the bacterium entered the body. Untreated infection leads to a secondary stage, often characterized by a rash that appears weeks later. The rash associated with secondary syphilis may take several forms, but it generally affects the palms and soles and does not itch (Richens and Mabey 2009). Mild fever, patchy hair loss, and weight loss are also common clinical signs in humans. If left untreated, tertiary-stage disease may occur, though sometimes not until decades after the infection was initially contracted. Destructive lesions called gummata may appear in virtually any organ of the body; neurosyphilis may result in psychiatric disorders, while cardiovascular syphilis killed many people in the pre-antibiotic era (Holmes et al. 2007).

The disease progression is similar for infection by all three of the *T. pallidum* subspecies, although a number of important differences have been reported (Table 1).

Table 1 Characteristics of natural infection with *T. pallidum in* different primate species

	Humans			Nonhuman primates[a]		
	Syphilis	Yaws	Bejel	Baboons-West Africa	Baboons-Tanzania	Gorillas
Genetic characteristics of strain responsible	*T. pallidum* subsp. *pallidum*	*T. pallidum* subsp. *pallidum*	*T. pallidum* subsp. *pallidum*	Genetic sequences from one strain collected in Guinea show most similarity to subsp. *pertenue*	Genetic sequences from strains at Lake Manyara National Park show most similarity to subsp. *pertenue*	Unknown
Primary-stage lesions	Hard, painless chancre develops at site where bacterium entered the body (usually anogenital region)	Papillomatous "mother yaw" develops at site where bacterium entered the body (most often legs)	Lesions at site where bacterium enters the body are rare. When present, they resemble chancres	When present at all: mild, small, keratotic lesions and ulcers affecting muzzle, eyelids, armpits	Moderate to severe genital ulceration; enlargement of inguinal lymph nodes	Scabby, dry raised lesions primarily affecting lips, nose, eyes, and cheeks. Large destructive lesions primarily affecting nose and mouth areas; lesions also found on wrists and ankles
Secondary-stage lesions	Non-itchy rash affecting palms and soles of feet; alopecia, mild fever, weight loss, enlargement of lymph nodes	Crustopapillomatous skin lesions, polydactylitis, osteoperiostitis, nodules, plaques, and papules on the skin	Angular stomatitis, mucous patches in mouth, rashes with hypertrophic lesions, pigmentary changes, osteoperiostitis			
Tertiary-stage lesions	Gummata can develop in virtually any organ of the body, neurological and cardiovascular involvement	Hyperkeratosis, ulceration around the nasal and maxillary areas (gangosa), saber tibia, gondou (hypertrophic osteitis of nasal process of maxilla)	Severe osteoperiostitis, sometimes resulting in saber tibia, gummata of the skin, palate, and nasal septum, gangosa	None reported		

(continued)

Table 1 (continued)

	Humans			Nonhuman primates[a]		
	Syphilis	Yaws	Bejel	Baboons-West Africa	Baboons-Tanzania	Gorillas
Primary transmission mode	Sexual	Skin to skin	Thought to be mouth to mouth. Fomites such as utensils and drinking vessels may be important	Unknown	Lesions targeting the genitals and involvement of only sexually active animals suggests sexual transmission	Unknown
Typical age of infection	Sexually mature individuals	Childhood	Childhood; 66 % acquired before the age of 16 in one study of 3,507 cases in Iraq	Unknown	Sexually mature animals	Lesions have been reported in almost 5 % of infants in the Parc National d'Odzala-Kokoua (Republic of Congo); peak incidence between infant and juvenile stages

Congenital transmission	Yes	Rarely, if ever	Rarely, if ever	Unknown	Unknown	Unknown
Geographic range	Global	Once very common in hot, humid regions of the world. Still reported in some countries in Western Africa and the South Pacific	Once very common in hot, arid regions such as the Middle East and the Eastern Mediterranean. Last reported case in Turkey, during the 1990s	West Africa: Guinea, Senegal, Cameroon	Tanzania	West Africa: Republic of Congo, Cameroon, Democratic Republic of Congo
Sources	Perine et al. (1984), Salazar et al. (2002a, b)	Noordhoek et al. (1991), Perine et al. (1984)	Yakinci et al. (1995), Pace and Csonka (1984), Csonka (1953), Perine et al. (1984)	Baylet et al. (1971), Fribourg-Blanc and Mollaret (1969), Harper et al. (2008), Smajs et al. (2011)	Wallis and Lee (1999), Mlengeya (2004), Knauf et al. (2012)	Cousins (1984), Karesh (2000), Levréro et al. (2007)

[a]Infection in NHPs is not well characterized enough to determine whether there are primary, secondary, and tertiary-stage lesions. Therefore, all lesions are described in one category

First, only syphilis is regularly transmitted congenitally. While there are several reports of possible yaws infection via *in utero* transmission in the literature (Engelhardt 1959; Wilson and Mathis 1930), it appears likely that congenital infection does not occur in the nonvenereal treponematoses (Antal et al. 2002). It has been hypothesized that this may be because yaws and bejel are typically acquired early in life, so active infection at sexual maturity is virtually nonexistent (Willcox 1955). It is also commonly believed that only subsp. *pallidum*, the agent of syphilis, affects the central nervous system (CNS). However, in-depth study of the progression of subsp. *pertenue* and *endemicum* infections has not been carried out, and one of the rare studies focusing on CNS involvement in yaws infection reported the presence of spirochetes in ocular fluid (Smith et al. 1971). In addition, Román and Román (1986) draw attention to a possible association between CNS complications and yaws infection. Finally, although for the most part syphilis transmission occurs via a venereal route and the other treponemal diseases are nonsexually transmitted, the potential for transmission via alternate routes is known for all *T. pallidum* infections. Cases of extragenital syphilis resulting from nonsexual contact have been reported several times (Luger 1972; Taylor 1954), for example, by human bites (Oh et al. 2008) or by mouth-to-mouth feeding in infants (Zhou et al. 2009). Similarly, infectious yaws and bejel lesions have been reported on the genitalia (Turner and Hollander 1957b; Wilson and Mathis 1930), suggesting that the opportunity for sexual transmission might be present to some extent in all of the *T. pallidum* subspecies. In summary, because so few in-depth studies have been performed on the nonsexually transmitted *T. pallidum* subspecies, the exact nature of the similarities and differences between the three diseases remains ambiguous.

Only recently have we gained the tools to differentiate between the three subspecies. For decades, a diagnosis of syphilis, yaws, or bejel was given based on the clinical presentation of the disease, the age of the patient, and the area of the world in which they lived. In areas where multiple treponemal diseases were present, the differential diagnosis could be quite puzzling (Lagarde et al. 2003). In the 1990s, single-nucleotide polymorphisms that reliably differentiated subsp. *pallidum* strains from the two nonsexually transmitted subspecies were identified for the first time (Centurion-Lara et al. 1998; Noordhoek et al. 1990). Subsequently, patterns of genetic variation that could be used to separate all three subspecies were identified (Centurion-Lara et al. 2006; Harper et al. 2008). Thus, while we still have no serological test that can differentiate between the three diseases, we can now use genetics to do so.

The multiple subspecies of *T. pallidum* present in humans raise the question: which selective pressures led to the emergence of multiple transmission modes? In the past, researchers have speculated that subsp. *pertenue* may represent the ancestral, nonsexually transmitted pathogen, perhaps infecting the earliest hominids (Cockburn 1967; Harper et al. 2008). In this view, subsp. *pallidum* is a relatively recent pathogen. It could have emerged in the Old World, as large cities began to appear and novel sexual behavior fostered sexual transmission (Hudson 1963). Or, according to a more conventionally held theory supported by recent genetic evidence (Harper et al. 2008), it may have arisen in the New World during Pre-Columbian times, introduced into the Old World by Columbus and his men during the voyages of discovery (Crosby 1969). Some researchers believe that the skeletal evidence indicates that subsp. *pallidum* may not have emerged until the voyages of discovery themselves, when Columbus and his crew took a nonvenereal form of *T. pallidum* back to the Old World, where it rapidly gained the ability to be sexually transmitted in its new environment (Baker and Armelagos 1988). Alternatively, one team of researchers has posited that all three *T. pallidum* subspecies are relatively recent additions to the family of human pathogens and arose at similar times in human history (Gray et al. 2006). Luckily, novel data have recently allowed us to reject yet another theory. The "Unitarian hypothesis" states that all human treponemal infections are caused by a single, protean pathogen with an opportunistic transmission mode, resulting in dramatically different clinical manifestations depending upon factors such as climate (Hudson 1963). Given the morphological and overall genetic similarity between the *T. pallidum* subspecies, this view did not seem unreasonable in the past. Recent genetic evidence stemming from intensive sequencing efforts by multiple groups has convincingly shown, however, that the three treponemal diseases, syphilis, yaws, and bejel, are caused by genetically distinct pathogens (Centurion-Lara et al. 2006; Harper et al. 2008; Smajs et al. 2011). Since an essential tenet of the Unitarian hypothesis is that the strains that cause the three diseases are interchangeable and thus cannot be distinguished, we can now discard this evolutionary explanation for *T. pallidum's* multiple transmission modes. As described above, however, many competing hypotheses remain.

In the remainder of this chapter, we will investigate what is known about *T. pallidum* in NHPs and will conclude by discussing what insights NHP infections may provide in understanding the history of human infection.

Treponemal Infection in Nonhuman Primates

Thanks to observational and serological surveys of wild NHPs, we now know that *Treponema* infection occurs naturally in many species. Based on the available sequences from the Fribourg-Blanc strain, collected from a wild baboon captured in Guinea in the 1960s (Fribourg-Blanc et al. 1966), we also know that the NHP strains responsible are very closely related to human *T. pallidum* strains. This baboon strain harbors 16S ribosomal subunit DNA that is 100 % identical to that found in *T. pallidum* human strains,[1] a level of identity which exceeds the threshold of 97–99 % often used to distinguish between bacterial species (Gevers et al. 2005). However, we do not yet know whether NHP strains represent a sister subspecies to *pertenue* or whether human subsp. *pertenue* and NHP strains represent one large clade.

[1]Sequences from the 16S ribosomal subunit (2nd operon) available from GenBank were obtained and aligned in ClustalX. They included the subsp. *pallidum* strains Dallas-1 (NC016844.1), Chicago (CP001752.1), Nichols (NC000919.1), Street Strain 14 (NC010741.1), and Mexico A (HM585252.1), the subsp. *pertenue* strains CDC-2 (NC016848.1), Gauthier (NC016843.1), Samoa D (NC016842.1), and the Fribourg-Blanc strain (HM165231.1).

Thus, although treponemal infection in NHPs is often referred to as yaws, we would like to raise the possibility that our understanding of the systematics involved may change when we know more about the relationship between human and NHP strains. Whole genome sequencing of different *T. pallidum* strains, especially the simian ones, is expected to shed further light on the small but potentially significant genetic differences between them.

Some background information on serological tests for *T. pallidum* is necessary before we review what is known about treponemal diseases in wild NHPs. Most surveys conducted in NHPs have relied heavily on serological screening methods, so the interpretation of their results requires some knowledge about the sensitivity and specificity of different tests. Serological tests for *T. pallidum* are basically divided into two types: non-treponemal and treponemal. Non-treponemal tests, often used for screening in humans, detect antibodies to phospholipid antigens, such as cardiolipid, that are not specific to *T. pallidum.* Because these antigens occur in a number of diseases, such as malaria, tuberculosis, viral fevers, leprosy, and trypanosomiasis, among others (Herring et al. 2006), they have a lower specificity than the other available tests and may not be particularly reliable when used on wild NHPs, especially those living in regions where multiple infections are common. In contrast, treponemal tests, which react to anti-*T. pallidum* antibodies, are quite specific (Herring et al. 2006). These tests include the *T. pallidum* immobilization (TPI) test, the fluorescent treponemal antibody absorption (FTA-ABS) test, and the microhemagglutination assay (MHA-TP). Despite their greater specificity, treponemal tests also have their limitations (Binnicker et al. 2011). Non-treponemal tests and the FTA-ABS test, which was used often in the surveys of NHPs discussed below, have been reported to cross-react with sera from patients with Lyme disease (Hunter et al. 1986; Magnarelli et al. 1990; Russell et al. 1984), which is caused by the *Borrelia* spec., spirochetes closely related to *T. pallidum.* Luckily, Lyme disease appears to be extremely rare in the Southern Hemisphere, where most NHPs live (Jowi and Gathua 2005; Yoshinari et al. 1993), so false positives from this disease are unlikely to be a significant problem when interpreting the results described here. In addition, autoimmune disorders are known to cause false-positive results when using non-treponemal serological tests, as well as the FTA-ABS assay when the latter is performed under certain conditions (Mackey et al. 1969; Seña et al. 2010). Luckily, this limitation does not appear to apply to the MHA-TP and TPI tests, which were also used in the surveys presented in this chapter, frequently in tandem with the FTA-ABS test (Mackey et al. 1969; Russell et al. 1984).

African Monkeys

NHP treponemal infection has probably been studied best in wild baboons (*Papio* spec.). Beginning in the 1960s, serological surveys demonstrated that treponemal disease was present at high prevalence in the baboons of Equatorial Guinea, Senegal, and Cameroon (Table 2) (Baylet et al. 1971; Fribourg-Blanc and Mollaret 1969). As described previously, one *T. pallidum* strain, the Fribourg-Blanc strain, was isolated from an infected baboon during the course of these surveys (Fribourg-Blanc et al. 1966).

Table 2 *T. pallidum* infection in African monkeys

Species	Origin	Seroprevalence in sample	Test	Source
Mangabey (*Cercocebus*, species not determined)	Laboratory animal caught in West Africa	0/1 (0.0 %)	FTA-ABS	Felsenfeld and Wolf (1971)
	Bangui region, Central African Republic	0/2 (0.0 %)	TPI and FTA-ABS	Fribourg-Blanc and Mollaret (1969)
Green monkey (*Chlorocebus sabaeus*)	Laboratory animals caught in West Africa	0/3 (0.0 %)	FTA-ABS	Felsenfeld and Wolf (1971)
Vervet monkey (*Chlorocebus pygerythrus*)	Laboratory animals caught in Kenya	1/2 (50.0 %), + result equivocal	FTA-ABS	Felsenfeld and Wolf (1971)
Chlorocebus, species not determined	Casamance region of Senegal	3/8 (37.5 %)	TPI and FTA-ABS	Fribourg-Blanc and Mollaret (1969)
	Fouta Toro Fleuve region of Senegal	28/45 (73.0 %)	FTA-ABS	Baylet et al. (1971)
	Niamey, Niger	0/1 (0.0 %)	TPI and FTA-ABS	Fribourg-Blanc and Mollaret (1969)
	Bangui, Central African Republic	0/3 (0.0 %)	TPI and FTA-ABS	Fribourg-Blanc and Mollaret (1969)
	Brazzaville region, Democratic Republic of Congo	0/3 (0.0 %)	TPI and FTA-ABS	Fribourg-Blanc and Mollaret (1969)
	Laboratory animals caught in East Africa	2/7 (28.6 %), + results equivocal	FTA-ABS	Felsenfeld and Wolf (1971)
Colobus monkey (*Colobus*, species not determined)	Casamance region of Senegal	1/1 (100.0 %)	TPI and FTA-ABS	Fribourg-Blanc and Mollaret (1969)
Patas monkey (*Erythrocebus patas*)	Laboratory animals caught in West Africa	1/13 (7.7 %)	FTA-ABS	Felsenfeld and Wolf (1971)
	Casamance region of Senegal	6/26 (23.1 %)	TPI and FTA-ABS	Fribourg-Blanc and Mollaret (1969)
	Mali	0/18 (0.0 %)	TPI and FTA-ABS	Fribourg-Blanc and Mollaret (1969)
	Bobo-Dioulasso region of Burkina Faso	0/17 (0.0 %)	TPI and FTA-ABS	Fribourg-Blanc and Mollaret (1969)
	Niamey, Niger	0/1 (0.0 %)	TPI and FTA-ABS	Fribourg-Blanc and Mollaret (1969)
	Fort-Lamy, Chad	1/14 (7.1 %)	TPI and FTA-ABS	Fribourg-Blanc and Mollaret (1969)
	Laboratory animals caught in East Africa	1/10 (10.0 %), + result equivocal	FTA-ABS	Felsenfeld and Wolf (1971)
	Fouta Toro Fleuve region, Senegal	1/4 (25.0 %)	FTA-ABS	Baylet et al. (1971)
	Koussanar region of Senegal	1/38 (2.6 %)	FTA-ABS	Baylet et al. (1971)

(continued)

Table 2 (continued)

Species	Origin	Seroprevalence in sample	Test	Source
Yellow baboon (*Papio cynocephalus*)	Kindia region of Guinea	164/216 (75.9 %)	TPI and FTA-ABS	Fribourg-Blanc and Mollaret (1969)
	Casamance region of Senegal	6/10 (60.0 %)	TPI and FTA-ABS	Fribourg-Blanc and Mollaret (1969)
	Kaolack region of Senegal	80/171 (46.8 %)	TPI and FTA-ABS	Fribourg-Blanc and Mollaret (1969)
	Bobo-Dioulasso region of Burkina Faso	0/159 (0.0 %)	TPI and FTA-ABS	Fribourg-Blanc and Mollaret (1969)
	Yaounde, Cameroon	3/3 (100.0 %)	TPI and FTA-ABS	Fribourg-Blanc and Mollaret (1969)
	Kenya	0/276 (0.0 %)	TPI and FTA-ABS	Fribourg-Blanc and Mollaret (1969)
Olive baboon (*Papio anubis*)	Casamance region of Senegal	49/82 (59.8 %)	FTA-ABS	Baylet et al. (1971)
	Koussanar region of Senegal	15/55 (27.3 %)	FTA-ABS	Baylet et al. (1971)
	Falémé region of Senegal	0/111 (0.0 %)	FTA-ABS	Baylet et al. (1971)
	Lake Manyara National Park, Tanzania	43/57 (75.4 %)	Serodia TP*PA	Knauf et al. (2012)
Baboon (*Papio*, species not determined)	Brazzaville region, Democratic Republic of Congo	0/2 (0.0 %)	TPI and FTA-ABS	Fribourg-Blanc and Mollaret (1969)

Experiments have shown that this strain is capable of causing infection in humans (Smith et al. 1971), and genetically it is closely related to, but possibly distinct from, human subsp. *pertenue* strains (Centurion-Lara et al. 2006; Harper et al. 2008; Smajs et al. 2011). Clinical signs in the affected populations were described as mild and included small keratotic lesions and ulcers on the muzzle, eyelids, and armpits (Baylet et al. 1971), though most infected animals did not appear to display any lesions at all (Table 1). It is interesting that in these same surveys, not a single animal from countries farther east, such as Burkina Faso ($n=159$) or Kenya ($n=276$), was found to be positive (Fribourg-Blanc and Mollaret 1969).

In the late 1980s, however, a form of treponemal disease with strikingly different clinical signs was described among the olive baboons (*P. anubis*) at Gombe Stream National Park (Table 1). "Penelope," an adult female baboon at Gombe, was the first recorded case displaying what would come to be recognized as the genital-associated lesions typical of the infection (Collins et al. 2011). Subsequently, researchers noticed the skin disease in several baboon troops in the national park (Wallis 2000b; Wallis and Lee 1999). At that time, laboratory analysis (dark field illumination) of skin lesions confirmed the presence of *T. pallidum*. However, DNA-based verification of the subspecies of *T. pallidum* involved was not performed, nor had other pathogens that cause genital ulceration in humans and NHPs been ruled out definitively. Because of the infection's association with genital lesions and the observation that it appeared in sexually mature animals, it was hypothesized that the disease might be sexually transmitted (Wallis and Lee 1999). Moreover, unlike the mild lesions described in West Africa, it was reported that lesions in a small proportion of the individuals affected at Gombe became so severe that urinary flow was obstructed and death resulted, with autopsy of one young male revealing widespread sepsis within the urogenital tract (Wallis and Lee 1999).

In the 1990s, similar lesions (Fig. 1) were reported for the first time in olive baboons at Lake Manyara National Park, also in Tanzania but 700 km away from Gombe (Mlengeya 2004). In contrast to other studies examining treponemal infection in NHPs, Knauf et al. (2012) were able to describe in detail the macroscopic clinical manifestations and histological findings that characterize *T. pallidum* infection in baboons. In addition, for the first time, they were able to demonstrate simian *T. pallidum* strains *in situ*, using immunohistochemistry (Fig. 2). Molecular biological tests such as qualitative and quantitative PCR were performed using skin tissue samples, and they demonstrated the presence of *T. pallidum* in infected animals while ruling out other pathogens. Qualitative and quantitative serological tests, including the Serodia TP*PA test, offered still more proof that *T. pallidum* infection was responsible for the lesions observed. Four informative genetic polymorphisms were used to demonstrate that, despite their predilection for causing anogenital lesions similar to those found in human syphilis, the simian strains collected from baboons at Lake Manyara National Park were most closely related to nonvenereal human *T. pallidum* subsp. *pertenue* strains. Finally, the study showed that many animals that looked healthy in the field had positive PCR, serological, or histological results for *T. pallidum*. For example, of 20 baboons with no clinical signs of infection, 15 showed histological abnormalities, of which six tested PCR positive

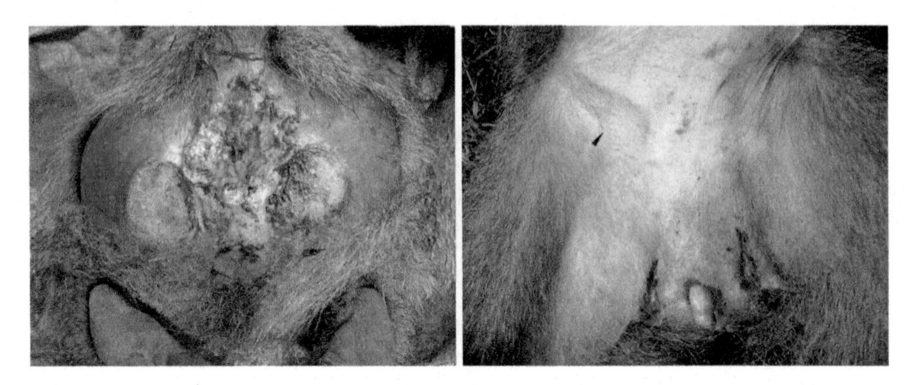

Fig. 1 Photos depicting clinical signs associated with genitotropic *T. pallidum* infection in baboons. The genitals of a severely affected female (*left*) and male (*right*) are shown

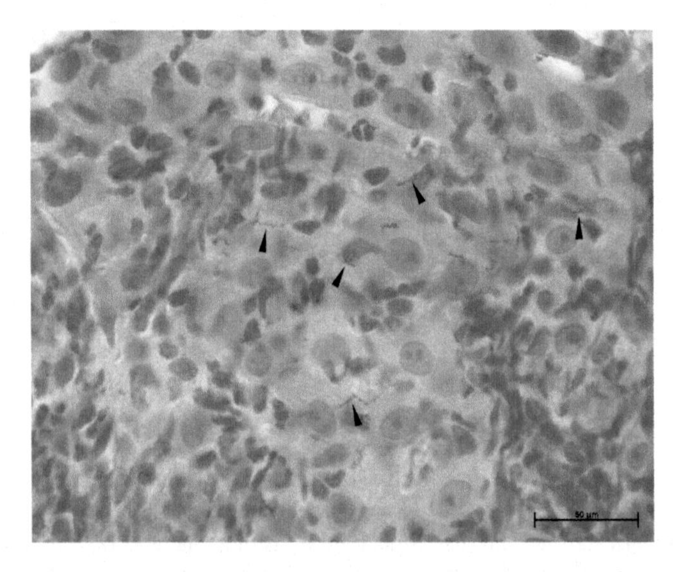

Fig. 2 Photos depicting T. pallidum in a skin tissue sample from the genitals of an Olive baboon at Lake Manyara National Park, Tanzania. Immunohistochemistry utilizing rabbit polyclonal antibodies against *T. pallidum*, was performed with epithelial cells counterstained using Mayer's hematoxylin. Bar 10 μm

for *T. pallidum*. Seven of the 20 clinically unaffected animals had anti-T. *pallidum* antibody titers ≥1:80 (Knauf et al. 2012). This indicates that the prevalence of the disease is much higher than originally suspected from field observations (Mlengeya 2004).

That this infection is prevalent in the baboons of Tanzania is also reinforced by a small survey we performed of baboons imported into the United States by laboratory suppliers (Harper, data not published). In 2006, we assayed 17 serum samples from olive baboons (*Papio anubis*) imported from Tanzania to the United States using the Sero-DIA TPPA test (Fujirebio Diagnostics, Malvern, PA), a highly sensitive and specific serological treponemal test. Sixteen samples came from adult males imported in 2003 ($n=6$) and 2005 ($n=10$). Pooled serum from six adult females imported in 2001 was also tested. Four of the 16 samples from adult males tested positive for *T. pallidum* antibodies, as did the pooled serum from the adult females. It is not clear where in Tanzania these animals were captured, but the relatively high prevalence of infection in these samples suggests that the infection has become established outside of Gombe and Lake Manyara National Parks.

Serological evidence of *T. pallidum* infection has also been found in other African monkey species (Table 2), though no gross-pathological lesions that appear to result from infection have yet been described. Antibodies have been detected in the *Chlorocebus* species [vervet (*C. pygerythrus)* and green monkeys (*C. sabaeus*)], patas monkeys (*Erythrocebus patas*), and in one Colobus monkey (*Colobus* spec.).

Gorillas

Reports that yaws infection is frequent in wild gorillas (*Gorilla gorilla*) in West Africa have appeared for some time. For example, stories from indigenous people as well as reports from the Service de Chasse in the former French Equatorial Africa region told of gorillas in the Ewo, Kelle, and Mekambo regions that suffered from a leprosy-like disease (Cousins 1984). In the 1950s, two researchers got the opportunity to examine four young gorillas captured in this area that displayed the "leprous" lesions described in earlier reports. The raised lesions, scabby and dry, primarily affected the lips, nose, eyes, and cheeks, with one animal also exhibiting lesions on the shoulder and forearm (Table 1). It should be noted that leprosy has been confirmed in an Asian wild-born macaque (Valverde et al. 1998), two African wild-born sooty mangabeys (Gormus et al. 1991), and wild-born chimpanzees from West Africa (Hubbard et al. 1991; Leininger et al. 1978; Suzuki et al. 2010). It is not yet clear whether these laboratory animals were infected while living in the wild or via contact with infected handlers while being captured and cared for prior to export in leprosy-endemic areas. However, in the case of the wild gorillas with "leprous" lesions, the researchers reported that the infection was successfully cleared up in 8 days with penicillin injections, which was consistent with *T. pallidum* infection rather than leprosy (Cousins 1984).

Mid-century, the scientist Armand Denis shared his observations on yaws in the wild gorillas of the Republic of Congo (ROC). He described the first case he saw, affecting an adult male, thus:

> It was like a mask eaten into by some flesh-consuming disease. The lips were gone. The nostrils were eaten almost away and the fangs of teeth were blackened and askew in what remained of the creature's lower jaw. Only the eyes were untouched... Whatever the disease was the wretched animal had caught I had no idea... (Denis 1963, p 181).

Although the clinical signs described again appear to overlap with those caused by leprosy, Denis learned that this disease was thought by the natives to be yaws, the same disease that was common in human populations that lived along the coast.

Table 3 *T. pallidum* infection in great apes

Origin	Seroprevalence in sample	Tests	Source
Chimpanzees (*Pan troglodytes*)			
Laboratory animals caught in western region of Central Africa	1/15 (6.7 %), + result equivocal	FTA-ABS	Felsenfeld and Wolf (1971)
Bukavu region, Republic of Congo	3/9 (33.3 %)	TPI and FTA-ABS	Fribourg-Blanc and Mollaret (1969)
Bangui region, Central African Republic	0/1 (0.0 %)	TPI and FTA-ABS	Fribourg-Blanc and Mollaret (1969)
Laboratory animals, origin not specified	48/250 (19.2 %)	FTA-ABS	Kuhn (1970)
Gorillas (*Gorilla gorilla*)			
Bukavu region, Republic of Congo	0/1 (0.0 %)	TPI and FTA-ABS	Fribourg-Blanc and Mollaret (1969)
Laboratory animals, origin not specified	0/14 (0.0 %)	FTA-ABS	Kuhn (1970)
Orangutans (*Pongo* spec.)			
Laboratory animals, origin not specified	0/39 (0.0 %)	FTA-ABS	Kuhn (1970)

Reports of yaws were not limited to gorillas living in the ROC. Lesions consistent with yaws were also described in the young gorillas of Rio Muni, in Equatorial Guinea (Cousins 1984). Similarly, yaws was diagnosed in two out of five young gorillas imported from the French Cameroons (modern day Cameroon and Nigeria) to the Lincoln Park Zoo, mid-century (Cousins 1972), though the method of *T. pallidum* confirmation is not given in the report. Sensitive and specific serological tests have confirmed that gorillas do indeed come into contact with *T. pallidum* (Table 3). Four blood samples from solitary gorillas in the Parc National d'Odzala-Kokoua in the Republic of Congo, two of which had skin lesions consistent with yaws, tested positive for treponemal antibodies (Karesh 2000).

Skeletal evidence consistent with (but not specific to) treponemal infection has also been found in the remains of wild gorillas (Lovell et al. 2000). Lovell and colleagues examined 126 gorilla skeletons collected from the Democratic Republic of Congo (DRC), the ROC, and Cameroon in the early twentieth century. Eighteen percent of gorillas were found to exhibit possible signs of treponemal disease, with active lesions found almost entirely among subadults. In addition, cranial deformities involving massive osseous tumors have been documented in gorillas, and some researchers have speculated that these may correspond to *goundou*, a rare manifestation of tertiary-stage yaws in humans (Cousins 2008). However, whereas in humans the lesions grow out of the nose and upper jaw, in gorillas, growth of goundou-like lesions seems to occur primarily over the cheekbones (Cousins 2008).

Recently, an evaluation of skin lesions in 377 gorillas living in the Parc National d'Odzala-Kokoua in the ROC gave us a much better understanding of treponemal disease in this population (Levréro et al. 2007). The macroscopic lesions identified were found to be similar to yaws in humans (Fig. 3) and affected 17 % of animals examined. However, since some animals that are infected do not display lesions (Karesh 2000), the actual prevalence of the infection is likely to be higher. Infection

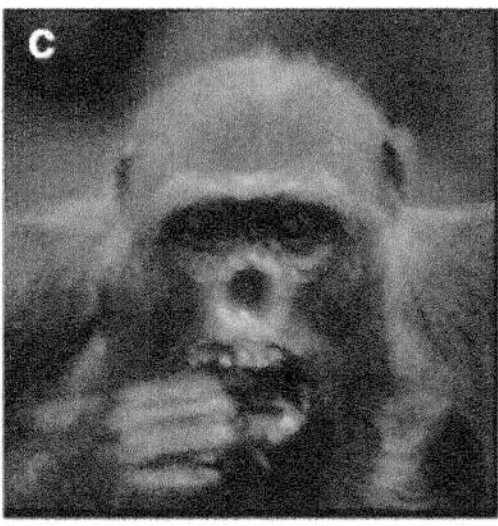

Fig. 3 Photos depicting skin lesions suspected to result from *T. pallidum* infection in gorillas. Skin lesions are shown on (**a**) an adult male, (**b**) a juvenile female, and (**c**) an adult female gorilla (Source: Levréro et al. 2007)

in the gorillas studied began early, with almost 5 % of infants displaying skin lesions compatible with yaws and peak incidence occurring between the infant and juvenile stages. Some animals were found to exhibit destruction of the nose and/or lips or to display deep lesions on their wrists or ankles. Possible social consequences of severe infection were hinted at in the observations of Levréro et al. One adult female with a nose that was completely destroyed was forced to leave her group; during the animal's seven subsequent visits to the clearing before disappearing, the researchers observed the males from groups she approached behaving antagonistically towards her. As the researchers note, the consequences of such rejection must certainly have implications for survival. Although lesions were also reported in gorillas at nearby sites, including Maya, Moba, and Lossi, infection may not be universal, since no such lesions were observed at the Mbeli clearing, also nearby.

Treponemal infection in gorillas has been identified with yaws, due to its gross-pathological manifestations and also the fact that the areas in which the gorillas reside tend to be places in which the human prevalence of this disease was quite high historically, affecting the vast majority of some human groups inhabiting the rain forest (Hackett 1953; Pampiglione and Wilkinson 1975) and still infecting more than 10 % of residents in some areas (Gersti et al. 2009). Unfortunately, not a single report of treponemal infection has been confirmed via molecular biological tests, such as *T. pallidum*-specific PCR or immunohistochemistry, capable of demonstrating the spirochetes *in situ*. Obtaining the samples needed to perform such tests in endangered great apes is extremely challenging, both politically and technically, and this explains why they have not yet been performed. However, without the information these tests can provide, it is impossible to determine with certainty that *T. pallidum* is responsible for the lesions and, if so, to which human subspecies the responsible strains are most closely related.

Although most reports of treponemal infection in wild gorillas have described a yaws-like disease, descriptions of animals with diseased sexual organs that may be linked to treponemal infection have also surfaced. For example, in Cameroon, one adult female exhibited genitalia covered with running sores, as well as deforming sores on the hands and wrists (Cousins 1984). This animal was said by the nearby residents to be suffering from "marjal" or "mebata," a local word for yaws. As with the yaws-like cases, providing an explanation for the etiology of such lesions is challenging, especially in the context of other sexually transmitted infections that can cause genital ulceration in NHPs. Definitively assigning an etiological agent in such cases will only be possible with advanced molecular biological techniques.

Chimpanzees

Descriptions of the manifestations of treponemal infection in chimpanzees (*Pan troglodytes*) are less common. That contact with *T. pallidum* occurs is clear from serological surveys of wild animals (Table 3). For example, one survey of more than 250 wild-born chimps, drawn from unknown locations, found that 48 (19 %) were positive for *T. pallidum* antibodies (Kuhn 1970). Serological evidence of infection has also been found in wild chimps from the DRC and Sierra Leone (Felsenfeld and Wolf 1971; Fribourg-Blanc and Mollaret 1969). In addition, the physical anthropologists who found evidence consistent with treponemal disease in gorillas also found that almost 20 % of 102 sets of wild chimp remains from Western Africa exhibited osseous lesions that could be due to treponemal disease (Lovell et al. 2000). However, only a single case of treponemal infection causing observable clinical manifestations has been reported in the literature (Edroma et al. 1997). This case occurred in a chimpanzee at Gombe, in Tanzania, who may have been infected as a result of the ongoing epidemic in neighboring baboons. The means via which *T. pallidum* was identified as the source of the disease was not described in the report.

Asian Monkeys

Some of the first yaws experiments were performed in macaques (*Macaca* spec.). However, it appears that natural infection of wild macaques is very rare, if it occurs at all; over 1,000 animals have been tested without one robust seropositive reaction (Table 4). Similarly, though Thivolet et al. found treponemal antibodies in 46 of 415 African monkeys, not a single one of the 152 Asian monkeys this research group tested were positive (Kuhn 1970).

Only one serologically reactive animal captured in Asia has been reported. Kuhn (1970) describes finding a seropositive reaction in a Celebes crested macaque (*Macaca nigra*), from Sulawesi, Indonesia. Given the paucity of positive infections among Asian monkeys, however, one must wonder whether this result represents one of the rare false positives that can stem from the FTA-ABS test or an infection acquired during the animal's time in captivity vs. during its life in the wild. Additional sampling of Asian NHPs will help answer this question.

Table 4 *T. pallidum* infection in Asian and South American monkeys

Species	Origin	Seroprevalence in sample	Tests	Source
Asia				
Macaques (*Macaca*, species not determined)	Laboratory animals caught in Southeast Asia	1/6 (16.7 %), + results equivocal	FTA-ABS	Felsenfeld and Wolf (1971)
Crab-eating Macaque (*Macaca fascicularis*)	Phnom-Penh region of Cambodia	0/1236 (0.0 %)	TPI and FTA-ABS	Fribourg-Blanc and Mollaret (1969)
Crab-eating Macaque (*Macaca fascicularis*)	Asia	0/5 (0.0 %)	TPI and FTA-ABS	Fribourg-Blanc and Mollaret (1969)
Rhesus Macaque (*Macaca mulatta)*	Laboratory animals caught in Southeast Asia	0/22 (0.0 %)	FTA-ABS	Felsenfeld and Wolf (1971)
Rhesus Macaque (*Macaca mulatta)*	India	0/5 (0.0 %)	TPI and FTA-ABS	Fribourg-Blanc and Mollaret (1969)
Rhesus Macaque (*Macaca mulatta)*	Asia	0/30 (0.0 %)	TPI and FTA-ABS	Fribourg-Blanc and Mollaret (1969)
Pig-tailed Macaque (*Macaca*, species not determined)	Asia	0/5 (0.0 %)	TPI and FTA-ABS	Fribourg-Blanc and Mollaret (1969)
South America				
Owl monkey (*Aotus*, species not determined)	Laboratory animals, origin not specified	2/84 (2.4 %), + results equivocal 0/9	FTA-ABS TPI	Levine et al. (1970)
Owl monkey (*Aotus trivirgatus*)	Laboratory animals caught in South America	0/3 (0.0 %)	FTA-ABS	Felsenfeld and Wolf (1971)
Red-bellied titi (*Callicebus moloch*)	Laboratory animals caught in South America	5/25 (20.0 %), + results equivocal	FTA-ABS	Felsenfeld and Wolf (1971)
Squirrel Monkey (*Saimiri sciureus*)	Laboratory animals caught in South America	0/18 (0.0 %)	FTA-ABS	Felsenfeld and Wolf (1971)
Squirrel monkey (*Saimiri sciureus*)	Laboratory animals, origin not specified	4/63 (6.3 %), + results equivocal 0/10	FTA-ABS TPI	Levine et al. (1970)
Marmoset (Callithrix spec.)	Laboratory animals, origin not specified	0/24 (0.0 %) 0/23	FTA-ABS TPI	Levine et al. (1970)

South American Monkeys

To date, not a single wild South American NHP has been found with a definite seropositive reaction to *T. pallidum,* although artificial infection of owl (*Aotus*) and squirrel (*Saimiri*) monkeys is possible (Elsas et al. 1968; Smith 1969). However, as is clear from Tables 2 and 4, sampling in African monkeys has been much more thorough. In order to conclude definitively that natural infection of South American monkeys does not occur, more extensive sampling would have to be performed. Sampling areas of Central and South America the prevalence of treponemal disease in human indigenous groups was once very high, such as French Guiana, Brazil, Venezuela, and Colombia (Black 1975; Hopkins and Flórez 1977; St John 1985), would be especially important.

Experimental Insights into *T. pallidum* Infection and Host Factors in Primates

NHPs played a pivotal role in the discovery of *T. pallidum.* In the early years of the twentieth century, Metchnikoff and Roux (1903, 1904, 1905) demonstrated that syphilis could be transferred from humans to chimpanzees and from one chimpanzee to another. Thus, apes became the first reliable animal model for the study of *T. pallidum* infection. Around the same time, Castellani (1907) demonstrated that monkeys could be infected with *T. pallidum* subsp. *pertenue.* Even so, our understanding of the differences in how infection progresses in humans versus NHPs is rudimentary. Here, we review what is known about NHP responses to experimental infection with *T. pallidum,* focusing on between-species similarities and differences (Table 5).

Turner and Hollander (1957b) did a series of experiments on rhesus macaques and African green monkeys. After inoculating animals in the thighs, eyebrows, and genital regions, the animals were followed for 14–17 months before postmortem examination; no gross changes suggesting syphilitic infection were identified, and a number of the animals remained seronegative throughout, though their organs contained infective spirochetes. On this basis, Turner and Hollander suggested that NHPs did not provide a suitable animal model for understanding *T. pallidum* pathogenesis. Similarly, although humans typically develop a primary chancre or "mother yaw" at the site where the bacterium enters the body, in a study utilizing ten macaques, Sepetjian et al. (1972) reported that only two animals developed significant lesions at the site of inoculation—the others exhibited no lesions at all or mild macular discoloration—and these lesions disappeared quickly. Like Turner and Hollander, Sepetjian et al. (1969) also observed no visceral lesions upon necroscopy after 4 months, although in half the animals treponemes were present in various organs.

Table 5 Comparison of NHP host response to experimental *T. pallidum* subsp. *pallidum* infection and human response to natural infection with syphilis

	Humans	Macaques	African Green Monkeys	Owl Monkeys (*Aötus trivirgatus*)
Geographic distribution of host	Global	Asia	Africa	South America
Lesion at site of inoculation	Yes	Small, mild, and quick to disappear when present at all after subsp. *pallidum* infection[a]	Only in half the animals studied; small and disappeared within several weeks	No
Treponemes disseminate through body	Yes	Yes	Yes	Yes
Lesions remote from inoculation site	Yes	None observed	None observed	Yes
Serologic response to infection	Yes	Sometimes but not always; response is often slow to develop	Appears rare (1 out of 4 animals seroconverted during study period)	Sometimes but not always
Robust CD4+ T-cell response	Yes	Yes	Unknown	Unknown
Sources	Salazar et al. (2002a, b), Perine et al. (1984)	Sepetjian et al. (1969, 1972), Turner and Hollander (1957b), Marra et al. (1998)	Turner and Hollander (1957b)	Elsas et al. (1968), Smith et al. (1965)

[a]Significant lesions at the site of inoculation were observed when the animals were injected with subsp. *pertenue* and Fribourg-Blanc strain (Sepetjian et al. 1969)

Other studies suggest that the NHP response to infection may offer parallels to that observed in humans, however. For example, experiments have demonstrated that owl monkeys and macaques develop a chronic, systemic infection in response to inoculation with strains of human-derived *T. pallidum* (Elsas et al. 1968; Marra et al. 1998). In some of these animals, a long-lasting serological response develops (Sepetjian et al. 1972), though, as Turner and Hollander and others have described, others remain seronegative for months on end, despite harboring infective treponemes (Smith et al. 1965; Turner and Hollander 1957b; Wells and Smith 1967). Similar to humans, NHPs can develop *T. pallidum*-related lesions weeks or months after infection, distant from the site of inoculation (Elsas et al. 1968; Sepetjian et al. 1969; Smith et al. 1965). Clark and Yobs (1968) have opined that the variation in terms of lesion development and serological response in NHP hosts such as the owl monkey might be viewed as typical of the varied response found in humans; in their view, the lesions observed are even consistent with primary and secondary stages of disease, as in humans.

Immunological studies suggest further similarities between the human and NHP response to *T. pallidum* infection. In humans, humoral, antibody, and $CD8^+$ cytotoxic T-cell responses have been shown to be relatively ineffective at clearing syphilitic infection or curbing the progression of lesions (Carlson et al. 2011). Instead, delayed-type hypersensitivity, which is mediated by $CD4^+$ T cells, appears to play a role of particular importance (Carlson et al. 2011). Similar findings have been reported in macaques infected with a subsp. *pallidum* strain. As in humans, CD4+ T cells appear to be responsible for clearing *T. pallidum* from the central nervous system during early infection (Marra et al. 1998). In terms of the components of the bacterium that stimulate an immune response, *T. pallidum* lacks lipopolysaccharides (LPS) in its cell wall and therefore does not cause an inflammation cascade by activating the toll-like receptor (TLR) 4 (Schroder et al. 2008). TLR4 activation is a common pathway induced by the LPS of Gram-negative bacteria and to a certain extent by the lipoteichoic acids of Gram-positive bacteria. This mechanism represents a major immunological host defense against bacterial infection, which is virtually absent in *T. pallidum* infection. Instead, the *T. pallidum* infected host responds to infection via the activation of TLR 2/1 by immunostimulatory lipoproteins in the outer membrane of the spirochete (Lien et al. 1999; Schroder et al. 2008). There is evidence that natural simian and human infections are similar in this respect. For example, in our laboratory, we performed immunoblots upon sera from wild baboons in Lake Manyara National Park infected with *T. pallidum*. The sera were tested for anti-*T. pallidum* IgM and IgG antibodies to recombinant *T. pallidum* proteins Tp15, Tp17, and Tp47, and *Treponema* membrane protein A (TmpA) obtained from human-infecting *T. pallidum* strains. The results consistently showed that the animals were producing antibodies against these proteins (Knauf unpublished data). On this basis, we predict that simian *T. pallidum* strains express immunostimulatory lipoproteins similar to the Tp15, Tp17, Tp47, and TmpA proteins of known human strains and that NHPs are responding to these antigens in a similar way to humans.

In humans, the relative distribution of the four immunoglobulin G subclasses changes over the course of infection (Baughn et al. 1988; Moskophidis 1989;

Salazar et al. 2002a, b). Whether similar changes occur in NHPs is not known, but field research on animals displaying signs of early- and late-stage disease will help answer this question.

Most of the results described above feature inoculation of NHPs with strains of *T. pallidum* obtained from human infections. In one study, however, five macaques were infected with the baboon-derived Fribourg-Blanc strain (Sepetjian et al. 1969). The lesions developed by the macaques were characteristic of yaws rather than syphilis. Rather than an indurated chancre, they displayed vegetative, hyperkeratotic papular lesions. Interestingly, the TPI serology was weak in one animal and nonreactive in the other three tested, one of which died 5 weeks into the study; the single weak reaction did not even begin to develop until the eleventh week of infection. FTA-ABS serology was similar, weak and slow to develop in two animals (one of which had the weak TPI test) and nonreactive in the other two. It remains unclear whether the fact that the antibody response of these macaques to an NHP-derived strain was so weak is significant. Is the NHP response to *T. pallidum* strains native to their own species similarly slow to their own species similarly slow to develop? At least for baboons naturally infected with *T. pallidum*, Knauf et al. (unpublished data) were able to demonstrate that anti-*T. pallidum* antibody titers in infected animals were generally high, with mean anti-*T. pallidum* IgM+IgG titers of 1:2.94E+04 ± SE9.87E+03 in the initial clinical stage of disease, 1:2.17E+05 ± SE1.83E+05 in the moderate stage, and 1.78E+06 ± SE1.38E+06 in the severe stage of infection. Moreover, comparison to a PCR-based test demonstrated that sensitivity (Sen) and positive predictive values (PP) for the antibody-based gelatin particle agglutination assay (Sen 100 %, PP 95 %), the FTA-ABS IgG (Sen 100 %, PP 98 %), and the immunoblot IgG (Sen 100 %, PP 93 %) serological tests in the baboons were reliable, although the specificity and negative prediction values need more testing. Of course in studies of wildlife, the major disadvantage is the paucity of information regarding the timing of infection. Thus, results regarding the serological response of wild NHPs to infection need to be considered carefully.

There is one report of inoculating human "volunteers" with a strain of *T. pallidum* obtained from an NHP infection. In The Caracas Project, a study of late-stage yaws and pinta carried out in 1969, five people were inoculated with the Fribourg-Blanc strain (Smith et al. 1971). Two of the five patients developed reactive FTA-ABS/TPI serology, while three had nonreactive tests. Furthermore, two of the three with nonreactive serology had elevated IgG levels in their cerebrospinal fluid, indicating that four of the five patients mounted some type of immunological response to infection. Finally, abnormal ophthalmological results consistent with late-stage yaws were found in three of the five patients, although it is not entirely clear that these abnormalities were due to infection with the Fribourg-Blanc strain. Thus, the Caracas Project showed that an NHP-derived *T. pallidum* strain could establish an infection in humans and possibly cause ocular abnormalities. Unfortunately, whether or not these infections yielded additional clinical signs in humans is unknown, as the project focused only on neuro-ophthalmologic manifestations. Nor was the length of experimental infection or a discussion of the ethics of infecting healthy adults with *T. pallidum* and allowing them to develop late-stage infection provided.

In sum, experimental studies of response to inoculation suggest that primates may mount an immune response against *T. pallidum* that shares some basic characteristics with that of humans while differing in other aspects. In all species examined, infection appears to be chronic, and the host uses a predominately CD4+ response to control pathogen growth. NHPs on the whole, though, appear to present much milder clinical signs than humans, often exhibiting a slow rise in antibody titers. Visceral signs of experimental infection have not been reported, in NHPs yet. Because most NHP studies have continued for only a few months or a year, it is not clear whether lesions akin to tertiary-stage symptoms in humans would emerge, given enough time. In addition, most experiments have been performed in species, such as Asiatic macaques or New World monkeys, which, for whatever reason, do not appear to be susceptible to infection in the wild. Therefore, it is not clear whether the same results would be found in species that have presumably evolved in tandem with their own *T. pallidum* strains. Finally, it should be noted that the results of the experiments described here are probably contingent to some degree upon the strain of *T. pallidum* used. For example, in one study, macaques inoculated with a subsp. *pallidum* strain did not develop significant lesions at the site of the injection; however, all macaques infected with a subsp. *pertenue* strain did develop considerable lesions (Sepetjian et al. 1969). While animal experiments studying *T. pallidum* pathogenesis are on the wane, especially those utilizing NHP hosts, field studies of treponemal infection in wild NHPs may help clarify the similarities and differences in response to infection that have evolved in various host species over the years.

Some Open Questions on Host Specificity and Strain Pathogenicity in Primates

As described above, even though *T. pallidum* infections have not been documented in species such as macaques, owl, and squirrel monkeys in the wild (Table 4), laboratory experiments show that these animals can be infected with *T. pallidum*. If these species are susceptible to infection, what prevents them from serving as hosts in the wild? Which is more important in explaining why NHP species in Africa but not Asia and South America are infected with *T. pallidum*: host geography (Fig. 4) or phylogeny (Fig. 5)? Is the reason for this inconsistency across primate species rooted in biological differences, the varied environments they inhabit, or chance playing out over evolutionary time? The genital ulcerative disease caused by *T. pallidum* in olive baboons, which has been reported exclusively in *Papio anubis* (Knauf et al. 2012; Wallis and Lee 1999), poses a similar puzzle. Olive and yellow baboons are known to hybridize (Alberts and Altmann 2001), which means that a significant level of sexual interaction occurs between the two species. However, even at *Treponema* hot spots in Tanzania where olive and yellow baboon subspecies overlap and transmission opportunities should arise frequently, yellow baboons have never been observed with genital lesions consistent with *T. pallidum* infection. Could

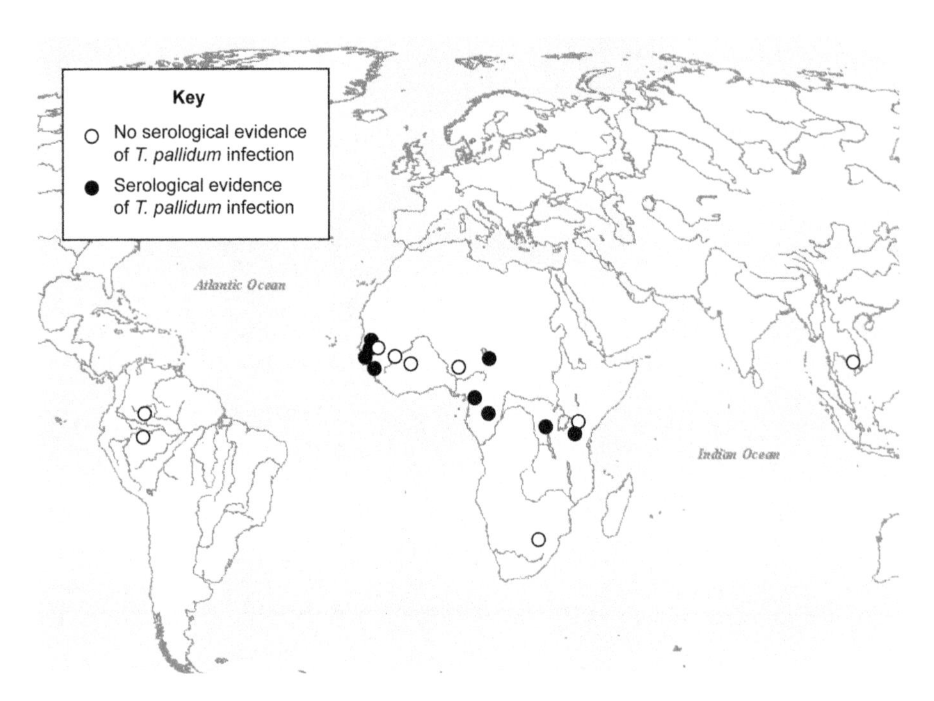

Fig. 4 Geographic distribution of *T. pallidum* infection in nonhuman primates. Drawn from sources cited in Tables 2–4; only areas in which more than ten animals had been tested were included

some constellation of host genetics in yellow baboons provide them with physiological protection against the disease? Or are differences in behavior between olive and yellow baboons responsible for the presence of the infection in one subspecies but not the other? The factors that determine whether or not an NHP clade serves as a natural host remain mysterious at this point.

Parallel questions exist about the importance of strain vs. host characteristics in determining clinical manifestations. For example, is the recently described infection in Tanzanian baboons, characterized by genital ulceration, caused by a strain or strains which are genetically distinct from the strains that cause milder, non-genitally ulcerating infections in the baboons of West Africa? Or is some characteristic of the Tanzanian baboons, or their environment, responsible for these novel manifestations? Previously, it has been demonstrated that climatic factors such as temperature and humidity play an important role in modulating clinical lesions in *T. pallidum* infection (Turner and Hollander 1957a). Could this important finding help explain the disease dynamics observed in some of the naturally occurring NHP epidemics in East Africa (Collins et al. 2011)? Detailed surveys of *T. pallidum* manifestations in different NHP groups may help begin to answer some of these questions, allowing us to begin to disentangle the effects of the evolution of host and pathogen from environment.

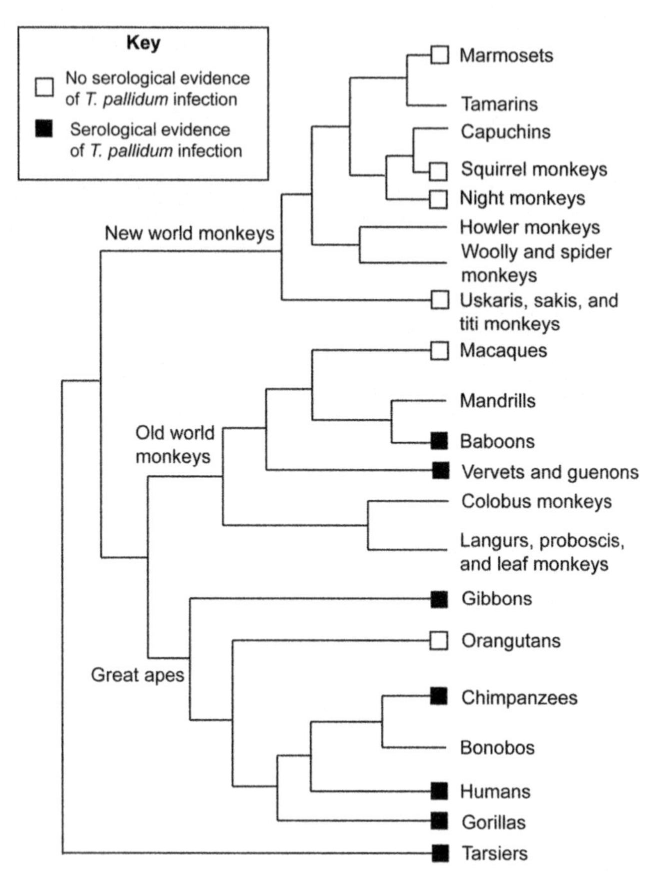

Fig. 5 Species distribution of *T. pallidum* infection in nonhuman primates. Drawn from sources cited in Tables 2–4; only species in which more than ten animals had been tested were included

Conclusion

Surveys of *T. pallidum* infection in NHPs performed thus far have provided fascinating insights into the ecology of treponemal disease in the wild. It appears that *T. pallidum* infection in wild NHPs is common in many parts of Africa, but very rare, if it occurs at all, in Asia and South America. Robust evidence of seropositive animals has only been found in Old World species of NHPs (Figs. 4 and 5). However, more comprehensive testing of Asian and South American NHPs must be performed before solid conclusions regarding the geographical and phylogenetic distribution of infection may be drawn.

Because the history of *T. pallidum* infection in humans remains controversial, studies of NHP strains may help elucidate the origin of our own. Thus far, only two NHP strains have been characterized genetically. These two strains are both from baboons, one from Guinea and the other from Tanzania, and both are closely related to human *T. pallidum* subsp. *pertenue* strains (Harper et al. 2008; Knauf et al. 2012), which are responsible for the disease yaws. However, it is not clear whether the NHP strains are merely closely related to subsp. *pertenue* or whether they are actually members of the subspecies. Sampling strains from other NHP species and geographic regions and sequencing their entire genomes using next-generation methods should help us better understand how NHP strains are related to human strains and to one another. If *T. pallidum* strains tend to diverge in the same order as their host species, this would indicate that the infection is ancient in primates, with each species possessing its own genetically distinct type of pathogen. It could also indicate that *T. pallidum* infected our earliest hominid ancestors. In contrast, if *T. pallidum* strains diverge in patterns that do not reflect their hosts' phylogenies, host switching may be implicated—and it is possible that *T. pallidum* infection may be a more recent addition to the human disease-scape.

The clinical signs that have been described in association with *T. pallidum* infection in NHPs are primarily consistent with mild, nonvenereal infection. However, the *T. pallidum* infection described in baboons in Tanzania is interesting in that it appears to be associated with the genitals and possibly spread via sexual transmission. Our field observations of hundreds of baboons at Lake Manyara (Knauf et al. 2012; Mlengeya 2004) echo the findings reported from Gombe (Wallis 2000a; Wallis and Lee 1999), in that the clinical lesions we observed are almost exclusively found in the anogenital region and appear to be limited to sexually mature individuals. Further epidemiological characterization of this infection should clarify the means of spread.

Given the fact that the human *T. pallidum* subspecies, transmitted via different modes, are genetically distinct (Centurion-Lara et al. 2006; Harper et al. 2008; Smajs et al. 2011), the same may be true of NHP strains transmitted via alternate routes. If future studies reveal that some NHP strains are consistently sexually transmitted while others are not, then comparison of closely related strains transmitted via different routes may help generate hypotheses about the genetic polymorphisms underlying sexual transmission. It is possible that comparison of genetic polymorphisms in *T. pallidum* associated with sexual transmission in both humans and NHPs could identify genetic regions that are especially important in adaptation to this mode of transmission. In addition, the selective pressures that favor sexual transmission of the pathogen in some NHPs but not others could be studied; environmental, social, immunological, and other factors might all be investigated as possible driving forces. In short, the spectrum of different transmission routes and types of clinical manifestations in NHPs may provide a unique opportunity to better understand the pathogenicity and evolution of *T. pallidum* in our own species.

Current efforts to study *T. pallidum* infection in wild NHPs are hampered by the difficulty of gathering adequate biological samples, especially in endangered species such as the great apes. A solid diagnosis of treponemal disease can only be obtained by demonstrating that the spirochete is present, and carefully documenting microscopic and macroscopic signs of infection, including molecular biological proof, is essential. Serology can be used to screen large numbers of wild NHPs for *T. pallidum* infection, but it requires that blood samples be taken. Moreover, because serology cannot aid in differentiating between the *T. pallidum* subspecies, nor can it provide information about clinical manifestations, disease progression, or even whether an infection is active, serum samples are not sufficient for a truly comprehensive examination. Though obtaining specimens from NHPs for histological and molecular biological tests is often very difficult, in order to reach a definitive diagnosis using the methods currently available, access to these types of samples is a necessity.

In the future, it is hoped that noninvasive means of studying treponemal infection in wild NHPs may be developed, which could greatly facilitate diagnosis and surveillance. Antibody-based tests have been successfully used on urine and fecal samples to study NHP pathogens such as simian immunodeficiency virus (SIV) (Santiago et al. 2003), simian foamy virus (SFV) (Liu et al. 2008), hepatitis B (Makuwa et al. 2005), and *Cryptosporidium* and *Giardia* (Salzer et al. 2007). In addition, urine and fecal samples have been used to obtain DNA from NHP pathogens such as SIV (Santiago et al. 2003), SFV (Liu et al. 2008), respiratory viruses (Köndgen et al. 2010), and *Plasmodium* spp. (Liu et al. 2010), paving the way for a flurry of recent phylogeographic studies of unprecedented size. Because it is not clear whether urine or feces are reasonable samples for *T. pallidum* antibody-based tests, or whether pathogen DNA is shed in urine or feces at detectable levels during the various stages of infection, the potential of noninvasive samples to revolutionize studies of *T. pallidum* in NHPs is unclear and currently under investigation (Knauf et al., unpublished data).

Understanding treponemal infection in NHPs is important for several reasons. First, it is possible that morbidity and mortality associated with *T. pallidum* infection has important conservation implications, especially for species such as the great apes (Levréro et al. 2007). Second, a dialogue about reinitiating the yaws eradication campaign that began mid-century is currently active (Asiedu et al. 2008; Rinaldi 2008). Before the feasibility of such a campaign may be assessed, we must understand the relevant disease ecology. Frequent transmission from an animal reservoir would have serious ramifications for an attempt to eradicate infection in humans, and phylogenetic studies can help determine whether or not this occurs. Third, understanding the evolution and transmission of *T. pallidum* among diverse primate species may help us better understand the forces that mold its evolution and transmission dynamics within our own species. Further study may shed light on the geographic distribution, the mode of transmission, and the NHP species affected by *T. pallidum* infection, as well as the relationship between human and NHP strains.

Acknowledgments We thank the Tanzania Wildlife Research Institute (TAWIRI), Tanzania National Parks (TANAPA), the Lake Manyara National Park headquarter staff, the NCA authority, and the Tanzania Commission for Science and Technology for making our study of *T. pallidum* infection in baboons possible. We also thank Jim Thomas for the use of his laboratory, Worldwide Primates and the Buckshire Corporation for providing samples, and Columbia University's Training Program in Cancer-Related Population Sciences (5-R25-CA 094061), the Robert Wood Johnson Foundation Health & Society Scholars program, NSF (Grant 0622399), the Wenner-Gren Foundation, and the Howard Hughes Medical Institute Predoctoral Fellowship program for their financial support.

References

Alberts SC, Altmann J (2001) Immigration and hybridization patterns of yellow and anubis baboons in and around Amboseli, Kenya. Am J Primatol 53(4):139–154

Antal G, Lukehart SA, Meheus AZ (2002) The endemic treponematoses. Microbes Infect 4:83–94

Arya OP, Bennett FJ (1976) Role of the medical auxiliary in the control of sexually transmitted disease in a developing country. Br J Vener Dis 52(2):116–121

Asiedu K, Amouzou B, Dhariwal A, Karam M, Lobo D, Patnaik S, Meheus A (2008) Yaws eradication: past efforts and future perspectives. Bull World Health Organ 86(7):499–500

Baker B, Armelagos G (1988) The origin and antiquity of syphilis: paleopathological diagnosis and interpretation. Curr Anthropol 29(5):703–737

Baughn RE, Jorizzo JL, Adams CB, Musher DM (1988) Ig class and IgG subclass responses to *Treponema pallidum* in patients with syphilis. J Clin Immunol 8(2):128–139

Baylet R, Thivolet J, Sepetjian M, Nouhouay Y, Baylet M (1971) La tréponématose naturelle ouverte du singe *Papio papio* en Casamance. Bulletin de la Sociêtê de Pathologie Exotique et de ses Filiales 64(6):842–846

Binnicker M, Jespersen D, Rollins L (2011) Treponema-specific tests for serodiagnosis of syphilis: comparative evaluation of seven assays. J Clin Microbiol 49(4):1313–1317

Black F (1975) Infectious diseases in primitive societies. Science 187:515–518

Carlson JA, Dabiri G, Cribier B, Sell S (2011) The immunopathobiology of syphilis: the manifestations and course of syphilis are determined by the level of delayed-type hypersensitivity. Am J Dermatopathol 33(5):433–460

Castellani A (1907) Experimental investigations on framboesia tropica (yaws). J Hyg 7(4):558–569

Centurion-Lara A, Castro C, Castillo R, Shaffer J, Wv V, Lukehart S (1998) The flanking region sequences of the 15-kDa lipoprotein gene differentiate pathogenic treponemes. J Infect Dis 177(4):1036–1040

Centurion-Lara A, Molini B, Godornes C, Sun E, Hevner K, Wv V, Lukehart S (2006) Molecular differentiation of *Treponema pallidum* subspecies. J Clin Microbiol 44(9):3377–3380

Clark J, Yobs A (1968) Observations of the pathogenesis of syphilis in *Aötus trivirgatus*. Br J Vener Dis 44:208–215

Cockburn T (1967) The evolution of human infectious diseases. In: Cockburn T (ed) Infectious diseases: their evolution and eradication. Charles C Thomas, Springfield, IL, pp 84–107

Collins A, Sindimwo A et al (2011) Reproductive disease in olive baboons (*Papio anubis*) of Gombe National Park: Outbreak, time-course and attempts to limit recurrence. In: The 8th TAWIRI Scientific Conference. Tanzania Wildlife Research Institute, Arusha, Tanzania

Cousins D (1972) Diseases and injuries in wild and captive gorillas. International Zoo Yearbook 12:211–218

Cousins D (1984) Notes on the occurrence of skin infections in gorillas. Der Zoologische Garten 54(4–5):333–338

Cousins D (2008) Possible goundou in gorillas. Gorilla J 37:22–24

Crosby A (1969) The early history of syphilis: a reappraisal. Am Anthropol 71(2):218–227

Csonka GW (1953) Clinical aspects of bejel. Br J Vener Dis 29:95–103

Denis A (1963) On safari: the story of my life. Collins, London

Diaz de Isla R (1539) Treatise on the Serpentine Malady, which in Spain is commonly called Bubas, which was drawn up in the Hospital of All Saints in Lisbon.

Edroma E, Rosen N, Miller P (1997) Conserving the Chimpanzees of Uganda

Elsas F, Smith J, Israel C, Gager W (1968) Late syphilis in the primate. Br J Vener Dis 44:267–273

Engelhardt H (1959) A study of yaws (does congenital yaws occur?). J Trop Med Hyg 62:238–240

Fegan D, Glennon M, Thami Y, Pakoa G (2010) Resurgence of yaws in Tanna, Vanuatu: time for a new approach? Trop Doct 40:68–69

Felsenfeld O, Wolf R (1971) Serological reactions with treponemal antigens in nonhuman primates and the natural history of treponematosis in man. Folia Primatol 16:294–305
Fribourg-Blanc A, Mollaret H (1969) Natural treponematosis of the African primate. Primates Med 3:113–121
Fribourg-Blanc A, Niel G, Mollaret H (1966) Confirmation serologique et microscopique de la treponemose du cynocephale de guinee. Bulletin de la Societe de Pathologie Exotique et de Ses Filiales 59(1):54–59
Gersti S, Kiwila G, Dhorda M, Lonlas S, Myatt M, Ilunga B, Lemasson D, Szumilin E, Guerin P, Ferradini L (2009) Prevalence study of yaws in the Democratic Republic of Congo using the lot quality assurance sampling method. PLoS One 4(7):e6338
Gevers D, Cohan F, Lawrence J, Spratt B, Coenye T, Feil E, Stackebrandt E, Peer YD, Vandamme P, Thompson F et al (2005) Re-evaluating prokaryotic species. Nat Rev Microbiol 3:733–739
Gormus B, Xu K, Alford P, Lee D, Hubbard G, Eichberg J, Meyers W (1991) A serologic study of naturally acquired leprosy in chimpanzees. Int J Lepr 59(3):450–457
Gray R, Mulligan C, Molini B, Sun E, Giacani L, Godornes C, Kitchen A, Lukehart S, Centurion-Lara A (2006) Molecular evolution of the *tprC, D, I, K, G* and *J* genes in the pathogenic genus *Treponema*. Mol Biol Evol 23(11):2220–2233
Guthe T, Reynolds FW, Krag P, Willcox RR (1953) Mass treatment of treponemal diseases, with particular reference to syphilis and yaws. Br Med J 1(4810):594–598
Guthe T, Ridet J, Vorst F, D'Costa J, Grab B (1972) Methods for the surveillance of endemic treponematoses and sero-immunological investigations of "disappearing" disease. Bull World Health Organ 46(1):1–14
Hackett C (1953) Extent and nature of the yaws problem in Africa. Bull World Health Organ 8(1–3):127–182
Harper K, Ocampo P, Steiner B, George R, Silverman M, Bolotin S, Pillay A, Saunders N, Armelagos G (2008) On the origin of the treponematoses: a phylogenetic approach. PLoS NTDs 2(1):e148
Harper K, Zuckerman M, Harper M, Kingston J, Armelagos G (2011) The origin and antiquity of syphilis revisited: an appraisal of Old World Pre-Columbian evidence for treponemal infection. Yearbook of Physical Anthropology 54:99–133
Herring A, Ballard R, Mabey D, Peeling R (2006) Evaluation of rapid diagnostic tests: syphilis. Nat Rev Microbiol 4(12):S33–S40
Holmes K, Sparling P, Stamm W, Piot P, Wasserheit J, Corey L, Cohen M (2007) Sexually transmitted diseases. McGraw Hill, New York
Hopkins D, Flórez D (1977) Pinta, yaws, and venereal syphilis in Colombia. Int J Epidemiol 6(4):349–355
Hubbard G, Lee D, Eichberg J, Gormus B, Xu K, Meyers W (1991) Spontaneous leprosy in a chimpanzee (*Pan troglodytes*). Vet Pathol 28:546–548
Hudson E (1963) Treponematosis and anthropology. Ann Intern Med 58:1037–1048
Hunter E, Russell H, Farshy C, Sampson J, Larsen S (1986) Evaluation of sera from patients with lyme disease in the fluorescent treponemal antibody-absorption test for syphilis. Sex Transm Dis 13(4):232–236
Jowi J, Gathua S (2005) Lyme disease: report of two cases. East Afr Med J 82(5):267–269
Karesh W (2000) Suivi de la santé des gorilles au Nord-Congo. Canopée 18:16–17
Karp G, Schlaeffer F, Jotkowitz A, Riesenberg K (2009) Syphilis and HIV co-infection. Eur J Intern Med 20(1):9–13
Knauf S, Batamuzi E, Mlengeya T, Kilewo M, Lejora I, Nordhoff M, Ehlers B, Harper K, Fyumagwa R, Hoare R et al (2012) *Treponema* infection associated with genital ulceration in wild baboons. Vet Pathol 49(2):292–303
Köndgen S, Schenk S, Pauli G, Boesch C, Leendertz F (2010) Noninvasive monitoring of respiratory viruses in wild chimpanzees. Ecohealth 7:332–341
Kuhn U (1970) The treponematoses. The Chimpanzee 3:71–81
Kumm H, Turner T (1936) The transmission of yaws from man to rabbits by an insect vector, *Hippelates pallipes* loew. Am J Trop Med 16(3):245–271

Lagarde E, Guyavarch E, Plau J, Gueye-Ndiaye A, Seck K, Enel C, Pison G, Ndoye I, Mboup S, MECORA (2003) Treponemal infection rates, risk factors and pregnancy outcome in a rural area of Senegal. Int J STD AIDS 14(3):208–215

Lahariya C, Pradhan S (2007) Can Southeast Asia eradicate yaws by 2010? Some lessons from the yaws eradication programme of India. Natl Med J India 20(2):1–6

Leininger J, Donham K, Rubino M (1978) Leprosy in a chimpanzee: morphology of the skin lesions and characterization of the organism. Vet Pathol 15:339–346

Levine B, Lawton Smith J, Israel CW (1970) Serology of normal primates. Br J Vener Dis 46:307–310

Levréro F, Gatti S, Gautier-Hion A, Ménard N (2007) Yaws disease in a wild gorilla population and its impact on the reproductive status of males. Am J Phys Anthropol 132:568–575

Lien E, Sellati TJ, Yoshimura A, Flo TH, Rawadi G, Finberg RW, Carroll JD, Espevik T, Ingalls RR, Radolf JD et al (1999) Toll-like receptor 2 functions as a pattern recognition receptor for diverse bacterial products. J Biol Chem 274(47):33419–33425

Liu W, Worobey M, Li Y, Keele B, Bibollet-Ruche F, Guo Y, Goepfert P, Santiago M, Ndjango J, Neel C et al (2008) Molecular ecology and natural history of simian foamy virus infection in wild-living chimpanzees. PLoS Pathog 4(7):e1000097

Liu W, Li Y, Learn G, Rudicell R, Robertson J, Keele B, Ndjango J, Sanz C, Morgan D, Locatelli S et al (2010) Origin of the human malaria parasite *Plasmodium falciparum* in gorillas. Nature 467:420–425

Lovell N, Jurmain R, Kilgore L (2000) Skeletal evidence of probable treponemal infection in free-ranging African apes. Primates 41:275–290

Luger A (1972) Non-venereally transmitted 'endemic' syphilis in Vienna. Br J Vener Dis 48:356–360

Mackey D, Price E, KNox J, Scotti A (1969) Specificity of the FTA-ABS test for syphilis: an evaluation. JAMA 207(9):1683–1685

Magnarelli L, Miller J, Anderson J, Riviere G (1990) Cross-reactivity of nonspecific treponemal antibody in serologic tests for Lyme disease. J Clin Microbiol 28(6):1276–1279

Makuwa M, Souquière S, Clifford S, Mouinga-Ondeme A, Bawe-Johnson M, Wickings E, Latour S, Simon F, Roques P (2005) Identification of hepatitis B virus genome in faecal sample from wild living chimpanzee (*Pan troglodytes troglodytes*) in Gabon. J Clin Virol 34(1):S83–88

Marra C, Castro C, Kuller L (1998) Mechanisms of clearance of *Treponema pallidum* from the CSF in a nonhuman primate model. Neurology 51:957–961

Metchnikoff E, Roux E (1903) Études experimentales sur la syphilis. Ann Inst Pasteur 17(808–821)

Metchnikoff E, Roux E (1904) Études experimentales sur la syphilis. Ann Inst Pasteur 18:1–6

Metchnikoff E, Roux E (1905) Études expérimentales sur la syphilis. Ann Inst Pasteur 19:673–698

Mitjà O, Hays R, Ipai A, Wau B, Bassat Q (2011) Osteoperiostitis in early yaws: case series and literature review. Clin Infect Dis 52(6):771–774

Mlengeya TDK (2004) Distribution pattern of a sexually transmitted disease (STD) of Olive Baboon in Lake Manyara National Park, Tanzania. College of African Wildlife Management, Moshi

Moskophidis M (1989) Analysis of the humoral immune response to *Treponema pallidum* in the different stages of untreated human syphilis. Zentralbl Bakteriol 271(2):171–179

Noordhoek G, Wieles B, Jvd S, Jv E (1990) Polymerase chain reaction and synthetic DNA probes: a means of distinguishing the causative agents of syphilis and yaws. Infect Immun 58(6):2011–2013

Noordhoek G, Engelkens H, Judanarso J, Jvd S, Aelbers G, Jvd S, Jv E, Stolz E (1991) Yaws in West Sumatra, Indonesia: Clinical manifestations, serological findings and characterization of new *Treponema* isolates by DNA probes. Eur J Clin Microbiol Infect Dis 10(1):12–19

Oh Y, Ahn SY, Hong SP, Bak H, Ahn SK (2008) A case of extragenital chancre on a nipple from a human bite during sexual intercourse. Int J Dermatol 47(9):978–980

Pace J, Csonka GW (1984) Endemic non-venereal syphilis (bejel) in Saudi Arabia. Br J Vener Dis 60:293–297

Pampiglione S, Wilkinson A (1975) A study of yaws among pygmies in Cameroon and Zaire. Br J Vener Dis 51:165–169

Perine P, Hopkins D, Niemel P, John RS, Causse G, Antal G (1984) Handbook of endemic Treponematoses: yaws, endemic syphilis, and pinta. World Health Organization, Geneva

Richens J, Mabey CW (2009) Sexually transmitted infections (Excluding HIV). In: Cook GC, Zumla AI (eds) Manson's tropical diseases, 22nd edn. Saunders-Elsevier, Philadelphia, pp 403–434

Rinaldi A (2008) Yaws: a second (and maybe last?) chance for eradication. PLoS NTDs 2(8):e275

Rolfs R, Goldberg M, Sharrar R (1990) Risk factors for syphilis: cocaine use and prostitution. Am J Public Health 80(7):853–857

Román G, Román L (1986) Occurrence of congenital, cardiovascular, visceral, neurologic, and neuro-ophthalmologic complications in late yaws: a theme for future research. Rev Infect Dis 8(5):760–770

Rothschild B, Turnbull W (1987) Treponemal infection in a Pleistocene bear. Nature 329:61–62

Rothschild B, Hershkovitz I, Rothschild C (1995) Origin of yaws in the Pleistocene. Nature 378:343–344

Russell H, Sampson J, Schmid G, Wilkinson H, Plikaytis B (1984) Enzyme-linked immunosorbent assay and indirect immunofluorescence assay for Lyme disease. J Infect Dis 149(3):465–470

Salazar J, Hazlett KRO, Radolf JD (2002) The immune response to infection with *Treponema pallidum*, the stealth pathogen. Microbes Infect 4:1133–1140

Salomone G (1999) Le pian chez les peuples de la forêt équatoriale du Nord-Congo et du sud de la République Centrafricaine. In: Bahuchet S, Bley D, Pagezy H, Vernazza-Licht N (eds) L'Homme et la Forêt Tropicale. University of Provence Press, Marseilles, pp 675–688

Salzer J, Rwego I, Goldberg T, Kuhlenschmidt M, Gillespie T (2007) *Giardia* sp. and *Cryptosporidium* sp. infections in primates in fragmented and undisturbed forest in Western Uganda. J Parasitol 93(2):439–440

Santiago M, Lukasik M, Kamenya S, Li Y, Bibollet-Ruche F, Bailes E, MUller M, Emery M, Goldenberg D, Lwanga J et al (2003) Foci of endemic simian immunodeficiency virus infection in wild-living Eastern chimpanzees (*Pan troglodytes schweinfurthii*). J Virol 77(13):7545–7562

Satchell G, Harrison R (1953) Experimental observations on the possibility of transmission of yaws by wound-feeding Diptera, in Western Samoa. Trans R Soc Trop Med Hyg 47(2):148–153

Satter E, Tokarz V (2010) Secondary yaws: an endemic treponemal infection. Pediatr Dermatol 27(4):364–367

Schroder NW, Eckert J, Stubs G, Schumann RR (2008) Immune responses induced by spirochetal outer membrane lipoproteins and glycolipids. Immunobiology 213(3–4):329–340

Seña A, White B, Sparling P (2010) Novel *Treponema pallidum* serologic tests: a paradigm shift in syphilis screening for the 21st century. Clin Infect Dis 51(6):700–708

Sepetjian M, Guerraz F, Salussola D, Thivolet J, Monier J (1969) Contribution à l'étude du tréponème isolé du singe par A. Fribourg-Blanc. Bull WHO 40:141–151

Sepetjian M, Thivolet J, Salussola D, Guerraz F, Monier J (1972) Étude comparative du FTA et du FTA ABS quantitatifs chez l'homme et l'animal a différents stades de la syphilis. Pathologie Biologie 20(9):449–455

Smajs D, Norris S, Weinstock G (2011) Genetic diversity in *Treponema pallidum*: implications for pathogenesis, evolution and molecular diagnostics of syphilis and yaws. Infect Genet Evol 12(2):191–202

Smith J (1969) Late ocular syphilis: transfer of infection from man to experimental animals. Trans Am Opthalmol Soc 67:658–697

Smith J, Singer J, Reynolds D, Moore M, Yobs A, Clark J (1965) Experimental ocular syphilis and neurosyphilis. Br J Vener Dis 41:15–23

Smith J, David N, Indgin S, Israel C, Levine B, Justice J, McCrary J, Medina R, Paez P, Santana E et al (1971) Neuro-ophthalmological study of late yaws and pinta II. The Caracas Project. Br J Vener Dis 47:226–251

St John R (1985) Yaws in the Americas. Rev Infect Dis 7(S2):S266–S272

Stamm L (2010) Global challenge of antibiotic-resistant *Treponema pallidum*. Antimicrob Agents Chemother 54(2):583–589
Suzuki K, Udono T, Fujisawa M, Tanigawa K, Idani G, Ishii N (2010) Infection during infancy and long incubation period of leprosy suggested in a case of a chimpanzee used for medical research. J Clin Microbiol 48(9):3432–3434
Taylor W (1954) Endemic syphilis in a South African coloured community. S Afr Med J 28(9): 176–178
Turner AB, Hollander DH (1957a) Factors affecting the evolution of experimental treponematosis. In: Turner AB, Hollander DH (eds) Biology of the Treponematoses. World Health Organization, Geneva, pp 70–94
Turner T, Hollander D (1957b) Biology of the Treponematoses. World Health Organization, Geneva
Valverde C, Canfield D, Tarara R, Esteves M, Gormus B (1998) Spontaneous leprosy in a wild-caught cynomolgus macaque. Int J Lepro Other Mycobact Dis 66(2):140–148
Wallis J (2000) Prevention of disease transmission in primate conservation. Ann NY Acad Sci 916:691–693
Wallis J, Lee D (1999) Primate conservation: the prevention of disease transmission. Int J Primatol 20(6):803–825
Wells J, Smith J (1967) Experimental ocular and neurosyphilis in the primate. Br J Vener Dis 43:10–17
WHO (2001) Global prevalence and incidence of selected curable sexually transmitted infections: Syphilis
Willcox R (1955) The non-venereal treponematoses. Br J Obstet Gynaecol 62(6):853–862
Wilson P, Mathis M (1930) Epidemiology and pathology of yaws. J Am Med Assoc 94: 1289–1292
Yahya-Malima KI, Evjen-Olsen B, Matee MI, Fylkesnes K, Haarr L (2008) HIV-1, HSV-2 and syphilis among pregnant women in a rural area of Tanzania: prevalence and risk factors. BMC Infect Dis 8:75
Yakinci C, Özcan A, Aslan T, Demirhan B (1995) Bejel in Malatya, Turkey. J Trop Pediatr 41(2): 117–120
Yoshinari N, Oyafuso L, Monteiro F, Pd B, Fd C, Ferreira L, Bonasser F, Baggio D, Cossermelli W (1993) Lyme disease. Report of a case observed in Brazil. Rev Hosp Clin Fac Med Sao Paulo 48(4):170–174
Zhou P, Qian Y, Lu H, Guan Z (2009) Nonvenereal transmission of syphilis in infancy by mouth-to-mouth transfer of prechewed food. Sex Transm Dis 36(4):216–217

Molecular Mimicry by γ-2 Herpesviruses to Modulate Host Cell Signaling Pathways

Lai-Yee Wong, Zsolt Toth, Kevin F. Brulois, Kyung-Soo Inn, Sun Hwa Lee, Hye-Ra Lee, and Jae U. Jung

Introduction

Herpesviruses are large double-stranded DNA viruses that can establish life-long infection in their respective hosts and can undergo two different phases in their life cycle: lytic or latent (Fig. 1a) (Pellett and Roizman 2007). Lytic replication is characterized by the expression of most viral genes in an ordered cascade (immediate early, early, and late), leading to the production of infectious virions. Latency is marked by minimal viral gene expression and the maintenance of the viral genome in the nucleus. Reactivation to lytic replication from latency can be triggered by multiple factors, such as stress or chemical reagents. The ability of herpesviruses to utilize these two very distinct modes of replication is an excellent survival strategy as establishment of latency with periodic reactivation may facilitate persistent infection in the host while allowing evasion from the immune system.

Herpesviruses are prevalent in nature, with most animal species being infected by at least one herpesvirus (Knipe and Howley 2007). Out of more than 200 herpesviruses identified to date, only eight are endemic to humans and are grouped into three subfamilies (α, β, γ) based on their structural and biological properties (Table 1). γ-Herpesviruses are lymphotropic and are further divided into two subgroups: γ-1 (Lymphocryptovirus) and γ-2 (Rhadinovirus) (Fig. 2). Epstein-Barr virus (EBV) was the first γ-herpesvirus to be discovered and is the prototype member of the γ-1 group (Diehl et al. 1968). Kaposi's sarcoma-associated herpesvirus (KSHV) is so far the only human virus assigned to the Rhadinovirus family, which also includes herpesvirus saimiri (HVS), rhesus rhadinovirus (RRV), and mouse herpesvirus 68 (mHV68) (Table 2) (Blaskovic et al. 1980; Chang et al. 1994; Desrosiers et al. 1997; Melendez et al. 1968; Moore et al. 1996).

L.-Y. Wong (✉) • Z. Toth • K.F. Brulois • K.-S. Inn • S.H. Lee • H.-R. Lee • J.U. Jung
Department of Molecular Microbiology and Immunology, University of Southern California, Keck School of Medicine, Los Angeles, CA 90033, USA
e-mail: LaiYee.Wong@usc.edu

J.F. Brinkworth and K. Pechenkina (eds.), *Primates, Pathogens, and Evolution*, Developments in Primatology: Progress and Prospects,
DOI 10.1007/978-1-4614-7181-3_8, © Springer Science+Business Media New York 2013

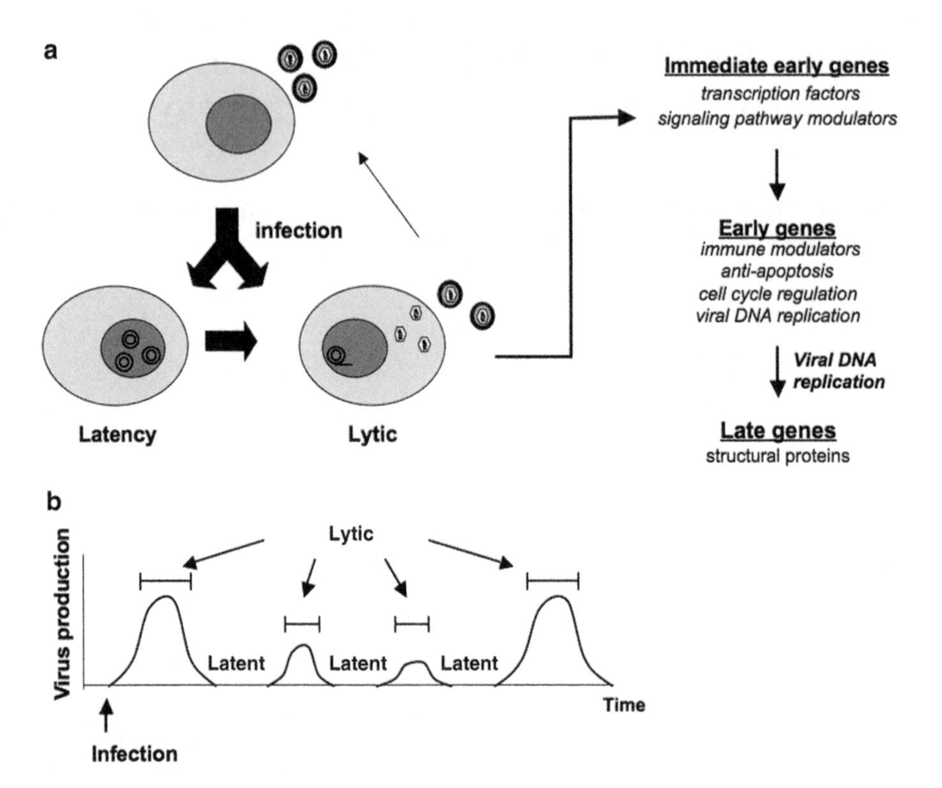

Fig. 1 (**a**) Diagram showing the two phases of a herpesvirus life cycle. Upon infection, the virus can enter latency where the viral genome is typically kept in an episomal form in the cell nucleus and only a few viral genes are expressed. Alternatively, the virus may undergo lytic replication to produce infectious virus that go on to infect new cells. Viral genes are expressed in an ordered cascade during lytic replication, starting with immediate-early genes, early genes, and late genes. Common functions of the proteins encoded by these lytic genes are listed underneath each group. Late gene expression only occurs after viral DNA replication. (**b**) A burst of lytic replication occurs after infection by a herpesvirus, after which the virus will establish latency in the now infected host. Periodically the virus will reactivate to produce new virus, followed by another round of latency where no virus is produced

γ-Herpesviruses are oncogenic viruses capable of causing neoplasia in the infected host (Table 2) and contain multiple open reading frames (ORFs) that contribute to virus-induced tumorigenesis. In addition to being the only oncoviruses in the herpesvirus family, the pathogenesis of γ-herpesviruses is largely associated with their latency program (Fig. 1b). In contrast, the disease symptoms of α- and β-herpesviruses are due to the lytic replication and the resulting host immune responses to contain the virus.

A striking feature of γ-herpesviruses is the high degree of molecular piracy among their viral genes that are postulated to have been "appropriated" from the host cell during virus-host evolution. This chapter focuses on several intriguing ORFs of KSHV, RRV, and HVS that are similar in structure (membrane localization

Table 1 Human herpesviruses

Name	Subfamily	Synonym	
HHV-1	α	Herpes simplex virus 1	HSV-1
HHV-2	α	Herpes simplex virus 2	HSV-2
HHV-3	α	Varicella-zoster virus	VZV
HHV-4	γ	Epstein-Barr virus	EBV
HHV-5	β	Cytomegalovirus	CMV
HHV-6	β	Human herpesvirus 6	
HHV-7	β	Human herpesvirus 7	
HHV-8	γ	Kaposi's sarcoma-associated herpesvirus	KSHV

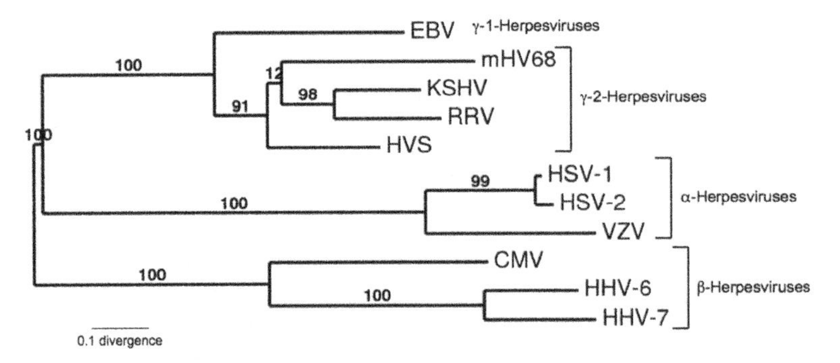

Fig. 2 Phylogenetic tree showing the three subfamilies of herpesviruses (α, β, γ). The tree was constructed using sequences of the viral DNA polymerase catalytic subunit. The program MUSCLE was used for alignment and PhyML for phylogeny analysis (Dereeper et al. 2008)

and signaling motifs), function (modulating lymphocyte activation events), and genomic position despite not having any sequence homology. Furthermore, we will also touch on the burgeoning field of virus-encoded microRNAs (miRNAs), which is another mechanism used by herpesviruses to manipulate host cell signaling through the repression of gene expression.

Kaposi Sarcoma-Associated Herpesvirus

KSHV was first discovered in 1994 as the etiological agent in Kaposi's sarcoma (KS) lesions and has since been further implicated in primary effusion lymphoma (PEL) and multicentric Castleman's disease (MCD) (Cesarman et al. 1995; Chang et al. 1994; Soulier et al. 1995). KSHV DNA has been detected in all cases of KS using multiple assays such as polymerase chain reaction and immunohistochemistry (Boshoff et al. 1995; Dupin et al. 1999). The first case of KS was described in 1872 by Dr. Moritz Kaposi as an indolent tumor affecting elderly men of Mediterranean and Jewish origins (Iscovich et al. 2000; Kaposi 1872). This later became known as classical KS after three other epidemiological forms were

Table 2 The respective host and associated malignancies of γ-herpesviruses

Group	Virus	Name	Host species	Associated malignancies
γ-1 Lymphocryptovirus	Epstein-Barr virus	EBV	*Homo sapiens* (human)	Burkitt's lymphoma, infectious mononucleosis, nasopharyngeal carcinoma (NPC)
	Kaposi's sarcoma-associated herpesvirus	KSHV	*Homo sapiens* (human)	Kaposi's sarcoma (KS), primary effusion lymphoma (PEL), multicentric Castleman's disease (MCD)
γ-2 Rhadinovirus	Herpesvirus saimiri	HVS	*Saimiri sciureus* (squirrel monkey)	Lymphoma (nonnatural host[a])
	Rhesus rhadinovirus	RRV	*Macaca mulatta* (rhesus monkey)	Lymphoproliferative disorder[b]
	Murine herpesvirus 68	mHV68	*Mus musculus* (mouse)	

[a]HVS-C in common marmosets and rabbits
[b]Naïve rhesus macaques

reported, which includes iatrogenic KS in immunosuppressed patients after organ transplants, endemic KS in sub-Saharan Africa, and acquired immunodeficiency syndrome (AIDS)-related epidemic KS (Ambroziak et al. 1995; Foreman et al. 1997; Gao et al. 1996; Kedes et al. 1996; Regamey et al. 1998; Schalling et al. 1995; Simpson et al. 1996). Unlike the classical form, endemic KS in Africa is extremely aggressive and rapidly fatal and occurs in both sexes with equal frequency (Dourmishev et al. 2003; Hengge et al. 2002). It also strikes children and in fact accounted for 4 % of all childhood cancer in Cameroon from 1986 to 1993 (Kasolo et al. 1997; Wabinga et al. 1993). Likewise, AIDS-related epidemic KS is a very aggressive, fulminant, and disseminated form of the disease (Schwartz 1996). It occurs predominantly in young homosexual and bisexual men with AIDS and was recognized as one of the first signs of the AIDS epidemic (Friedman-Kien 1981; Gottlieb et al. 1981).

Despite the various epidemiological forms of KS with different clinical courses, all KS lesions are characterized by spindle-shaped cells of endothelial origin with infiltration of inflammatory cells (Niemi and Mustakallio 1965). Analysis of KS lesions showed that most of these spindle cells are latently infected, with a very low number of cells undergoing lytic replication at any time (Staskus et al. 1997; Zhong et al. 1996). KS spindle cells are unique as these cells are not fully transformed as with conventional tumor cells (Aluigi et al. 1996; Benelli et al. 1996). Explanted spindle cells gradually lose viral episomes in vitro and cannot form tumors in nude mice

(Aluigi et al. 1996; Ganem 2006; Salahuddin et al. 1988). These spindle cells secrete many proinflammatory and angiogenic factors, which contribute to the inflammatory and neovascular characteristics of KS (Salahuddin et al. 1988). The driving force behind spindle cells has largely been attributed to the expression of a viral latent protein; FLICE-inhibitory protein (vFLIP); an antiapoptotic, anti-autophagy factor; and a strong inducer of the nuclear factor kappa-light-chain-enhancer of activated B-cell (NF-κB) pathway (Grossmann et al. 2006; Lee et al. 2009).

KSHV is also associated with PEL and MCD, rare lymphoproliferative disorders of B-cell origin usually diagnosed in AIDS patients. Coinfection with EBV is commonly found in cases of PEL, an aggressive subtype of non-Hodgkin B-cell lymphoma. PEL is characterized by proliferation of cells that lack most B-cell markers but have features reminiscent of plasmablasts (Jenner et al. 2003). MCD is a rare, polyclonal lymphocyte hyperplasia in which KSHV-infected cells are found to be IgM-λ positive B cells located in the mantle zone in the lymph node (Du et al. 2001). In vitro infection of human tonsillar B cells with KSHV revealed that the virus has a propensity to infect IgM-λ positive B cells and subsequently acquire characteristics similar to infected cells in MCD (Hassman et al. 2011). It is unclear why KSHV shows favoritism towards a specific subtype of B cells or how this specificity is linked to the pathogenesis seen in MCD.

In nature, KSHV infection is restricted to humans. In an experimental setting, KSHV could infect a variety of primary cells and cell lines from different species such as human, mouse, and rat (Lagunoff et al. 2002; McAllister and Moses 2007; Renne et al. 1998). Despite the promiscuous ability of KSHV to infect multiple cell types, these infected cells are not immortalized or transformed and, without any selective pressure such as antibiotic selection, tend to lose the viral genome after several passages (Lagunoff et al. 2002). Recently (2012), Jones et al. successfully developed a KSHV-induced tumorigenesis in vitro model using primary rat mesenchymal cell (Jones et al. 2012).

The in vivo study of KSHV and its malignancies uses mostly nude, transgenic, or humanized mouse model systems (Dittmer et al. 1999; Wu et al. 2006). KSHV-positive PEL cells are transplantable into nude mice lacking functional B and T cells, but the virus does not spread to murine cells (Picchio et al. 1997). The study of vFLIP using transgenic mice showed increased incidence of lymphomas, and when this viral protein expression is restricted to B cells, the mice acquired phenotypes similar to MCD abnormalities (Ballon et al. 2011; Chugh et al. 2005). The first nonhuman primate model of KSHV infection was established using common marmosets (*Callithrix jacchus*) (Chang et al. 2009). Infection with recombinant KSHV led to persistent infection in these animals, and one marmoset showed development of KS-like lesion, which shared similar histopathological characteristics as human AIDS-associated KS lesions (such as presence of spindle cells with detectable expression of KSHV proteins). It would be interesting to determine if the KS-like lesion progresses similarly in this common marmoset animal model compared to human KS.

Table 3 KSHV genes and cellular homologues

KSHV ORF	Name	Function	Cellular homologue
ORF4	KCP	Complement	CR2
K2	vIL-6	Cytokine	IL-6
K4	vMIP-II	Chemokine	CCL3
K4.1	vMIP-III	Chemokine	CCL3
K6	vMIP-I	Chemokine	CCL3
K9	vIRF-1	Interferon	IRF4 and 8
K10	vIRF-4	Interferon	IRF4 and 8
K10.5	vIRF-3	Interferon	IRF4 and 8
K11	vIRF-2	Interferon	IRF4 and 8
ORF16	vBcl-2	Antiapoptotic and anti-autophagy	Bcl2
ORF71	vFLIP	Antiapoptotic and anti-autophagy	cFLIP
ORF72	v-cyclin	Cell cycle	Cyclin D2
ORF74	vGPCR	Signaling	Interleukin-8 receptor
K14	v-OX2	Cytokine	OX2

Molecular Piracy in KSHV

Sequencing of the complete KSHV genome revealed that KSHV encodes multiple ORFs with noticeable homology to cellular genes (Neipel et al. 1997). These pirated genes are involved in multiple pathways such as antiapoptosis, autocrine/paracrine signaling, immune responses, and cell cycle regulation. Most of these "pilfered" genes are also found in other Rhadinoviruses at the equivalent genomic position, suggesting that the piracy event may have happened early during the virus evolution. KSHV v-cyclin, a latent gene, is a type D cyclin with 53 % similarity to both HVS and cellular cyclin (Li et al. 1997). In addition, some of the genes first identified as unique to KSHV turned out to have cellular equivalents that were discovered after the characterization of the viral counterparts. Two notable examples are the KSHV K3 and K5 genes, also known as the modulator of immune recognition MIR-1 and MIR-2, respectively. K3 and K5 are ubiquitin E3 ligases and potent immune dysregulators that downregulate a variety of surface immune receptors, such as MHC class I (Coscoy and Ganem 2000; Ishido et al. 2000). Studies on K3 and K5 led to the identification of a family of cellular ubiquitin E3 ligases called membrane-associated RING CH domain (MARCH) proteins which share a common enzymatic motif and domain organization (Goto et al. 2003). A list of pirated viral genes and their cellular homologues is provided in Table 3.

While the few examples of viral molecular mimicry discussed so far contain cellular homologues that are conserved not only in function but also in sequence, the γ-2-herpesviruses also contain unique genes that have no cellular counterparts but functionally mimic a cellular protein. The first and last ORFs of these viruses (but the first and second ORFs for HVS) encode transmembrane proteins involved in modulating lymphocyte activation events and are unique to the respective viruses (K1 and K15 for KSHV, R1 and R15 for RRV, Tip and STP for HVS) (Fig. 3).

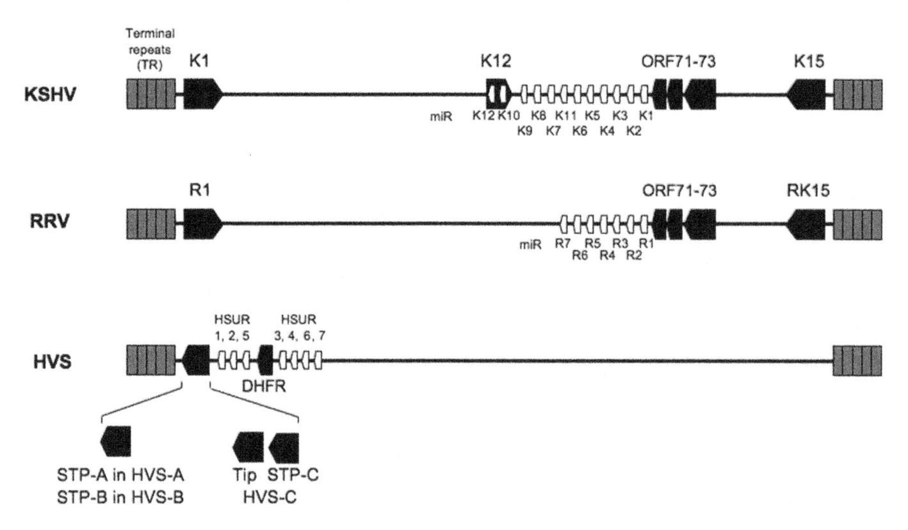

Fig. 3 Genomic location of the viral transmembrane proteins and viral pre-miRNAs of KSHV, RRV, and HVS. In HVS strains A and B, STP-A and STP-B are the first ORFs of the genome, but in HVS subgroup C, Tip is the first ORF followed by STP-C as the second ORF. The pre-miRNAs are located in the latency-associated region. The U-rich RNAs of HVS are located at the left end region of the viral genome. Dihydrofolate reductase (DHFR)

The ability of γ-herpesviruses to subordinate the lymphocyte activation pathways is particularly important to ensure its survival and concealment from the host immune surveillance since they establish a latent reservoir within circulating lymphocytes. These viral proteins control multiple key cellular pathways involved in cellular transformation and thus are likely to be important contributors to viral-dependent lymphoproliferation phenotypes seen in infected organisms.

Sequence comparison between these distinct ORFs show low sequence similarity, but they share structural and functional conservation. Some of the proteins even show a high level of sequence divergence between different strains of the same virus. It is possible that the proximity of these genes to the terminal repeats, an area of high homologous recombination activity, contributed to this sequence variation. Despite not having cellular homologues, these viral proteins possess domains commonly found in cellular proteins. It is unclear whether these genes evolved from a common ancestor that pirated a cellular gene and later acquired unique functions due to their different tropisms and high mutation frequency, or whether the different viruses acquired these genes independently during the course of their evolution.

KSHV K1

K1, the first ORF of KSHV, is located at the 5′ end of the KSHV genome after the terminal repeats (Fig. 2) (Lagunoff and Ganem 1997; Lee et al. 1998b; Russo et al. 1996). K1 is expressed at low levels during latency, and its presence has been

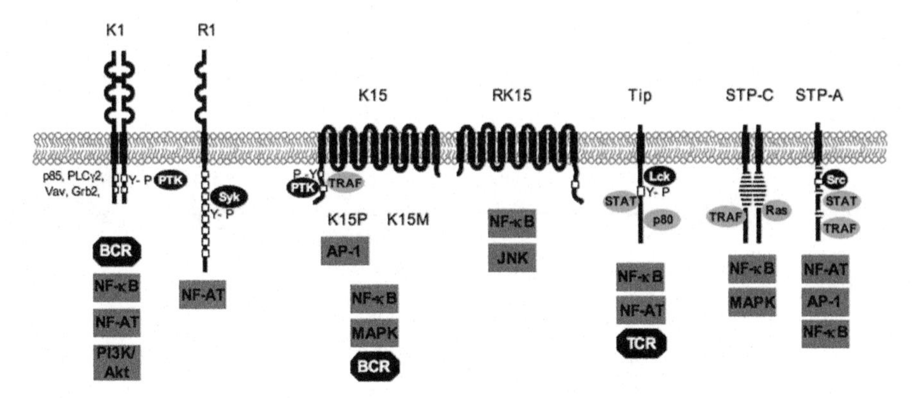

Fig. 4 Structures of the transmembrane signaling proteins of KSHV (K1 and K15), RRV (R1 and RK15), and HVS (Tip, STP-C and STP-A). The pathways depicted by squares indicate they are activated by the respective viral proteins, while those in octagons indicate inhibition of signaling. P-Y denotes phosphorylated tyrosine residues

detected in PEL, MCD, and KS tumors (Lagunoff and Ganem 1997; Lee et al. 2003; Samaniego et al. 2001). *K1* encodes a type-I transmembrane glycoprotein that mimics the signaling activity of the cellular B-cell receptor (BCR) (Fig. 4) (Lee et al. 1998a). Its extracellular domain is highly divergent between different KSHV strains, yet they all maintain the ability to oligomerize via this region (Lagunoff et al. 1999; Lee et al. 1998b). On the other hand, the cytoplasmic tails of K1 variants are highly conserved and contain an immunoreceptor tyrosine-based activation (ITAM) motif, which is involved in cell activation signals (Cambier 1995; Lee et al. 1998a). The structure of K1 is similar to the BCR, and indeed K1 expression has been shown to generate a signaling profile reminiscent of BCR activation. At the same time, K1 deregulates BCR by retaining the newly synthesized μ chain of the BCR complex in the endoplasmic reticulum (ER) as well as promoting the internalization of surface BCRs via clathrin-dependent endocytosis (Lee et al. 2000; Tomlinson and Damania 2008). Thus, K1 can be localized to the cell surface, ER, and endosomes (early and recycling) (Tomlinson and Damania 2008). The presence of K1 in recycling endosomes strongly suggests that K1 is recycled after endocytosis; however, direct evidence for this scenario has yet to be presented (Table 4).

Through its ITAM motif, K1 induces calcium mobilization and increases tyrosine phosphorylation of cellular proteins leading to the activation of nuclear factor for activated T cells (NFAT) and NF-κB, events reminiscent of BCR activation (Lagunoff et al. 1999; Lee et al. 1998a). A prerequisite step for initiation of BCR signaling is the clustering of multiple BCRs, which only occurs after ligand binding. In contrast, K1 is constitutively active and does not require extracellular signals. It has been postulated that this intrinsic activation of K1 is due to its extracellular domain, which mediates multimerization even in the absence of ligand binding (Lagunoff et al. 1999). K1 interacts with a variety of cellular signal transduction

Table 4 Summary of the ORFs discussed

	KSHV K1	KSHV K15	RRV R1	RRV RK15	HVS STP	HVS Tip
Interaction partners	Syk, Lyn, Vav, PLCγ2, Grab, PTK-1 and –2	Src, Hck, Lyn, Fyn, Yes	Syk		Src, TRAFs, Ras (STP-C), STAT3	Lck, p80, STAT-1 and -3
Signaling motifs	ITAM—phosphorylated by Src	SH2—phosphorylated by Src kinases, SH3, TRAF-binding motif	ITAM—phosphorylated by Syk	SH3 only	TRAF-binding motif (STP-A and STP-C), collagen-like repeats (STP-A and STP-C), SH2 (STP-A and STP-B)	SH3 and CSKH (phosphorylated by Lck)
Ligand dependent	No. Multimerization of extracellular domain				No. Oligomerization of collagen-like repeats	
Ca^{2+} mobilization	Increased upon αK1 stimulation	Inhibited upon BCR stimulation	Increased upon stimulation			Decreased
Cellular pTyr level	Increased	Decreased	Increased			Decreased
Activated pathways	NFAT, NF-κB, AP-1, VEGF, PI3K/Akt	NF-κB (SH2 dependent), AP-1 (K15P only), MAPK (TRAF and SH2 dependent), JNK (K15M only), OncomiRs (only K15M tested)	NFAT	NF-κB (SH2 dependent), JNK	NF-κB, NFAT, MAPK, AP-1	NF-κB, NFAT
Downregulate	BCR surface levels	BCR signaling and surface levels				Lck activity, TCR and CD4 surface levels

(continued)

Table 4 (continued)

	KSHV K1	KSHV K15	RRV R1	RRV RK15	HVS STP	HVS Tip
Transformation	Yes. Rat-1 fibroblasts and common marmoset T cells induces lym-phoma in transgenic mice		Yes. Rat-1 fibroblast and common marmoset T cells		STP-C—T cells from rabbit, primates, and human—induces lym-phoma in rhesus monkeys. STP-A –lymphocytes from common marmosets	
Consequences	Cell survival, inflammatory cytokines, angiogenesis, blocks BCR signaling, and suppressing lytic replication	Angiogenesis, metastasis			Antiapoptosis, IL-2-independent growth	Blocks TCR signaling

molecules termed the K1 "signalosome" and characterized in a study by Lee et al. (Lee et al. 2005). These cellular factors include SH2-binding factors such as spleen tyrosine kinase (Syk), Lyn, phospholipase C-γ2 (PLCγ2), Vav, Grab2, and protein tyrosine phosphatases 1 and 2 (Lee et al. 2005). These SH2-binding proteins participate in a variety of signaling pathways such as tyrosine phosphorylation, intracellular calcium fluxes, and transcription factor activation, ultimately leading to the upregulation of inflammatory cytokine and angiogenic factor genes (Lee et al. 2005). KSHV-infected cells are highly dependent on autocrine and paracrine signaling, so a cytokine-rich environment is important for KSHV-induced oncogenesis (Ensoli et al. 1989, 2000). K1 expression results in the production of inflammatory cytokines and angiogenic factors such as vascular endothelial growth factor (VEGF), highlighting the central role of K1 in promoting the paracrine model of KSHV tumorigenicity (Lee et al. 2005; Prakash et al. 2002, 2005; Wang et al. 2004). In addition, K1 also protects infected cells from apoptosis via multiple mechanisms such as blocking the binding of Fas ligand to the Fas receptor and activation of the phosphoinositide 3-kinase (PI3K) and AKT pathways (Berkova et al. 2009; Tomlinson and Damania 2004). The ability of K1 to modulate multiple cellular signal transduction pathways underlies its oncogenic potential, as shown by the ability of rodent fibroblast cells expressing K1 to undergo transformation and foci formation (Lee et al. 1998b).

Furthermore, when the saimiri transforming protein (STP) oncogene of HVS is replaced by K1, this recombinant virus retains the ability to immortalize common marmoset (*Callithrix jacchus*) T cells in vitro independent of exogenous cytokines and induced lymphoma in infected common marmosets, albeit at a delayed time point compared to HVS wild-type infected animals (Lee et al. 1998b). Results from K1 transgenic mice highlight the transforming potential of K1 in vivo as these mice develop spindle cell sarcoma and plasmablastic lymphoma after a year (Prakash et al. 2002). The authors showed that K1 constitutively activated the NF-κB and VEGF pathways, and, indeed, serums levels of VEGF were three times higher in K1 transgenic mice compared to control mice (Prakash et al. 2002). Thus, it is likely that K1 is an important contributor to the angiogenic characteristic seen in KS tumors, where most of the cells are latently infected.

Given the ability of K1 to activate so many different pathways, it is interesting to consider its affect on viral lytic replication. Lagunoff et al. showed that inhibition of K1 by expressing a dominant-negative mutant (with tyrosine to phenylalanine substitutions in the ITAM) actually reduces the efficiency of viral lytic replication (Lagunoff et al. 2001).

Their data suggests that K1 contributes to efficient lytic replication (Lagunoff et al. 2001). Since the ITAM is essential for the signaling role of K1, it is hard to tell if the signaling activity of K1 is intimately linked to its role in enhancing viral lytic replication.

KSHV K15

K15 is another gene unique to KSHV and is found at the 3′-end of the KSHV genome, adjacent to the terminal repeats (Fig. 3). This gene contains eight exons and multiple splicing variants with the dominant spliced form encompassing all the exons to express a protein with 12 transmembrane domains followed by a short cytoplasmic tail (Fig. 4) (Brinkmann et al. 2003; Choi et al. 2000b; Glenn et al. 1999). While *K15* is widely considered to be a lytic gene, low levels of expression can be detected in KSHV-positive PEL cell lines explanted from patients (Sharp et al. 2002; Wong and Damania 2006). It is uncertain if this low level of K15 expression is due to lytic reactivation in a minority of infected cells (1–5 %) or if the latently infected cells periodically express a low amount of K15. Two variants of K15 exist: the predominant (P) form and the minor (M) form, termed K15P and K15M, respectively (Glenn et al. 1999; Poole et al. 1999). These two variants show low sequence identity at the protein level, but are almost structurally (splicing pattern and protein domains) and functionally identical (Glenn et al. 1999; Poole et al. 1999). The cytoplasmic tails of the K15 variants are highly conserved, suggesting that loss of function in this domain is deleterious to the virus. So far, most studies have been done with the K15P form.

K15P

Expression of K15P interferes with BCR signaling but induces the activation of NF-κB, activator protein-1 (AP-1), and mitogen-activated protein kinase (MAPK) pathways, which ultimately leads to expression of genes involved in cell survival, proliferation, and inflammation. Specifically, K15P activates the extracellular signal-regulated kinases (ERK), c-Jun amino-terminal kinases (JNK), but not the p38 cascade of the MAPK superfamily (Brinkmann et al. 2003, 2007; Cho et al. 2008; Pietrek et al. 2010, Tsai et al. 2009; Wang et al. 2009). The cytoplasmic domain of K15 contains multiple motifs important in cellular signaling cascades, including a tumor necrosis factor (TNF), receptor-associated factors (TRAF)-interacting site, a Src homology 2 (SH2) domain, and a proline-rich SH3 motif (Choi et al. 2000b). These domains mediate interactions with a variety of cellular signaling molecules to modulate the pathways mentioned above. The SH2 and SH3 motifs of K15 interact with multiple Src kinases such as Src, Hck, Lck, Fyn, and Yes, with the Y481 in the SH2 motif being constitutively phosphorylated by tyrosine kinases (Brinkmann et al. 2003). This phosphorylation is required for activation of NF-κB and MAPK (Brinkmann et al. 2003).

The K15 cytoplasmic tail is sufficient to downregulate BCR signaling, as shown in a study using a chimeric protein composed of the K15 cytoplasmic portion fused to the CD8α transmembrane domain and expressed in a B-cell line (Choi et al. 2000b). The inhibition of the calcium cascade upon BCR stimulation requires intact SH2 and SH3 motifs of K15 (Choi et al. 2000b). It is hypothesized that the recruitment of Lyn by K15 sequesters this major B-cell kinase from BCR signaling events (Cho et al. 2008). Indeed, mutating the SH3 motif of K15 interferes with its ability to deregulate BCR signaling (Cho et al. 2008). Furthermore, K15 increases the rate of BCR internalization from the cell surface; thus, it is likely that K15 uses multiple mechanisms to inhibit BCR activation (Lim et al. 2007).

K15P Versus K15M

Despite the conservation of signaling motifs in the cytoplasmic tails of the K15P and K15M variants, there are phenotypical differences between these two variants. For one, even though both are predominantly cytoplasmic, the P variant contains a mitochondria localization sequence and can thus localize to the host mitochondria and other organelles such as ER and Golgi, while K15M is found on lysosomal membranes (Choi et al. 2000b; Sharp et al. 2002; Tsai et al. 2009; Wang et al. 2007). Although both K15 variants activate NF-κB to the same degree, only the P form can positively regulate AP-1 activity (Tsai et al. 2009). Whether this distinct effect is due to the difference in localization remains to be tested.

The importance of the SH2 motif in K15 to initiate the expression of cellular cytokines involved in angiogenesis and inflammation (e.g., IL-8, IL-6, CXCL2) also varies slightly between the two K15 variants. While mutation of the amino acids tyrosine-glutamic acid-glutamic acid-valine at position 481–484 (Y^{481}EEV) of the K15P SH2 domain abolishes this phenotype, the K15M Y^{490}EEV mutant retained residual activity in cytokine upregulation, which may be due to the ability of K15M to activate the JNK pathway independently of its SH2 motif (Pietrek et al. 2010; Wang et al. 2007). Furthermore, K15M can upregulate miR-21 and miR-23, cellular cancer-related miRNAs (oncomiRs) that are involved in tumor cell proliferation, metastasis, and angiogenesis (Tsai et al. 2009). This ability is dependent on the tyrosine residue in the SH2 domain. It is unclear whether K15P, which also has an SH2 motif, can upregulate the same oncomiRs.

Despite the higher prevalence of the K15P isoform in KSHV isolates, the two variants seem to function equivalently in most of the in vitro assays used to study them (Pietrek et al. 2010; Wang et al. 2007). Whether the P variant has an in vivo survival advantage over the M isolates (due to activation of AP-1 pathway) remains to be tested.

Rhesus Rhadinovirus

RRV was first discovered in rhesus monkeys (*Macaca mulatta*) housed at the New England Primate Research Center (NEPRC) (Desrosiers et al. 1997). The animals reacted to HVS antigens in serology tests, hinting that they were infected by an agent related to γ-herpesviruses. A further survey after the characterization of RRV revealed that at least 90 % of rhesus monkeys housed at both the NEPRC and

Oregon Regional Primate Research Center (ORPRC) were seropositive for RRV (Mansfield et al. 1999; Wong et al. 1999). This uniformly high prevalence of infection at both centers suggested that RRV is endemic in rhesus monkeys. Sequencing of its viral genome revealed that RRV is a close cousin of KSHV with high sequence conservation. Indeed, both these Old World Rhadinoviruses share homologous genes that are not found in New World Rhadinoviruses like HVS (e.g., viral interleukin 6 (*vIL-6*)) (Alexander et al. 2000; Searles et al. 1999). RRV has been detected in B cells of rhesus macaques, suggesting that, similar to KSHV, lymphocytes are the major reservoir of viral persistence (Bergquam et al. 1999; Desrosiers et al. 1997; Wong et al. 1999). Unlike KSHV, which primarily enters latency upon de novo infection, RRV can undergo lytic replication and produce high viral titer in primary rhesus fibroblasts (Desrosiers et al. 1997).

To recapitulate the pathogenesis of KSHV often seen in HIV-coinfected patients, two groups from NEPRC and ORPRC have infected rhesus monkeys with both RRV and simian immunodeficiency viruses (SIVmac239, a virus used to model HIV-like infection in rhesus monkeys) (Mansfield et al. 1999; Wong et al. 1999). One group reported that transient lymphadenopathy followed by persistent infection has been detected in the rhesus macaques infected with RRV alone (Mansfield et al. 1999). At 3 months postinfection, these pathologies disappeared as the immune systems of the rhesus macaques were able to dampen the pathogenesis of the virus (Mansfield et al. 1999). However, when the animals were coinfected with both RRV and SIVmac239, they showed weaker antibody response to both viruses and had shorter survival times compared to monkeys infected with only one of the agents. The other group found that coinfected animals displayed symptoms of lymphoproliferative disorder such as splenomegaly and hypergammaglobulinemia in coinfected animals, symptoms reminiscent of MCD induced by KSHV in humans (Wong et al. 1999). These results demonstrate the utility of RRV/SIVmac239 coinfection in monkeys as a model to recapitulate the KSHV-related lymphoproliferative disorders in HIV-positive patients.

RRV R1

Like *K1*, the *R1* gene encodes a transmembrane protein with an extracellular immunoglobulin domain. However, the cytoplasmic tail of R1 contains multiple potential ITAM motifs and is considerably longer than K1 (171 and 38 amino acids for R1 and K1, respectively) (Fig. 4) (Damania et al. 1999, 2000). Nonetheless, R1 is able to transduce a signaling profile similar to that of K1-activated B cells (Damania et al. 2000). R1 specifically interacts with the tyrosine kinase Syk and is itself a substrate of Syk (Damania et al. 2000). Fusion of the cytoplasmic tail of R1 to the extracellular and transmembrane domains of CD8 results in increased cellular tyrosine phosphorylation, intracellular calcium influx, and NFAT activation upon stimulation with an αCD8 antibody, events that are reminiscent of B-cell activation (Damania et al. 2000). In fact, expression of R1 is sufficient to constitutively activate NFAT, a key family of transcription factors important for lymphocyte activation, without any stimulation (Damania et al. 2000). Thus R1 is a close cousin of K1 with respect to its ability to control B-cell signaling independently of BCR (Fig. 4). Interestingly, the duration of calcium signaling was longer and stronger with R1 than K1 upon stimulation with αCD8 antibody (Damania et al. 2000). Whether this difference in activity strength is due to the longer cytoplasmic tail with more potential ITAM motifs is unknown.

Likewise, R1 displays oncogenic traits similar to K1 (Damania et al. 1999). When expressed in Rat-1 cells, R1 can induce foci formation in vitro and tumors in vivo when inoculated into nude mice (Damania et al. 1999). Furthermore, recombinant HVS with R1, expressed in place of STP, could still immortalize T cells from common marmosets, highlighting the ability of R1 to activate cellular pathways leading to cell growth and survival, even in the absence of exogenous signals (Damania et al. 1999). The authors did not mention if there were any differences in the transforming abilities of HVS/K1 or HVS/R1 versus wild-type HVS (Damania et al. 1999).

RRV RK15

The *RK15* gene is situated at the same genomic location as KSHV *K15* and is the homologue of K15 in RRV (Fig. 3) (Alexander et al. 2000). It is termed RK15 as not to cause confusion with an RRV gene already named R15 of strain 17577 that is the RRV homologue of cellular OX2 and KSHV ORF74 (Pratt et al. 2005; Wang et al. 2009). Similar to K15, RK15 displays a complex splicing pattern with the most common transcript encoding all the 12 transmembranes (Wang et al. 2009). While RK15 does not have the SH2 YEEV motif that is important for the signaling activities of K15, it retains the SH3 motif, albeit at a different location (K15P–P^{387}PLP, K15M–P^{396}PLP versus RK15–P^{492}PLP) (Brinkmann et al. 2003; Wang et al. 2007, 2009). Nonetheless, RK15 can still activate the NF-κB and JNK pathways to a similar degree as K15, but only weakly activates the ERK cascade compared to K15. This suggests that RK15 utilizes an SH2-independent mechanism to target NF-κB and JNK activity. Furthermore this also implies that RK15 require an SH2 domain to robustly activate the ERK pathway. A study by Wang et al. using microarray analysis to compare the spectrum of cellular genes regulated by K15 and RK15 found that there was little overlap between the cellular genes regulated by these two viral proteins (Wang et al. 2009). While both induced the expression of proinflammatory cytokines and chemokines such as IL-8 and IL-6, the extent of induction was more robust in the presence of K15 compared to RK15. On the other hand, RK15 preferentially upregulates cellular factors such as fibroblast growth factor 21 (FGF21), Kruppel-like factor 15 (KLF15), and complement component 5a receptor 1 (C5aR1), genes whose expression is unchanged in the presence of K15 (Wang et al. 2009). This differential display of gene induction activities between K15 and RK15 could be due to the inability of RK15 to activate the ERK pathway, leading to weaker induction of proinflammatory signaling pathways by RK15. One possible reason why K15 preferentially activates a wider range of cytokines and chemokines genes compared to RK15 could be due to the proinflammatory signature of KS, and so far, no KS-like symptoms have been seen in RRV and SIV infected macaques (Mansfield et al. 1999; Wong et al. 1999).

Herpesvirus Saimiri

HVS is the prototype of γ-2-herpesviruses and belongs to the New World Rhadinovirus group (Roizmann et al. 1992). This virus is nonpathogenic in its natural host, the squirrel monkey *(Saimiri sciureus)*, but infection of other New World primates such as common marmosets *(Callithrix jacchus)* results in fatal T-cell lymphoma and leukemia (Fleckenstein and Ensser 2007; Melendez et al. 1968, 1969). In addition, certain strains of HVS are capable of immortalizing T cells from humans, rhesus monkeys, common marmosets, and rabbits to interleukin-2 (IL-2)-independent growth (Ablashi et al. 1985; Biesinger et al. 1992; Daniel et al. 1975a; Duboise et al. 1998a). IL-2 is required for T-cell growth and function; thus, the ability of certain strains of HVS to induce IL-2-independent transformation demonstrates its potent oncogenicity and gives these infected cells a growth advantage when the cytokine supply is limited.

The genome organization of HVS is similar to other Rhadinoviruses, but HVS lacks certain genes that are shared between the Old World Rhadinoviruses (such as vIRFS encoded by KSHV and RRV). Three strains of HVS named A, B, and C have been discovered so far, each of which display different oncogenic potentials (Biesinger et al. 1990; Daniel et al. 1975b; Desrosiers and Falk 1982; Falk et al. 1972; Medveczky et al. 1984). Subgroup C is the most potent; it is able to immortalize T cells from humans, rabbits, and primates in vitro and causes lymphoma in rhesus monkeys upon inoculation (Duboise et al. 1998a). Strain A is able to induce cytokine-independent transformation of lymphocytes from common marmosets. Finally, subgroup B is the least oncogenic. The oncogenic dissimilarities between these HVS strains are due to the 5′ ORFs at the location corresponding to K1 and R1 of KSHV and RRV, respectively (Fig. 3) (Desrosiers et al. 1985; Koomey et al. 1984; Murthy et al. 1989). Similar to low sequence conservation of K1 and R1 in different strains, these HVS ORFs also have low homology between the different HVS strains (Fig. 4).

Saimiri Transforming Protein

In HVS subgroups A and B, the first ORF encodes a protein named STP A and B (STP-A and STP-B), respectively (Fig. 3) (Murthy et al. 1989). In HVS subgroup C (strain C488), however, STP-C is the second ORF from the left end of the genome, as the first ORF encodes a protein called tyrosine-interacting protein (Tip) (Fig. 3) (Duboise et al. 1998a). STP-C and Tip are translated from a bicistronic mRNA; thus, both genes are very likely to be expressed at the same time (Biesinger et al. 1995). The STPs and Tip are not required for viral replication but are essential for the transformation of lymphocytes in vitro and the induction of lymphoma in vivo. Expression of STP-A or STP-C alone in rat fibroblasts results in transformation of the cells and tumor formation following inoculation in nude mice (Jung et al. 1991).

Even though the STPs are structurally similar, they do not share high sequence homology, with STP-B being 28 % and 22 % identical to STP-A and STP-C, respectively. STP-A and STP-C both have collagen-like repeats (glycine-X-Y with X or Y being proline or glutamine) after a highly acidic amino terminus, while STP-B lacks these repeats. STP-C has 18 of these repeats arranged in tandem, while STP-A only has 9 repeats that are scattered throughout the protein. The importance of these collagen-like repeats for STP oligomerization and ultimately the oncogenicity of HVS has been shown in several studies. Indeed, a chimeric version of STP-B containing the 18 copies of repeats from STP-C is now able to transform cells (Choi et al. 2000a). Oligomerization of STP-C through its collagen repeats may mimic a ligand-independent, constitutively active receptor, which then activates the NF-κB pathway to induce expression of genes important for survival and transformation. Not surprisingly, mutating the repeats in STP-C abolishes its oncogenic activity (Choi et al. 2000a).

STP-C was the first reported virus-encoded protein to interact with cellular rat sarcoma (Ras), a molecular switch for MAPK cascade important for oncogenesis and cell growth (Jung and Desrosiers 1995). Expression of STP-C activates the Ras signaling cascade, and disrupting the STP-C-Ras interaction interferes with the transformation of cells by this viral oncogene (Fig. 4). In fact, a recombinant virus that encodes Ras in lieu of STP-C can still transform lymphocytes, albeit with lower efficiency compared to the wild-type virus (Jung and Desrosiers 1995). Thus, Ras is an important player in the transformation of lymphocytes by STP-C.

STPs also contain other motifs involved in cellular signaling cascades, such as a TRAF-interacting domain and an SH2 motif (in STP-A and STP-B only). TRAFs are a family of cellular proteins involved in relaying signaling from initiation by the external stimuli to downstream molecules and thus regulate diverse functions. The TRAF-binding site of STP-C is required for its interaction with TRAF-1, TRAF-2, and TRAF-3 and subsequent activation of NF-κB (Lee et al. 1999). Mutation of the TRAF-binding motif abolishes the ability of STP-C to immortalize primary human T cells in vitro. However, this STP-C TRAF-binding mutant protein can still transform lymphocytes of common marmoset in vivo and in vitro, suggesting that while host TRAFs are important for the oncogenicity of HVS STP-C, this multifunctional protein can utilize other mechanisms to achieve the same result (Lee et al. 1999).

Residue Y^{115} in the SH2 motif of STP-A is responsible for binding to Src and is itself a substrate for this tyrosine kinase (Chung et al. 2004; Garcia et al. 2007). This interaction leads to strong activation of the transcription factors, AP-1 and NFAT, while the binding of STP-A to TRAF-6 elicits a vigorous NF-kB response. The activation of all three pathways contributes to IL-2 promoter activity (Garcia et al. 2007).

Another binding partner of STP-A is signal transducer and activator of transcription 3 (STAT3), and a triple complex is formed with STAT3, STP-A, and Src. Even though the interaction between STP-A and STAT3 is independent of its interaction with Src, the close proximity of these molecules results in phosphorylation and activation of STAT3 by Src (Chung et al. 2004). STP-A-induced STAT3-mediated signal transduction leads to upregulation of proteins involved in antiapoptosis and cell cycle such as Bcl-XL and cyclin D1, ultimately allowing STP-A-expressing cells to survive and proliferate in the absence of serum (Chung et al. 2004). Together, these studies suggest that STP activates these key cellular pathways to prevent apoptosis and contribute to IL-2 -independent growth.

Tyrosine-Interacting Protein

Tip is encoded by the first ORF of HVS subgroup C (Fig. 3) (Biesinger et al. 1995). Like all the viral proteins discussed so far, Tip also contains multiple motifs for interacting with cellular proteins. First and foremost is the interaction of Tip with Lck, the major Src tyrosine kinase in T cells (Biesinger et al. 1995). This binding is mediated by the SH3 motif and the C-terminal Src-related kinase homology (CSKH) domain of Tip (Jung et al. 1995a). The interaction between Tip and Lck results in the phosphorylation of Tip by the kinase as well as deregulation of the downstream T-cell receptor (TCR) signaling cascade, likely by sequestering Lck away from the TCR (Cho et al. 2004).

In addition, Tip expression also decreases the cell surface level of TCR and CD4 molecules, which is dependent on the localization of Tip to the membrane rafts via its transmembrane and an amphipathic helix motif just preceding the transmembrane domain (Cho et al. 2006; Min et al. 2008). Tip also interacts with a cellular endosomal protein, p80, through its serine-rich (SR) domain (Park et al. 2002). This interaction in turn mediates formation of large vesicles that help traffic the internalized TCR and CD4 molecules to the lysosome for degradation (Park et al. 2003).

The functional consequences of Tip binding to Lck are controversial. In certain contexts, expression of Tip suppresses the activation of Lck. For example, expression of Tip in the human Jurkat T-cell line results in lowered levels of tyrosine phosphorylation, while in fibroblasts Tip is able to inhibit the activity of an oncogenic mutant of Lck (F505) (Jung et al. 1995b). On the other hand, infection of human peripheral blood T lymphocytes (PBL) with a Tip-deleted mutant virus results in a significant decrease in Lck activation compared to wild-type virus, strongly suggesting that the presence of Tip increases the activity of Lck (Lund et al. 1997). The enzymatic activity of Lck is also enhanced by Tip in an in vitro assay (Kjellen et al. 2002; Lund et al. 1997; Wiese et al. 1996). Duboise et al. mutated the Lck-binding SH3 motif (proline to alanine mutations) of Tip in the context of the virus and surprisingly found that this mutant virus can still immortalize common marmoset lymphocytes in vivo and in vitro (Duboise et al. 1998b). This data suggests that deregulation of Lck by Tip is not required for its ability to immortalize. Interestingly, this mutant virus was even more pathogenic compared to the wild-type virus, with increased lymphoid infiltration upon necropsy of the infected animals (Duboise et al. 1998b). However, it was reported that mutating just one of the dual Lck-interacting sites in Tip does not completely abrogate the interaction between Tip and Lck (Heck et al. 2006). In summary, the oncogenic ability of HVS and the modulation of Tip on Lck are complex and multifaceted.

Another function of Tip is its activation of several transcription factors including NF-κB, NFAT, STAT1, and STAT3 (Merlo and Tsygankov 2001). The modulation of STATs by Tip is Lck dependent, but STAT activation is dispensable for the transformative function of Tip (Heck et al. 2005).

Summary of Viral Signaling Molecules

The studies on these six ORFs from KSHV, RRV, and HVS suggest conservation of functions despite the lack of sequence homology and divergence of cell tropism. K1 and R1 both activate pathways that are also induced by BCR activation and both have transforming potentials. In addition, the abilities of K1 and R1 to replace STP in recombinant HVS and still immortalize T cells from common marmosets suggest that these three viral proteins have a conserved capability to activate the same pathways leading to cell survival and transformation (Damania et al. 1999; Lee et al. 1998b). Comparison of K15 and RK15 revealed a weaker propensity for RK15 to activate inflammatory pathways, perhaps reflecting the more cytokine-dependent nature of KSHV pathogenesis as compared with RRV. Nonetheless, all of these viral proteins activate cellular pathways important for cell survival. The in vivo contribution of K1, K15, R1, and RK15 to virus-induced tumorigenesis is still unknown and hopefully will be elucidated in the near future.

γ-2-Herpesvirus miRNAs

Another mechanism used by herpesviruses to manipulate host cell signaling pathways is through the regulation of gene expression by virus-encoded miRNAs. miRNAs are noncoding RNAs 19–24 nucleotides in length that regulate gene expression by binding to the 3′ untranslated region (UTR) of the target mRNAs to either repress protein translation or induce mRNA degradation. The sequence of the 5′ nucleotides from positions 2–7 of the miRNA determines the target and is termed the seed sequence. Perfect complementation between the miRNA and the target mRNA leads to proteolytic cleavage of the target, while imperfect complementation results in repression of protein translation. The biogenesis pathway of miRNAs will not be covered here, but more in depth reading on these regulatory RNAs can be found in these reviews (Garzon et al. 2009; Krol et al. 2010; Winter et al. 2009).

Herpesviruses, in general, prove to be abundantly rich in miRNAs (reviewed in (Boss et al. 2009)). miRNAs are ideal tools for viruses to modulate gene expression as they are non-immunogenic, economical (miRNAs require less coding capacity compared to ORFs), capable of acting on multiple targets and can undergo rapid evolution to target new transcripts (Skalsky and Cullen 2010). Furthermore, viruses can utilize the highly conserved cellular pathway for miRNA biogenesis without having to encode any specific viral factors dedicated for miRNA processing.

Herpesvirus miRNAs are Expressed During Latency

KSHV encodes 12 pre-miRNAs clustered between ORF K12 and ORF71, an area termed the latency-associated region (Fig. 3) (Cai et al. 2005; Grundhoff et al. 2006; Pfeffer et al. 2005; Samols et al. 2005). From these 12 pre-miRNAs, 17–25 mature miRNAs can be predicted to form following maturation (Abend et al. 2010; Lin et al. 2010, 2011). As these miRNAs are located within the major latency-associated region of the viral genome, it is not surprising that these miRNAs are highly enriched in latent PEL cells (Cai et al. 2005; Samols et al. 2005). With the exception of miR-K10 and miR-K12, all other miRNAs levels remain constant during lytic replication, indicating that KSHV miRNAs primarily function during viral latency (Cai et al. 2005; Grundhoff et al. 2006).

KSHV miRNAs have been shown to target both viral and cellular genes, exerting control over both the viral life cycle and host immune responses. miR-K12-11 is an ortholog of cellular miR-155, an important regulator of lymphocyte differentiation and innate immunity (Faraoni et al. 2009; Gottwein et al. 2007; Skalsky et al. 2007). miR-155 is a potent antiapoptotic miRNA when expressed in breast cancer cells and has been implicated in tumorigenesis of lymphoid and myeloid cancers (Ovcharenko et al. 2007). PEL cells lack the expression of miR-155, but express high levels of miR-K12-11 (Skalsky et al. 2007). Since it has an identical seed sequence and regulates the same set of genes as miR-155, miR-K12-11 may contribute to the KSHV-induced oncogenesis (Gottwein et al. 2007). A recent study (2011) showed that miR-K12-11 is indeed the viral mimic of the cellular miR-155 (Boss et al. 2011). Ectopic expression of either miRNAs in humanized NSG (NOD-scid interleukin-2 receptor γ-chain null (IL2Rγ^{null})) mouse model led to hyperproliferation of human B cells in the spleen. This phenotype could be traced to the dysregulation of CCAAT enhancer-binding protein β (C/EBPβ), a regulator of IL-6 transcription, by these miRNAs (Boss et al. 2011). Since IL-6 is a key driver of inflammation and proliferation of lymphocytes, activation of IL-6 by miR-K12-11 is likely to be an important aspect of KSHV pathogenesis.

The example of miR-K1, which targets multiple signaling pathways, highlights the versatility of miRNA-mediated regulation of gene expression. First, miR-K1 has been shown to be an important regulator of latency by positively regulating the NF-κB pathway through reducing the level of the inhibitor IκBα to maintain the viral latency state (Lei et al. 2010). Furthermore, by manipulating this key pathway,

KSHV can suppress host immunity and enhance cell survival. Second, miR-K1 specifically prevents the expression of p21, an important regulator of cell cycle (Gottwein and Cullen 2010).

Lytic replication is critical for gammaherpesviruses to maintain a disseminated latent infection in vivo but has unwanted consequences such as immune activation and cell death to release infectious virion. Thus, tight control of lytic reactivation may allow gammaherpesviruses to maintain persistent infection in the presence of host immune surveillance. Rta, encoded by ORF50, is the master transcriptional regulator of lytic replication in KSHV and is directly targeted by at least two viral miRNAs (miR-K9, miR-K12-7) (Bellare and Ganem 2009; Lin et al. 2011; Lu et al. 2010b). As a consequence of miR-K3 targeting the nuclear factor I/B (NFIB), a transcriptional activator for the Rta promoter, miR-K3 also has a role in repressing expression of Rta and thus contributes to the maintenance of latency (Lu et al. 2010a). The number of viral miRNAs devoted to targeting Rta, either directly or indirectly through inhibition of other mRNAs, suggests that KSHV tightly maintains latency to prevent inappropriate induction of lytic replication.

Sequencing of KSHV miRNAs from PEL cell lines and tissue samples from patients with different KSHV-induced diseases revealed high conservation among most of the viral miRNAs, suggesting the importance of miRNAs in KSHV biology (Marshall et al. 2007). Interestingly, a few miRNAs (e.g., miR-K12) showed polymorphisms between the different isolates (Marshall et al. 2007). Since it has been shown that a single nucleotide polymorphism (SNP) in a viral miRNA could affect its maturation and function, it is tempting to speculate that these "minor" changes in viral miRNA sequences may impact the virus biology in a significant manner (Gottwein et al. 2006). These viral miRNA SNPs could reflect a relatively rapid adaptation strategy by the virus to outwit the host response.

RRV also encodes 15 miRNAs processed from 7 pre-miRNAs located in the same genomic region as the KSHV miRNAs (Fig. 3) (Schafer et al. 2007; Umbach et al. 2010). Umbach et al. (2010) used a deep sequencing assay to analyze the expression of viral miRNAs in RRV-induced tumors. They showed that all but one RRV miRNA lack sequence homology with KSHV miRNAs. The targets and functions of RRV miRNAs have yet to be characterized. The observation that the miRNAs of KSHV and RRV are located in the same viral genomic region but do not share sequence homology suggests that these viruses may have adapted the use of miRNA before evolutionary diverging, but have since altered the seed sequence to target their respective host.

Analysis of the HVS genome failed to predict with sufficient certainty the presence of miRNAs (Walz et al. 2010). Several potential miRNA candidates are predicted to be encoded at the location where miRNAs are found in the KSHV and RRV genomes, but have yet to be verified (Walz et al. 2010). HVS, however, does encode seven viral U-rich RNAs called HSURs in its 3′ terminal L-DNA region (Fig. 3) (Biesinger et al. 1990; Lee et al. 1988). These noncoding RNAs are not required for viral replication in cell culture (Ensser et al. 1999). Steiz and colleagues found that HSUR-1 and HSUR-2 bind to cellular miR-27, miR-142-3p, and miR-16, leading to degradation of miR-27 while not affecting the other two miRNAs

(Cazalla et al. 2010). The functional consequences of miR-27 downregulation and miR-142-3p and miR-16 binding by HVS is an exciting area that warrants further investigation.

As the examples described above illustrate, there is little conservation in the sequence and function among the known miRNAs expressed by these γ-herpesviruses. However, as this area is still under intense investigation, there may be more targets that are yet to be discovered. Due to the flexibility of the miRNA seed sequence, each miRNA may target many mRNAs, enabling these viruses to target the same pathways either by using different seed sequences to bind to a different target site of the same mRNA or by targeting different mRNAs that encode components of the same pathway. That miR-K11 and miR-155 have the same seed sequence suggests that KSHV may have pirated cellular miRNA for its own advantage.

Conclusion

In summary, herpesviruses are remarkably well adapted to their respective hosts as even the oncogenic γ-2-herpesviruses rarely cause fatalities unless the host is immunocompromised or contribution from other factors (e.g., genetic or environment). As life-long "passengers" in their infected hosts, herpesviruses must strike a delicate balance with the host immune response. To achieve this, γ-2-herpesviruses devote multiple resources to manipulate the key signaling pathways described here, ultimately promoting the survival of virus-infected cell, immune evasion, and tumorigenesis. The molecular piracy strategy used by these viruses has been remarkably successful for their adaptation to specific niches. The abilities of the viral terminal transmembrane proteins (K1 and K15 of KSHV, R1 and RK 15 of RRV, STP and Tip of HVS) and miRNAs to manipulate host cell signaling pathways and drive viral replication at appropriate times very likely contribute to the pathogenesis of these viruses. Moreover, the high frequency of molecular piracy by γ-herpesviruses allows researchers to study these viral mimics and gain insights into the mechanisms viruses use to divert cellular pathways, even leading to the discovery of new classes of cellular proteins.

References

Abend JR, Uldrick T, Ziegelbauer JM (2010) Regulation of tumor necrosis factor-like weak inducer of apoptosis receptor protein (TWEAKR) expression by Kaposi's sarcoma-associated herpesvirus microRNA prevents TWEAK-induced apoptosis and inflammatory cytokine expression. J Virol 84:12139–12151

Ablashi DV, Schirm S, Fleckenstein B, Faggioni A, Dahlberg J, Rabin H, Loeb W, Armstrong G, Peng JW, Aulahk G et al (1985) Herpesvirus saimiri-induced lymphoblastoid rabbit cell line: growth characteristics, virus persistence, and oncogenic properties. J Virol 55:623–633

Alexander L, Denekamp L, Knapp A, Auerbach MR, Damania B, Desrosiers RC (2000) The primary sequence of rhesus monkey rhadinovirus isolate 26–95: sequence similarities to Kaposi's sarcoma-associated herpesvirus and rhesus monkey rhadinovirus isolate 17577. J Virol 74:3388–3398

Aluigi MG, Albini A, Carlone S, Repetto L, De Marchi R, Icardi A, Moro M, Noonan D, Benelli R (1996) KSHV sequences in biopsies and cultured spindle cells of epidemic, iatrogenic and Mediterranean forms of Kaposi's sarcoma. Res Virol 147:267–275

Ambroziak JA, Blackbourn DJ, Herndier BG, Glogau RG, Gullett JH, McDonald AR, Lennette ET, Levy JA (1995) Herpes-like sequences in HIV-infected and uninfected Kaposi's sarcoma patients. Science 268:582–583

Ballon G, Chen K, Perez R, Tam W, Cesarman E (2011) Kaposi sarcoma herpesvirus (KSHV) vFLIP oncoprotein induces B cell transdifferentiation and tumorigenesis in mice. J Clin Invest 121:1141–1153

Bellare P, Ganem D (2009) Regulation of KSHV lytic switch protein expression by a virus-encoded microRNA: an evolutionary adaptation that fine-tunes lytic reactivation. Cell Host Microbe 6:570–575

Benelli R, Albini A, Parravicini C, Carlone S, Repetto L, Tambussi G, Lazzarin A (1996) Isolation of spindle-shaped cell populations from primary cultures of Kaposi's sarcoma of different stage. Cancer Lett 100:125–132

Bergquam EP, Avery N, Shiigi SM, Axthelm MK, Wong SW (1999) Rhesus rhadinovirus establishes a latent infection in B lymphocytes in vivo. J Virol 73:7874–7876

Berkova Z, Wang S, Wise JF, Maeng H, Ji Y, Samaniego F (2009) Mechanism of Fas signaling regulation by human herpesvirus 8 K1 oncoprotein. J Natl Cancer Inst 101:399–411

Biesinger B, Trimble JJ, Desrosiers RC, Fleckenstein B (1990) The divergence between two oncogenic Herpesvirus saimiri strains in a genomic region related to the transforming phenotype. Virology 176:505–514

Biesinger B, Muller-Fleckenstein I, Simmer B, Lang G, Wittmann S, Platzer E, Desrosiers RC, Fleckenstein B (1992) Stable growth transformation of human T lymphocytes by herpesvirus saimiri. Proc Natl Acad Sci U S A 89:3116–3119

Biesinger B, Tsygankov AY, Fickenscher H, Emmrich F, Fleckenstein B, Bolen JB, Broker BM (1995) The product of the herpesvirus saimiri open reading frame 1 (tip) interacts with T cell-specific kinase p56lck in transformed cells. J Biol Chem 270:4729–4734

Blaskovic D, Stancekova M, Svobodova J, Mistrikova J (1980) Isolation of five strains of herpesviruses from two species of free living small rodents. Acta Virol 24:468

Boshoff C, Schulz TF, Kennedy MM, Graham AK, Fisher C, Thomas A, McGee JO, Weiss RA, O'Leary JJ (1995) Kaposi's Sarcoma-associated herpesvirus infects endothelial and spindle cells. Nat Med 1:1274–1278

Boss IW, Plaisance KB, Renne R (2009) Role of virus-encoded microRNAs in herpesvirus biology. Trends Microbiol 17:544–553

Boss IW, Nadeau PE, Abbott JR, Yang Y, Mergia A, Renne R (2011) A Kaposi's sarcoma-associated herpesvirus-encoded ortholog of microRNA miR-155 induces human splenic B-cell expansion in NOD/LtSz-scid IL2Rgammanull mice. J Virol 85:9877–9886

Brinkmann MM, Glenn M, Rainbow L, Kieser A, Henke-Gendo C, Schulz TF (2003) Activation of mitogen-activated protein kinase and NF-kappaB pathways by a Kaposi's sarcoma-associated herpesvirus K15 membrane protein. J Virol 77:9346–9358

Brinkmann MM, Pietrek M, Dittrich-Breiholz O, Kracht M, Schulz TF (2007) Modulation of host gene expression by the K15 protein of Kaposi's sarcoma-associated herpesvirus. J Virol 81:42–58

Cai X, Lu S, Zhang Z, Gonzalez CM, Damania B, Cullen BR (2005) Kaposi's sarcoma-associated herpesvirus expresses an array of viral microRNAs in latently infected cells. Proc Natl Acad Sci U S A 102:5570–5575

Cambier JC (1995) Antigen and Fc receptor signaling. The awesome power of the immunoreceptor tyrosine-based activation motif (ITAM). J Immunol 155:3281–3285

Cazalla D, Yario T, Steitz JA (2010) Down-regulation of a host microRNA by a Herpesvirus saimiri noncoding RNA. Science 328:1563–1566

Cesarman E, Chang Y, Moore PS, Said JW, Knowles DM (1995) Kaposi's sarcoma-associated herpesvirus-like DNA sequences in AIDS-related body-cavity-based lymphomas. N Engl J Med 332:1186–1191

Chang Y, Cesarman E, Pessin MS, Lee F, Culpepper J, Knowles DM, Moore PS (1994) Identification of herpesvirus-like DNA sequences in AIDS-associated Kaposi's sarcoma. Science 266:1865–1869

Chang H, Wachtman LM, Pearson CB, Lee JS, Lee HR, Lee SH, Vieira J, Mansfield KG, Jung JU (2009) Non-human primate model of Kaposi's sarcoma-associated herpesvirus infection. PLoS Pathog 5:e1000606

Cho NH, Feng P, Lee SH, Lee BS, Liang X, Chang H, Jung JU (2004) Inhibition of T cell receptor signal transduction by tyrosine kinase-interacting protein of Herpesvirus saimiri. J Exp Med 200:681–687

Cho NH, Kingston D, Chang H, Kwon EK, Kim JM, Lee JH, Chu H, Choi MS, Kim IS, Jung JU (2006) Association of herpesvirus saimiri tip with lipid raft is essential for downregulation of T-cell receptor and CD4 coreceptor. J Virol 80:108–118

Cho NH, Choi YK, Choi JK (2008) Multi-transmembrane protein K15 of Kaposi's sarcoma-associated herpesvirus targets Lyn kinase in the membrane raft and induces NFAT/AP1 activities. Exp Mol Med 40:565–573

Choi JK, Ishido S, Jung JU (2000a) The collagen repeat sequence is a determinant of the degree of herpesvirus saimiri STP transforming activity. J Virol 74:8102–8110

Choi JK, Lee BS, Shim SN, Li M, Jung JU (2000b) Identification of the novel K15 gene at the rightmost end of the Kaposi's sarcoma-associated herpesvirus genome. J Virol 74:436–446

Chugh P, Matta H, Schamus S, Zachariah S, Kumar A, Richardson JA, Smith AL, Chaudhary PM (2005) Constitutive NF-kappaB activation, normal Fas-induced apoptosis, and increased incidence of lymphoma in human herpes virus 8 K13 transgenic mice. Proc Natl Acad Sci USA 102:12885–12890

Chung YH, Cho NH, Garcia MI, Lee SH, Feng P, Jung JU (2004) Activation of Stat3 transcription factor by Herpesvirus saimiri STP-A oncoprotein. J Virol 78:6489–6497

Coscoy L, Ganem D (2000) Kaposi's sarcoma-associated herpesvirus encodes two proteins that block cell surface display of MHC class I chains by enhancing their endocytosis. Proc Natl Acad Sci USA 97:8051–8056

Damania B, Li M, Choi JK, Alexander L, Jung JU, Desrosiers RC (1999) Identification of the R1 oncogene and its protein product from the rhadinovirus of rhesus monkeys. J Virol 73:5123–5131

Damania B, DeMaria M, Jung JU, Desrosiers RC (2000) Activation of lymphocyte signaling by the R1 protein of rhesus monkey rhadinovirus. J Virol 74:2721–2730

Daniel MD, Hunt RD, Dubose D, Silva D, and Melendez LV (1975a) Induction of herpesvirus saimiri lymphoma in New Zealand white rabbits inoculated intravenously. IARC Sci Publ (11 pt 2):205–208

Daniel MD, Silva D, Jackman D, Sehgal P, Baggs RB, Hunt RD, King NW, Melendez LV (1975b) Reactivation of squirrel monkey heart isolate (Herpesvirus saimiri strain) from latently infected human cell cultures and induction of malignant lymphoma in marmoset monkeys. Bibl Haematol 43:392–395

Dereeper A, Guignon V, Blanc G, Audic S, Buffet S, Chevenet F, Dufayard JF, Guindon S, Lefort V, Lescot M, Claverie JM, Gascuel O (2008) Phylogeny.fr: robust phylogenetic analysis for the non-specialist. Nucleic Acids Res 36:W465–W469

Desrosiers RC, Falk LA (1982) Herpesvirus saimiri strain variability. J Virol 43:352–356

Desrosiers RC, Bakker A, Kamine J, Falk LA, Hunt RD, King NW (1985) A region of the Herpesvirus saimiri genome required for oncogenicity. Science 228:184–187

Desrosiers RC, Sasseville VG, Czajak SC, Zhang X, Mansfield KG, Kaur A, Johnson RP, Lackner AA, Jung JU (1997) A herpesvirus of rhesus monkeys related to the human Kaposi's sarcoma-associated herpesvirus. J Virol 71:9764–9769

Diehl V, Henle G, Henle W, Kohn G (1968) Demonstration of a herpes group virus in cultures of peripheral leukocytes from patients with infectious mononucleosis. J Virol 2:663–669
Dittmer D, Stoddart C, Renne R, Linquist-Stepps V, Moreno ME, Bare C, McCune JM, Ganem D (1999) Experimental transmission of Kaposi's sarcoma-associated herpesvirus (KSHV/HHV-8) to SCID-hu Thy/Liv mice. J Exp Med 190:1857–1868
Dourmishev LA, Dourmishev AL, Palmeri D, Schwartz RA, Lukac DM (2003) Molecular genetics of Kaposi's sarcoma-associated herpesvirus (human herpesvirus-8) epidemiology and pathogenesis. Microbiol Mol Biol Rev 67:175–212
Du MQ, Liu H, Diss TC, Ye H, Hamoudi RA, Dupin N, Meignin V, Oksenhendler E, Boshoff C, Isaacson PG (2001) Kaposi sarcoma-associated herpesvirus infects monotypic (IgM lambda) but polyclonal naive B cells in Castleman disease and associated lymphoproliferative disorders. Blood 97:2130–2136
Duboise SM, Guo J, Czajak S, Desrosiers RC, Jung JU (1998a) STP and Tip are essential for herpesvirus saimiri oncogenicity. J Virol 72:1308–1313
Duboise SM, Lee H, Guo J, Choi JK, Czajak S, Simon M, Desrosiers RC, Jung JU (1998b) Mutation of the Lck-binding motif of Tip enhances lymphoid cell activation by herpesvirus saimiri. J Virol 72:2607–2614
Dupin N, Fisher C, Kellam P, Ariad S, Tulliez M, Franck N, van Marck E, Salmon D, Gorin I, Escande JP, Weiss RA, Alitalo K, Boshoff C (1999) Distribution of human herpesvirus-8 latently infected cells in Kaposi's sarcoma, multicentric Castleman's disease, and primary effusion lymphoma. Proc Natl Acad Sci U S A 96:4546–4551
Ensoli B, Nakamura S, Salahuddin SZ, Biberfeld P, Larsson L, Beaver B, Wong-Staal F, Gallo RC (1989) AIDS-Kaposi's sarcoma-derived cells express cytokines with autocrine and paracrine growth effects. Science 243:223–226
Ensoli B, Sturzl M, Monini P (2000) Cytokine-mediated growth promotion of Kaposi's sarcoma and primary effusion lymphoma. Semin Cancer Biol 10:367–381
Ensser A, Pfinder A, Muller-Fleckenstein I, Fleckenstein B (1999) The URNA genes of herpesvirus saimiri (strain C488) are dispensable for transformation of human T cells in vitro. J Virol 73:10551–10555
Falk LA, Wolfe LG, Deinhardt F (1972) Isolation of Herpesvirus saimiri from blood of squirrel monkeys (Saimiri sciureus). J Natl Cancer Inst 48:1499–1505
Faraoni I, Antonetti FR, Cardone J, Bonmassar E (2009) miR-155 gene: a typical multifunctional microRNA. Biochim Biophys Acta 1792:497–505
Fleckenstein B, Ensser A (2007) Gammaherpesviruses of New World primates. Human herpesviruses: biology, therapy, and immunoprophylaxis. Cambridge: Cambridge University Press.
Foreman KE, Bacon PE, Hsi ED, Nickoloff BJ (1997) In situ polymerase chain reaction-based localization studies support role of human herpesvirus-8 as the cause of two AIDS-related neoplasms: Kaposi's sarcoma and body cavity lymphoma. J Clin Invest 99:2971–2978
Friedman-Kien AE (1981) Disseminated Kaposi's sarcoma syndrome in young homosexual men. J Am Acad Dermatol 5:468–471
Ganem D (2006) KSHV infection and the pathogenesis of Kaposi's sarcoma. Annu Rev Pathol 1:273–296
Gao SJ, Kingsley L, Li M, Zheng W, Parravicini C, Ziegler J, Newton R, Rinaldo CR, Saah A, Phair J, Detels R, Chang Y, Moore PS (1996) KSHV antibodies among Americans, Italians and Ugandans with and without Kaposi's sarcoma. Nat Med 2:925–928
Garcia MI, Kaserman J, Chung YH, Jung JU, Lee SH (2007) Herpesvirus saimiri STP-A oncoprotein utilizes Src family protein tyrosine kinase and tumor necrosis factor receptor-associated factors to elicit cellular signal transduction. J Virol 81:2663–2674
Garzon R, Calin GA, Croce CM (2009) MicroRNAs in cancer. Annu Rev Med 60:167–179
Glenn M, Rainbow L, Aurade F, Davison A, Schulz TF (1999) Identification of a spliced gene from Kaposi's sarcoma-associated herpesvirus encoding a protein with similarities to latent membrane proteins 1 and 2A of Epstein-Barr virus. J Virol 73:6953–6963

Goto E, Ishido S, Sato Y, Ohgimoto S, Ohgimoto K, Nagano-Fujii M, Hotta H (2003) c-MIR, a human E3 ubiquitin ligase, is a functional homolog of herpesvirus proteins MIR1 and MIR2 and has similar activity. J Biol Chem 278:14657–14668

Gottlieb GJ, Ragaz A, Vogel JV, Friedman-Kien A, Rywlin AM, Weiner EA, Ackerman AB (1981) A preliminary communication on extensively disseminated Kaposi's sarcoma in young homosexual men. Am J Dermatopathol 3:111–114

Gottwein E, Cullen BR (2010) A human herpesvirus microRNA inhibits p21 expression and attenuates p21-mediated cell cycle arrest. J Virol 84:5229–5237

Gottwein E, Cai X, Cullen BR (2006) A novel assay for viral microRNA function identifies a single nucleotide polymorphism that affects Drosha processing. J Virol 80:5321–5326

Gottwein E, Mukherjee N, Sachse C, Frenzel C, Majoros WH, Chi JT, Braich R, Manoharan M, Soutschek J, Ohler U, Cullen BR (2007) A viral microRNA functions as an orthologue of cellular miR-155. Nature 450:1096–1099

Grossmann C, Podgrabinska S, Skobe M, Ganem D (2006) Activation of NF-kappaB by the latent vFLIP gene of Kaposi's sarcoma-associated herpesvirus is required for the spindle shape of virus-infected endothelial cells and contributes to their proinflammatory phenotype. J Virol 80:7179–7185

Grundhoff A, Sullivan CS, Ganem D (2006) A combined computational and microarray-based approach identifies novel microRNAs encoded by human gamma-herpesviruses. RNA 12:733–750

Hassman LM, Ellison TJ, Kedes DH (2011) KSHV infects a subset of human tonsillar B cells, driving proliferation and plasmablast differentiation. J Clin Invest 121:752–768

Heck E, Lengenfelder D, Schmidt M, Muller-Fleckenstein I, Fleckenstein B, Biesinger B, Ensser A (2005) T-cell growth transformation by herpesvirus saimiri is independent of STAT3 activation. J Virol 79:5713–5720

Heck E, Friedrich U, Gack MU, Lengenfelder D, Schmidt M, Muller-Fleckenstein I, Fleckenstein B, Ensser A, Biesinger B (2006) Growth transformation of human T cells by herpesvirus saimiri requires multiple Tip-Lck interaction motifs. J Virol 80:9934–9942

Hengge UR, Ruzicka T, Tyring SK, Stuschke M, Roggendorf M, Schwartz RA, Seeber S (2002) Update on Kaposi's sarcoma and other HHV8 associated diseases. Part 1: epidemiology, environmental predispositions, clinical manifestations, and therapy. Lancet Infect Dis 2:281–292

Iscovich J, Boffetta P, Franceschi S, Azizi E, Sarid R (2000) Classic kaposi sarcoma: epidemiology and risk factors. Cancer 88:500–517

Ishido S, Wang C, Lee BS, Cohen GB, Jung JU (2000) Downregulation of major histocompatibility complex class I molecules by Kaposi's sarcoma-associated herpesvirus K3 and K5 proteins. J Virol 74:5300–5309

Jenner RG, Maillard K, Cattini N, Weiss RA, Boshoff C, Wooster R, Kellam P (2003) Kaposi's sarcoma-associated herpesvirus-infected primary effusion lymphoma has a plasma cell gene expression profile. Proc Natl Acad Sci U S A 100:10399–10404

Jones T, Ye F, Bedolla R, Huang Y, Meng J, Qian L, Pan H, Zhou F, Moody R, Wagner B, Arar M, Gao SJ (2012) Direct and efficient cellular transformation of primary rat mesenchymal precursor cells by KSHV. J Clin Invest 122:1076–1081

Jung JU, Desrosiers RC (1995) Association of the viral oncoprotein STP-C488 with cellular ras. Mol Cell Biol 15:6506–6512

Jung JU, Trimble JJ, King NW, Biesinger B, Fleckenstein BW, Desrosiers RC (1991) Identification of transforming genes of subgroup A and C strains of Herpesvirus saimiri. Proc Natl Acad Sci U S A 88:7051–7055

Jung JU, Lang SM, Friedrich U, Jun T, Roberts TM, Desrosiers RC, Biesinger B (1995a) Identification of Lck-binding elements in tip of herpesvirus saimiri. J Biol Chem 270:20660–20667

Jung JU, Lang SM, Jun T, Roberts TM, Veillette A, Desrosiers RC (1995b) Downregulation of Lck-mediated signal transduction by tip of herpesvirus saimiri. J Virol 69:7814–7822

Kaposi M (1872) Idiopathic multiple pigmented sarcoma of the skin. Archiv fur Dertmatologie und Syphilis 4:265–273
Kasolo FC, Mpabalwani E, Gompels UA (1997) Infection with AIDS-related herpesviruses in human immunodeficiency virus-negative infants and endemic childhood Kaposi's sarcoma in Africa. J Gen Virol 78(Pt 4):847–855
Kedes DH, Operskalski E, Busch M, Kohn R, Flood J, Ganem D (1996) The seroepidemiology of human herpesvirus 8 (Kaposi's sarcoma-associated herpesvirus): distribution of infection in KS risk groups and evidence for sexual transmission. Nat Med 2:918–924
Kjellen P, Amdjadi K, Lund TC, Medveczky PG, Sefton BM (2002) The herpesvirus saimiri tip484 and tip488 proteins both stimulate lck tyrosine protein kinase activity in vivo and in vitro. Virology 297:281–288
Knipe DM, Howley PM (eds) (2007) Field virology. Lippincott Williams & Wilkins, Philadelphia, PA
Koomey JM, Mulder C, Burghoff RL, Fleckenstein B, Desrosiers RC (1984) Deletion of DNA sequence in a nononcogenic variant of Herpesvirus saimiri. J Virol 50:662–665
Krol J, Loedige I, Filipowicz W (2010) The widespread regulation of microRNA biogenesis, function and decay. Nat Rev Genet 11:597–610
Lagunoff M, Ganem D (1997) The structure and coding organization of the genomic termini of Kaposi's sarcoma-associated herpesvirus. Virology 236:147–154
Lagunoff M, Majeti R, Weiss A, Ganem D (1999) Deregulated signal transduction by the K1 gene product of Kaposi's sarcoma-associated herpesvirus. Proc Natl Acad Sci USA 96: 5704–5709
Lagunoff M, Lukac DM, Ganem D (2001) Immunoreceptor tyrosine-based activation motif-dependent signaling by Kaposi's sarcoma-associated herpesvirus K1 protein: effects on lytic viral replication. J Virol 75:5891–5898
Lagunoff M, Bechtel J, Venetsanakos E, Roy AM, Abbey N, Herndier B, McMahon M, Ganem D (2002) De novo infection and serial transmission of Kaposi's sarcoma-associated herpesvirus in cultured endothelial cells. J Virol 76:2440–2448
Lee SI, Murthy SC, Trimble JJ, Desrosiers RC, Steitz JA (1988) Four novel U RNAs are encoded by a herpesvirus. Cell 54:599–607
Lee H, Guo J, Li M, Choi JK, DeMaria M, Rosenzweig M, Jung JU (1998a) Identification of an immunoreceptor tyrosine-based activation motif of K1 transforming protein of Kaposi's sarcoma-associated herpesvirus. Mol Cell Biol 18:5219–5228
Lee H, Veazey R, Williams K, Li M, Guo J, Neipel F, Fleckenstein B, Lackner A, Desrosiers RC, Jung JU (1998b) Deregulation of cell growth by the K1 gene of Kaposi's sarcoma-associated herpesvirus. Nat Med 4:435–440
Lee H, Choi JK, Li M, Kaye K, Kieff E, Jung JU (1999) Role of cellular tumor necrosis factor receptor-associated factors in NF-kappaB activation and lymphocyte transformation by herpesvirus Saimiri STP. J Virol 73:3913–3919
Lee BS, Alvarez X, Ishido S, Lackner AA, Jung JU (2000) Inhibition of intracellular transport of B cell antigen receptor complexes by Kaposi's sarcoma-associated herpesvirus K1. J Exp Med 192:11–21
Lee BS, Connole M, Tang Z, Harris NL, Jung JU (2003) Structural analysis of the Kaposi's sarcoma-associated herpesvirus K1 protein. J Virol 77:8072–8086
Lee BS, Lee SH, Feng P, Chang H, Cho NH, Jung JU (2005) Characterization of the Kaposi's sarcoma-associated herpesvirus K1 signalosome. J Virol 79:12173–12184
Lee JS, Li Q, Lee JY, Lee SH, Jeong JH, Lee HR, Chang H, Zhou FC, Gao SJ, Liang C, Jung JU (2009) FLIP-mediated autophagy regulation in cell death control. Nat Cell Biol 11:1355–1362
Lei X, Bai Z, Ye F, Xie J, Kim CG, Huang Y, Gao SJ (2010) Regulation of NF-kappaB inhibitor IkappaBalpha and viral replication by a KSHV microRNA. Nat Cell Biol 12:193–199
Li M, Lee H, Yoon DW, Albrecht JC, Fleckenstein B, Neipel F, Jung JU (1997) Kaposi's sarcoma-associated herpesvirus encodes a functional cyclin. J Virol 71:1984–1991

Lim CS, Seet BT, Ingham RJ, Gish G, Matskova L, Winberg G, Ernberg I, Pawson T (2007) The K15 protein of Kaposi's sarcoma-associated herpesvirus recruits the endocytic regulator intersectin 2 through a selective SH3 domain interaction. Biochemistry 46:9874–9885
Lin YT, Kincaid RP, Arasappan D, Dowd SE, Hunicke-Smith SP, Sullivan CS (2010) Small RNA profiling reveals antisense transcription throughout the KSHV genome and novel small RNAs. RNA 16:1540–1558
Lin X, Liang D, He Z, Deng Q, Robertson ES, and Lan K (2011) miR-K12-7-5p encoded by Kaposi's sarcoma-associated herpesvirus stabilizes the latent state by targeting viral ORF50/RTA. PLoS One 6:e16224
Lu CC, Li Z, Chu CY, Feng J, Feng J, Sun R, Rana TM (2010a) MicroRNAs encoded by Kaposi's sarcoma-associated herpesvirus regulate viral life cycle. EMBO Rep 11:784–790
Lu F, Stedman W, Yousef M, Renne R, Lieberman PM (2010b) Epigenetic regulation of Kaposi's sarcoma-associated herpesvirus latency by virus-encoded microRNAs that target Rta and the cellular Rbl2-DNMT pathway. J Virol 84:2697–2706
Lund T, Medveczky MM, Medveczky PG (1997) Herpesvirus saimiri Tip-484 membrane protein markedly increases p56lck activity in T cells. J Virol 71:378–382
Mansfield KG, Westmoreland SV, DeBakker CD, Czajak S, Lackner AA, Desrosiers RC (1999) Experimental infection of rhesus and pig-tailed macaques with macaque rhadinoviruses. J Virol 73:10320–10328
Marshall V, Parks T, Bagni R, Wang CD, Samols MA, Hu J, Wyvil KM, Aleman K, Little RF, Yarchoan R, Renne R, Whitby D (2007) Conservation of virally encoded microRNAs in Kaposi sarcoma–associated herpesvirus in primary effusion lymphoma cell lines and in patients with Kaposi sarcoma or multicentric Castleman disease. J Infect Dis 195:645–659
McAllister SC, Moses AV (2007) Endothelial cell- and lymphocyte-based in vitro systems for understanding KSHV biology. Curr Top Microbiol Immunol 312:211–244
Medveczky P, Szomolanyi E, Desrosiers RC, Mulder C (1984) Classification of herpesvirus saimiri into three groups based on extreme variation in a DNA region required for oncogenicity. J Virol 52:938–944
Melendez LV, Daniel MD, Hunt RD, Garcia FG (1968) An apparently new herpesvirus from primary kidney cultures of the squirrel monkey (Saimiri sciureus). Lab Anim Care 18:374–381
Melendez LV, Daniel MD, Garcia FG, Fraser CE, Hunt RD, King NW (1969) Herpesvirus saimiri. I. Further characterization studies of a new virus from the squirrel monkey. Lab Anim Care 19:372–377
Merlo JJ, Tsygankov AY (2001) Herpesvirus saimiri oncoproteins Tip and StpC synergistically stimulate NF-kappaB activity and interleukin-2 gene expression. Virology 279:325–338
Min CK, Bang SY, Cho BA, Choi YH, Yang JS, Lee SH, Seong SY, Kim KW, Kim S, Jung JU, Choi MS, Kim IS, Cho NH (2008) Role of amphipathic helix of a herpesviral protein in membrane deformation and T cell receptor downregulation. PLoS Pathog 4:e1000209
Moore PS, Gao SJ, Dominguez G, Cesarman E, Lungu O, Knowles DM, Garber R, Pellett PE, McGeoch DJ, Chang Y (1996) Primary characterization of a herpesvirus agent associated with Kaposi's sarcomae. J Virol 70:549–558
Murthy SC, Trimble JJ, Desrosiers RC (1989) Deletion mutants of herpesvirus saimiri define an open reading frame necessary for transformation. J Virol 63:3307–3314
Neipel F, Albrecht JC, Fleckenstein B (1997) Cell-homologous genes in the Kaposi's sarcoma-associated rhadinovirus human herpesvirus 8: determinants of its pathogenicity? J Virol 71:4187–4192
Niemi M, Mustakallio KK (1965) The fine structure of the spindle cell in Kaposi's sarcoma. Acta Pathol Microbiol Scand 63:567–575
Ovcharenko D, Kelnar K, Johnson C, Leng N, Brown D (2007) Genome-scale microRNA and small interfering RNA screens identify small RNA modulators of TRAIL-induced apoptosis pathway. Cancer Res 67:10782–10788
Park J, Lee BS, Choi JK, Means RE, Choe J, Jung JU (2002) Herpesviral protein targets a cellular WD repeat endosomal protein to downregulate T lymphocyte receptor expression. Immunity 17:221–233

Park J, Cho NH, Choi JK, Feng P, Choe J, Jung JU (2003) Distinct roles of cellular Lck and p80 proteins in herpesvirus saimiri Tip function on lipid rafts. J Virol 77:9041–9051
Pellett PE, Roizman B (2007) Fields virology. Lippincott-Raven, Philadelphia, PA
Pfeffer S, Sewer A, Lagos-Quintana M, Sheridan R, Sander C, Grasser FA, van Dyk LF, Ho CK, Shuman S, Chien M, Russo JJ, Ju J, Randall G, Lindenbach BD, Rice CM, Simon V, Ho DD, Zavolan M, Tuschl T (2005) Identification of microRNAs of the herpesvirus family. Nat Methods 2:269–276
Picchio GR, Sabbe RE, Gulizia RJ, McGrath M, Herndier BG, Mosier DE (1997) The KSHV/HHV8-infected BCBL-1 lymphoma line causes tumors in SCID mice but fails to transmit virus to a human peripheral blood mononuclear cell graft. Virology 238:22–29
Pietrek M, Brinkmann MM, Glowacka I, Enlund A, Havemeier A, Dittrich-Breiholz O, Kracht M, Lewitzky M, Saksela K, Feller SM, Schulz TF (2010) Role of the Kaposi's sarcoma-associated herpesvirus K15 SH3 binding site in inflammatory signaling and B-cell activation. J Virol 84:8231–8240
Poole LJ, Zong JC, Ciufo DM, Alcendor DJ, Cannon JS, Ambinder R, Orenstein JM, Reitz MS, Hayward GS (1999) Comparison of genetic variability at multiple loci across the genomes of the major subtypes of Kaposi's sarcoma-associated herpesvirus reveals evidence for recombination and for two distinct types of open reading frame K15 alleles at the right-hand end. J Virol 73:6646–6660
Prakash O, Tang ZY, Peng X, Coleman R, Gill J, Farr G, Samaniego F (2002) Tumorigenesis and aberrant signaling in transgenic mice expressing the human herpesvirus-8 K1 gene. J Natl Cancer Inst 94:926–935
Prakash O, Swamy OR, Peng X, Tang ZY, Li L, Larson JE, Cohen JC, Gill J, Farr G, Wang S, Samaniego F (2005) Activation of Src kinase Lyn by the Kaposi sarcoma-associated herpesvirus K1 protein: implications for lymphomagenesis. Blood 105:3987–3994
Pratt CL, Estep RD, Wong SW (2005) Splicing of rhesus rhadinovirus R15 and ORF74 bicistronic transcripts during lytic infection and analysis of effects on production of vCD200 and vGPCR. J Virol 79:3878–3882
Regamey N, Tamm M, Wernli M, Witschi A, Thiel G, Cathomas G, Erb P (1998) Transmission of human herpesvirus 8 infection from renal-transplant donors to recipients. N Engl J Med 339:1358–1363
Renne R, Blackbourn D, Whitby D, Levy J, Ganem D (1998) Limited transmission of Kaposi's sarcoma-associated herpesvirus in cultured cells. J Virol 72:5182–5188
Roizmann B, Desrosiers RC, Fleckenstein B, Lopez C, Minson AC, Studdert MJ (1992) The family Herpesviridae: an update. The Herpesvirus Study Group of the International Committee on Taxonomy of Viruses. Arch Virol 123:425–449
Russo JJ, Bohenzky RA, Chien MC, Chen J, Yan M, Maddalena D, Parry JP, Peruzzi D, Edelman IS, Chang Y, Moore PS (1996) Nucleotide sequence of the Kaposi sarcoma-associated herpesvirus (HHV8). Proc Natl Acad Sci U S A 93:14862–14867
Salahuddin SZ, Nakamura S, Biberfeld P, Kaplan MH, Markham PD, Larsson L, Gallo RC (1988) Angiogenic properties of Kaposi's sarcoma-derived cells after long-term culture in vitro. Science 242:430–433
Samaniego F, Pati S, Karp JE, Prakash O, and Bose D (2001) Human herpesvirus 8 K1-associated nuclear factor-kappa B-dependent promoter activity: role in Kaposi's sarcoma inflammation? J Natl Cancer Inst Monogr (28):15–23
Samols MA, Hu J, Skalsky RL, Renne R (2005) Cloning and identification of a microRNA cluster within the latency-associated region of Kaposi's sarcoma-associated herpesvirus. J Virol 79:9301–9305
Schafer A, Cai X, Bilello JP, Desrosiers RC, Cullen BR (2007) Cloning and analysis of microRNAs encoded by the primate gamma-herpesvirus rhesus monkey rhadinovirus. Virology 364:21–27
Schalling M, Ekman M, Kaaya EE, Linde A, Biberfeld P (1995) A role for a new herpes virus (KSHV) in different forms of Kaposi's sarcoma. Nat Med 1:707–708

Schwartz RA (1996) Kaposi's sarcoma: advances and perspectives. J Am Acad Dermatol 34:804–814
Searles RP, Bergquam EP, Axthelm MK, Wong SW (1999) Sequence and genomic analysis of a Rhesus macaque rhadinovirus with similarity to Kaposi's sarcoma-associated herpesvirus/ human herpesvirus 8. J Virol 73:3040–3053
Sharp TV, Wang HW, Koumi A, Hollyman D, Endo Y, Ye H, Du MQ, Boshoff C (2002) K15 protein of Kaposi's sarcoma-associated herpesvirus is latently expressed and binds to HAX-1, a protein with antiapoptotic function. J Virol 76:802–816
Simpson GR, Schulz TF, Whitby D, Cook PM, Boshoff C, Rainbow L, Howard MR, Gao SJ, Bohenzky RA, Simmonds P, Lee C, de Ruiter A, Hatzakis A, Tedder RS, Weller IV, Weiss RA, Moore PS (1996) Prevalence of Kaposi's sarcoma associated herpesvirus infection measured by antibodies to recombinant capsid protein and latent immunofluorescence antigen. Lancet 348:1133–1138
Skalsky RL, Cullen BR (2010) Viruses, microRNAs, and host interactions. Annu Rev Microbiol 64:123–141
Skalsky RL, Samols MA, Plaisance KB, Boss IW, Riva A, Lopez MC, Baker HV, Renne R (2007) Kaposi's sarcoma-associated herpesvirus encodes an ortholog of miR-155. J Virol 81:12836–12845
Soulier J, Grollet L, Oksenhendler E, Cacoub P, Cazals-Hatem D, Babinet P, d'Agay MF, Clauvel JP, Raphael M, Degos L et al (1995) Kaposi's sarcoma-associated herpesvirus-like DNA sequences in multicentric Castleman's disease. Blood 86:1276–1280
Staskus KA, Zhong W, Gebhard K, Herndier B, Wang H, Renne R, Beneke J, Pudney J, Anderson DJ, Ganem D, Haase AT (1997) Kaposi's sarcoma-associated herpesvirus gene expression in endothelial (spindle) tumor cells. J Virol 71:715–719
Tomlinson CC, Damania B (2004) The K1 protein of Kaposi's sarcoma-associated herpesvirus activates the Akt signaling pathway. J Virol 78:1918–1927
Tomlinson CC, Damania B (2008) Critical role for endocytosis in the regulation of signaling by the Kaposi's sarcoma-associated herpesvirus K1 protein. J Virol 82:6514–6523
Tsai YH, Wu MF, Wu YH, Chang SJ, Lin SF, Sharp TV, Wang HW (2009) The M type K15 protein of Kaposi's sarcoma-associated herpesvirus regulates microRNA expression via its SH2-binding motif to induce cell migration and invasion. J Virol 83:622–632
Umbach JL, Strelow LI, Wong SW, Cullen BR (2010) Analysis of rhesus rhadinovirus microRNAs expressed in virus-induced tumors from infected rhesus macaques. Virology 405:592–599
Wabinga HR, Parkin DM, Wabwire-Mangen F, Mugerwa JW (1993) Cancer in Kampala, Uganda, in 1989–91: changes in incidence in the era of AIDS. Int J Cancer 54:26–36
Walz N, Christalla T, Tessmer U, Grundhoff A (2010) A global analysis of evolutionary conservation among known and predicted gammaherpesvirus microRNAs. J Virol 84:716–728
Wang L, Wakisaka N, Tomlinson CC, DeWire SM, Krall S, Pagano JS, Damania B (2004) The Kaposi's sarcoma-associated herpesvirus (KSHV/HHV-8) K1 protein induces expression of angiogenic and invasion factors. Cancer Res 64:2774–2781
Wang L, Brinkmann MM, Pietrek M, Ottinger M, Dittrich-Breiholz O, Kracht M, Schulz TF (2007) Functional characterization of the M-type K15-encoded membrane protein of Kaposi's sarcoma-associated herpesvirus. J Gen Virol 88:1698–1707
Wang L, Pietrek M, Brinkmann MM, Havemeier A, Fischer I, Hillenbrand B, Dittrich-Breiholz O, Kracht M, Chanas S, Blackbourn DJ, Schulz TF (2009) Identification and functional characterization of a spliced rhesus rhadinovirus gene with homology to the K15 gene of Kaposi's sarcoma-associated herpesvirus. J Gen Virol 90:1190–1201
Wiese N, Tsygankov AY, Klauenberg U, Bolen JB, Fleischer B, Broker BM (1996) Selective activation of T cell kinase p56lck by Herpesvirus saimiri protein tip. J Biol Chem 271:847–852
Winter J, Jung S, Keller S, Gregory RI, Diederichs S (2009) Many roads to maturity: microRNA biogenesis pathways and their regulation. Nat Cell Biol 11:228–234
Wong EL, Damania B (2006) Transcriptional regulation of the Kaposi's sarcoma-associated herpesvirus K15 gene. J Virol 80:1385–1392

Wong SW, Bergquam EP, Swanson RM, Lee FW, Shiigi SM, Avery NA, Fanton JW, Axthelm MK (1999) Induction of B cell hyperplasia in simian immunodeficiency virus-infected rhesus macaques with the simian homologue of Kaposi's sarcoma-associated herpesvirus. J Exp Med 190:827–840

Wu W, Vieira J, Fiore N, Banerjee P, Sieburg M, Rochford R, Harrington W Jr, Feuer G (2006) KSHV/HHV-8 infection of human hematopoietic progenitor (CD34+) cells: persistence of infection during hematopoiesis in vitro and in vivo. Blood 108:141–151

Zhong W, Wang H, Herndier B, Ganem D (1996) Restricted expression of Kaposi sarcoma-associated herpesvirus (human herpesvirus 8) genes in Kaposi sarcoma. Proc Natl Acad Sci USA 93:6641–6646

Neotropical Primates and Their Susceptibility to *Toxoplasma gondii*: New Insights for an Old Problem

José Luiz Catão-Dias, Sabrina Epiphanio, and Maria Cecília Martins Kierulff

Introduction

Toxoplasmosis is a zoonosis caused by *Toxoplasma gondii*, an obligate intracellular coccidium belonging to the Phylum Apicomplexa. This coccidium is considered one of the most competent known parasites, and it is believed that it can virtually parasitize all species of birds and mammals. In most immunocompetent hosts, infection with *T. gondii* is subclinical or causes discrete clinical and pathological alterations. However, in different groups of mammals, including strepsirrhines from Madagascar, marsupials from Oceania, hares from Europe, and, in particular, neotropical nonhuman primates, toxoplasmosis can cause disease processes associated with high mortality rates. In our experience, it is possibly the most important cause of acute infectious disease death affecting Platyrrhini in captivity in Brazil. The reasons for the high susceptibility of New World Primates (NWPs) to toxoplasmosis are not fully understood, but it is assumed that platyrrhines have evolved for over 20 million years without the presence of felines and, thus, these did not develop an efficient immune response to the parasite. In addition, the arboreal habits of platyrrhines would have contributed to minimize the contact of NWPs with the protozoan after its arrival to the American continent during the Pleistocene, about one to three million years ago.

J.L. Catão-Dias (✉)
Laboratório de Patologia Comparada de Animais Selvagens,
Departamento de Patologia, Faculdade de Medicina Veterinária e Zootecnia,
Universidade de São Paulo, São Paulo, Brazil
e-mail: zecatao@usp.br

S. Epiphanio
Departamento de Ciências Biológicas, Universidade Federal de São Paulo,
Diadema, São Paulo, Brazil

M.C.M. Kierulff
Departamento de Ciências Biológicas e Agrárias, Centro Universitário Norte do Espírito
Santo, Universidade Federal do Espírito Santo, São Mateus, São Paulo, Brazil

J.F. Brinkworth and K. Pechenkina (eds.), *Primates, Pathogens, and Evolution*,
Developments in Primatology: Progress and Prospects,
DOI 10.1007/978-1-4614-7181-3_9, © Springer Science+Business Media New York 2013

On the other hand, available clinical, serological, pathological, and experimental data suggest that the response of NWPs to toxoplasmosis is not homogeneous. There is a clear difference between the genus *Cebus*, which apparently reacts mildly to the disease, and most other groups, for which toxoplasmosis can be devastating. The understanding of the reason for this variability of responses of NWPs to *T. gondii* is not clear. The aim of this chapter is to present a brief summary of existing knowledge about toxoplasmosis in NWP and review some possibilities that justify them.

New World Primate Ecology

New World primates (NWPs, Platyrrhini) are distributed from South and Central America to North America, in Mexico. Table 1 presents a distribution of NWPs according to their diet and forest strata preferences. They are all small- to medium-sized primates, weighing from 120 g to approximately 10 kg. The different species live in sympatry throughout most of their ranges with up to 13 species at some Amazonian sites. One of the characteristics of Platyrrhini is the absence of terrestrial species, and few species occasionally forage on the ground or travel short distances between trees, but none spend most of the day feeding on the ground (Feagle 1999). New World primates include species that feed on gums, fruits, leaves, seeds, a variety of invertebrates, and small vertebrates. Some of the smaller species rely heavily on nectar during the dry periods of the year. Some NWPs show morphological adaptations for more specialized diets, while others are omnivorous, and most species show preference for one of the arboreal strata, where they stay most part of the time.

Perelman et al. (2011) divided Platyrrhini into three families and listed 17 genera, but did not include the two monotypic genera: *Callibella* (Aguiar and Lacher 2003; Van Roosmalen and Van Roosmalen 2003) and *Oreonax* (Groves 2001), although some authors do not separate *Oreonax* from *Lagothrix* (Rosenberger and Matthews 2008). According to Rylands and Mittermeier (2009), there are 19 genera and 199 species and subspecies of primates in the Neotropics.

Pitheciidae (*Cacajao,* uakaris; *Calicebus,* titis; *Chiropotes,* bearded sakis; *Pithecia*, sakis) have an unusual dental specialization for processing fruits and seeds encased in a hard outer covering that are generally too hard for other monkeys to bite through (Kinzey 1992). Compared to *Cacajao* and *Chiropotes*, *Pithecia* seems to eat fruits with relatively softer outer covering, has a more diverse diet, and concentrates its activity in the middle and upper canopy but can use the lower canopy for foraging (Feagle 1999).

Chiropotes prefers high rain forests, being usually in the middle and upper levels of the main canopy (Feagle 1999). They feed on hard, often unripe fruits and on seeds with very hard shells, which they open with their large canines. They occasionally feed on insects (Norconk et al. 2009). *Cacajao* are also specialized on fruits with hard outer shells and immature seeds and include in their diet small amounts

Table 1 Diet and forest strata preferences by Neotropical primates

Family	Genus[a]	Species and subspecies[b]	Common names	Diet	Forest strata preference	References
Cebidae	*Mico*	14	Amazonian marmosets	Fruits, arthropods, flowers, exudates	Middle of the canopy to the ground, lower levels of the forest	Bicca-Marques et al. (2011), Feagle (1999)
	Cebuella	2	Pygmy marmoset	Exudates, arthropods, fruits, and small vertebrates	Middle of the canopy to the ground, lower levels of the forest	Norconk et al. (2009), Feagle (1999), Soini (1988), Bicca-Marques et al. (2011).
	Callithrix	6	Marmosets	Exudates, arthropods, fruits, small vertebrates, eggs, seeds, molluscs	Middle of the canopy to the ground, lower levels of the forest	Norconk et al. (2009), Feagle (1999), Bicca-Marques et al. (2011)
	Callimico	1	Goeldi's monkey	Arthropods, fruits, fungi, exudates	Middle of the canopy to the ground, lower levels of the forest	Norconk et al. (2009), Feagle (1999)
	Leontopithecus	4	Lion tamarins	Fruits, arthropods, exudates, flowers, small vertebrates	Middle of the canopy to the ground, lower levels of the forest	Norconk et al. (2009), Feagle (1999), Kierulff et al. (2002)
	Saguinus	33	Tamarins	Arthropods, fruits, exudates, young leaves, small vertebrates	Middle of the canopy to the ground, lower levels of the forest	Norconk et al. (2009), Feagle (1999), Snowdon and Soini (1988), Bicca-Marques et al. (2011)
	Aotus	12	Owl monkeys	Fruits, young leaves, flowers, arthropods	No preference for a particular canopy level	Norconk et al. (2009), Feagle (1999), Cunha (2008), Bicca-Marques et al. (2011)
	Saimiri	10	Squirrel monkeys	Arthropods, fruits, young leaves, flowers, seeds, small vertebrates, and eggs	Middle and lower levels of the forest, eventually come down to the ground.	Norconk et al. (2009), Feagle (1999), Defler (2005), Baldwin and Baldwin (1981), Boinski (1987), Ingberman et al. (2008), Bicca-Marques et al. (2011)

(continued)

Table 1 (continued)

Family	Genus[a]	Species and subspecies[b]	Common names	Diet	Forest strata preference	References
	Cebus	26	Capuchin monkeys	Fruits, arthropods, young leaves, seeds, flowers, small vertebrates, eggs. In the wild, use large stones as tools to open hard palm fruits	Main canopy levels but frequently come down to the understory or to the ground during both travel and feeding	Norconk et al. (2009), Izawa (1979), Freese and Oppehheimer (1981), Fragaszy et al. (2004)
Atelidae	*Lagothrix*	5	Woolly monkeys	Fruits, insects, young leaves, flowers, seeds	Upper levels of the main canopy, rarely coming down to the ground	Norconk et al. (2009), Feagle (1999), Bicca-Marques et al. (2011)
	Brachyteles	2	Muriquis	Young leaves, fruits (pulp), flowers, mature leaves, seeds	Upper levels of the main canopy, rarely coming down to the ground	Norconk et al. (2009), Feagle (1999), Mendes et al. (2010)
	Ateles	15	Spider monkeys	Fruits (pulp), young leaves, flowers	Highest levels of the forest but eventually come down to the ground	Norconk et al. (2009), Feagle (1999), Zanon et al. (2008), Bicca-Marques et al. (2011).
	Alouatta	19	Howler monkeys	Young leaves, fruits (pulp), mature leaves, flowers	Most species prefer the main canopy and emergent levels; species from dry areas regularly come down to the ground	Norconk et al. (2009), Feagle (1999), Bicca-Marques et al. (2011)
Pitheciidae	*Cacajao*	6	Uakaris	Seeds, fruits (pulp), flowers	Upper parts of the forest but eventually come down to the ground to forage	Norconk et al. (2009), Rickli and Reis (2008)

Chiropotes	5	Bearded saki monkeys	Seeds, fruit (pulp), arthropods, flowers	Middle and upper levels of the main canopy, rarely coming down to the ground	Norconk et al. (2009), Feagle (1999), Bicca-Marques et al. (2011).
Pithecia	9	Saki monkeys	Seeds, fruits (pulp, whole, arils), young leaves, arthropods, and flowers	Middle and upper canopy, eventually come down to the ground to forage	Norconk et al. (2009), Feagle (1999)
Callicebus	29	Titi monkeys	Fruits (pulp), seeds, young leaves, flowers, arthropods	Main canopy to the understory and rarely coming down to the ground	Norconk et al. (2009), Bordignon et al. (2008), Feagle (1999), Bicca-Marques et al. (2011)

[a]Taxonomy of Perelman et al. (2011)
[b]Number of species and subspecies according to Rylands and Mittermeier (2009)

of leaves, flowers, nectar, and insects (Rickli and Reis 2008). They use more frequently the upper parts of the forest but can eventually come down to the ground to forage (Rickli and Reis 2008).

Calicebus have very short canine teeth in comparison with other species of the family Pitheciidae. They are mainly frugivorous but can supplement their diet with leaves, seeds, flowers, and insects (Bordignon et al. 2008; Norconk et al. 2009). The different species are distributed in different habitats from mature forest to the dry scrub forest (caatinga) or bamboo thickets, where they use the main canopy or the understory or low levels in the forest (Feagle 1999; Bordignon et al. 2008).

The family Atelidae includes two predominantly folivorous genera (*Alouatta,* howler monkey, and *Brachyteles,* muriqui) that supplement their diet with fruits, flowers, and seeds and two frugivorous genera, the spider (*Ateles*) and woolly monkeys (*Lagothrix* and *Oreonax*), that feed mainly on fruits but also leaves, buds, flowers, and insects (Groves 2001; Zanon et al. 2008; Norconk et al. 2009; Bicca-Marques et al. 2011). All atelines have a long, prehensile tail, and spider monkeys, woolly monkeys, and muriquis are largely restricted to high primary rain forests where they prefer the upper levels of the main canopy, rarely coming down to the ground (Mendes et al. 2010; Feagle 1999). *Alouatta* are found in a variety of habitats, including primary and secondary forest, dry deciduous forest, and habitats containing patches of relatively low trees in open savannah. Most species seem to prefer the main canopy and emergent levels, but some species that live in drier areas (*A. caraya*) regularly come down to the ground and cross-open areas between patches of forest (Feagle 1999).

The family Cebidae is composed of four subfamilies: Cebinae that includes capuchin monkeys (*Cebus*); Saimirinae, represented by squirrel monkeys (*Saimiri*); Callitrichinae, the smallest and most distinctive New World primates, separated in Goeldi's monkey (*Callimico*), tamarins (*Saguinus* and *Leontopithecus*), and marmosets (*Mico, Callithrix, Cebuella, Callibella*); and the subfamily Aotinae (owl monkeys), the only nocturnal Platyrrhini (Fernandez-Duque 2006; Norconk et al. 2009; Perelman et al. 2011).

Aotus are found in a variety of forest habitats, and there are no indications that they prefer any particular canopy level. They are primarily frugivorous with a diet that is supplemented by flowers, leaves, insects, and occasionally small vertebrates and eggs (Wright 1981; Feagle 1999; Cunha 2008). Callitrichines use different types of forest but seem to be characterized by the ability to exploit marginal and disturbed habitat, and their diet is composed of fruits, flower, arthropods (mainly insects), exudates, fungus, and small vertebrates (lizards, birds, frogs, and small rodents). All species spend most of the time in the middle levels of the forest and forage for fruits and insects in the middle of the canopy to the ground in the lower levels of the forest (Snowdon and Soini 1988; Soini 1988; Feagle 1999; Kierulff et al. 2002; Bicca-Marques et al. 2011).

Capuchin monkeys (*Cebus*) and squirrel monkeys (*Saimiri*) are the two most omnivorous Platyrrhini (Feagle 1999). *Saimiri* occupy a variety of rain forest habitats but seem to prefer riverine and secondary forests, where they are commonly found in the middle and lower levels (Feagle 1999; Defler 2005). They are frugivores and insectivores and supplement their diet with leaves, seeds, small vertebrates, nectar, and eggs (Baldwin and Baldwin 1981; Boinski 1987; Feagle 1999; Defler

2005; Ingberman et al. 2008). The insect component of their diets is the highest of any non- Callitrichinae, but the differences between these genera lie in mandibular and dental robusticity, which is much stronger in *Cebus* and could determine differences in their diets (Norconk et al. 2009).

Capuchin monkeys are divided into tufted (*Cebus apella*, *C. macrocephalus*, *C. libidinosus*, *C. nigritus*, *C. robustus*, *C. cay*, *C. flavius*, *C. xanthosternos*) and untufted (*C. albifrons*, *C. olivaceus*, *C. kaapori*) (Bicca-Marques et al. 2011)[1]. *Cebus* are found in virtually all types of neotropical forest including humid and dry forest, swamp forests, seasonally flooded forest, as well as more open vegetation types in savannahs and caatingas, where rainfall is absent for 5–6 months each year (Fragaszy et al. 2004; Bicca-Marques et al. 2011). They seem to prefer the main canopy levels but frequently come down to the understory or to the ground during both travel and feeding (Feagle 1999).

Their diet includes many types of fruits and other vegetal parts and animal matter (invertebrates and small vertebrates), and they are considered omnivores and also classified as frugivore–insectivore (Izawa 1979; Freese and Oppehheimer 1981; Feagle 1999). Fragaszy et al. (2004) characterized them as innovative and extreme foragers due to their ability to acquire sustenance from a variety of potentially dangerous sources that require special foraging skills and also for trying to eat almost anything remotely edible. It gives them three types of adaptive advantages: first, the flexibility to switch from more accessible foods such as fruits to more inaccessible ones at time of food scarcity; second, the capacity to exploit habitats with different structure and phenological characteristics (such as secondary or disturbed forests); and third, it reduces the degree of dietary overlap between capuchins and other arboreal vertebrates, mainly other primate species, more specialized in fruit or insects (Brown and Zunino 1990; Fragaszy et al. 1990, 2004).

Toxoplasma gondii

Taxonomy and Trophism

Toxoplasmosis is a zoonosis of global occurrence caused by the protozoan *Toxoplasma gondii*, an obligate intracellular coccidium belonging to Phylum Apicomplexa, Class Sporozoasida. The first descriptions of *T. gondii* occurred almost simultaneously in 1908, and the history of the agent's discovery reveals the independent work of researchers in two different continents. In Tunisia, Nicolle and Manceaux, investigating the

---

[1]Recent phylogeographic analysis has shown that capuchins contain two well-supported monophyletic clades, the morphologically distinct "gracile" (or untufted) and "robust" (or tufted) groups, and placed the age of the split at 6.7 Ma (95 % highest posterior density 4.1–9.4 Ma) (Alfaro et al. 2012a). Morphological and behavioral–ecological data also support a division of capuchins into the same two distinct groups. As a consequence Alfaro et al. (2012b) have argued for a division of capuchin monkeys into two genera: *Sapajus* Kerr, 1792, for the robust capuchins and *Cebus* Erxleben, 1777, for the gracile capuchins.

participation of rodent of *Ctenodactylus gundi* species in the epidemiological cycle of leishmaniasis, described arc-shaped protozoa, which were given the name from the Greek "toxo" (arc) "plasma" (life) (Nicolle and Manceaux 1909). One week later, in Brazil, Alfonso Splendore, investigating histological lesions in rabbits resembling human leishmaniasis, described a new protozoan, morphologically similar to that reported by Nicolle and Manceaux (Splendore 1909). In subsequent years, several reports describing similar parasites in other hosts were observed; however, immunobiological evidence showed that the different strains isolated from humans and animals represented the same agent (Innes 2010).

As the only species of the genus, until recently, it was believed that *T. gondii* was a clonal organism with minimal genetic variation, a condition that allowed classifying it into three lineages, namely, I, II, and III. However, newly published work has shown that strains isolated from the free-living chicken (*Gallus domesticus*) from different regions of Brazil have phenotypic and genetic characteristics markedly different from strains from other countries, indicating that the genetic variability of the parasite is higher than originally thought (Dubey 2009; Dubey and Su 2009).

In the 100 years since its discovery, *T. gondii* has proven to be a remarkably competent parasite, since it appears to be able to parasitize virtually all species of mammals and birds (Innes 2010). Recent work suggests that any cell of these vertebrates can be infected by the protozoan (Elmore et al. 2010). Moreover, *T. gondii* likely emerged in the manner that all other Coccidia have, using only one host and being transmitted via fecal–oral route (Dubey 2009). However, the acquired skill of being transmitted through other routes, such as ingestion of meat or through the placenta, gave the parasite the opportunity to occupy new geographical areas and move into new biomes and corresponding hosts (Dubey and Su 2009). These features make *T. gondii* the archetype of the ideal parasite and be considered the most successful existing parasite by some authors (Innes 2010).

On the other hand, the genetic basis of the virulence mechanisms involved in *Toxoplasma* infection is only beginning to be unraveled. Recent data obtained from the proteomic analysis of rhoptries show that several proteins in this organelle such as ROP18 and ROP16 are released within parasitophorous vacuoles in the exact moment of the host cell invasion by the protozoan and may represent, therefore, the communication interface in the host–parasite relationship (Bradley et al. 2005; Boothroyd 2009; Elmore et al. 2010).

Toxoplasma gondii Life Cycle

The life cycle of *T. gondii* was elucidated throughout the 1960s and 1970s of the twentieth century with the identification of the parasite's sexual cycle (Frenkel et al. 1969, 1970; Hutchison et al. 1970; Ferguson et al. 1974). Figure 1 summarizes the *T. gondii* life cycle (Gardiner et al. 1998). Three infective forms are recognized: sporozoites in sporulated oocysts and asexual fast- and slow-replicating forms tachyzoites and bradyzoites. Felines are the only known natural hosts and become infected through meat consumption or intake of water and/or other food contaminated with infective forms. However, there is a marked difference in competence

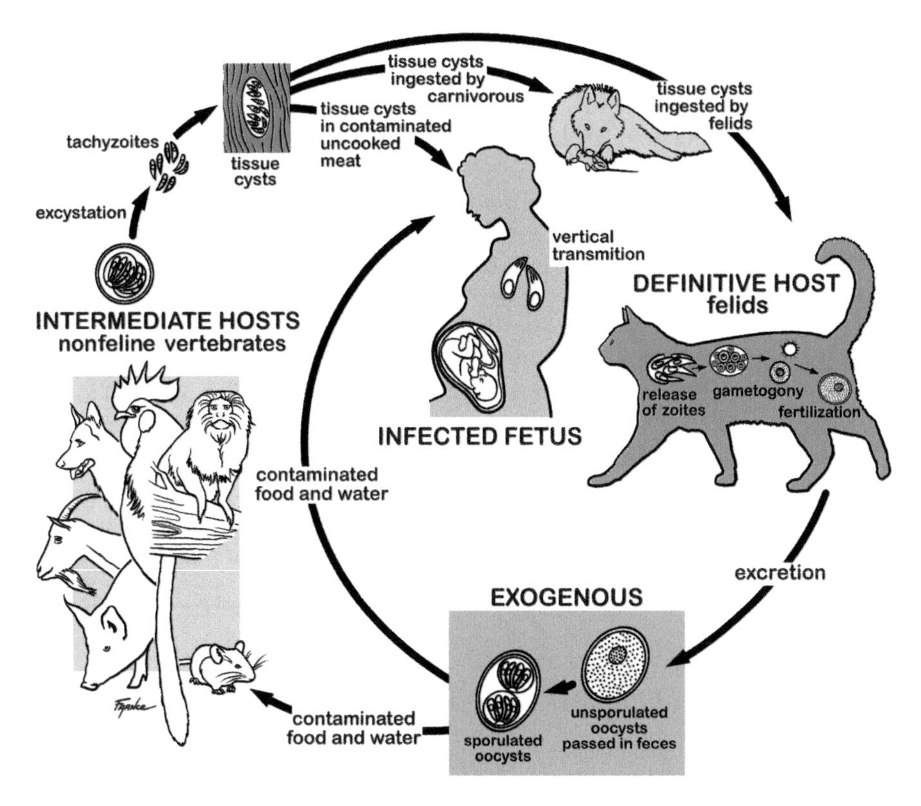

Fig. 1 *Toxoplasma gondii* life cycle (Adapted from Gardiner et al. 1998)

between the various types of zoites in terms of how effectively they infect susceptible felines. In this sense, bradyzoites are exceptionally efficient when compared to tachyzoites and sporozoites. The ingestion of only one bradyzoite is sufficient to cause infection in felines, while the intake of at least 1,000 oocysts is required to produce the same effect (Dubey 2009). Besides the domestic cat (*Felis catus*), 17 species of wild felines, including five New World species (*Puma concolor*, cougar; *Puma yagouaroundi*, jaguarundi; *Leopardus geoffroyi*, Geoffroy's cat; *L. pardalis*, ocelot; and *L. colocolo*, pampas cat) have been identified as natural hosts (Silva 2007; Elmore et al. 2010).

Once ingested by cats, the infective form invades the enterocyte of the small intestine and undergoes asexual reproduction cycles, followed by sexual reproduction with the formation of non-sporulated oocyst with eight sporozoites. These oocysts are excreted in feces and, depending on environmental conditions, undergo sporulation and become infective. Fecal excretion of non-sporulated oocysts by immunocompetent felines occurs 1–2 weeks after infection and, once established, generates large amounts of oocysts. However, recently published data showed that the incidence of immunosuppressive events can promote new episodes of oocyst excretion in affected animals (Malmasi et al. 2009). Sporulated oocysts are very resistant to environmental conditions such as desiccation and freezing, and to the action of disinfectants, and can survive in the environment for many months. It is believed that the favorable conditions of humidity and temperature, as those found in several tropical biomes, such as Amazon and Atlantic forest, greatly improve the viability of sporulated oocysts (Silva 2007).

The infection of intermediate hosts may occur in many ways, either through meat consumption/ingestion of prey infected by asexual zoites, the consumption of other food and water contaminated with sporulated oocysts, or congenitally (Dubey et al. 1998). However, there are reports of infection caused by blood transfusions, transplantation, and laboratory accidents in humans (Dubey and Jones 2008). The possibility of transmission occurring through inhalation of aerosols or ingestion of secretions containing *T. gondii* in Neotropical primates has also been suggested (Furuta et al. 2001; Carme et al. 2009). In most vertebrate homeotherm hosts, zoites multiply asexually within the infected cells through endodyogeny and tend to form perennial cysts viable for many years in multiple tissues and organs, particularly in skeletal and cardiac muscle, and in the central nervous system (Dubey 2009).

In summary, *T. gondii* life cycle is characterized by sexually reproducing and developing oocysts in the small intestine of a natural feline host, before transmitting to an intermediate homeotherm host. Within the intermediate host, the parasite develops, forming cysts in various tissues.

Toxoplasma gondii and Host Immune Response

Although host immune responses to *T. gondii* have been studied since the pathogen's discovery more than a century ago, their interactions are not fully understood (Boothroyd 2009; Tait and Hunter 2009). Experimental studies have shown that more than 1,000 host genes are modulated in *Toxoplasma*-infected cells, among them genes encoding proteins implicated in several processes including inflammation, apoptosis, metabolism, and cell growth and differentiation (Blader and Saeij 2009).

The development of toxoplasmosis in mice and humans is determined by mechanisms involving the pathogenicity of the parasite strain, in addition to the host immune status (Hill et al. 2011; Pifer and Yarovinsky 2011). Surface antigen 1 (SAG1), the major surface protein of *T. gondii,* has been recognized as an essential target of adaptive immune response. However, the parasite has developed strategies to avoid the powerful immune response (Buzoni-Gatel and Werts 2006). In addition, it is believed that three main *T. gondii* genotypes responsible for infection in humans (types I, II, and III) may induce different patterns of the disease (Saeij et al. 2005). Clonal lineages differ in growth, migration, and transmigration. In laboratory mice, it is known that type I strains are very virulent (lethal dose – LD100 of one parasite), while type II and III strains are much less virulent (lethal dose – LD50 $\sim10^3$ and $\sim10^5$). In humans, type I strains are frequently associated with postnatally acquired ocular infections, whereas type II strains are more related with congenital infections and toxoplasmic encephalitis (Blader and Saeij 2009; Sibley et al. 2009).

Typically, the natural and intermediate host response to *T. gondii* infection is capable of holding back parasite dissemination, reducing mortality rates. The early stimulation of the immune system following infection is an essential step in establishing a balanced host–parasite relationship, once both host and parasite survive the initial phases of infection and, in this way, the process progresses towards chronic disease (Aliberti 2005).

Immune System: How Cells React

It is known that both humoral and cellular immune responses are involved in the resolution of acute toxoplasmosis, but the cellular response is considered the most important mechanism responsible for the host defense (Däubener and Hadding 1997; Innes 1997). When the parasites invade the intestinal mucosa of intermediate hosts (the main infection route), the infected enterocytes suffer morphological and physiological changes and secret chemokines and cytokines that attract polymorphonuclear leukocytes, macrophages, and dendritic cells (DCs) (Buzoni-Gatel and Werts 2006).

Proinflammatory cytokines produced by lymphocytes, macrophages, DCs, and neutrophils are crucial for controlling *T. gondii.* In the first stages of infection, *T. gondii* activates cells such as macrophages, DCs, and neutrophils to produce high levels of IL-12 (Gazzinelli et al. 1993b; Johnson and Sayles 1997). Mitogen-activated protein kinase p38 is required for IL-12 production by macrophages in response to soluble tachyzoite antigen (STAg) (Mason et al. 2004).

Many studies have demonstrated that IFN-γ, the hallmark of the inflammatory response, is the major defense component against *T. gondii* infection that inhibits parasite replication in various human and mouse cells (Suzuki et al. 1988; Sharma 1990; Däubener and Hadding 1997; Buzoni-Gatel and Werts 2006). Additionally, IFN-γ, which is produced by natural killer cells (NK cells) in response to IL-12 secretion, contributes to the differentiation of lymphocytes into the Th1 phenotype. IFN-γ, produced by activated T lymphocytes, natural killer cells, and natural killer T cells, stimulates macrophages to produce reactive oxygen intermediates (ROI), leading to the death of *T. gondii* (Nathan et al. 1983). Hence, IFN-γ is an essential cytokine for resistance to acute and chronic *Toxoplasma* infections (Suzuki et al. 1988; Gazzinelli et al. 1993a). Similar to IFN-γ, TNF-α, IL-6, and IL-1 have synergistic effects on the induction of an adequate immune response against *T. gondii* (Lang et al. 2007).

In addition to IFN-γ, other cytokines, such as TNF-α, IL-2, and lymphotoxin-α, are cofactors important to host responses to *T. gondii* infection. Natural killer cells and $CD4^+$ and $CD8^+$ T cells are three lymphocyte subsets that produce these cytokines and have been suggested to influence *T. gondii* immunity in mice and humans (reviewed in Blanchard et al. 2008). IL-2, produced by $CD4^+$ T cells, is an important T cell mitogen (reviewed in Tait and Hunter 2009). $CD4^+$ and $CD8^+$ T cell activation prevent reactivation of infection, probably by IFN-γ production (Gazzinelli et al. 1992). $CD8^+$ T cells are known to be crucial for protection against the intracellular parasite *T. gondii*, because they are involved in the capacity to induce apoptosis in infected cells (Däubener and Hadding 1997). $CD8^+$ T cells have also been reported to directly kill the extracellular and intracellular *T. gondii*, independent or dependent of major histocompatibility complexes (MHC), respectively. In addition, perforin present in the granules of $CD8^+$ T cells and NK cells has been implicated in the parasite death, and it has been found to play a critical role in chronic toxoplasmosis (Denkers et al. 1997).

The production of nitric oxide (NO) by macrophages limits the growth and replication of *T. gondii*, and this secretion depends on the expression of the inducible NO synthase (iNOS). IFN-γ through signal transducer and activator of transcription 1 (STAT1) regulates effector mechanisms, including the iNOS overexpression. NO and reactive nitrogen intermediates have been recognized as the main effector molecules of microbicidal and microbiostatic activities in activated macrophages. Experimentally, it has been shown that in acute infection, the lethality of the pathogen is associated with high NO levels in serum (reviewed by Denkers and Gazzinelli 1998; Tait and Hunter 2009). In addition, IFN-γ induces indoleamine 2,3-dioxygenase, resulting in L-tryptophan depletion and inhibition of parasite growth (Fujigaki et al. 2002).

CD40 (in T cells) and CD154 (in macrophages) pathways promote killing of *T. gondii* through induction of vacuole–lysosomal fusion. In addition, these pathways promote proinflammatory cytokines and active autophagic mechanisms (Subauste 2009). The upregulation of p47 GTPases, in response to IFN-γ production, is also involved in the autophagy process (reviewed by Tait and Hunter 2009).

In vivo studies indicate that the activation of the host innate immunity plays a crucial role in the early resistance against infection and pathogenesis of toxoplasmosis (Gazzinelli et al. 1994). Toll-like receptors (TLR) are a family of innate immune receptors focused on recognition of "pathogen-associated molecular patterns" (PAMPs), which are molecules that are critical for microorganisms and naturally not expressed by host cells (Pifer and Yarovinsky 2011). However, ligands from the parasite that stimulate these receptors are not completely known. Toll-like receptors (TLR) such as TLR2, TLR4, TLR9, and TLR11 bind to *Toxoplasma*-derived factors, and TLR11 is considered to be a major innate immune receptor that regulates IL-12 response to *T. gondii* infection (Yarovinsky et al. 2005). It has already been reported that TLR9 is required for Th1 immune response after oral infections with *T. gondii* (reviewed in Oykhman and Mody 2010). Glycosylphosphatidylinositols (GPIs) of *Toxoplasma*, as well as other apicomplexan protozoa, stimulate the production of TNF-α in macrophages through NF-κB activation, via both TLR4 and TLR2 (reviewed in Debierre-Grockiego and Schwarz 2010). However, only the loss of MyD88, an adaptor protein that mediates TLR signaling, is essential for the survival in parasite-infected animals (Scanga et al. 2002; Yarovinsky et al. 2005; Debierre-Grockiego et al. 2007).

Humoral Immunity

Humoral immunity has been considered of minor importance in toxoplasmosis protection and resistance (Sharma 1990). However, antigens from the parasite induce the production of antibodies such as IgM, IgG, and IgA (Correa et al. 2007).

These immunoglobulins may lead to neutralization, inhibition of parasite cell invasion, activation of the classical pathway of complement, and inflammation (Däubener and Hadding 1997; Correa et al. 2007). Anti-*T. gondii* antibodies are mainly specific to tachyzoites (reviewed in Hegab and Al-Mutawa 2003). However, some studies have shown that IgA is important against cyst formation in mucosa during oral infection (reviewed in Denkers and Gazzinelli 1998).

IgG is the most important immunoglobulin involved in the humoral immune response against *T. gondii* infection. Specific IgG isotypes, IgG1 in humans and IgG2a in mice, can play an important role in resistance to different pathogens through mechanisms such as complement fixation, opsonization, or antibody-dependent cell cytotoxicity. In humans, these antibodies reach a peak around 6–14 months postinfection, decreasing afterwards, with the host remaining reactive forever. On the other hand, very high IgG titers are indicative of acute infection (Sharma 1990; Denkers and Gazzinelli 1998).

The detection of specific IgM in host serum indicates a recently acquired infection and does not indicate reinfection, because a minimum titer of IgG suppresses IgM production (reviewed in Hegab and Al-Mutawa 2003). In the newborn, IgM is diagnostic of congenital infection, since maternal IgM cannot cross the placental barrier (reviewed in Hegab and Al-Mutawa 2003). Intriguingly, *T. gondii* IgM in adults commonly persists well over 6 months. A microreactivation of cysts and the generation of cross-reactive hetero- or autoantibodies can be involved in long-lasting IgM (Correa et al. 2007).

In *T. gondii* infection, IgA is produced during the digestive stage. However, IgA can be produced in the eye during intraocular disease in acute or recurrent infection, but not in chronic toxoplasmosis (reviewed in Hegab and Al-Mutawa 2003). IgA response appears prior to any IgG production and is upregulated by IL-10 and TGF-β (Correa et al. 2007). Moreover, IgA can be present in the colostrum, but does not cross the placenta (reviewed in Hegab and Al-Mutawa 2003). In addition, a recent study has shown that IgE induces elimination of intracellular parasites by human macrophages through its ability to trigger CD23 signaling, and this capability is dependent on nitric oxide (NO) and controlled by IL-10 (Vouldoukis et al. 2011).

The Balance Between Proinflammatory and Anti-Inflammatory Response

The balance between the production of proinflammatory (IFN-γ, TNF-α, IL-6, IL-1) and anti-inflammatory (TGF- β and IL-10) cytokines appears to be decisive for the outcome of *T. gondii* infection (reviewed by Lang et al. 2007). IL-10, an anti-inflammatory cytokine, traditionally inhibits proinflammatory responses, controlling cytokine and chemokine production. Neutralization of IL-10 in murine models increases central nervous system inflammation. IL-10-deficient mice lose control of their immune response and die in the acute phases of toxoplasmosis, after uncontrolled IFN-gamma and TNF-α production (reviewed by Aliberti 2005). In addition, IL-10 contributes to the suppression of the T cell function, and it is considered to play a vital role in the control of *T. gondii* immunopathology (reviewed by Tait and Hunter 2009).

Lipoxin, an anti-inflammatory eicosanoid, plays an important role in regulating the immune response to *T. gondii* (reviewed by Machado and Aliberti 2009). Soluble tachyzoites antigen (STAg) initiates lipoxin A4 production, inhibits dendritic cell migration, and hinders in vivo and in vitro IL-12 production (Aliberti et al. 2002a, b).

Recently, it has been determined that IL-27 is a new anti-inflammatory cytokine involved in the regulatory mechanism modulating infection-induced pathology (reviewed by Tait and Hunter 2009). An in vivo study shows that IL-27R-deficient mice generate aberrant IL-2 responses that are associated with fatal toxoplasmosis. Depletion of IL-2 was found to prolong the survival of infected *IL-27R*$^{-/-}$ mice, strongly suggesting that IL-27 limits IL-2 production during Th1 differentiation (Villarino et al. 2006).

The P2X(7) receptor is a transmembrane receptor that is expressed on the surface of a broad range of immune cells, but also in parenchymal cells. The activation of this receptor by extracellular ATP in infected cells can kill intracellular pathogens or may stimulate the production of proinflammatory cytokines in immune cells. The P2X(7) receptor can mediate *T. gondii* death by human and murine macrophages (Jamieson et al. 2010; Lees et al. 2010). However, *T. gondii* infection in the murine model showed that the absence of the P2X(7) receptor did not affect IFN-γ, IL-12, IL-1β, monocyte chemoattractant protein-1 (MCP-1), or TNF production. However, significant and prolonged production of nitric oxide and delayed production of IL-10 in P2X(7) R-deficient mice lead to more susceptibility and weight loss (Miller et al. 2011).

Immune Evasion Strategies

Protozoans have developed mechanisms to escape the immune system of immunocompetent hosts. Some of these mechanisms include antigenic masking and variation, serum factor blocking, intracellular location, and immunosuppression (Seed 1996). Figure 2 summarizes the immune evasion mechanisms involved in *T. gondii* infection in an immunocompetent host.

It has been demonstrated that *T. gondii* interferes with macrophages and the signaling pathways of dendritic cells, where it blocks the nuclear import of transcription factors such as Stat1 and NF-κB. This leads to, among other consequences, inhibition of TNF-α and production of IL-12 by dendritic cells and macrophages, blockade of MHC class II upregulation, and defects in the production of reactive oxygen intermediates (ROI), reactive nitrogen intermediates (RNI), and costimulatory molecules such as CD80 and CD86 (Denkers and Butcher 2005; Luder et al. 2009).

T. gondii infection induces IL-10, lipoxin A4, TGF-β, and IFN-α and IFN-β upregulation and inhibits NO production and p47 GTPases (Lang et al. 2007; Luder et al. 2009). In addition, *T. gondii* could interfere with NO production at

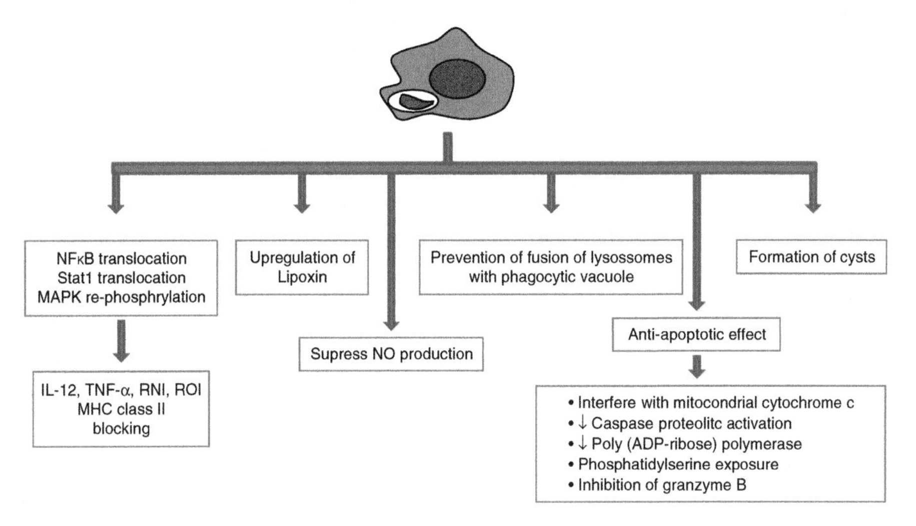

Fig. 2 Different escape mechanisms of *Toxoplasma*-infected cells. Nuclear factor-κB (NF-κB), signal transducers and activator of transcription 1 (Stat1), mitogen-activated protein kinase (MAPK), interleukin 12 (IL-12), tumor necrosis factor-α (TNF-α), reactive nitrogen intermediate (RNI), reactive oxygen intermediate (ROI), major histocompatibility complex class II (MHC class II), nitric oxide (NO), and adenosine diphosphate (ADP) (Based on Seed 1996; Denkers and Butcher 2005; Lang et al. 2007; Luder et al. 2009; Santos et al. 2011)

transcriptional level by reduction of mRNA and protein levels of iNOS. The reduction of NO production in serum has been implicated in triggering a conversion stage in the parasite and protecting the host from immunopathological effects of infection (reviewed in Lang et al. 2007).

In addition, the parasite can suppress the host immune response through induction of apoptosis in CD4+ T cells while inhibiting the apoptosis of infected cells by interfering with the mitochondrial cytochrome c protein, decreasing caspase proteolytic activation, and exposing of phosphatidylserine, among others (reviewed by Denkers and Butcher 2005; Lang et al. 2007; Luder et al. 2009; Santos et al. 2011). Finally, *T. gondii* can avoid the fusion of lysosomes with the phagocytic vacuole and form cysts, thus escaping the host immune system (Seed 1996; Denkers and Butcher 2005).

Toxoplasmosis and the Usual Response in Mammals

The tachyzoite, due to its rapid replication, is the protozoan form involved in triggering *Toxoplasma*-induced tissue necrosis. Thus, the host's ability to prevent, inhibit, or minimize the spread of tachyzoites will determine, in large part, the clinical changes observed in toxoplasmosis. In most hosts, particularly in immunocompetent adult mammalians, *T. gondii* infection has no clinically relevant implications, since the immune response is usually effective in limiting the multiplication of tachyzoites. In the natural feline host, infection is also predominantly subclinical, and clinically important changes are mainly observed in congenitally infected offspring or in adults debilitated by immunosuppressive processes (i.e., animals positive for feline immunodeficiency virus). In these cases, the main clinical signs are fever, anorexia, lethargy, abdominal pain, and neurological and ocular dysfunctions (Elmore et al. 2010). In farm animals, toxoplasmosis is an important cause of abortion in sheep and to a lesser extent in pigs and goats worldwide, particularly in Europe, the UK, and Oceania (Brown and Barker 2007; Klevar 2007; Innes 2010). Infected offspring from several domestic species, including cats, dogs, and pigs, may present widespread interstitial pneumonia, myocarditis, necrotizing hepatitis, meningoencephalomyelitis, lymphadenitis, and myositis (Brown and Barker 2007; Klevar 2007).

Most immunocompetent humans that are infected after birth do not develop relevant clinical symptoms, while few have fever and generalized lymphadenopathy. In immunocompromised people, toxoplasmosis is an important cause of retinochoroiditis, encephalitis, and pneumonia (Elmore et al. 2010).

Toxoplasma gondii Infection, Toxoplasmosis, and New World Primates

Toxoplasma *Epidemiology in NWP*

The first report of toxoplasmosis in New World primates was made in 1916, affecting a specimen of *Stentor seniculus* (*Alouatta seniculus*) (reviewed by Nery-Guimaraes et al. 1971). Since that time, many other reports of this disease in NWP have been made, predominantly affecting captive animals (Hessler et al. 1971; Anderson and McClure 1982; Borst and Vanknapen 1984; Cunningham et al. 1992; Dietz et al. 1997; Pertz et al. 1997; Juan-Salles et al. 1998; Bouer et al. 1999; Epiphanio et al. 2000, 2001, 2003; Andrade et al. 2007; Carme et al. 2009; Cedillo-Pelaez et al. 2011). Although it has been known that NWPs are susceptible to toxoplasmosis for almost 100 years, very little is known about the immune response induced by *T. gondii* in these animals, and only serological data are available in the literature.

Table 2 presents a comparison between major clinical and pathological manifestations of toxoplasmosis in NWP, felids, and humans. In general, toxoplasmosis in NWPs is a disease with a hyperacute clinical course. In most cases, the animals are found dead without prior clinical history. When present, the main clinical findings reported are prostration, dyspnea, hypothermia, nasal foamy serum–bloody exudation, anorexia, and vomiting. At necropsy, in most animals, the macroscopic changes are diverse and occur in multiple organs and tissues and are characterized by severe pulmonary edema and congestion, hepatic congestion with hepatomegaly, splenomegaly, spleen lymphoid hyperplasia, mesenteric and mediastinal fibrin-hemorrhagic

Table 2 Major clinical and pathological manifestations of toxoplasmosis in New Word Primates, felids, and humans

	New Word Primates			Felids		Humans		
	Pattern I (Callitrichinae)	Pattern II (most Cebidae and Atelidae)	Pattern III (*Cebus*)	Immune competent	Immune suppressed	Immune competent	Immune suppressed	Congenitally infected
Clinical manifestations	Hyperacute, markedly severe; mortality close to 100 %. Prostration, dyspnea, nasal foamy exudation, anorexia	Acute, severe, and mortality from 20 % to 80 %. Prostration, anorexia, hypothermia, dyspnea, vomiting	Subacute, mild, with very low mortality rate	Subclinical or mild self-limiting diarrhea	Congenitally or lactationally infected offspring; stillborn or lethargy depression, hypothermia, ascites, hepatomegaly, chorioretinitis, and sudden death Older cats with immunosuppressive processes: anorexia, lethargy, dyspnea, persistent/intermittent fever	Most subclinical; few develop fever and lymphadenopathy. Debilitating ocular disease has been reported in few cases	Fever, headache, myalgia, anorexia, fatigue, abdominal pain, vomiting, nausea, dyspnea, arthralgia. Lymphadenopathy	Asymptomatic to fatal. Prematurity, intrauterine growth retardation, debilitating ocular disease, strabismus, psychomotor impairment, prostration, microcephalus, hydrocephalus, convulsion, icterus, hypotonia, and hepatomegaly
Pathological findings	Severe fibrin-hemorrhagic necrotizing pneumonia, hepatitis, splenitis, enteritis, and lymphadenitis	Severe fibrin-hemorrhagic necrotizing pneumonia, hepatitis, splenitis, enteritis, and lymphadenitis	Mild and nonspecific	Nonspecific	Moderate to severe necrotizing hepatitis, lymphadenitis, and pneumonia. Neuronal necrosis and meningitis	Retinochoroiditis	Encephalitis	Retinochoroiditis, anterior uveitis, and encephalitis

Based on Anderson and McClure (1982), Elmore et al. (2010) Epiphanio et al. (2003), Silva (2007), Epiphanio and Catão-Dias (unpublished data)

lymphadenitis, gastric ulceration, hemorrhagic enteritis, brain edema, and congestion. Microscopically, the main reported changes are fibrin-hemorrhagic interstitial pneumonia, fibrin-necrotic splenitis, multifocal necrotizing hepatitis, fibrin-hemorrhagic lymphadenitis, and multifocal necrotic-hemorrhagic enteritis (Anderson and McClure 1982; Cunningham et al. 1992; Dietz et al. 1997; Juan-Salles et al. 1998; Epiphanio et al. 2000, 2001, 2003; Andrade et al. 2007; Carme et al. 2009; Cedillo-Pelaez et al. 2011). However, detailed studies of the macro- and microscopic pathological findings resulting from toxoplasmosis in NWPs showed that there is significant variation in the types and intensity of lesions, depending on the Platyrrhini species affected. Necroscopy, histopathological, histochemical, immunohistochemical, morphometry, and ultrastructural analyses of 33 cases of toxoplasmosis in NWPs in captivity (11 Atelidae, i.e., 6 *A. fusca* and 5 *L. lagotricha*; 22 Cebidae, i.e., 18 Callitrichinae, 3 Saimirinae, and 1 Aotinae) showed that Atelidae presented pathological changes more variable and pleomorphic than Cebidae, making the preliminary necroscopic diagnosis more difficult. In particular, jaundice was observed only in *A. fusca*, affecting 50 % (3/6) of animals evaluated (Epiphanio et al. 2003).

A relevant aspect that stands out in the analysis of several reports of toxoplasmosis in Platyrrhini is the scarcity of information on the occurrence of fatal clinical outcomes involving the genus *Cebus*. The cases described include the death of one specimen of *C. apella*[2] and one of *C. capucinus* due to natural infection (De Rodaniche 1954; Nery-Guimaraes and Franken 1971).

Serological investigation shows that, similar to other intermediate hosts, Platyrrhini species apparently exhibit distinct humoral responses against infection by *T. gondii*. Table 3 presents a summary of epidemiological surveys of NWPs for *Toxoplasma*. When surveys conducted with apparently healthy animals are considered, and excluding data for a very small number of individuals from genera *Aotus* and *Ateles* (Bouer et al. 2010), the frequency of animals with positive titer of anti-*Toxoplasma gondii* antibodies is generally higher for genus *Cebus*. In *Cebus* kept in captivity, the frequencies ranged from 28.7 % to 79 %, while a frequency of 30.2 % in free-living *Cebus* has been reported (Garcia et al. 2005; Leite et al. 2008; Bouer et al. 2010).

[2]The capuchin monkeys are particularly complex in their taxonomy. For many years, taxonomic arrangements reduced all tufted capuchin monkeys to just one species, *Cebus apella*, with 11 and 16 (Hill 1960) subspecies. The most recent revisions, by Groves (2001) and Silva (2001), both based on morphology, differently recognized, as species, the following: *apella* and *macrocephalus* in the Amazon and *libidinosus*, *nigritus*, *robustus*, *cay*, and *xanthosternos* to the south. Groves (2001) presented an alternative as follows: Amazon forms *C. apella apella*, *C. a. fatuellus*, *C. a. macrocephalus*, *C. a. peruanus*, and *C. a. tocantinus* and southern forms *C. libidinosus libidinosus*, *C. l. pallidus*, *C. l. paraguayanus*, *C. l. juruanus* (Amazonian), *C. nigritus nigritus*, *C. n. robustus*, *C. n. cucullatus*, and *C. xanthosternos* (see Fragaszy et al. (2004) and Rylands et al. (2005)). In Brazil, captive capuchins (independently of origin) are generally named as *Cebus apella*, and in many occasions and in different institutions, individuals of different subspecies or species are kept together generating hybrid groups (M.C.M. Kierulff, personal observation). Because of all these problems, we decided to maintain the original names used for *Cebus* species cited in the references, mostly named just as *Cebus apella* with no distinction to subspecies, even known that it refers to species other than the Amazonians.

Table 3 Seroprevalence of *Toxoplasma gondii* in New World Primates

Genus/species	Local	Animal tested (*n*)	Test	Captive/free ranging	Positive test (%)	Reference
S. oedipus oedipus	South America	100	SFR	FR	0	Werner et al. (1969), Nery-Guimaraes and Franken (1971)
S. sciureus	Brazil	17	SFR	Captive/FR	17,6	Nery-Guimaraes and Franken (1971)
C. apella		26			15,3	
A. belzebuth		1			0	
Aotus sp.		1			0	
A. geoffroyi		1			0	
C. jacchus		2			0	
C. penicillata		5			0	
L. lagotricha		1			0	
C. apella	Brazil	5	SFR	FR	60	Sogorb et al. (1972)
A. fusca		12			42,1	
Saimiri spp.	Brazil	49	IHT	FR	63,3	Ferraroni and Marzochi (1980)
C. apella	Colombia	10	NA	Captive	0/0	Cadavid et al. (1991)
C. capucinus		15			13,3	
C. albifrons		22			40,9	
Saimiri sciureus	England	4[a]	IFA	Captive	100/75	Cunningham et al. (1992)
		11#			100/54,5	
A. seniculus	French Guiana	50	DA	FR	4	De Thoisy et al. (2003)
S. midas		50			0	
C. apella	Brazil	43	MAT	FR	30,2	Garcia et al. (2005)
A. caraya		17			17,6	
L. lagotricha	USA	2[a]	LA	Captive	100	Gyimesi et al. (2006)
		13	IHT MAT		0/0/0	
C. apella	Brazil	14	IFAT	Captive	28.7	Leite et al. (2008)
		13	MAT		30.8	

(continued)

Table 3 (continued)

Genus/species	Local	Animal tested (*n*)	Test	Captive/free ranging	Positive test (%)	Reference
Alouatta caraya	USA	1	MAT	Captive	0	de Camps et al. (2008)
Ateles geoffroyi		4			0	
Callicebus moloch		5			0	
donacophilus		3			0	
Callimico goeldii		1			0	
Callithrix kuhlii		4			0	
L. lagotricha		1			0	
L. chrysomelas		3			0	
L. rosalia		8			0	
Pithecia pithecia		6			0	
Saguinus geoffroyi						
S. sciureus	Israel	24[a]	MAT	Captive	83,3	Salant et al. (2009)
C. apella	Brazil	105	IFAT	Captive	79	Bouer et al. (2010)
Callithrix sp.		42			26,2	
Alouatta sp.		20			50	
Leontopithecus		15			20	
sp. *Ateles* sp.		7			57,14	
Saimiri sp.		6			33,33	
Saguinus sp.		5			0	
Aotus sp.		3			66,66	
Lagothrix sp.		3			0	

C. jacchus	Brazil	25	LA	Captive	0	Epiphanio & Catão-Dias, unpublised data
C. penicillata		18			0	
C. geoffroyi		8			0	
C. aurita		2			0	
C. kuhlii		1			0	
Callithrix sp.		1			0	
S. bicolor		2			0	
S. midas		3			0	
S. niger		2			0	
L. chrysomelas		46			2,2	
L. rosalia		1			0	
L. chrysopygus		1			0	
C. apella		100			73	
A. marginatus		2			100	
A. belzebuth		1			0	
A. paniscus		5			20	
Ateles sp.		3			100	
Saimiri sciureus		6			0	
L. lagotricha		2			0	
A. caraya		7			42,9	
A. fusca		5			0	
Alouatta sp.		5			0	
Aotus sp.		2			50	

IFAT Indirect immunofluorescence, *MAT* Modified Agglutination Test, *IHT* indirect heamagglution test, *LA* latex agglutination, *PCR* polymerase chain reaction, *SFR* Sabin-Feldman reaction, *DA* direct agglutination method

[a]died; # survivors

Moreover, the data linking the evolution of the serological profile with the occurrence of toxoplasmosis outbreaks in NWPs can provide interesting information for understanding this process. At the London Zoo, one-third of the *S. sciureus* colony died of toxoplasmosis, and serological surveys showed that most animals had titers indicative of recent infection. IgG was detected in 11 surviving animals and IgM in six individuals from this group (Cunningham et al. 1992). In another outbreak that caused the death of 24 *S. sciureus* in Israel, 83.3 % of the animals had positive serology, suggesting that humoral immunity is not an effective defense mechanism for sudden toxoplasmosis or reinfection in this NWP species (Salant et al. 2009). Similarly, *Lagothrix lagotricha* individuals who died of sudden toxoplasmosis showed positive serology for anti-*Toxoplasma gondii* antibodies in three distinct types of tests (latex agglutination, indirect hemagglutination, modified agglutination), besides identification of *T. gondii* by PCR (Gyimesi et al. 2006).

Experimental infections have been performed in an attempt to better understand the role of the humoral response in the development of toxoplasmosis in NWPs. In one experiment, 28 *Saguinus* sp. were infected and died within few days, without the detection of specific antibodies anti-*Toxoplasma gondii* (Werner et al. 1969). In another study, five *C. apella* individuals were infected with *T. gondii* (N strain, type II tachyzoite suspension—1×10^5 mL—by intraperitoneal route) and exhibited nonspecific and mild clinical signs for only 3 days postinfection and were euthanized 102 days postinfection. Macroscopic and microscopic lesions observed were mild and were not correlated with toxoplasmosis. However, anti-*Toxoplasma gondii* IgG titers were detected by IFA and ELISA 9 days postinfection and lasted until the end of the investigation (Bouer et al. 2010). In another experiment, *S. sciureus* orally infected died approximately 1 week postinfection; however, anti-*Toxoplasma* immunoglobulin titers were not detected by immunoblot (Furuta et al. 2001).

Response Patterns to T. gondii *Infections in NWP*

Serological, clinical, pathological, and experimental data available on toxoplasmosis in NWPs suggest three distinct response patterns to *T. gondii* infection: patterns I, II, and III. Pattern I is that observed mainly in Callitrichinae. In these animals (*Saguinus, Leontopithecus, Callithrix*), the disease is markedly severe, with mortality close to 100 %, which makes the occurrence of animals with positive serology to be very low or zero. Pattern II involves a diverse group of NWPs from families Cebidae (*Saimiri, Aotus*) and Atelidae (*Alouatta, Ateles, Lagothrix*), being characterized by the occurrence of severe outbreaks, with variable mortality, but with the survival of a reasonable number of individuals with positive serology in the population (from 15 % to 66 %), particularly *Saimiri* and *Alouatta*. In both patterns, despite the pleomorphism observed in some Atelidae (Epiphanio et al. 2003), the lesions are predominantly consistent and characterized by multifocal, coalescing, severe, and multisystemic necrosis, associated with the presence of large amounts of tachyzoites. In turn, pattern III is observed in the genus *Cebus*, which differs greatly from those seen in most other platyrrhines. In *Cebus*, the infection tends to induce high and persistent IgG titers, the animals rarely die, and the morphological changes seen in experimental cases are mild and nonspecific (Bouer et al. 2010).

The variable prevalence of anti-*Toxoplasma gondii* antibodies in NWPs, and especially the very low frequency of antibodies in Callitrichinae, may explain the very high susceptibility of these animals to infection. Callitrichinae, specifically, die rapidly during the acute phase of infection, before IgG production is initiated. On the other hand, the high IgG titers observed in *Cebus*, associated with rare reports of death in this genus, suggests that these animals are able to establish an efficient antibody immune response against the infection.

Over the course of 20 years at the Laboratory of Comparative Pathology of Wildlife (LAPCOM—Laboratório de Patologia Comparada de Animais Selvagens) of the FMVZ–USP, the authors have witnessed at least six major toxoplasmosis outbreaks in NWP. These outbreaks mainly occurred among the genera *Callithrix, Leontopithecus, Lagothrix, Alouatta, Saimiri,* and *Saguinus*. In these outbreaks, all individuals who exhibited clinical signs subsequently died. In at least two of the outbreaks, diseased animals (*Leontopithecus chrysopygus, L. rosalia, L. chrysomelas*) were submitted to recommended therapeutic procedures (Osborn and Lowenstine 1998), but did not survive. Significantly, *Cebus* specimens were not affected in any of the outbreaks followed by LAPCOM, although these animals constitute the large majority of the NWPs populations kept in captivity in the zoos involved.

The clinical and serological aspects of toxoplasmosis (described above) for most NWPs, especially Callitrichinae, resemble those reported for the mountain hare (*Lepus timidus*). These Eurasian animals are especially susceptible to the disease and develop severe acute symptoms that are often fatal. Experimental studies have shown that mountain hares have very low antibody titers and inefficient proliferation of T lymphocytes when exposed to *T. gondii*, compared with those observed in domestic rabbits (*Oryctolagus cuniculus*) (Gustafsson et al. 1997), suggesting that the hares are incompetent to establish an efficient adaptive immune response against the parasite.

Other aspects that deserve to be addressed in understanding differing susceptibility of NWP species to *T. gondii* include the variability of strains, infective doses, and the possibility of recrudescence of cysts. The recent characterization of different *T. gondii* strains shows that there is significant genetic diversity that was previously unknown (Dubey 2009; Dubey and Su 2009). In addition, it is known that the main three known strains can induce different patterns of the disease in humans (Saeij et al. 2005).

Furthermore, studies investigating the types of strains involved in cases of toxoplasmosis in NWPs are rare and are restricted to four outbreaks in *Saimiri*: two in French Guiana, with the identification of strains II and III and atypical alleles (Carme et al. 2009); one in Israel involving strain III (Salant et al. 2009); and one in Mexico involving strain I (Cedillo-Pelaez et al. 2011). The high susceptibility of *Saimiri* to the three main *T. gondii* strains suggests that, at least for this NWP species, the genetic variability of the parasite may not be a determining factor for disease manifestation.

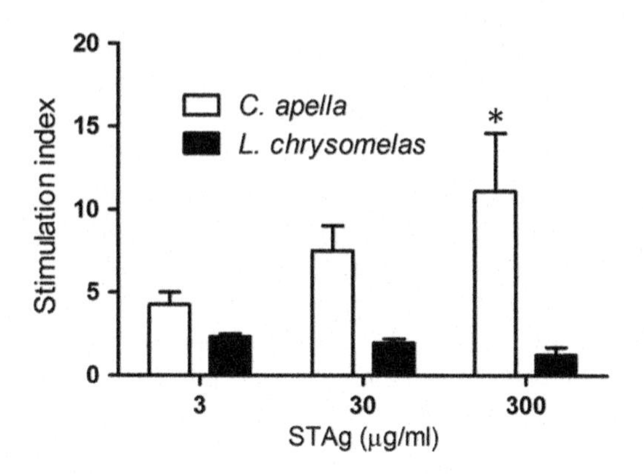

Fig. 3 Peripheral blood mononuclear cells proliferation in response to soluble *Toxoplasma gondii* tachyzoite antigen (STAg). *C. apella* $n=2$ and *L. chrysomelas* $n=2$; $^{*}p \leq 0.05$, two-way ANOVA test

In humans, the immunocompromised host (patients with AIDS, organ transplantation, cancer, or taking immunosuppressants) may experience cyst reactivation, with bradyzoites transforming back into tachyzoites, leading to infection recrudescence and life-threatening encephalitis (Montoya and Liesenfeld 2004). Toxoplasmosis recrudescence has been demonstrated in other hosts, such as in immunosuppressed *Rattus norvegicus* after experimental treatment with dexamethasone (Silva 2007) and in dogs and cats undergoing renal transplantation (Bernsteen et al. 1999). It has already been suggested that captivity stress, as well as immunosuppression, may lead to reactivation of *Toxoplasma* cysts in NWPs (Gyimesi et al. 2006). Thus, one possibility to explain, at least partially, the existing data is that part of Atelidae and Cebidae (except *Cebus*) with positive serology for *T. gondii* could be chronically infected and, in response to diverse immunosuppressant stimuli, could undergo infection reactivation with the development of clinical disease and death. Another possibility is that animals were reinfected with a strain more virulent than the previous one, or with higher infective doses, resulting in acute toxoplasmosis. In these cases, it is possible to suggest that these animals would respond rapidly with the production of antibodies, progressing to resistance or death.

To investigate differences in cellular immune response against *T. gondii*, proliferation assays were performed with peripheral blood mononuclear cells (PBMC) from two *Cebus apella* and two *L. chrysomelas*. These species were selected because they represent the extremes of the NWP susceptibility to *T. gondii,* taking into account the epidemiological, serological, clinical, and pathological records of LAPCOM. PBMCs were stimulated with varying concentrations (3, 30, and 300 μg/ml) of soluble antigens of *T. gondii* tachyzoites (STAg) and analyzed by liquid scintigraphy. Positive controls were stimulated with phytohemagglutinin (PHA). Our preliminary stimulation indices (SI) show that stimulation with STAg is dose dependent in *C. apella,* but not in *L. chrysomelas* (Fig. 3). Moreover, the SI were 1.8-, 5.28-, and 6.64-fold higher in *C. apella* when compared with *L. chrysomelas* (Epiphanio and Catão-Dias, unpublished data).

These results, although preliminary and involving a small number of samples, suggest that the immune system of *Cebus* is more able to respond to infection by

T. gondii, by more efficiently controlling the spread of pathogens and allowing the establishment of a more balanced host–parasite relationship similar to the vast majority of warm-blooded hosts infected with *T. gondii*. Moreover, the inadequate proliferative response observed in *L. chrysomelas* would prevent the emergence of an efficient cellular response against the infection, favoring the development of fatal infections.

Our results, associated with those published by other researchers, raise several questions about the mechanisms involved in the unsatisfactory cellular immune response shown by the vast majority of NWPs against *T. gondii,* such as:

1. Would the inflammatory response induced by *T. gondii* in NWPs be intensified to the point of leading to the overproduction of IFN-γ, TNF-α, IL-6, IL-12, or IL-1β?
2. Could there be overstimulation of TLR, with consequent amplification of the expression of the transcription factor NF-kB?
3. Could regulatory mechanisms, such as anti-inflammatory cytokines TGF-β, IL-10, and lipoxins, be downregulated?
4. Could there be disturbances in the fusion of the parasitophorous vacuole with lysosomes, favoring the survival of *T. gondii* in cells infected by *T. gondii* in NWPs?
5. Could the apoptosis mechanisms of *Toxoplasma*-infected cells be deficient?
6. Could the mechanisms of NO production be lacking?

Currently, many sequences of genes encoding proteins related to the immune system of platyrrhines are known (Table 4). Hopefully in the near future, these tools can be used to clarify some of the many unanswered questions regarding the toxoplasmosis immunology in NWPs.

Toxoplasma gondii Infection and NWPs' Response: Approaching the Susceptibility Differences from Ecological and Behavioral Perspectives

Approximately three million years ago, Panamanian land bridge formed and allowed the immigration of many intermittent "invaders" from the North America into the South America, including carnivores (e.g., felids, canids, and mustelids) over the course of the Pleistocene (Webb 1976; Simpson 1980; Marshall 1988). At that time, *T. gondii* may also have migrated into the continent with carnivore species (reviewed in Sibley et al. 2009). In fact, *T. gondii* strains from North and South America share a common ancestry, and it was estimated that they last shared a common ancestor one million years ago (Sibley et al. 2009).

Platyrrhines, however, have a long evolutionary history preceding the migration of carnivore species into South America. The early platyrrhine fossils come from the Late Oligocene in Bolivia (24–28 million years ago). Indeed, species related to the living Neotropical primates were present in Colombia, during the Middle to Late Miocene, suggesting a common platyrrhine ancestor in the Late Oligocene or Early Miocene (14–24 million years ago) (Fleagle and Tejedor 2002). Recent

Table 4 List of selected genes related to immune response in New World Primates

Species	Gene	GenBank accession
Aotus infulatus	IL-12B	DQ989359.1
Aotus lemurinus	TNF-a	AF097329
Aotus lemurinus	IFN-γ	AF097327.1
Aotus nancymaae	CD4	FJ623078.1
Aotus nancymaae	TLR9	AY788894.1
Aotus nancymaae	IFN-γ	AF014512.1
Aotus nigriceps	TNF-a	AF097328
Aotus trivirgatus	CD40 ligand	AF344860.1
Aotus trivirgatus	MHCI	AB113205.1
Aotus vociferans	TNF-a	AF014508
Aotus vociferans	IL-10	AAD01532.1
Aotus vociferans	IFN-γ	AF014507.1
Ateles belzebuth	TLR4	AB446521
Ateles belzebuth	MHCI	AB113112
Ateles geoffroyi	IL-10	ABM65916.1
Ateles geoffroyi	TLR4	AB446522
Callicebus moloch	MHCII	AF197231.1
Callithrix jacchus	IL-12B	AB539805.1
Callithrix jacchus	iNOS	AM712438
Callithrix jacchus	CD4	AF452616.1
Callithrix jacchus	CD8	DQ189217
Callithrix jacchus	IL-1a	AB539804
Callithrix jacchus	TNF-a	DQ520835
Callithrix jacchus	TLR4	AB446516
Callithrix jacchus	TLR9	XM_002758237
Callithrix jacchus	IL-27	XM_002756059.1
Callithrix jacchus	CD40	DQ189221.1
Callithrix jacchus	p47 GTPase	XM_002762275
Callithrix jacchus	MyD88	XM_002759734
Callithrix jacchus	P2X(7)	XM_002753098
Callithrix jacchus	MHCII	AF197230.1
Cebus apella	TLR4	AB446520.1
Leontopithecus rosalia	TLR4	AB446518
Saguinus imperator	TLR5	FJ542217
Saguinus labiatus	MHCII	JF414576.1
Saguinus mystax	IFN-γ	FJ598592.1
Saguinus oedipus	TNF	AY091968
Saguinus oedipus	TLR4	AB446517.1
Saguinus oedipus	TLR2	EU488857.1
Saguinus oedipus	MHCII	AF197226.1
Saimiri sciureus	IL-12B	DQ989358.1
Saimiri sciureus	CD4	AF452617
Saimiri sciureus	CD8	AJ130819
Saimiri sciureus	IL-1b	AF294754
Saimiri sciureus	TNF-a	AJ437697
Saimiri sciureus	TNF	DQ989365
Saimiri sciureus	TLR4	AB446519
Saimiri sciureus	IL-10	Q8MKG9.1
Saimiri sciureus	IFN-γ	AF414102.1

phylogenetic analysis using DNA samples of primate species showed that Platyrrhini diverged from a last common ancestor with Catarrhini 43.5 million years ago during the Eocene. The common ancestor to Pitheciidae originated 20.2 million years ago, and the Cebidae radiation initiated with the emergence of Cebinae and Saimirinae approximately 20 million years ago (Perelman et al. 2011). These data indicate that nonhuman primates were already in South America at the time of the felid (possibly infected with *T. gondii*) invasion, from the north hemisphere.

The high susceptibility of NWPs to toxoplasmosis has been known for almost 100 years. The main hypotheses proposed to explain this condition can be summarized as follows: (a) NWPs have evolved for over 20 million years without the presence of felids and therefore would not have acquired adaptations to the pathogen over this time, especially cellular responses to *T. gondii* (Cunningham et al. 1992), and (b) even after the arrival of felids (possibly infected with *T. gondii*) to the Neotropics, the arboreal habits of NWPs would have restricted their contact with the feces of felids infected by the protozoan oocysts, limiting the development of efficient immune response (Innes 1997). The available data compiled in this review corroborates many of the assumptions covered above to justify the high susceptibility of some NWPs to toxoplasmosis. On the other hand, they do not explain the significant differences observed between groups of NWPs, in particular the high resistance reported for *Cebus*.

To some extent, it is possible to assess how a particular animal explores and occupies an environment in relation to the intensity and diversity of the parasite load it carries. A study conducted in Costa Rica showed *C. capucinus* with higher parasite infestation in fecal samples than *Alouatta* and *Ateles* (Stuart et al. 1998). Fragaszy et al. (2004) suggested that at least three characteristics of *Cebus* behavior may lead to higher parasite infestation in these monkeys than in sympatric *Alouatta* and *Ateles*: *Cebus* drinks from water holes, frequently forages on the ground, and eats a wider variety of foods. These behaviors may bring *Cebus* into contact with greater variety of parasites. Considering the epidemiological characteristics of *T. gondii*, it is possible to speculate that the aspects described above could justify a diverse exposure of NWPs to the protozoan.

Contaminated water can be a source of toxoplasmosis, and this behavior has already been recorded for other *Alouatta* species that drink water in holes in branches or trunks or in bromeliads (Glander 1978; Gilbert and Stouffer 1989; Bicca-Marques 1992; Giudice and Mudry 2000) or go to the ground to drink water (Almeida-Silva et al. 2005). Other Platyrrhini such as *Aotus* (Wright 1981), *Saimiri* (Baldwin and Baldwin 1981), *Cebuella* (Soini 1988), *Saguinus* (Snowdon and Soini 1988), and *Brachyteles* (Mendes et al. 2010) have also been observed drinking on the banks of streams and/or rivers during the dry season.

Cebus sp. comes down to the ground more frequently in comparison to other NWPs where they may be more frequently exposed to excreted *T. gondii*. Recent studies have reported the use of tools (stones used as hammers and anvils) to open hard nuts by wild *Cebus* in places where the groups (from different species) use the ground more frequently (Fragaszy et al. 2004; Canale et al. 2009). Robinson (1984) found that many of the invertebrates consumed by *C. olivaceus* were found on the

Fig. 4 Group of yellow-breasted capuchin monkeys (*Cebus xanthosternos*) eating a bristle-spined rat pup (*Chaetomys subspinosus*) they had just caught, Una Biological Reserve, Bahia, Brazil (Photo by Jean Marc Lernould)

ground. They visually search the leaf litter and sweep the leaves to reveal the insects hidden underneath (Fragaszy et al. 2004).

Additionally, in our opinion, an important behavioral characteristic of *Cebus* is that it is the most carnivorous of the platyrrhines. It has been often noted that the consumption of invertebrates and small warm-blooded vertebrates plays a role in the transmission of *T. gondii* to NWPs (Epiphanio et al. 2003; Carme et al. 2009; Salant et al. 2009). Most of the protein in *Cebus* diets comes from invertebrates (insects and other arthropods), but while hunting for invertebrates, they sometimes find vertebrates that they capture and consume. It has been widely reported that they capture and consume a variety of relatively large vertebrates that may weigh up to one third the *Cebus* body weight and may constitute up to 3 % of their feeding time (Fragaszy et al. 2004). Along with chimpanzees (*Pan troglodytes*), *Cebus* are one of the few nonhuman primate species that have been reported to hunt vertebrate prey in more than an occasional, incidental manner (Fragaszy et al. 2004).

The types of vertebrate prey that *Cebus* has been reported to consume include birds and their nestlings and eggs, lizards, frogs, rodents, bats, squirrels (*Sciurus variegatoides*), coati pups (*Nasua narica*), and infant titi monkeys (*Callicebus moloch*) (Izawa 1978; Terborgh 1983; Fedigan 1990; Galetti 1990; Rose 1997; Sampaio and Ferrari 2005). *Cebus xanthosternos* from the Una Reserve, Bahia, Brazil, have been seen preying upon bristle-spined rat pups (*Chaetomys subspinosus*) on four occasions (Priscilla G. Suscke, personal communication) (Fig. 4). In a forest in Rio de Janeiro, a bamboo rat (*Kannabateomys amblyonyx*) was found preyed upon immediately following a passage of a *Cebus nigritus* group

(M.C.M. Kierulff, personal observation). Ferreira et al. (2002) described predation on birds by a group of *Cebus* at the Tietê Ecological Park, São Paulo. Resende et al. (2003) reported the same semi-free-ranging group eating an adult male rat (*Rattus rattus*) and an infant opossum (*Didelphis* sp.). These and all other warm-blooded vertebrates that may be consumed by *Cebus* are potential source of *T. gondii* infection.

As previously described, other NWP species show a variety of items in their diet, and sometimes they may even prey on small vertebrates such as birds and rodents. Neotropical primates, despite their preference for forest strata, do forage or move around for short distances on the ground. However, none have the sophisticated hunting techniques, so carnivorous diet, and spend as much time on the ground as the genus *Cebus*. Due to their behavior and ecology, *Cebus* seems to be the platyrrhine with the greatest access to *T. gondii* in nature. We believe that the hunting and exploratory habits of *Cebus* promotes frequent interactions with the protozoa and may have led this monkey genus to select and develop a more effective immune response and consequent resistance to *T. gondii*.

Final Comments and Research Perspectives

Most NWPs are very susceptible to toxoplasmosis, and according to the authors' experience, this is possibly the most important cause of acute death of infectious origin affecting Platyrrhini in captivity in Brazil. However, the susceptibility of NWPs to *T. gondii* is variable, with Callitrichinae (*Callithrix*, *Saguinus*, and *Leontopithecus*) showing mortality rate close to 100 %, Atelidae and some Cebidae (*Saimiri*, *Aotus*) showing variable mortality patterns, and genus *Cebus* showing high resistance, with rare deaths reported due to toxoplasmosis.

The reasons for the high susceptibility of most NWPs to *T. gondii* are not clear. We believe that it may be due, at least in part, to ecological and behavioral characteristics of different NWPs that led to different degrees of exposure to *T. gondii* over evolutionary time. It is possible to speculate that such variable exposure to the protozoa may have led, along the evolutionary process of NWPs, to differentiated immunological features culminating in the relative ability to resist the infection.

Naturally, there are many unanswered questions to investigate and novel areas to research regarding *Toxoplasma*–NWP interactions. To better understand Toxoplasmosis manifestation in NWPs, we believe certain studies are very important including further research on NWP cellular (proliferative assays, role and measurement of cytokines) and humoral immune responses to the pathogen (more comprehensive serological surveys, both in captivity and in the wild; use and validation of different techniques) as well as molecular epidemiology of *T. gondii* (characterization of strains and their environmental distribution). Obtaining new information in these areas will certainly help clarify questions about NWP–*T. gondii* interactions.

Finally, considering our laboratory has witnessed on several occasions the devastating effect that toxoplasmosis can have on ex situ conservation programs for NWPs, we would like to emphasize the importance of curatorial/zoological institutions adopting the best management practices. We see such policies as the only effective option, currently, for the prevention of new outbreaks that can, otherwise, decimate genetically invaluable populations of NWPs.

Acknowledgments We are grateful to the staff at LAPCOM–FMVZ/USP, UNIFESP, and UFES; without their efforts, this chapter would not have been possible. In particular, we would like to acknowledge the continuous financial support from *Fundação de Amparo à Pesquisa do Estado de São Paulo* (FAPESP) and *Conselho Nacional de Desenvolvimento Científico e Tecnológico* (CNPq). José Luiz Catão-Dias is a recipient of a scholarship by the CNPq (301517/2006-1).

References

Aguiar JM, Lacher TE Jr (2003) On the morphological distinctiveness of *Callithrix humilis* Van Roosmalen et al. 1998. Neotrop Primates 11:11–18

Alfaro JWL, Boubli JP, Olson LE, Di Fiore A, Wilson B, Gutierrez-Espeleta GA, Chiou KL, Schulte M, Neitzel S, Ross V, Schwochow D, Nguyen MTT, Farias I, Janson CH, Alfaro ME (2012a) Explosive Pleistocene range expansion leads to widespread Amazonian sympatry between robust and gracile capuchin monkeys. J Biogeogr 39:272–288

Alfaro JWL, Silva JS Jr, Rylands AB, Boubli JP (2012b) How different are robust and gracile capuchin monkeys? An argument for the use of *Sapajus* and *Cebus*. Am J Primatol 74(4):273–286. doi:10.1002/ajp.22007:1-14

Aliberti J (2005) Host persistence: exploitation of anti-inflammatory pathways by *Toxoplasma gondii*. Nat Rev Immunol 5:162–170

Aliberti J, Hieny S, Reis e Sousa C, Serhan CN, Sher A (2002a) Lipoxin-mediated inhibition of IL-12 production by DCs: a mechanism for regulation of microbial immunity. Nat Immunol 3:76–82

Aliberti J, Serhan C, Sher A (2002b) Parasite-induced lipoxin A4 is an endogenous regulator of IL-12 production and immunopathology in *Toxoplasma gondii* infection. J Exp Med 196:1253–1262

Almeida-Silva B, Guedes PG, Boubli JP, Strier KB (2005) Deslocamento terrestre e o comportamento de beber em um grupo de barbados (*Alouatta guariba clamitans* Cabrera, 1940) em Minas Gerais, Brasil. Neotrop Primates 13:1–3

Anderson DC, McClure HM (1982) Acute disseminated fatal toxoplasmosis in a squirrel monkey. J Am Vet Med Assoc 181:1363–1366

Andrade MCR, Coelho JMCO, Amendoeira MRR, Vicente RT, Cardoso CVP, Ferreira PCBF, Marchevsky RS (2007) Toxoplasmosis in squirrel monkeys: histological and immunohistochemical analysis. Ciência Rural 37:1724–1727

Baldwin JD, Baldwin JI (1981) The squirrel monkeys, genus *Saimiri*. In: Coimbra-Filho AF, Mittermeir RA (eds) Ecology and behavior of neotropical primates. Academia Brasileira de Ciências, Rio de Janeiro, pp 241–276

Bernsteen L, Gregory CR, Aronson LR, Lirtzman RA, Brummer DG (1999) Acute toxoplasmosis following renal transplantation in three cats and a dog. J Am Vet Med Assoc 215:1123–1126

Bicca-Marques JC (1992) Drinking behavior in the black howler monkey (*Alouatta caraya*). Folia Primatol (Basel) 58:107–111

Bicca-Marques JC, Silva VM, Gomes DF (2011) In: Reis NR, Perachi AL, Pedro WA, Lima IP (eds) Mamíferos do Brasil, 2nd edn. Londrina, PR. pp. 107–150.

Blader IJ, Saeij JP (2009) Communication between *Toxoplasma gondii* and its host: impact on parasite growth, development, immune evasion, and virulence. APMIS 117:458–476
Blanchard N, Gonzalez F, Schaeffer M, Joncker NT, Cheng T, Shastri AJ, Robey EA, Shastri N (2008) Immunodominant, protective response to the parasite *Toxoplasma gondii* requires antigen processing in the endoplasmic reticulum. Nat Immunol 9:937–944
Boinski S (1987) Mating patterns in squirrel-monkeys (Saimiri-Oerstedi) - implications for seasonal sexual dimorphism. Behav Ecol Sociobiol 21:13–21
Boothroyd JC (2009) *Toxoplasma gondii*: 25 years and 25 major advances for the field. Int J Parasitol 39:935–946
Bordignon MO, Setz EZF, Caselli CB (2008) Gênero *Callicebus* Thomas 1903. In: Reis NR, Perachi AL, Andrade FR (eds) Primatas brasileiros. Technical Books, Londrina, pp 153–166
Borst GHA, Vanknapen F (1984) Acute acquired toxoplasmosis in primates in a zoo. J Zoo Wildl Med 15:60–62
Bouer A, Werther K, Catao-Dias JL, Nunes AL (1999) Outbreak of toxoplasmosis in *Lagothrix lagotricha*. Folia Primatol (Basel) 70:282–285
Bouer A, Werther K, Machado RZ, Nakaghi AC, Epiphanio S, Catao-Dias JL (2010) Detection of anti-*Toxoplasma gondii* antibodies in experimentally and naturally infected non-human primates by Indirect Fluorescence Assay (IFA) and indirect ELISA. Rev Bras Parasitol Vet 19:26–31
Bradley PJ, Ward C, Cheng SJ, Alexander DL, Coller S, Coombs GH, Dunn JD, Ferguson DJ, Sanderson SJ, Wastling JM, Boothroyd JC (2005) Proteomic analysis of rhoptry organelles reveals many novel constituents for host-parasite interactions in *Toxoplasma gondii*. J Biol Chem 280:34245–34258
Brown CCB, Barker DC (2007) Alimentary system. In: Maxie MG (ed) Pathology of domestic animals. Elsevier, Philadelphia, PA, pp 1–296
Brown AD, Zunino GE (1990) Dietary variability in Cebus apella in extreme habitats—evidence for adaptability. Folia Primatol 54:187–195
Buzoni-Gatel D, Werts C (2006) *Toxoplasma gondii* and subversion of the immune system. Trends Parasitol 22:448–452
Cadavid AP, Canas L, Estrada JJ, Ramirez LE (1991) Prevalence of anti-*Toxoplasma gondii* antibodies in Cebus spp in the Santa Fe Zoological Park of Medellin, Colombia. J Med Primatol 20: 259–261
Canale GR, Guidorizzi CE, Kierulff MC, Gatto CA (2009) First record of tool use by wild populations of the yellow-breasted capuchin monkey (*Cebus xanthosternos*) and new records for the bearded capuchin (*Cebus libidinosus*). Am J Primatol 71:366–372
Carme B, Ajzenberg D, Demar M, Simon S, Darde ML, Maubert B, de Thoisy B (2009) Outbreaks of toxoplasmosis in a captive breeding colony of squirrel monkeys. Vet Parasitol 163:132–135
Cedillo-Pelaez C, Rico-Torres CP, Salas-Garrido CG, Correa D (2011) Acute toxoplasmosis in squirrel monkeys (*Saimiri sciureus*) in Mexico. Vet Parasitol 80:368–371
Correa D, Canedo-Solares I, Ortiz-Alegria LB, Caballero-Ortega H, Rico-Torres CP (2007) Congenital and acquired toxoplasmosis: diversity and role of antibodies in different compartments of the host. Parasite Immunol 29:651–660
Cunha RGT (2008) Gênero *Aotus* Illiger 1811. In: Reis NR, Perachi AL, Andrade FR (eds) Primatas brasileiros. Technical Books, Londrina, pp 115–125
Cunningham AA, Buxton D, Thomson KM (1992) An epidemic of toxoplasmosis in a captive colony of squirrel monkeys (*Saimiri sciureus*). J Comp Pathol 107:207–219
Däubener W, Hadding U (1997) Cellular immune reactions directed against *Toxoplasma gondii* with special emphasis on the central nervous system. Med Microbiol Immunol 185:195–206
de Camps S, Dubey JP, Saville WJ (2008) Seroepidemiology of Toxoplasma gondii in zoo animals in selected zoos in the midwestern United States. J Parasitol 94: 648–653
De Rodaniche E (1954) Spontaneous toxoplasmosis in the whiteface monkey, *Cebus capucinus*, in Panama. Am J Trop Med Hyg 3:1023–1025

de Thoisy B, Demar M, Aznar C, Carme B (2003) Ecologic correlates of *Toxoplasma gondii* exposure in free-ranging neotropical mammals. J Wildl Dis 39: 456–459
Debierre-Grockiego F, Schwarz RT (2010) Immunological reactions in response to apicomplexan glycosylphosphatidylinositols. Glycobiology 20:801–811
Debierre-Grockiego F, Campos MA, Azzouz N, Schmidt J, Bieker U, Resende AG, Santos Mansur D, Weingart R, Schmidt RR, Golenbock DT, Gazzinelli RT, Schwarz RT (2007) Activation of TLR2 and TLR4 by glycosylphosphatidylinositols derived from *Toxoplasma gondii*. J Immunol 179:1129–1137
Defler TR (2005) Primates of Colombia: conservation international. 550 p
Denkers EY, Butcher BA (2005) Sabotage and exploitation in macrophages parasitized by intracellular protozoans. Trends Parasitol 21:35–41
Denkers EY, Gazzinelli RT (1998) Regulation and function of T-cell-mediated immunity during *Toxoplasma gondii* infection. Clin Microbiol Rev 11:569–588
Denkers EY, Yap G, Scharton-Kersten T, Charest H, Butcher BA, Caspar P, Heiny S, Sher A (1997) Perforin-mediated cytolysis plays a limited role in host resistance to *Toxoplasma gondii*. J Immunol 159:1903–1908
Dietz HH, Henriksen P, Bille-Hansen V, Henriksen SA (1997) Toxoplasmosis in a colony of new world monkeys. Vet Parasitol 68:299–304
Dubey JP (2009) History of the discovery of the life cycle of *Toxoplasma gondii*. Int J Parasitol 39:877–882
Dubey JP, Jones JL (2008) *Toxoplasma gondii* infection in humans and animals in the United States. Int J Parasitol 38:1257–1278
Dubey JP, Su CL (2009) Population biology of *Toxoplasma gondii*: what's out and where did they come from. Mem Inst Oswaldo Cruz 104:190–195
Dubey JP, Lindsay DS, Speer CA (1998) Structures of *Toxoplasma gondii* tachyzoites, bradyzoites, and sporozoites and biology and development of tissue cysts. Clin Microbiol Rev 11:267–299
Elmore SA, Jones JL, Conrad PA, Patton S, Lindsay DS, Dubey JP (2010) *Toxoplasma gondii*: epidemiology, feline clinical aspects, and prevention. Trends Parasitol 26:190–196
Epiphanio S, Guimaraes MA, Fedullo DL, Correa SH, Catao-Dias JL (2000) Toxoplasmosis in golden-headed lion tamarins (*Leontopithecus chrysomelas*) and emperor marmosets (*Saguinus imperator*) in captivity. J Zoo Wildl Med 31:231–235
Epiphanio S, Sa LR, Teixeira RH, Catao-Dias JL (2001) Toxoplasmosis in a wild-caught black lion tamarin (*Leontopithecus chrysopygus*). Vet Rec 149:627–628
Epiphanio S, Sinhorini IL, Catao-Dias JL (2003) Pathology of toxoplasmosis in captive new world primates. J Comp Pathol 129:196–204
Feagle JG (1999) Primate adaptation and evolution. Academic, San Diego, CA
Fedigan LM (1990) Vertebrate predation in Cebus capucinus: meat eating in a neotropical monkey. Folia Primatol (Basel) 54:196–205
Ferguson DJ, Hutchison WM, Dunachie JF, Siim JC (1974) Ultrastructural study of early stages of asexual multiplication and microgametogony of *Toxoplasma gondii* in the small intestine of the cat. Acta Pathol Microbiol Scand B Microbiol Immunol 82:167–181
Fernandez-Duque E (2006) Aotinae: social monogamy in the only nocturnal haplorhines. In: Campbell CJ, Fuentes A, MacKinnon KC, Panger M, Bearder SK (eds) Primates in perspective. Oxford University Press, New York, NY, pp 139–154
Ferraroni JJ, Marzochi MC (1980) [Prevalence of *Toxoplasma gondii* infection in domestic and wild animals, and human groups of the Amazonas region]. Mem Inst Oswaldo Cruz 75: 99–109
Ferreira RG, Resende BD, Mannu M, Ottoni EB, Izar P (2002) Bird predation and prey-transfer in brown capuchin monkeys (*Cebus apella*). Neotrop Primates 10:84–89
Fleagle JG, Tejedor MF (2002) Early platyrrhines of southern South America. Cambridge Stud Biol Evolut Anthropol 33:161–173
Fragaszy DM, Visalberghi E, Robinson JG (1990) Variability and adaptability in the genus *Cebus*. Folia Primatol 54:114–118
Fragaszy DM, Visalberghi E, Fedigan LM (2004) The complete capuchin—the biology of the genus *Cebus*. Cambridge University Press, Cambridge, UK

Freese CH, Oppehheimer JR (1981) The Capuchin monkeys, Genus *Cebus*. In: Coimbra-Filho AF, Mittermeir RA (eds) Ecology and behavior of neotropical primates. Academia Brasileira de Ciências, Rio de Janeiro, pp 331–390
Frenkel JK, Dubey JP, Miller NL (1969) *Toxoplasma gondii*: fecal forms separated from eggs of the nematode *Toxocara cati*. Science 164:432–433
Frenkel JK, Dubey JP, Miller NL (1970) *Toxoplasma gondii* in cats: fecal stages identified as coccidian oocysts. Science 167:893–896
Fujigaki S, Saito K, Takemura M, Maekawa N, Yamada Y, Wada H, Seishima M (2002) L-tryptophan-L-kynurenine pathway metabolism accelerated by *Toxoplasma gondii* infection is abolished in gamma interferon-gene-deficient mice: cross-regulation between inducible nitric oxide synthase and indoleamine-2,3-dioxygenase. Infect Immun 70:779–786
Furuta T, Une Y, Omura M, Matsutani N, Nomura Y, Kikuchi T, Hattori S, Yoshikawa Y (2001) Horizontal transmission of *Toxoplasma gondii* in squirrel monkeys (*Saimiri sciureus*). Exp Anim 50:299–306
Galetti M (1990) Predation on squirrel (*Sciurus aestuans*) by Capuchin Monkey (*Cebus apella*). Mammalia 54:152–154
Garcia JL, Svoboda WK, Chryssafidis AL, de Souza ML, Shiozawa MM, de Moraes AL, Teixeira GM, Ludwig G, da Silva LR, Hilst C, Navarro IT (2005) Sero-epidemiological survey for toxoplasmosis in wild New World monkeys (*Cebus* spp.; *Alouatta caraya*) at the Parana river basin, Parana State, Brazil. Vet Parasitol 133:307–311
Gardiner CH, Fayer R, Dubey JP (1998) An atlas of protozoan parasites in animal tissues. United States Department of Agriculture, Agriculture Handbook, Washington, DC
Gazzinelli R, Xu YH, Hieny S, Cheever A, Sher A (1992) Simultaneous depletion of Cd4+ and $Cd8^{+}$ Lymphocytes-T is required to reactivate chronic infection with *Toxoplasma-gondii*. J Immunol 149:175–180
Gazzinelli RT, Eltoum I, Wynn TA, Sher A (1993a) Acute cerebral toxoplasmosis is induced by in vivo neutralization of TNF-alpha and correlates with the down-regulated expression of inducible nitric oxide synthase and other markers of macrophage activation. J Immunol 151: 3672–3681
Gazzinelli RT, Hieny S, Wynn TA, Wolf S, Sher A (1993b) Interleukin-12 is required for the T-lymphocyte-independent induction of interferon-gamma by an intracellular parasite and induces resistance in T-cell-deficient hosts. Proc Natl Acad Sci U S A 90:6115–6119
Gazzinelli RT, Hayashi S, Wysocka M, Carrera L, Kuhn R, Muller W, Roberge F, Trinchieri G, Sher A (1994) Role of Il-12 in the initiation of cell-mediated-immunity by *Toxoplasma-gondii* and Its regulation by Il-10 and nitric-oxide. J Eukaryot Microbiol 41:S9
Gilbert KA, Stouffer PC (1989) Use of a ground-water source by mantled howler monkeys (*Alouatta palliata*). Biotropica 21:380
Giudice AM, Mudry MD (2000) Drinking behavior in the black howler monkey (*Alouatta caraya*). Zoocriadores 3:11–19
Glander KE (1978) Drinking from arboreal water sources by mantled howling monkeys (Alouatta palliata Gray). Folia Primatol 29:206–217
Groves CP (2001) Primate taxonomy. Smithsonian Institution Press, Washington, DC
Gustafsson K, Wattrang E, Fossum C, Heegaard PM, Lind P, Uggla A (1997) *Toxoplasma gondii* infection in the mountain hare (Lepus timidus) and domestic rabbit (Oryctolagus cuniculus). II. Early immune reactions. J Comp Pathol 117:361–369
Gyimesi ZS, Lappin MR, Dubey JP (2006) Application of assays for the diagnosis of toxoplasmosis in a colony of woolly monkeys (*Lagothrix lagotricha*). J Zoo Wildl Med 37:276–280
Hegab SM, Al-Mutawa SA (2003) Immunopathogenesis of toxoplasmosis. Clin Exp Med 3:84–105
Hessler JR, Woodard JC, Tucek PC (1971) Lethal toxoplasmosis in a woolly monkey. J Am Vet Med Assoc 159:1588–1594
Hill WC (1960) Primates. Comparative anatomy and taxonomy IV. Cebidae Part A. University Press, Edinburgh, p xxii, 523
Hill RD, Gouffon JS, Saxton AM, Su C (2011) Differential gene expression in mice infected with distinct *Toxoplasma* strains. Infect Immun 80:968–974

Hutchison WM, Dunachie JF, Siim JC, Work K (1970) Coccidian-like nature of *Toxoplasma gondii*. Br Med J 1:142–144

Ingberman B, Stone AI, Cheida CC (2008) Gênero Saimiri (Voigt 1831). In: Reis NR, Perachi AL, Andrade FR (eds) Primatas Brasileiros. Technical Books, Londrina, pp 41–46

Innes EA (1997) Toxoplasmosis: comparative species susceptibility and host immune response. Comp Immunol Microbiol Infect Dis 20:131–138

Innes EA (2010) A brief history and overview of *Toxoplasma gondii*. Zoonoses Public Health 57:1–7

Izawa K (1978) Frog eating behavior of wild black-capped capuchin (*Cebus apella*). Primates 19:633–642

Izawa K (1979) Foods and feeding behavior of wild black-capped capuchin (*Cebus apella*). Primates 20:57–76

Jamieson SE, Peixoto-Rangel AL, Hargrave AC, de Roubaix LA, Mui EJ, Boulter NR, Miller EN, Fuller SJ, Wiley JS, Castellucci L, Boyer K, Peixe RG, Kirisits MJ, Elias LD, Coyne JJ, Correa-Oliveira R, Sautter M, Smith NC, Lees MP, Swisher CN, Heydemann P, Noble AG, Patel D, Bardo D, Burrowes D, McLone D, Roizen N, Withers S, Bahia-Oliveira LMG, McLeod R, Blackwell JM (2010) Evidence for associations between the purinergic receptor P2X(7) (P2RX7) and toxoplasmosis. Genes Immun 11:374–383

Johnson LL, Sayles PC (1997) Interleukin-12, dendritic cells, and the initiation of host-protective mechanisms against *Toxoplasma gondii*. J Exp Med 186:1799–1802

Juan-Salles C, Prats N, Marco AJ, Ramos-Vara JA, Borras D, Fernandez J (1998) Fatal acute toxoplasmosis in three golden lion tamarins (*Leontopithecus rosalia*). J Zoo Wildl Med 29:55–60

Kierulff MCM, Raboy BE, Procopio de Oliveira P, Miller K, Passos FC, Prado F (2002) Behavioral ecology of lion tamarins. In: Kleiman DG, Rylands AB (eds) Lion tamarins: biology and conservation. Smithsonian Institution Press, Washington, pp 157–187

Kinzey WG (1992) Dietary adaptations in the Pitheciinae. Am J Phys Anthropol 88:499–514

Klevar S (2007) Tissue cyst forming coccidia; *Toxoplasma gondii* and *Neospora* caninum as a cause of disease in farm animals. Acta Vet Scand 49:S1

Lang C, Gross U, Luder CG (2007) Subversion of innate and adaptive immune responses by *Toxoplasma gondii*. Parasitol Res 100:191–203

Lees MP, Fuller SJ, McLeod R, Boulter NR, Miller CM, Zakrzewski AM, Mui EJ, Witola WH, Coyne JJ, Hargrave AC, Jamieson SE, Blackwell JM, Wiley JS, Smith NC (2010) P2X7 receptor-mediated killing of an intracellular parasite, *Toxoplasma gondii*, by human and murine macrophages. J Immunol 184:7040–7046

Leite TN, Maja Tde A, Ovando TM, Cantadori DT, Schimidt LR, Guercio AC, Cavalcanti A, Lopes FM, Da Cunha IA, Navarro IT (2008) Occurrence of infection *Leishmania* spp. and *Toxoplasma gondii* in monkeys (*Cebus apella*) from Campo Grande, MS. Rev Bras Parasitol Vet 17(Suppl 1):307–310

Luder CG, Stanway RR, Chaussepied M, Langsley G, Heussler VT (2009) Intracellular survival of apicomplexan parasites and host cell modification. Int J Parasitol 39:163–173

Machado FS, Aliberti J (2009) Lipoxins as an immune-escape mechanism. Adv Exp Med Biol 666:78–87

Malmasi A, Mosallanejad B, Mohebali M, Sharifian Fard M, Taheri M (2009) Prevention of shedding and re-shedding of *Toxoplasma gondii* oocysts in experimentally infected cats treated with oral Clindamycin: a preliminary study. Zoonoses Public Health 56:102–104

Marshall LG (1988) Land mammals and the Great American Interchange. Am Sci 76:380–388

Mason NJ, Fiore J, Kobayashi T, Masek KS, Choi Y, Hunter CA (2004) TRAF6-dependent mitogen-activated protein kinase activation differentially regulates the production of interleukin-12 by macrophages in response to *Toxoplasma gondii*. Infect Immun 72:5662–5667

Mendes SL, Silva MP, Strier KB (2010) O Muriqui. Vitória, ES, Brasil, Instituto de Pesquisas da Mata Atlântica-IPEMA. 95 p

Miller CM, Zakrzewski AM, Ikin RJ, Boulter NR, Katrib M, Lees MP, Fuller SJ, Wiley JS, Smith NC (2011) Dysregulation of the inflammatory response to the parasite, *Toxoplasma gondii*, in P2X(7) receptor-deficient mice. Int J Parasitol 41:301–308

Montoya JG, Liesenfeld O (2004) Toxoplasmosis. Lancet 363:1965–1976
Nathan CF, Murray HW, Wiebe ME, Rubin BY (1983) Identification of interferon-gamma as the lymphokine that activates human macrophage oxidative metabolism and antimicrobial activity. J Exp Med 158:670–689
Nery-Guimaraes F, Franken AJ (1971) Toxoplasmosis in nonhuman primates. II. Attempts at experimental infection in Macacca mulata, *Cebus apella* and *Callithrix jacchus*; and search for antibodies in several species of platyrrhinus. Mem Inst Oswaldo Cruz 69:97–111
Nery-Guimaraes F, Franken AJ, Chagas WA (1971) Toxoplasmosis in nonhuman primates. I. Natural infection in *Macacca mulata* and *Cebus apella*. Mem Inst Oswaldo Cruz 69:77–87
Nicolle MC, Manceaux L (1909) On a new protozoan in gundis (*Toxoplasma* N. Gen). Arch Inst Pasteur Tunis 1:96–103
Norconk MA, Wright BW, Conklin-Brittain NL, Vinyard CJ (2009) Mechanical and nutritional properties of food as factors in platyrrhine dietary adaptations. In: Garber PA, Estrada A, Bicca-Marques JC, Heymann EW, Strier KB (eds) South American primates: comparative perspectives in the study of behavior, ecology and conservation. Developments in primatology: progress and prospect. Springer, New York, NY, pp 279–319
Osborn KG, Lowenstine LJ (1998) Respiratory diseases. In: Bennett BT, Abee CR, Henrickson R (eds) Nonhuman primates in biomedical research. Academic, San Diego, CA, pp 263–309
Oykhman P, Mody CH (2010) Direct microbicidal activity of cytotoxic T-lymphocytes. J Biomed Biotechnol 2010:249482
Perelman P, Johnson WE, Roos C, Seuanez HN, Horvath JE, Moreira MA, Kessing B, Pontius J, Roelke M, Rumpler Y, Schneider MP, Silva A, O'Brien SJ, Pecon-Slattery J (2011) A molecular phylogeny of living primates. PLoS Genet 7:e1001342
Pertz C, Dubielzig RR, Lindsay DS (1997) Fatal *Toxoplasma gondii* infection in golden lion tamarins (*Leontopithecus rosalia rosalia*). J Zoo Wildl Med 28:491–493
Pifer R, Yarovinsky F (2011) Innate responses to *Toxoplasma gondii* in mice and humans. Trends Parasitol 27:388–393
Resende BD, Greco VLG, Otonni EB, Izar P (2003) Some observations on the predation of small mammals by tufted capuchin monkeys (*Cebus apella*). Neotrop Primates 11:103–104
Rickli RI, Reis NR (2008) Gênero *Cacajao* Lesson 1840. In: Reis NR, Perachi AL, Andrade FR (eds) Primatas brasileiros. Technical Books, Londrina, pp 147–151
Robinson JG (1984) Diurnal variation in foraging and diet in the wedge-capped capuchin Cebus olivaceus. Folia Primatol 43:216–228
Rose L (1997) Vertebrate predation and food-sharing in *Cebus* and *Pan*. Int J Primatol 18:727–765
Rosenberger AL, Matthews LJ (2008) *Oreonax*-not a genus. Neotrop Primates 15:8–12
Rylands AB, Mittermeier RA (2009) The diversity of the New World primates (Platyrrhini): an annotated taxonomy. In: Garber PA, Estrada A, Bicca-Marques JC, Heymann EW, Strier KB (eds) South American primates: comparative perspectives in the study of behavior, ecology and conservation. Developments in primatology: progress and prospects. Springer, New York, NY, pp 23–54
Rylands AB, Kierulff MCM, Mittermeier RA (2005) Notes on the taxonomy and distributions of the tufted capuchin monkeys (*Cebus*, Cebidae) of South America. Lundiana 6:97–110
Saeij JP, Boyle JP, Boothroyd JC (2005) Differences among the three major strains of *Toxoplasma gondii* and their specific interactions with the infected host. Trends Parasitol 21:476–481
Salant H, Weingram T, Spira DT, Eizenberg T (2009) An outbreak of toxoplasmosis amongst squirrel monkeys in an Israeli monkey colony. Vet Parasitol 159:24–29
Sampaio DT, Ferrari SF (2005) Predation of an infant titi monkey (*Callicebus moloch*) by a tufted capuchin (Cebus apella). Folia Primatol 76:113–115
Santos TA, Portes Jde A, Damasceno-Sa JC, Caldas LA, Souza W, Damatta RA, Seabra SH (2011) Phosphatidylserine exposure by *Toxoplasma gondii* is fundamental to balance the immune response granting survival of the parasite and of the host. PLoS One 6:e27867

Scanga CA, Aliberti J, Jankovic D, Tilloy F, Bennouna S, Denkers EY, Medzhitov R, Sher A (2002) Cutting edge: MyD88 is required for resistance to *Toxoplasma gondii* infection and regulates parasite-induced IL-12 production by dendritic cells. J Immunol 168:5997–6001

Seed JR (1996) Protozoa: pathogenesis and defenses. In: Baron S (ed) Medical microbiology. University of Texas Medical Branch at Galveston, Galveston, TX

Sharma SD (1990) Immunology of toxoplasmosis. In: Wyler DJ (ed) Modern parasite biology: cellular, immunological and molecular aspects. W.H. Freeman and Company, New York, NY, pp 184–199

Sibley LD, Khan A, Ajioka JW, Rosenthal BM (2009) Genetic diversity of *Toxoplasma gondii* in animals and humans. Philos Trans R Soc Lond B Biol Sci 364:2749–2761

Silva JCR (2007) Toxoplasmose. In: Cubas ZS, Silva JCR, Catão-Dias JL (eds) Tratado de animais selvagens: medicina veterinária. Editora Roca, São Paulo, pp 768–784

Silva JS Jr (2001) Especiação nos macacos-prego e caiararas, gênero *Cebus* Erxleben, 1777 (Primates, Cebidae). Ph.D. Thesis, Universidade Federal do Rio de Janeiro, Rio de Janeiro

Simpson GG (1980) *Splendid isolation:* the curious history of South American mammals. Yale University, New Haven, CT

Snowdon CT, Soini P (1988) The tamarins, genus *Saguinus*. In: Mittermeier RA, Rylands AB, Coimbra-Filho AF, Fonseca GAB (eds) Ecology and behavior of neotropical primates. World Wildlife Fund, Washington, DC, pp 223–298

Sogorb S F, Jamra LF, Guimaraes EC, Deane MP (1972) Toxoplasmose espontanea em animais domesticos e silvestres, em Sao Paulo. Revista. Inst Med trop S Paulo 14: 314–320

Soini P (1988) The pygmy marmoset, genus *Cebuella*. In: Mittermeier RA, Rylands AB, Coimbra-Filho AF, Fonseca GAB (eds) Ecology and behavior of neotropical primates. World Wildlife Fund, Washington, DC, pp 79–129

Splendore A (1909) A new protozoan parasite of rabbit found in histological lesions similar to human Kala-Azar. Rev Soc Sci S Paulo 3:109–112

Stuart M, Pendergast V, Rumfelt S, Pierberg S, Greenspan L, Glander K, Clarke M (1998) Parasites of wild howlers (*Alouatta* spp.). Int J Primatol 19:493–512

Subauste CS (2009) CD40, autophagy and *Toxoplasma gondii*. Mem Inst Oswaldo Cruz 104:267–272

Suzuki Y, Orellana MA, Schreiber RD, Remington JS (1988) Interferon-gamma: the major mediator of resistance against *Toxoplasma gondii*. Science 240:516–518

Tait ED, Hunter CA (2009) Advances in understanding immunity to *Toxoplasma gondii*. Mem Inst Oswaldo Cruz 104:201–210

Terborgh J (1983) Five New world primates. Princeton University Press, Princeton, NJ

Van Roosmalen MGM, Van Roosmalen T (2003) The description of a new marmoset genus, *Callibella* (Callitrichinae, Primates), including its molecular phylogenetic status. Neotrop Primates 11:1–10

Villarino AV, Stumhofer JS, Saris CJ, Kastelein RA, de Sauvage FJ, Hunter CA (2006) IL-27 limits IL-2 production during Th1 differentiation. J Immunol 176:237–247

Vouldoukis I, Mazier D, Moynet D, Thiolat D, Malvy D, Mossalayi MD (2011) IgE mediates killing of intracellular *Toxoplasma gondii* by human macrophages through CD23-dependent, interleukin-10 sensitive pathway. PLoS One 6:e18289

Webb SD (1976) Mammalian faunal dynamics of the Great American Interchange. Paleobiology 2:220–234

Werner H, Janitschke K, Kijhler H (1969) Uber Beobachtungen an Marmoset-Affen *Saguinus* (*Oedipomidas*) *Oedipus* nach oraler und intraperitonealer infektion mit verschiedenen en zystenbildenden *Toxoplasma*-Stgmmen unterschiedlicher Virulenz. I. Mitteilung: Klinische, pathologisch-anatomische, histologische und parasitologische zentralblatt Bakteriologie Befunde. Parasitenkunde fir Infektionskrankheiten und Hygiene 209:553–569

Wright PC (1981) The night monkeys, genus *Aotus*. In: Coimbra-Filho AF, Mittermeir RA (eds) Ecology and behavior of neotropical primates. Academia Brasileira de Ciências, Rio de Janeiro, pp 211–240

Yarovinsky F, Zhang DK, Andersen JF, Bannenberg GL, Serhan CN, Hayden MS, Hieny S, Sutterwala FS, Flavell RA, Ghosh S, Sher A (2005) TLR11 activation of dendritic cells by a protozoan profilin-like protein. Science 308:1626–1629

Zanon CMV, Reis NR, Filho HO (2008) Gênero *Ateles* E.Geoffroy 1806. In: Reis NR, Perachi AL, Andrade FR (eds) Primatas brasileiros. Technical Books, Londrina, pp 169–173

The Evolution of SIV in Primates and the Emergence of the Pathogen of AIDS

Edward J.D. Greenwood, Fabian Schmidt, and Jonathan L. Heeney

Introduction

Three decades have passed since the first reports of opportunistic diseases in previously healthy individuals and the description of acquired immunodeficiency syndrome (AIDS) as a new human disease (CDC 1981a, b, c). Shortly after, the causative agent was identified as a T-cell tropic retrovirus (Barre-Sinoussi et al. 1983) that eventually came to be termed human immunodeficiency virus 1 (HIV-1). HIV-1 is a retrovirus of the Lentivirus genus and the first primate lentivirus to be discovered. A second human lentivirus, HIV-2, was later discovered in patients suffering from AIDS in West Africa (Clavel et al. 1986).

Lentiviruses have since been discovered to naturally infect over 40 different African primate species, termed simian immunodeficiency viruses (SIVs). Study of SIV infection of the majority of these species is difficult, as many are endangered and (rightly) protected in the wild, with limited or no captive populations available for study. However, some important facts have been established. Firstly, it is now clear that HIV-1 has originated from SIVcpz of the common chimpanzee (*Pan troglodytes*) and HIV-2 from SIVsmm of the sooty mangabey (*Cercocebus atys*). Secondly, some species of SIV-infected African primates are present in European and US research centers, and thus, the natural history of their infection has been studied in detail. In particular, the SIV infection of two African green monkey species (*Chlorocebus sabaeus* and *C. pygerythrus*) and sooty mangabeys has been studied intensively. In these species, it is clear that the vast majority of individuals do not progress to AIDS.

In contrast, Asian macaques are not infected with SIV in the wild, but can develop AIDS when experimentally infected with SIV from other species,

Edward JD Greenwood and Fabian Schmidt contributed equally to this work

E.J.D. Greenwood • F. Schmidt (✉) • J.L. Heeney
Department of Veterinary Medicine, University of Cambridge, Cambridge, UK

J.F. Brinkworth and K. Pechenkina (eds.), *Primates, Pathogens, and Evolution*, Developments in Primatology: Progress and Prospects,
DOI 10.1007/978-1-4614-7181-3_10, © Springer Science+Business Media New York 2013

providing a model for disease in humans caused by HIV. Rhesus macaques (*Macaca mulatta*) and pig-tailed macaques (*Macaca nemestrina*) are the two species principally used in these studies. These models were first established after macaques in numerous American primate centers developed AIDS-like clinical signs and were also found to be infected with T-cell tropic retroviruses, which collectively came to be termed SIVmac (Benveniste et al. 1986; Daniel et al. 1984). Due to the similarity between SIVmac and SIVsmm, it was hypothesized early on that SIVmac could have its origins in SIVsmm infection of sooty mangabeys (Murphey-Corb et al. 1986), which has since been confirmed. It is likely that SIVsmm was unknowingly transmitted from sooty mangabeys into macaques during invasive experiments for the study of prion diseases in American primate centers and subsequently spread within captive macaque populations (reviewed by Apetrei et al. 2006).

Comparison of the pathogenic infection of humans and macaques with HIV/SIV with the nonpathogenic infection of sooty mangabeys and African green monkeys has provided key insight into the pathways most important in the development of AIDS in susceptible species and the host mechanisms that have evolved in African primate species to avoid disease as a result of lentivirus infection.

In this chapter, we will first discuss the age and diversity of primate/human lentiviruses, the outcome of SIV transmission into humans, and the mechanisms proposed to have promoted the pandemic spread of HIV-1. Next, we will compare the natural history of pathogenic HIV infection of humans and SIV infection of Asian macaques (the best available animal model of human HIV/AIDS) with the nonpathogenic SIV infection of sooty mangabeys and African green monkeys. We will discuss in depth the mechanisms that have been proposed to explain the dichotomous outcome of lentivirus infection between these groups. We will then examine the specific differences between HIV-1 and other primate lentiviruses, including potential mechanisms for the high pathogenicity of HIV-1. Finally, we will review what is known of the host–virus relationship in the species in which the HIV-1 lineage evolved and suggest that examination of this relationship is of special importance to HIV-1 and SIV research.

Age and Diversity of the SIV Lineage

SIVs observed in wild African primates are generally species-specific: multiple isolates of virus from one primate species generally form monophyletic lineages in phylogenetic trees. The degree to which SIVs differ within one species varies, but is largely biased by the number of isolates sequenced and their geographical distribution (Bibollet-Ruche et al. 2004; Liegeois et al. 2012). Species with highly divergent SIVs have also been observed, which are usually the result of cross-species transmissions, sometimes followed by recombination between distant SIVs (Aghokeng et al. 2007; Liegeois et al. 2012; Souquiere et al. 2001). High frequencies of recombination are a characteristic of primate lentiviruses (Chen et al. 2006), but such recombination events are most apparent when heterologous viruses recombine. These so-called mosaic or chimeric viruses are described for African green monkeys, mandrills (*Mandrillus sphinx*), and chimpanzees (Bailes et al. 2003; Jin et al. 1994; Takemura and Hayami 2004) and can be identified when comparing phylogenetic trees created using alignments from different parts of the genome, as in Fig. 1.

A precise age of primate lentiviruses has been difficult to ascertain, but has been estimated in several studies. Such dating methods normally require the use of a "molecular clock," in which sequences are compared, and a known or estimated rate of genetic change is applied to estimate the time to the most recent common ancestor. Estimates resulting from molecular clock methods are dependant on how this rate of change is estimated. While this is relatively simple for eukaryotic species, as the rate of genetic change is both slow and well established for different species, calibrating a molecular clock for retroviruses, which change extremely rapidly and often recombine, is much more challenging.

Using only relatively modern SIV and HIV sequences of known dates (from 1975 to 2005) to calibrate a molecular clock resulted in a estimate that the primate lentivirus lineage is only centuries old (Wertheim and Worobey 2009). However, this dating was controversial as it was already suspected that using only modern sequences to extrapolate the history of a possibly ancient lineage would lead to erroneous estimates (Sharp et al. 2000).

However, a recent study of SIV infection of primates on the African island of Bioko has allowed for a new calibration of the molecular clock estimate of the age of this lineage (Worobey et al. 2010). The island has been separated from the African mainland for a period of 10,000–12,000 years and accommodates a number of primate species. Individuals from four of these species were found to be infected with SIV. Most importantly, the Bioko drill (*Mandrillus leucophaeus poensis*) is infected with an SIV similar to that isolated from the mainland drill (*Mandrillus leucophaeus leucophaeus*). The time to the most recent common ancestor of SIVdrl from Bioko and SIVdrl from mainland Africa is therefore known to be at least 10,000 years old, providing a new method for calibrating the molecular clock. The resulting estimate is that SIVs have been present in African primates for 76,000 years.

Finally, evidence exists that the primate lentivirus lineage is ancient. The genomes of a number of lemur species of genera *Cheirogaleus* and *Microcebus* contain sequences of a lentivirus that has at some point infected germ line cells and become integrated into the genome—an endogenous lentivirus. After integration into the genome, the sequences of endogenous retroviruses are expected to be subject to the same rate of mutation as other host genomic sequences. This rate is well established for eukaryotic species and is much slower than the rate of mutation of exogenous retroviruses. Endogenous retroviruses are therefore ideal for estimating dates on ancient timescales. It seems that there were two independent integration events, both estimated to have occurred around 4 million years ago (Gifford et al. 2008; Gilbert et al. 2009). Unless SIV was introduced to Madagascar independently and prior to the introduction of SIV to the African mainland, this would indicate that the African primate lentivirus lineage is at least equally as ancient.

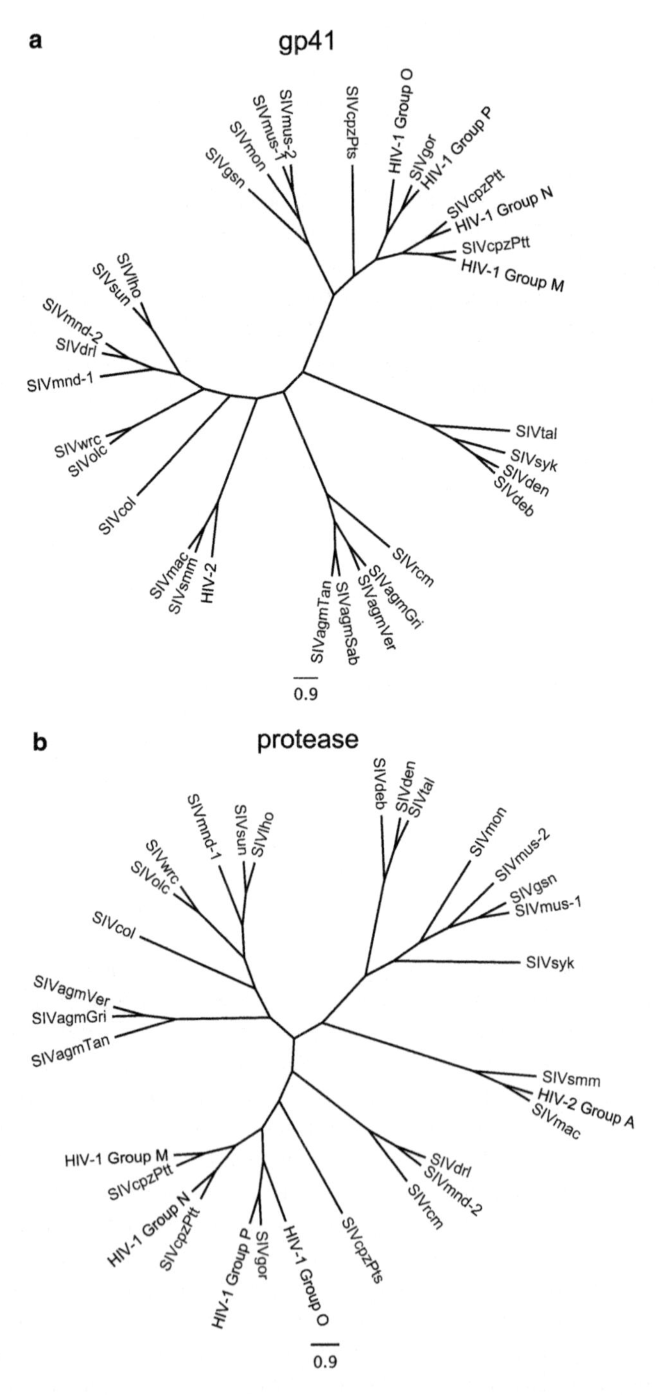

Fig. 1 Phylogenetic trees demonstrating the relationships between different SIVs and HIV-1 and HIV-2. Trees were created from alignments using nucleic acid sequences from the (**a**) gp41 and (**b**) protease genes. *Scale bar* indicates 0.9 substitutions per site. SIVagmSab is excluded from the protease tree as a recombination event has occurred in the region used for this alignment

In addition, in general, the relationship between SIVs of different species mirrors the relationships between their primate hosts, with SIVs from closely related primates being the most similar to one another. It has therefore been suggested that species-specific SIVs could be the result of concurrent host diversification, with splits in the SIV lineage occurring at the same time as splits in the primate lineage (host-dependant evolution) (Sharp et al. 2000). The SIVs of African green monkeys show particularly strong evidence of this. Each of the four species of African green monkey (*Chlorocebus aethiops, C. tantalus, C. pygerythrus, C. sabaeus*) is infected with a species-specific SIV, and the genetic divergence between viruses mirrors the divergence of the host species (Jin et al. 1994; Muller et al. 1993; Sharp et al. 2000). For host-dependant evolution to occur, the most recent common ancestor of the four African green monkey species must have been infected with SIV. The time to the most recent common ancestor of these species is approximately 3 million years (Fabre et al. 2009).

Distribution of Accessory Genes in the Primate Lentivirus Lineage

HIV/SIV are retroviruses of the Lentivirus genus. The genomes of all retroviruses include the genes *gag, pol, and env*, the major structural and enzymatic proteins of the virus. All known members of the Lentivirus genus also possess the regulatory proteins Rev and Tat, including the oldest recognized lentivirus, "RELIK," an endogenous retrovirus estimated to have integrated into the genome of the European rabbit (*Oryctolagus cuniculus*) over 12 million years ago (Katzourakis et al. 2007; Keckesova et al. 2009). All extant lentiviruses found in primates, cats, sheep, and cattle possess an additional gene, *vif*.

In contrast, four genes are unique to extant primate lentiviruses; *nef* and *vpr* are found in all primate lentiviruses, while *vpx* and *vpu* are found in nonoverlapping subsets of these viruses. Our understanding of the functions of *vif*, *nef*, *vpr*, *vpx,* and *vpu* has increased dramatically during recent years, and it can be concluded that all factors have evolved to play a role in counteracting host defense mechanisms called restriction factors (Ayinde et al. 2010; Kirchhoff 2010).

Vpx, one of the two genes unique to primate lentiviruses, evolved in the *Papionini* tribe of primates, and the distribution of this gene within primate SIVs is likely to have been increased as a result of several recombination events between SIVs of different primate species (Takemura and Hayami 2004). It is found in SIVsmm (of sooty mangabeys), SIVrcm (red-capped mangabey, *Cercocebus torquatus*), SIVmnd-2 (mandrill), and SIVdrl (drill). The other gene unique to primate lentiviruses, *vpu*, was acquired by an SIV within a subset of guenon species (Bailes et al. 2003). Guenons (tribe *Cercopithecini*) are a species-rich group of primates; however, only four guenon species, which associate in the wild, carry SIVs with the *vpu* gene. These are the mona monkey (*Cercopithecus mona*), greater spot-nosed monkey (*C. nictitans*), and mustached monkey (*C. cephus*), infected with SIVmon,

SIVgsn, and SIVmus, respectively. A fourth *Cercopithecus* species, Dent's mona monkey (*C. mona denti*), was found to be infected with an SIV harboring the *vpu* gene (SIVden), but with a shorter coding region (Dazza et al. 2005; Schmokel et al. 2010). Two other nonhuman primate species are infected with a vpu-carrying virus—the chimpanzee, *Pan troglodytes*, infected with SIVcpz, and the western gorilla (*Gorilla gorilla*) infected with SIVgor. SIVcpz has been identified in two subspecies of chimpanzees, *Pan troglodytes troglodytes* and *P. t. schweinfurthii*, while surveys of wild *P. t. verus* and *P. t. ellioti* (previously known as *P. t. vellerosus*) have demonstrated with some confidence that they are not infected (Sharp and Hahn 2011). SIVgor has been identified in the western lowland gorilla subspecies (*Gorilla gorilla gorilla*) and has thus far been found exclusively in Cameroon (Neel et al. 2010). SIVgor is highly related to SIVcpz and gorillas have most likely become infected more recently through cross-species transmission of SIVcpz from *Pan troglodytes troglodytes*.

The recombination events involved in the genesis of SIVcpz/SIVgor are of particular interest. Phylogenetic analyses suggest that a *vpx*-expressing virus found in red-capped mangabeys was transmitted into chimpanzees where it recombined with an SIV expressing the accessory gene *vpu*. The chimpanzee mosaic virus shows homology with SIVs found in a subset of guenons within the 3′ half of its genome (Courgnaud et al. 2002). The 5′ region of the genome and possibly the *nef* gene at the 3′ end are closely related to SIVrcm (Beer et al. 2001; Kirchhoff 2009). One recombination crossover is therefore likely to have occurred in the short region between *vpr* and *vpu* and another between *env* and *nef* (Kirchhoff 2009; Sharp et al. 2005). It is believed that the *vpx* gene was transferred within the initial recombination, but must have subsequently been lost as no remnant of this gene is apparent in contemporary SIVcpz. The *vpu* gene however remained conserved.

Transmission of SIV into Humans and the Spread of HIV

There are four recognized HIV-1 groups: M, N, O, and P. As each of these HIV-1 groups is closest in sequence homology to different isolates of SIV in chimpanzees or gorillas, it is clear that there have been at least four cross-species transmission events into humans, with M and N group viruses having their source in chimpanzees of the *Pan troglodytes troglodytes* subspecies and O and P group most closely related to SIV of gorillas (Keele et al. 2006; Plantier et al. 2009; Van Heuverswyn et al. 2006). To prevent confusion, it should also be noted that there is a separate alphabetical nomenclature for describing different subtypes (also referred to as clades) of HIV-1 group M viruses. Based on sequence homology, HIV-1 group M viruses are assigned to subtypes from A to K or identified to have been generated through recombination between viruses of previously identified subtypes.

Non-M HIV-1 viruses are mostly restricted to Cameroon. Group O infection accounts for around 1 % of all HIV-1 infection in Cameroon (Vergne et al. 2003), while N has been identified in less than 20 individuals (Vallari et al. 2010a) and group P in only two cases so far (Plantier et al. 2009; Vallari et al. 2010b). Clinical data regarding non-M infections is limited, but it is important to note that group O and group N infections have been identified in patients with AIDS (Ayouba et al. 2000; Gurtler et al. 1994), while the first group P infection to be identified was associated with depleted CD4+ T cells and a high viral load (Plantier et al. 2009). There is therefore no evidence that HIV-1 non-M infections are less pathogenic than HIV-1 M infections, despite their more limited spread.

In addition to HIV-1, there is a second human immunodeficiency virus, HIV-2. In contrast to HIV-1, HIV-2 infects globally only between 1 and 2 million individuals and is predominantly confined to West Africa (Campbell-Yesufu and Gandhi 2011). In common with HIV-1, HIV-2 consists of several groups, each the result of a different cross-species transmission event. Eight groups, A–H, have been recognized, with only groups A and B having spread more extensively within the human population. The remaining groups are limited to infections of single individuals. HIV-2 is the result of transmission of SIV from the sooty mangabey, and because of this different source, HIV-2 has a different genetic structure to HIV-1. Viral genomes of several primate lentiviruses, along with HIV-1 and HIV-2, are compared in Fig. *2*. There are critical differences between HIV-1 and HIV-2 infection of humans that are discussed at the end of this chapter.

As with dating the age of the primate lentivirus lineage, dating the cross-species transmission events leading to HIV has been difficult. However, the availability of two HIV-1 group M sequences attained from archived samples from 1959 and 1960, from Kinshasa (formally Leopoldville), Democratic Republic of Congo (DRC, formally Belgian Congo and, later, Zaire), allows for greater confidence in dating the origin of the HIV-1 group M pandemic. While the chimpanzees infected with SIVcpz isolates most closely related to HIV-1 group M are found in Cameroon, it seems that Kinshasa is a candidate for the epicenter of the HIV-1 pandemic. A study conducted in 1997 indicated that there is unparalleled genetic diversity of HIV-1 M in the DRC, with all subtypes represented, along with recombinant forms not represented outside of the DRC (Vidal et al. 2000). Notably, HIV-1 extracted from the two samples from 1959 and 1960 show a high degree of genetic divergence—comparable to the genetic difference between two contemporary isolates of different subtypes. This suggests that by 1960, HIV-1 had already circulated extensively in humans in this region (Worobey et al. 2008).

Using sequences from these two early samples, along with later samples to calibrate a molecular clock, leads to an estimate that the cross-species transmission event leading to HIV-1 group M most likely occurred near the start of the twentieth century, between 1873 and 1924 (Worobey et al. 2008). Dating the cross-species transmission events that resulted in the other HIV-1 and HIV-2 groups is more problematic due to the markedly fewer available sequences and the lack of sequences from older achieved material. Estimates for the date of the four cross-species transmission events leading to HIV-1 groups O and N and HIV-2 groups A and B are all also within or near the beginning of the twentieth century (de Sousa et al. 2010; Wertheim and Worobey 2009).

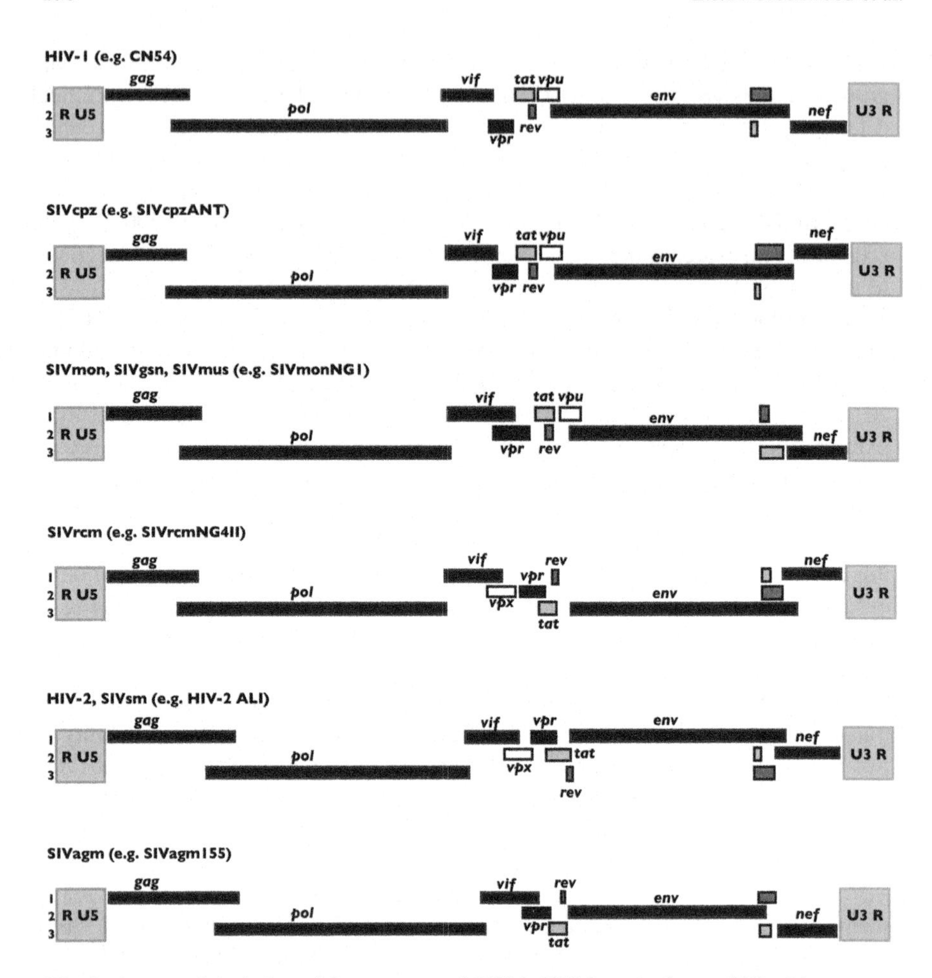

Fig. 2 Annotated depiction of the genomes of HIV-1, HIV-2, and relevant SIVs. *Diagrams* are based on annotated sequences, with the sequence used named in *brackets*. *Numbers* 1–3 at *left* indicate reading frames

Pandemic Spread of HIV

Human exposure to SIV-infected primates is unlikely to be novel to the late nineteenth and early twentieth century. Notably, some strains of another human retrovirus, human T-cell lymphotropic virus (HTLV), present in Africa seem to be the result of cross-species transmission events from primates occurring thousands of years ago (Switzer et al. 2006; Van Dooren et al. 2001), also most likely through bushmeat. Several authors have therefore attempted to identify mechanisms to explain why only the cross-species transmission events in this recent time frame have resulted in the several HIV-1 and HIV-2 epidemics and the HIV-1 group M pandemic—i.e., mechanisms that promoted the spread of HIV that are unique to the twentieth century.

Firstly, Worobey and colleagues have suggested that the development of large urban centers, not present in western Central Africa prior to 1900, facilitated the spread of HIV-1 M. They assert that prior to 1910, there was not a single site with a population greater than 10,000 people in western Central Africa (Worobey et al. 2008). Kinshasa underwent particularly rapid growth, from only a few thousand people in 1905, to around 40,000 in 1940, and over 400,000 in 1961 (Chitnis et al. 2000).

However, this rapid urbanization, combined with the (often-forced) movement of workers, led to disruption of social norms that may have been a greater contributing factor than the size of the urban centers alone. De Sousa and colleagues carried out extensive analysis of colonial medical articles and reports regarding Kinshasa from the start of the twentieth century to the country's independence (and general expulsion of Belgian authorities) in 1960 (Chin 2007; de Sousa et al. 2010). They note that in 1929 there were a large number of commercial sex workers, possibly due to the heavily male biased population, at 4:1 males to females. They also find that sexually transmitted diseases, including syphilis and other genital ulcerative diseases, were highly prevalent, with one survey in 1930–1932 finding that 5 % of the female population had active genital ulcers (de Sousa et al. 2010). While the low risk of HIV-1 transmission from heterosexual sex, estimated to be between 0.01 and 1.1 % per act (Boily et al. 2009), could be viewed as a barrier to early spread of HIV-1 through purely sexual contact, the risk of transmission is considerably increased when genital ulcers (and other non-ulcerative sexually transmitted diseases) are present in either the HIV-positive or HIV-negative sexual partner, with most studies finding a 2–5-fold increase in risk of infection, but some finding the risk of transmission to be almost 20-fold higher (Boily et al. 2009; Fleming and Wasserheit 1999).

The role of non-sterile injections throughout sub-Saharan Africa in the twentieth century has also been proposed as a major factor allowing the early spread of HIV-1. Marx and colleagues propose that initial "serial human passage" of the virus through non-sterile injections allowed the adaptation to humans which they suggest would be required for subsequent spread by sexual transmission (Marx et al. 2001). They point to numerous mass injection campaigns in sub-Saharan Africa and evidence for large numbers of injections involving the reuse of non-sterilized injection materials in the twentieth century, especially since the discovery of penicillin and its use in Africa from the 1950s onwards.

The mechanisms reviewed above that are proposed to have facilitated the spread of HIV-1 span the entirety of the twentieth century. Interestingly, in their 2008 analysis, Worobey et al. propose a model of the growth of HIV-1 that suggests relatively slow growth prior to 1960, followed by much more rapid expansion after this date. Although their analysis is likely to be subject to bias due to the limited availability of samples available, this computational analysis has some support from the early reports of AIDS in Africa, as it correlates with marked increases in the opportunistic diseases that now define AIDS, such as Kaposi's sarcoma and esophageal candidiasis, occurring in Kinshasa, Uganda, Zambia, and Rwanda from the late 1970s to the early 1980s (Quinn et al. 1986). This would be consistent with aggressive spread of

HIV-1 in the 1960s and early 1970s, due to the approximately decade-long incubation period between HIV-1 infection and the development of AIDS (see below).

Together, these lines of evidence tend not to favor extensive spread in urban centers in the early part of the twentieth century. May and colleagues have proposed an elegant model of rural spread among "loosely interlinked villages" (May et al. 2001). One significant attraction of this model is that it predicts that HIV-1 could persist at a very low level for decades within rural villages, with only limited spread within each village compensated for by introduction to new villages, preventing the virus from becoming extinct. In this model, the period of extremely slow spread is then followed by much more rapid expansion of HIV, without requiring mechanisms postulated above. Of course, this does not exclude a role for the other factors reviewed, especially as urbanization, high levels of other sexually transmitted diseases, and the reuse of non-sterilized injection materials could all have contributed to the introduction of HIV-1 to urban centers and extensive spread within these centers—which was almost certainly a requirement for HIV dissemination throughout and beyond Africa.

While HIV-1 group M subtype B causes only a minority of HIV-1 infections in Africa, it is the most prevalent subtype in a large number of countries outside of Africa, including the USA and Europe. The disparity between the prevalence of subtype B outside and within Africa could indicate that the pandemic beyond Africa was the result of a single transmission from Africa. Gilbert and colleagues examined sequences from archived samples collected in 1982 and 1983 of Haitian nationals that had immigrated to the USA after 1975 and hospitalized with AIDS prior to 1981 (Gilbert et al. 2007). They focused on these patients as shortly after the recognition of AIDS as a novel syndrome firstly primarily in homosexual men in the USA in 1981 (Gottlieb et al. 1981), and intravenous drug users in 1982 (CDC 1982b), there were reports of Haitian nationals in the USA suffering from AIDS (CDC 1982c). The majority of these Haitian patients indicated they had practiced neither homosexual sex nor intravenous drug use, suggesting that being a recent Haitian immigrant was an independent risk factor for the development of AIDS.

In their analysis of sequences from these patients, in addition to other sequences of HIV-1 isolates from Haiti, and sequences representing HIV-1 subtype B isolates from multiple other countries, Gilbert and colleagues find that subtype B is the most genetically diverse within Haiti, indicating an earlier introduction of HIV-1 subtype B into this nation than any other, and that the HIV-1 epidemic in Haiti was the result of a single introduction of the virus from Africa. They also show that with few exceptions, the pandemic spread of HIV-1 subtype B beyond Haiti (including North American, South American, European, and Asian countries) is the result of another single founder event linked to Haiti, which they postulate to be most likely the introduction of HIV-1 to the USA through immigration.

The authors estimate that the introduction of HIV-1 from Africa into Haiti occurred between 1962 and 1970. Interestingly, it seems that there was a large movement of Haitians to the DRC (then Zaire) in the early 1960s following two events: firstly, political problems appearing in Haiti after Francois Duvalier

('Papa Doc') took power in 1957 and, secondly, the vacancies for "executive" positions (administrators, healthcare workers, and teachers) in Zaire following the expulsion of the Belgian authorities in 1960. Many Haitians returned to Haiti or to other nations in the late 1960s and early 1970s (Chin 2007; Molez 1998; Piot et al. 1984), possibly indicating a potential mechanism by which HIV-1 was introduced into Haiti. Gilbert et al. also estimate that the single founder event leading to the spread of HIV-1 subtype B from Haiti occurred between 1966 and 1972 (Gilbert et al. 2007), which is constant with the earliest cases of AIDS in the USA, retrospectively diagnosed as occurring in 1978 (CDC 1982a).

Natural History of HIV/SIV Infection of AIDS-Resistant and AIDS-Susceptible Species

The clinical progression of HIV-1 infection of humans is normally monitored by assessing the number of peripheral blood CD4+ T cells and the peripheral blood viral load. In the acute stage of infection, the highest levels of virus replication are reached, with viral loads frequently in the region of 10^6–10^7 copies of the RNA genome per ml of blood (Kaufmann et al. 1998; Rieder et al. 2010). This occurs concurrently with a sudden drop in the CD4+ T-cell count that is partly recovered after the first few months of infection. In the chronic phase of infection, a lower, relatively stable "set point" viral load is established, generally in the region of 10^4–10^6 copies/ml, with the chronic viral load correlating with the rate of disease progression (Mellors et al. 1996; Rodriguez et al. 2006). Levels of peripheral blood CD4+ T cells slowly decline, until they are depleted to the point that the immune system can no longer function. This depletion of CD4+ T cells is thought to be due to a number of factors, including, but certainly not limited to, the direct infection of CD4+ T cells by HIV-1. The mean CD4+ T-cell count in healthy HIV-negative adult humans has been found to be between 700 cells/μl and 1,200 cells/μl in studies of numerous different populations (Aina et al. 2005; Hazenberg et al. 2000; Kibaya et al. 2008; Tugume et al. 1995). The risk of developing opportunistic infections—the clinical manifestations of AIDS—is greatly increased when the CD4+ count drops below 200 cells/μl (Begtrup et al. 1997; Phair et al. 1990; Phillips et al. 1989). The median survival time following untreated HIV-1 infection is approximately 8–12 years (Collaborative Group on AIDS Incubation and HIV Survival 2000; Morgan et al. 2002).

Two primate species are commonly used in modeling the pathogenesis of HIV: the rhesus macaque and the pig-tailed macaque. SIVmac/SIVsmm infection of rhesus macaques follows a similar, albeit accelerated, course to HIV-1 infection of humans, with progression to AIDS normally occurring within 6 months to 3 years of infection (Brown et al. 2007). Pig-tailed macaques seem to have an even higher susceptibility to developing AIDS. In addition to developing AIDS as a result of SIVsmm or SIVmac infection, they also can develop AIDS following infection with SIVagm, which does not cause disease in rhesus macaques (Favre et al. 2009; Hirsch et al. 1995).

The natural history of the SIV infection of African green monkeys or sooty mangabeys with SIVagm and SIVsmm, respectively, is very different. These animals generally do not develop disease despite peripheral blood viral loads similar to those of humans and rhesus macaques. A recent study of sooty mangabeys has demonstrated that SIV infection does result in an extremely slow depletion of CD4+ T cells (Taaffe et al. 2010), but unlike in humans and rhesus macaques, the magnitude of CD4+ T-cell loss does not correlate with viral load. CD4+ T-cell loss has not been reported in African green monkeys. However, in both African green monkeys and sooty mangabeys, measurements of CD4+ T cells are complicated by factors relating to expression of CD4 (see below). For both of these species, only a single case of an AIDS-like condition in an SIV-infected animal has been described (Ling et al. 2004; Traina-Dorge et al. 1992). The mechanisms by which these species circumvent disease development are of great interest and are discussed below.

Limiting Target Cell Availability

Entry of HIV-1 into the cell is mediated by the use of CD4 as a receptor, along with a coreceptor. CD4 is expressed principally on CD4+ T cells, and it binds to MHC-II on antigen-presenting cells. These T cells, upon activation, become "T helper cells," which direct the adaptive immune response primarily through cytokine production. The coreceptor used by HIV-1 is principally the chemokine receptor C-C chemokine receptor type 5 (CCR5), but in some cases a different chemokine receptor, C-X-C chemokine receptor type 4 (CXCR4). Interestingly, humans lacking CCR5 expression through an inherited mutation are almost completely resistant to infection by both sexual and parenteral routes (Dean et al. 1996; Wilkinson et al. 1998). Partly because of the restriction of CCR5 expression to memory T cells, memory CD4+ T cells are preferentially infected compared to naïve CD4+ T cells. In humans and rhesus macaques, quantitative PCR detection of the HIV-1 genome in different CD4 T-cell subsets consistently finds that the majority of infected cells have a memory CD4+ cell phenotype, though infected naïve T cells are also found (Brenchley et al. 2004a; Mattapallil et al. 2005; Ostrowski et al. 1999). Macrophages and dendritic cells, the primary antigen-presenting cells of the immune system, also express CD4 and CCR5 and can be productively infected by HIV-1. An obvious potential mechanism by which species may adapt to SIV infection would therefore be to restrict the expression of the viral receptor and coreceptor.

Interesting observations have been made regarding CD4 expression in sooty mangabeys and African green monkeys. Adult uninfected African green monkeys have a much lower frequency of peripheral blood CD4+ T cells than other primates, including humans, sooty mangabeys, and rhesus macaques (Beaumier et al. 2009). They have a large peripheral blood T-cell population with a unique phenotype of CD4-CD8α^{dim}. Interestingly, it seems that while naïve T cells express normal levels of CD4, activation and development of a subset of these into a memory phenotype is associated with down-modulation of CD4 and upregulation of CD8α. Curiously, these cells retain the functionality of T helper cells and are restricted to antigen presented on MHC-II, despite the lack of CD4 expression (Beaumier et al. 2009). As in humans and rhesus macaques, CD4+ T cells with a memory phenotype make up the majority of SIVagm-infected CD4+ T cells. However, while the limited number of memory T cells that maintain CD4 expression has the highest level of infection, the CD4-CD8α^{dim} population contains relatively low numbers of infected cells. Thus, the loss of CD4 expression in differentiation from naïve to memory apparently protects a subset of CD4+ memory T cells from infection.

Similarly, a large population of CD4-CD8- ("double-negative") peripheral blood T cells has been identified in both infected and uninfected sooty mangabeys (Milush et al. 2011). These cells were first identified when it was discovered that some strains of SIVsmm are in fact capable of causing very rapid and severe loss of CD4+ T cells in sooty mangabeys, to levels that would be associated with AIDS in humans and macaques. Despite this, these animals maintain a competent immune system, and this led to the discovery that the CD4–CD8- T cells seem to be capable of performing functions normally associated with CD4+ T cells. In both species, the segregation of functions normally carried out by CD4+ T cells to cells resistant to infection means that regardless of the extent of CD4+ T-cell depletion, a minimum level of immune function can be maintained.

Expression of the coreceptor CCR5 has also been examined in primate species, in the context of SIV infection. One report has compared the frequency of CCR5+ CD4+ T cells in the blood and lymph nodes of uninfected individuals of multiple primate species—comparing those species that are known to be SIV infected with the wild (including African green monkeys and sooty macaques) with humans and rhesus macaques. This report demonstrates that sooty mangabeys, African green monkeys, and numerous other African species that are host to SIV have a much lower frequency of CD4+ T cells that express CCR5 (Pandrea et al. 2007a). The low frequency of CCR5 expression has therefore been assumed to be convergent evolution by SIV-exposed species to restrict target cell availability.

A recent study has further demonstrated that upon in vitro stimulation, naïve T cells from sooty mangabeys show more restricted CCR5 upregulation compared to naïve T cells from rhesus macaques. This failure to upregulate CCR5 is especially pronounced in cells that take on a central memory phenotype after activation. This may protect central memory T cells from infection; as in sooty mangabeys, this population shows a much lower frequency of infection than the Tcm population of rhesus macaques (Paiardini et al. 2011). However, it should be stressed that in this comparison, different viral isolates were used to infect the two different species, and the possibility of minor differences in viral tropism cannot be excluded.

Further examination of sooty mangabeys has demonstrated a relatively high frequency of CCR5 deletion mutant alleles, with 8 % homozygous for CCR5 deletions in a large captive population (Riddick et al. 2010). Interestingly, in contrast to HIV-1-infected humans, these animals are not resistant to SIV infection. In the captive population studied, prevalence of naturally acquired SIV infection was similar in the CCR5 deletion homozygous animals as in heterozygous animals or those with two functional CCR5 alleles (Riddick et al. 2010). Viruses isolated from both homozygous CCR5 deletion animals and those expressing CCR5 were able to use a variety of additional coreceptors, including CXCR6, GPR15, and GPR1, in addition to maintaining CCR5 utilization. Similarly, in red-capped mangabeys, two studies have shown a high frequency of one CCR5 deletion (CCR5 Δ24), with approximately 60–70 % of animals being homozygous for this deletion (Beer et al. 2001; Chen et al. 1998). Cells from red-capped mangabeys with the homozygous CCR5 deletion were completely resistant to infection by CCR5-using viruses from other species. Interestingly, SIV of red-capped mangabeys has apparently adapted to this selection pressure, as this virus uses principally CCR2b as a coreceptor and cannot use CCR5 (Beer et al. 2001; Chen et al. 1998).

Damage to the Mucosal Immune System and Bacterial Translocation

The clinical picture of HIV-1 infection of humans suggests that damage to the immune system is a chronic, gradual process, resulting in the cumulative immunological collapse that allows the onset of opportunistic disease. In contrast, a number of studies suggest that a great deal of irreversible damage occurs early in infection in the secondary lymphoid organs and gut-associated immune system.

The gut-associated lymphoid system (GALT) can be divided into inductive sites (gut-draining mesenteric lymph nodes and Peyer's patches), at which T-cell responses are generated through interaction with antigen-presenting cells and effector sites, primarily lymphocytes residing in the lamina propria (lamina propria lymphocytes, LPLs) and between the epithelial cells of the mucosal surface (intraepithelial lymphocytes, IELs). There are a large number of CD4+ T cells in the Peyer's patches and the lamina propria, with expression of CCR5 at a much higher frequency in these compartments than CD4+ T cells in the peripheral blood or other secondary lymphoid tissues.

Several studies have demonstrated a dramatic depletion of CD4+ T cells in one or more of these compartments from the very earliest stage of HIV-1 infection, with little or no recovery of these cells in the chronic phase of infection (Brenchley et al. 2004b; Estes et al. 2008; Guadalupe et al. 2003; Mehandru et al. 2004). The result of this T-cell loss has been the topic of much discussion in the field. Some authors have stated that the gut mucosal immune system contains the majority of lymphocytes in the body (Guadalupe et al. 2003; Mehandru et al. 2004; Veazey et al. 1998), with some further stating that the acute loss of CD4+ T cells in the gut therefore "reflects the loss of most CD4+ T-cells in the body" (Brenchley and Paiardini 2011). However, available literature on this subject (Ganusov and De Boer 2007) finds that it is more likely that at most around 20 % of all lymphocytes are resident in the human gut, with a similar percentage of the total CD4+ T cells found there.

Despite this, there does seem to be at least one important consequence of damage to the mucosal immune system by HIV-1 infection, as it appears to result in the loss of integrity of the mucosal barrier. This allows translocation of bacteria and bacterial products and their systemic dissemination, which in turn causes the production of pro-inflammatory cytokines by cells of the innate immune system. The level of bacterial translocation, as measured by concentration of lipopolysaccharide (LPS) in the peripheral blood, is significantly upregulated in chronically infected HIV-1 patients when compared to uninfected controls in European and US cohorts (Brenchley et al. 2006). LPS translocation also shows a significant correlation with the level of peripheral blood activated CD8+ T cells, as measured by expression of the markers CD38 and HLA-DR (Brenchley et al. 2006). This acute damage to the gut mucosa therefore seems to be a major contributing factor to the chronic immune activation that is thought to drive the progression from chronic HIV-1 infection to AIDS (see below). The damage to the gut mucosal immune system in SIVmac-infected rhesus macaques is similar to humans, with profound depletion of CD4+ T cells occurring within weeks of infection at multiple sites of the GI tract and increased levels of plasma LPS (Brenchley et al. 2006; Ling et al. 2007; Mattapallil et al. 2005; Veazey et al. 1998).

The importance of LPS translocation in HIV-1 pathogenesis, supported by data from HIV-1+ patients in the USA and Europe, is not entirely supported by two studies carried out in sub-Saharan Africa. One study finds no difference in LPS levels between patient samples taken from pre-infection to AIDS (Redd et al. 2009), while another finds a significant difference only between uninfected patients and patients that have already progressed to AIDS—not earlier in the disease process (Nowroozalizadeh et al. 2010). In this second study, there is however a nonsignificant trend for increased plasma LPS in HIV-1-positive patients compared to negative control individuals and a significant inverse correlation between peripheral blood CD4+ T-cell counts and plasma LPS. However, these findings would be compatible with a theory that LPS translocation is allowed to occur due to loss of CD4+ T cells by other mechanisms and becomes progressively worse throughout the course of disease—rather than necessarily indicating that LPS translocation is a driving factor in immune activation and loss of CD4+ T cells from the outset of infection.

However, the importance of loss of the integrity of the gut mucosal barrier in pathogenic lentivirus infections is emphasized by the different outcome in African nonhuman primates. Surprisingly, two studies have demonstrated that in nonpathogenic infection of African green monkeys and sooty mangabeys, there is also massive depletion of gut mucosal T cells, comparable with depletion seen in SIVmac-infected rhesus macaques (Gordon et al. 2007; Pandrea et al. 2007b). Recovery of CD4+ T cells in this compartment after acute infection is variable, with some animals restoring numbers of cells to near pre-infection levels and some maintaining very low levels throughout the chronic phase of infection. Despite this, the mucosal barrier appears to remain intact, as SIV-infected African green monkeys and sooty mangabeys do not show increased levels of plasma LPS (Brenchley et al. 2006; Gordon et al. 2007; Pandrea et al. 2007b).

The critical difference between the different outcomes may involve a specific subset of CD4+ T cells, Th17 cells. These cells are postulated to be involved in defense against bacterial pathogens through the production of the cytokines IL-17 and IL-22 (Brenchley et al. 2008; Liang et al. 2006) and are found at higher frequency in the gut mucosa compared with peripheral blood or lungs (Brenchley et al. 2008; Cecchinato et al. 2008). In HIV-1-infected humans and SIVmac-infected rhesus macaques, in addition to the general destruction of gut CD4+ T cells, there is a specific greater loss of Th17 cells—they are underrepresented in the residual CD4+ T cells (Brenchley et al. 2008; Cecchinato et al. 2008). In contrast, in the SIV infection of sooty mangabeys, total gut CD4+ T cells are depleted in SIV infection but the percentage of Th17 cells within the remaining CD4+ cells is not altered (Brenchley et al. 2008). Furthermore, in a study directly comparing SIVagm infection of pig-tailed macaques and African green monkeys, significant specific depletion of Th17 only occurs in pig-tailed macaques, which progress to AIDS after SIVagm infection (Favre et al. 2009). Thus, maintenance of the Th17 cell population seems to play an important role in sustaining a competent mucosal barrier, which in turn prevents systemic spread of bacteria from the gut and the subsequent immune activation which follows in AIDS-susceptible species. How natural host species have evolved to avoid this specific depletion of Th17 cells in the face of massive general loss of CD4+ T cells in the gut has yet to be elucidated.

Persistent or Resolving Innate and Adaptive Immune Activation

As mentioned above, elevated levels of immune activation in the chronic phase of infection is seen as an integral part of HIV-1 pathogenesis. Numerous markers of activation of the innate and adaptive immune system are elevated in HIV-1-infected humans, with many showing correlation with the rate of disease progression, especially markers of activation of CD4+ and CD8+ T cells. High levels of immune activation are thought to drive CD4+ T-cell depletion by inducing multiple rounds of expansion, followed by activation-induced cell death (AICD), or infection of activated cells by HIV-1 (as they are more susceptible than naive/resting cells). The level of increased expression of markers of immune activation on CD8+ T cells such as CD38 can be used to predict disease progression with power similar to that of set point viral load measurement (Deeks et al. 2004; Hazenberg et al. 2003). These activated cells are unlikely to represent exclusively CD8+ T cells that are responding to HIV-1 antigens, as the proportion of HIV-1-specific CD8+ T cells in the peripheral blood is generally of a lower percentage than the proportion of CD8+ T cells expressing immune activation markers (depending on the markers measured) (Doisne et al. 2004; Gea-Banacloche et al. 2000; Saez-Cirion et al. 2007; Sieg et al. 2005). In addition, while increased frequency and level of CD38 expression on CD8+ T cells correlates with increased viral load (Deeks et al. 2004) and predicts faster disease progression in HIV-1-infected humans, the frequency of HIV-1-specific CD8+ cells has been shown to either correlate inversely with viral load (Edwards et al. 2002; Ogg et al. 1998) or show no correlation with viral load and survival (Addo et al. 2003; Schellens et al. 2008).

Studies using both flow cytometry to measure levels of activated T cells and microarray analysis to measure gene expression in the peripheral blood and lymphoid tissues, throughout the course of infection, have demonstrated striking differences in immune activation in SIV-infected African green monkeys and sooty mangabeys compared to disease susceptible HIV-1-infected humans and SIV-infected rhesus and pig-tailed macaques. In the acute stage of infection, in all species, there is a dramatic increase in the number of circulating activated T cells, expression of genes that are induced by type I interferon, and soluble markers of immune activation in the plasma (Harris et al. 2010; Kornfeld et al. 2005; Li et al. 2009; Meythaler et al. 2009; Stacey et al. 2009). In all species, the level of immune activation is reduced in the transition from acute to chronic phase of infection. However, the key difference between pathogenic and nonpathogenic infections is the magnitude of this reduction in immune activation after the acute phase. In humans and macaques, the reduction is only partial, and multiple markers of immune activation remain highly elevated compared to pre-infection levels or uninfected controls.

In contrast, in sooty mangabeys and African green monkeys, the reduction in immune activation after acute infection is more complete. In sooty mangabeys, a small number of markers remain slightly (but significantly) upregulated when chronically SIV-infected animals are compared to uninfected controls (Meythaler et al. 2009; Silvestri et al. 2003). Interestingly, the extent of low-level chronic immune activation does correlate inversely with CD4+ T-cell counts (Silvestri et al. 2003). Chronically infected African green monkeys are indistinguishable from uninfected animals in the majority of markers, as immune activation is completely suppressed (Kornfeld et al. 2005; Lederer et al. 2009; Lozano Reina et al. 2009).

Of particular interest is the strong induction of interferon-sensitive genes in the lymph nodes of all species, which are then reduced to near-normal levels in sooty mangabeys and African green monkeys, while remaining high in humans and rhesus and pig-tailed macaques. Continuous immune activation has severe consequences for the secondary lymphoid environment in pathogenic infections (see below). In addition, the type I interferon response is likely connected to the other elements of immune activation. Purified activated CD4+ T cells from HIV-1-infected humans show much higher expression of interferon-sensitive genes than activated cells from uninfected humans (Sedaghat et al. 2008), suggesting that type I interferon plays a role in the excessive T-cell activation seen in HIV-1 infection. In addition, the down-modulation of a type I interferon receptor on monocytes (which is likely to be a measure of previous type I interferon exposure) correlates with lower CD4+ T-cell counts, higher viral loads, and higher expression of CD38 on CD8+ T cells (Hardy et al. 2009).

The mechanism by which sooty mangabeys and African green monkeys restrict the interferon response and other elements of immune activation after the acute stage of infection has been difficult to elucidate and may well be different for each species. Down-modulation of the interferon response in the lymphoid tissue occurs in natural host species despite similar levels of viral replication in secondary lymphoid tissues (Gueye et al. 2004). One study has shown that in vitro, plasmacytoid dendritic cells (pDCs), a major interferon-producing cell population, from sooty mangabeys produce much less interferon in response to a range of stimuli, compared to humans or rhesus macaques, and suggested this results in reduced immune activation in SIV infection (Mandl et al. 2008). However, this is somewhat at odds with the high levels of expression of interferon-sensitive genes in the acute stage of infection in this species, which does not suggest a reduced capacity for interferon production (Bosinger et al. 2009). Furthermore, immunohistochemical stainings of lymph node sections from rhesus macaques, African green monkeys, and sooty mangabeys during acute infection found pDCs to be strongly producing type I IFNs in all three species (Harris et al. 2010). Crucially, in sooty mangabeys and African green monkeys, the interferon response is downregulated after the acute phase of infection, while it remains elevated throughout the infection of rhesus macaques.

Several studies have analyzed gene expression throughout infection, comparing species with different outcomes. In African green monkeys, there is an early induction of the immunosuppressive cytokine *IL-10* (Jacquelin et al. 2009; Kornfeld et al. 2005; Lederer et al. 2009) when compared to rhesus macaques and pig-tailed macaques, but this is not found in sooty mangabeys. Several other immunosuppressive genes are upregulated in sooty mangabeys, such as the enzyme indoleamine-pyrrole 2,3-dioxygenase (*IDO/INDO*) (Bosinger et al. 2009). However, other studies using similar methods of analysis at the gene expression level, in addition to immunohistochemistry to analyze expression at the protein level, have found increased expression of IL-10 and IDO, along with other immunosuppressive molecules postulated to restrict immune activation in natural host species, in pathogenic infections of humans (Li et al. 2009) and rhesus macaques (Estes et al. 2006; Jacquelin et al. 2009). Thus, no convincing mechanism has been postulated by which the immune system is consistently able to bring itself under control in nonpathogenic infections of these African primates.

Destruction of the Lymph Node Environment

One consequence of high and persistent levels of immune activation appears to be severe damage to the secondary lymphoid environment. The structure of secondary lymphatic tissue is of vital importance in mediating interactions between antigen-presenting cells, T cells, and B cells, which are required to induce an immune response. The T-cell zones of secondary lymphoid tissues are also the major anatomical site in which naïve T cells may reside. Pathological changes have been recognized in the lymph nodes of HIV-1 patients since the first description of the disease, with severe lymphadenopathy being a symptom of HIV-1 infection. Lymph nodes from HIV-1-infected patients usually show hyperplasia and, in the later stages of disease, follicular lysis and involution. Damage to the secondary lymph node environment has long been postulated to be an important underlying factor in AIDS pathogenesis (Heeney 1995).

Important changes occur in the T-cell zone, causing dramatic effects on the whole body T-cell population. Firstly, infected humans and rhesus macaques show significantly higher levels of collagen deposition in this area compared to uninfected controls (Diaz et al. 2010; Estes et al. 2007). As might be expected, the level of collagen deposition correlates inversely with the number of naïve (and total) CD4+ T cells resident in the lymph node, suggesting that collagen deposition discourages habitation of this niche. In addition, the degree of collagen deposition before the start of antiretroviral treatment inversely correlates with the increase in peripheral blood T cells after 6 months of treatment (Schacker et al. 2002) or after an even longer duration of treatment (Kumarasamy et al. 2009). Deposition of collagen may therefore physically restrict the size of the niche available for CD4+ T cells to occupy. In addition, a study in rhesus macaques has demonstrated that collagen deposition restricts access of naïve T cells to the fibroblastic reticular cell (FRC) network. This network provides the framework for migration of T cells through the T-cell zone, and cells of this network also produce IL-7, a key stimulus for naïve T-cell survival. Loss of access to this network in these animals leads to apoptosis of naïve T cells in the node (Zeng et al. 2011).

The cause of this damage to the secondary lymphoid tissues has been connected to two factors. First, in HIV-1-infected humans, there is abnormal accumulation of CD4+ and CD8+ effector memory T cells in lymphoid tissues. Effector memory cells are rare in the lymphoid tissue of uninfected humans, as their normal role is to respond to antigen in non-lymphoid tissue. The extent of the infiltration correlates with the area of collagen deposition in the tissue (Brenchley et al. 2004b). These cells are presumably recruited to the lymph nodes as a result of the pro-inflammatory environment induced by HIV-1, and so the presence of such cells could either be a cause of damage or only a consequence of the inflammation ongoing in the lymph node.

The second factor is the increased prevalence of TGFβ1-producing regulatory T cells (Tregs) in the lymphatic tissue of infected humans and rhesus macaques. Tregs are CD4+ T cells with an immunosuppressive function, including the production of the cytokine TGFβ1. This cytokine has multiple functions, most prominently as an immunosuppressant and in mediating wound repair. However, inappropriate expression is known to have a role in tissue fibrosis in other disease pathways (Branton and Kopp 1999). In both humans and rhesus macaques, increased numbers of TGFβ1-expressing cells and increased collagen deposition occur in the acute stage of infection and become exacerbated through the chronic stage of infection. Interestingly, increased numbers of T cells producing the immunosuppressive factors IDO and IL-10 are also found in infected rhesus macaques (Estes et al. 2006). It seems likely that TGFβ1 is produced as part of a negative feedback response to the high levels of immune activation, with the side effect of promoting damaging fibrosis in the immune microenvironment.

The damaging effect of this immunosuppressive response is somewhat paradoxical, especially given that greater or similar levels of IL-10, IDO, and even TGFβ1 are found in gene expression analyses of SIV-infected African green monkeys and sooty mangabeys when compared to rhesus and pig-tailed macaques (Bosinger et al. 2009; Jacquelin et al. 2009; Lederer et al. 2009). However, when using immunohistology to analyze expression at the protein level, SIV-infected sooty mangabeys do not show increased numbers of TGFβ1-producing T cells in the lymph node, and increased collagen deposition is not found in the T-cell zone. It seems likely that the pathological aspect of TGFβ1 production in the lymph nodes of humans and macaques is due to the failure, despite an immunosuppressive response, to bring levels of immune activation down after acute infection. The ability mentioned above of sooty mangabeys and African green monkeys to resolve innate and adaptive immune activation after chronic infection therefore likely saves the lymph node environment from destruction.

The destruction of the lymph node environment in pathogenic infection leads to loss of naïve T cells and the inability to generate de novo CD4+ and CD8+ T-cell responses by preventing normal interactions between T cells and antigen-presenting cells. Both naïve and memory T cells may become inappropriately activated due to the high levels of interferon and are lost through subsequent apoptosis or direct HIV infection. The combination of loss of memory T cells and the environment required to generate new responses presumably drives the immune system to the critical point at which AIDS occur (Heeney 1995). Existing memory cells to a specific antigen/pathogen are sufficiently depleted that there is no longer a meaningful memory response to opportunistic infections, and the generation of new responses is sufficiently delayed by the loss of proper lymph node environment and depletion of naïve T cells that an immune response cannot be mounted in time to prevent illness and death. The key pathways leading to disease in humans and macaques and how they are prevented in African green monkeys and sooty mangabeys are outlined in Fig. 3.

The HIV-1 Viral Lineage Has Increased Pathogenic Potential: Limitations of Available Animal Models

As discussed, SIVsmm generally does not cause disease in sooty mangabey but has the potential to cause an AIDS-like disease upon direct inoculation of Asian macaques (McClure et al. 1989; Silvestri et al. 2005). The disease caused in rhesus macaques is a commonly used animal model for AIDS in humans caused by HIV-1. SIVsmm has also been transmitted into humans, causing the HIV-2 epidemic. Numerous lines of evidence demonstrate that in humans, HIV-2 is not as pathogenic as HIV-1. In surveys taking place in Guinea-Bissau, Holmgren and colleagues found only a twofold increase in mortality in individuals infected with HIV-2, comparable with previous findings within the same geographical region (Holmgren et al. 2007; Poulsen et al. 1997; Ricard et al. 1994). In contrast, HIV-1 is associated with a 10–15-fold increase in mortality rate (reviewed in (Jaffar et al. 2004). Longitudinal investigations with statistically meaningful cohorts indicate that about 80 % of HIV-2-infected individuals do not develop disease and maintain a normal life expectancy (van der Loeff et al. 2010). One-third of all HIV-2-infected

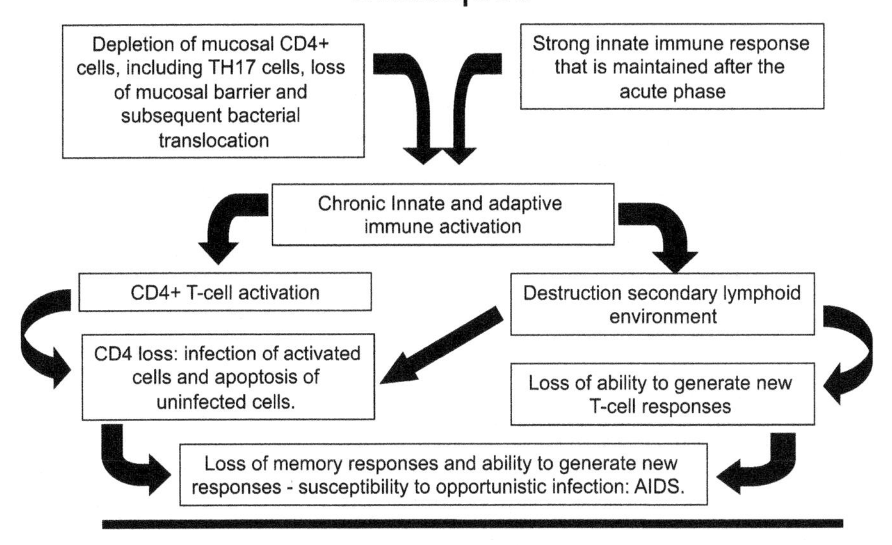

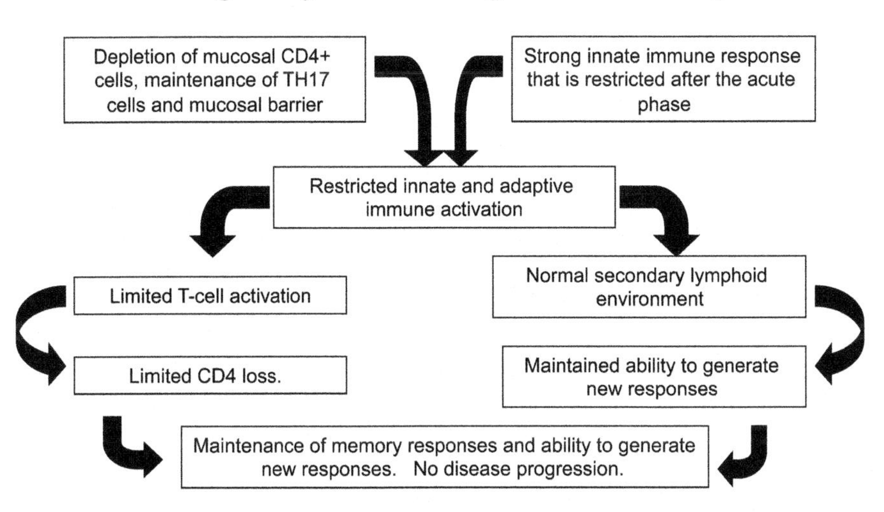

Fig. 3 Pathways leading to the development of AIDS in progressive SIV/HIV infection compared with the nonprogressive SIV infection of African primates

individuals have an undetectable plasma viral load (van der Loeff et al. 2010). Nevertheless, progressive HIV-2 infections show clinical features that are indistinguishable from AIDS caused by HIV-1 (Martinez-Steele et al. 2007).

Given the considerable difference in the outcome of HIV-1 and HIV-2 infection of humans, is it appropriate to use the disease of rhesus macaques infected with viruses of the SIVsm/SIVmac/HIV-2 lineage as a model for disease caused by HIV-1 in humans? If it is the case that there are specific aspects of the HIV-1 lineage that make this virus more pathogenic than HIV-2 in humans, then this model becomes less attractive in understanding HIV-1 pathogenesis. We will now examine the differences between these viruses to establish if this is the case.

Despite their different origins, HIV-1 and HIV-2 are relatively closely related retroviruses with 60 % similarity at the amino acid level in the capsid and polymerase proteins and 30 % similarity in the envelope protein (Guyader et al. 1987). However, the genomes of both viral lineages differ by two genes. While *vpx* is present in the HIV-2/SIVsmm lineage (in addition to SIVrcm, SIVdrl, and SIVmnd2), the accessory gene *vpu*, not *vpx*, is found in HIV-1, in addition to SIVcpz and the SIVs of four species of the *Cercopithecus* genus (Barlow et al. 2003; Courgnaud et al. 2002, 2003; Huet et al. 1990; Strebel et al. 1988).

Both *vpu* and *vpx* seem to have evolved to allow lentiviruses to escape host restriction mechanisms. In macaques, the accessory viral gene *vpx* has been directly shown to contribute to virulence. Animals infected with an SIVmac mutant lacking the *vpx* gene show lower virus burdens, delayed declines in CD4 lymphocytes, and either a lack of disease or delayed progression to disease (Gibbs et al. 1995; Hirsch et al. 1998). The Vpx of the SIVsmm/SIVmac/HIV-2 viral lineage enables the virus to efficiently replicate in primate macrophages by antagonizing a cellular restriction factor termed SAM domain and HD domain 1 (SAMHD1) that is expressed in this cell type (Laguette et al. 2011; Sharova et al. 2008). That vpx can provide a fitness advantage to the virus was further confirmed in vivo when a wild-type and a vpx-deleted SIVsmm isolate were inoculated into pig-tailed macaques, resulting in the vpx mutant showing a strong competitive disadvantage in the early virus dissemination (Hirsch et al. 1998).

vpu, the accessory gene uniquely expressed in the SIVcpz/HIV-1 lineage, including the small subset of guenon SIVs, also seems to posess various functions that may impact pathogenicity. Vpu helps HIV-1 to escape host restriction and enhances release of virus particles (Klimkait et al. 1990). Vpu contributes to HIV-1-induced CD4 receptor downregulation, which enhances virus replication (Willey et al. 1992). In addition, it increases the release of progeny virions from infected cells and cell-to-cell spread of viral particles by antagonizing tetherin, an interferon-induced host restriction factor that directly cross-links virions on the host cell surface (Neil et al. 2008). However, both CD4 and tetherin downregulation are not exclusive to the lineage of HIV-1 and are facilitated by the nef or env gene in other lentiviruses (Gupta et al. 2009; Le Tortorec and Neil 2009; Zhang et al. 2009). However, two functions of vpu have recently been discovered that are thus far unique to HIV-1. Firstly, Vpu inhibits the recycling of CD1d receptor from endosomal compartments, strongly inhibiting the ability of infected DC to activate CD1d-restricted natural killer T cells (NKT cells) (Moll et al. 2010). Secondly, Vpu downregulates the NKT and B cell coactivator (NTB-A) at the surface of infected cells and as a result interferes with the degranulation of NK cells that recognize the infected cells (Shah et al.

2010). Importantly, HIV-2-infected cells do not downregulate NTB-A and are killed more efficiently by NK cells than HIV-1-infected cells (Shah et al. 2010). Furthermore, *vpu* seems to have influenced the evolution of the other viral genes. A mechanism encoded in the viral gene *nef* is believed to have evolved to restrict T-cell activation, through down modulation of the T-cell receptor (also referred to as CD3) in infected cells. However, in all viruses that carry *vpu* this function of nef is reported to be lost (Schindler et al. 2006). Immunodeficiency viruses depend on activated T cells for replication, but accelerated immune activation induces a range of host defense mechanisms. It has been suggested that acquisition of the *vpu* gene by this viral lineage has allowed the virus to trigger immune activation, increasing viral replication, without the consequence of host-mediated restriction (Kirchhoff 2009). There are therefore reasons to believe that *vpu* plays an important role in HIV-1-induced disease, and models of human AIDS should include this factor.

While pig-tailed macaques can become transiently infected with HIV-1 (Agy et al. 1992), the infection does not persist. Other macaque species are resistant to infection, as macaque-encoded restriction factors provide dominant-acting blocks the establishment of infection (Stremlau et al. 2004). Using the macaque model to investigate influence of the *vpu* gene in vivo therefore requires more complex approaches. Our increased understanding of host restriction factors has led to novel approaches in which HIV-1 isolates have been artificially equipped with viral gene variants capable of antagonizing the macaque's host restriction factors (Hatziioannou et al. 2009). However, to date, these artificial HIV variants are not pathogenic. An alternative approach has been to infect macaques with partial recombinants between SIVs and HIV-1 isolates, called SHIVs. The first SHIVs were created by exchanging the SIV envelope for an HIV-1 envelope sequence in order to test HIV-1 vaccine candidates (Stremlau et al. 2004). Some SHIVs are equipped with the HIV-1 *vpu* gene in addition to the HIV envelope, and a few studies indicate that the presence of *vpu* can influence the pathogenic outcome in pig-tailed macaques (Hout et al. 2005; Singh et al. 2001, 2003; Stephens et al. 2002). However, the mechanistic background remains unknown, especially as the various described functions of *vpu* are yet not characterized for these Asian primate hosts. The lack of a suitable model for HIV-1 has driven the search for alternative animal models such as the rodent model in which mice were "humanized" with bone marrow/liver/thymus grafts from human donors (Melkus et al. 2006). To date, these models are still in a premature state and only further research will establish whether insights into HIV-1 virulence can be gained with such models.

Examining the Virus–Host Relationships Responsible for Generating HIV-1

As stated previously, other than HIV-1, only SIVcpz of chimpanzees and SIVmon/mus/gsn/den of four primate species of the guenon genus carry the *vpu* gene. Given the difficulties in assessing the role of vpu in existing in vivo models described above, it would be extremely informative to understand the outcome of SIV infection in these species. Here, we will review what is known of SIV infection of chimpanzees and in these guenon species.

Before the origin of HIV-1 was traced to chimpanzees, it was found that chimpanzees can be persistently infected with HIV-1 (Alter et al. 1984). As no other species can be persistently infected with HIV-1, the lack of alternatives made this animal model initially attractive to study HIV-1. However, as the vast majority of animals infected with HIV-1 failed to develop clinical signs, and due to cost and ethical reasons, the use of the chimpanzee model for HIV-1 study has been generally abandoned. Nevertheless, by 1996, over 200 chimpanzees had been infected with HIV-1 (Committee on Long-Term Care of Chimpanzees 1997). Many of these animals have now been infected for over 20 years. Of these, only 5 animals, all at the Yerkes Primate Centre, have been described as developing AIDS and subsequently euthanized (Juompan et al. 2008). The majority of these 200 HIV-1-infected chimpanzees were infected with the CXCR4 coreceptor using HIV-1 virus isolate, IIIb. It would be tempting to postulate that due to the expansion of this virus in vitro and the artificial nature of infection with a CXCR4 using virus, this isolate is inherently less able to induce disease. Unfortunately, data exists that this strain remains able to induce disease in humans, as a laboratory worker was accidentally infected with HIV-1_{IIIb} and progressed to AIDS after 8 years of infection (Beaumont et al. 2001). In general, captive chimpanzees infected with HIV-1 rapidly control viral load to undetectable levels. There is therefore direct evidence that HIV-1 is generally not pathogenic in chimpanzees.

The vast majority of chimpanzees held in European and US primate centers and therefore the majority of HIV-1-infected chimpanzees were of the *Pan troglodytes verus* subspecies. This subspecies is not naturally infected with SIVcpz—SIV infection has only been found in the *troglodytes* and *schweinfurthii* subspecies despite extensive surveys of captive and wild populations of the *verus* subspecies. It is worth noting, however, that while SIV infection has not been found in the *verus* subspecies, these animals must have a significant history of exposure to the SIV of the western red colobus monkey. The red colobus monkey is frequently hunted by chimpanzees (Boesch and Boesch-Achermann 2000) and has a high prevalence of SIV infection (Locatelli et al. 2008), though this virus has not been found in *verus* chimpanzees (Leendertz et al. 2011). This could suggest that some resistance mechanisms shared with naturally SIV-infected primate species, such as low frequency of CCR5 expression on CD4+ T cells (Pandrea et al. 2007a), have evolved in the *verus* subspecies due to historic exposure to SIVwrc.

Given the close relationship between HIV-1 and SIVcpz, it would seem reasonable to assume that SIVcpz would also not be pathogenic in the chimpanzee. However, a recent report studying wild chimpanzees has found the opposite. A study of wild, habituated animals (wild animals accustomed to being observed very closely) of the *P. t. schweinfurthii* subspecies has found that SIVcpz-infected animals within the group showed greater risk of mortality of the study period (Keele et al. 2009). In addition, the bodies of three infected animals were recovered for necropsy, along with the bodies of two uninfected animals. Significant depletion of

CD4+ cells in the periarteriolar lymphoid sheaths (PALS) of the spleens of these animals was shown in the SIVcpz-infected animals. Particularly severe depletion was shown in an animal that had become infected during the study, three years prior to death, whose death was attributed to an "AIDS-like" disease by the group reporting this finding. A more recent article reports also identified a *Pan troglodytes troglodytes* chimpanzee with AIDS-like clinical signs in a Cameroonian sanctuary (Etienne et al. 2011). A pathogenic outcome of SIVcpz infection would lend further support to the importance of *vpu* in HIV-1 infection.

From 1989 to 2005, seven SIVcpz-infected chimpanzees were housed in primate centers in Europe and the USA (Heeney et al. 2006): one naturally infected *schweinfurthii* animal, two experimentally infected *schweinfurthii* animals, and four experimentally infected *verus* animals. Of these seven animals, four are still alive, including Noah, a naturally infected animal that first tested positive in 1989, aged around 2 years old, and remains asymptomatic after over 20 years of infection (Greenwood et al., unpublished data). Three animals have died of cardiac conditions, which are common in captive chimpanzees (Seiler et al. 2009).

Two of these chimpanzees have already been studied in some detail as part of a cohort of HIV-1-infected chimpanzees. As with the HIV-1-infected chimpanzees, these two SIVcpz-infected chimpanzees seem to lack the profound changes to the immune system that are found in HIV-1-infected humans (Gougeon et al. 1997; Rutjens et al. 2008, 2010). Given the apparent pathogenic outcome of SIVcpz infection of wild chimpanzees, it should be possible to identify, in these captive animals, disease mechanisms that are shared with HIV-1-infected humans. Further study of samples from these and other captive SIVcpz-infected animals is therefore warranted, especially if this is possible using previously archived samples and/or non-invasive methods.

As previously mentioned, SIVcpz is the result of recombination between SIVrcm and SIVgsn/mon/mus. The importance of the *vpu* gene seems to be emphasized by the fact that it is retained in the chimeric virus in favor of *vpx*. The small subset of *Cercopithecus* species infected with these *vpu*-carrying viruses have not been investigated for their outcome of infection. The only knowledge we have of these viruses is their genomic sequences and their prevalence in the wild. Interestingly, the few studies that investigated significant numbers of these *Cercopithecus* animals found an exceptionally low prevalence of the *vpu*-harboring SIVs in mustached and greater spot-nosed guenon populations in Cameroon (2–4 %). This is in stark contrast to the seroprevalence seen in primate populations that carry an SIV lacking the *vpu* gene, which reported seroprevalence of 50–90 % (Aghokeng et al. 2006, 2009; Ellis et al. 2004). It therefore seems that the outcome of SIVgsn/SIVmon infection is likely to differ from other SIV infection of natural hosts. It is possible to postulate two possible mechanisms for this low prevalence. SIV infection in these species could be much more pathogenic than in other species, perhaps due to the presence of *vpu*—which leads to infected animals being rapidly removed from the population. Alternatively, these species could have evolved novel mechanisms to prevent the spread of the virus within the population, though presumably the evolutionary pressure for this to occur would again have to be the result of increased pathogenicity of SIV infection of these specific species. The opportunity of further study for research on these monkeys is restricted, as they are not found in primate research centers and are (rightly) protected in the wild. While surveys of bushmeat samples from African countries have allowed some limited analysis, the low prevalence of SIV carrying will make further studies a major challenge. Nevertheless, due to their possible key role in our understanding of HIV-1 pathogenicity, such efforts may be justified.

Conclusion

The African primate origins of HIV-1 and HIV-2 are now well established, and a relatively thorough history of how HIV-1 has spread from western Central Africa throughout the world has been proposed. However, the mechanisms by which HIV causes AIDS in humans remain heavily debated. It should be noted that it is not possible to review here the entirety of the vast and conflicting literature describing the observed alterations to the immune system in HIV infection and the various mechanisms proposed to cause CD4+ T-cell loss and AIDS in HIV in infected humans.

Instead, we have highlighted the key differences found throughout the course of infection between the pathogenic HIV/SIV infection of humans and macaques and the nonpathogenic SIV infection of two well-studied African primate species and used these differences to propose a relatively straightforward model by which lentivirus infection induces AIDS in humans and macaques but not in sooty mangabeys and African green monkeys.

Finally, we have reviewed the possible limitation of current animal models that specifically the HIV-1/SIVcpz viral lineage may have greater potential to cause disease than other primate lentiviruses, which is not well addressed by current primate models. Specifically, we have noted the functions of the *vpu* gene. While some of the functions of *vpu* are carried out by other genes in primate lentiviruses (such as down-modulation of CD4 and tetherin), other functions are so far described uniquely for this gene. It is also noted that the virus–host relationship in primate species infected with a *vpu*-harboring virus seems to differ from that of other African primates. While further examination of primate species infected with *vpu*-carrying viruses—chimpanzees, gorillas, and members of the *Cercopithecus* genus—is clearly complicated by ethical and practical considerations, noninvasive studies of these species have already proven to be fruitful. Expansion of such noninvasive studies may lead to new insights in the subject of lentivirus pathogenesis and host adaptation not available in other animal models.

Acknowledgments We would like to thank Joel Wertheim for creating the HIV/SIV alignments used to create Fig. 1.

References

Addo MM, Yu XG, Rathod A, Cohen D, Eldridge RL, Strick D, Johnston MN, Corcoran C, Wurcel AG, Fitzpatrick CA et al (2003) Comprehensive epitope analysis of human immunodeficiency virus type 1 (HIV-1)-specific T-cell responses directed against the entire expressed HIV-1 genome demonstrate broadly directed responses, but no correlation to viral load. J Virol 77(3):2081–2092

Aghokeng AF, Liu W, Bibollet-Ruche F, Loul S, Mpoudi-Ngole E, Laurent C, Mwenda JM, Langat DK, Chege GK, McClure HM et al (2006) Widely varying SIV prevalence rates in naturally infected primate species from Cameroon. Virology 345(1):174–189

Aghokeng AF, Bailes E, Loul S, Courgnaud V, Mpoudi-Ngolle E, Sharp PM, Delaporte E, Peeters M (2007) Full-length sequence analysis of SIVmus in wild populations of mustached monkeys (Cercopithecus cephus) from Cameroon provides evidence for two co-circulating SIVmus lineages. Virology 360(2):407–418

Aghokeng AF, Ayouba A, Mpoudi-Ngole E, Loul S, Liegeois F, Delaporte E, Peeters M (2009) Extensive survey on the prevalence and genetic diversity of SIVs in primate bushmeat provides insights into risks for potential new cross-species transmissions. Infect Genet Evol 10(3):386–396

Agy MB, Frumkin LR, Corey L, Coombs RW, Wolinsky SM, Koehler J, Morton WR, Katze MG (1992) Infection of *Macaca nemestrina* by human immunodeficiency virus type-1. Science 257(5066):103–106

Aina O, Dadik J, Charurat M, Amangaman P, Gurumdi S, Mang E, Guyit R, Lar N, Datong P, Daniyam C et al (2005) Reference values of CD4 T lymphocytes in human immunodeficiency virus-negative adult Nigerians. Clin Diagn Lab Immunol 12(4):525–530

Alter HJ, Eichberg JW, Masur H, Saxinger WC, Gallo R, Macher AM, Lane HC, Fauci AS (1984) Transmission of HTLV-III infection from human plasma to chimpanzees: an animal model for AIDS. Science 226(4674):549–552

Apetrei C, Lerche NW, Pandrea I, Gormus B, Silvestri G, Kaur A, Robertson DL, Hardcastle J, Lackner AA, Marx PA (2006) Kuru experiments triggered the emergence of pathogenic SIVmac. AIDS 20(3):317–321

Ayinde D, Maudet C, Transy C, Margottin-Goguet F (2010) Limelight on two HIV/SIV accessory proteins in macrophage infection: is Vpx overshadowing Vpr? Retrovirology 7:35

Ayouba A, Souquieres S, Njinku B, Martin PM, Muller-Trutwin MC, Roques P, Barre-Sinoussi F, Mauclere P, Simon F, Nerrienet E (2000) HIV-1 group N among HIV-1-seropositive individuals in Cameroon. AIDS 14(16):2623–2625

Bailes E, Gao F, Bibollet-Ruche F, Courgnaud V, Peeters M, Marx PA, Hahn BH, Sharp PM (2003) Hybrid origin of SIV in chimpanzees. Science 300(5626):1713

Barre-Sinoussi F, Chermann JC, Rey F, Nugeyre MT, Chamaret S, Gruest J, Dauguet C et al (1983) Isolation of a T-lymphotropic retrovirus from a patient at risk for acquired immune deficiency syndrome (AIDS). Science 220(4599):868–871

Barlow KL, Ajao AO, Clewley JP (2003) Characterization of a novel simian immunodeficiency virus (SIVmonNG1) genome sequence from a mona monkey (Cercopithecus mona). J Virol 77(12):6879–6888

Beaumier CM, Harris LD, Goldstein S, Klatt NR, Whitted S, McGinty J, Apetrei C, Pandrea I, Hirsch VM, Brenchley JM (2009) CD4 down-regulation by memory CD4+ T cells in vivo renders African green monkeys resistant to progressive SIVagm infection. Nat Med 15(8):879–885

Beaumont T, van Nuenen A, Broersen S, Blattner WA, Lukashov VV, Schuitemaker H (2001) Reversal of human immunodeficiency virus type 1 IIIB to a neutralization-resistant phenotype in an accidentally infected laboratory worker with a progressive clinical course. J Virol 75(5):2246–2252

Beer BE, Foley BT, Kuiken CL, Tooze Z, Goeken RM, Brown CR, Hu J, St Claire M, Korber BT, Hirsch VM (2001) Characterization of novel simian immunodeficiency viruses from red-capped mangabeys from Nigeria (SIVrcmNG409 and -NG411). J Virol 75(24):12014–12027

Begtrup K, Melbye M, Biggar RJ, Goedert JJ, Knudsen K, Andersen PK (1997) Progression to acquired immunodeficiency syndrome is influenced by CD4 T-lymphocyte count and time since seroconversion. Am J Epidemiol 145(7):629–635

Benveniste RE, Arthur LO, Tsai CC, Sowder R, Copeland TD, Henderson LE, Oroszlan S (1986) Isolation of a lentivirus from a macaque with lymphoma: comparison with HTLV-III/LAV and other lentiviruses. J Virol 60(2):483–490

Bibollet-Ruche F, Bailes E, Gao F, Pourrut X, Barlow KL, Clewley JP, Mwenda JM, Langat DK, Chege GK, McClure HM et al (2004) New simian immunodeficiency virus infecting De Brazza's monkeys (Cercopithecus neglectus): evidence for a cercopithecus monkey virus clade. J Virol 78(14):7748–7762

Boesch C, Boesch-Achermann H (2000) The chimpanzees of the Taï Forest: behavioural ecology and evolution. Oxford University Press, Oxford, p 316, viii

Boily MC, Baggaley RF, Wang L, Masse B, White RG, Hayes RJ, Alary M (2009) Heterosexual risk of HIV-1 infection per sexual act: systematic review and meta-analysis of observational studies. Lancet Infect Dis 9(2):118–129

Bosinger SE, Li Q, Gordon SN, Klatt NR, Duan L, Xu L, Francella N, Sidahmed A, Smith AJ, Cramer EM (2009) Global genomic analysis reveals rapid control of a robust innate response in SIV-infected sooty mangabeys. J Clin Invest 119(12):3556–3572

Branton MH, Kopp JB (1999) TGF-beta and fibrosis. Microbes Infect 1(15):1349–1365

Brenchley JM, Paiardini M (2011) Immunodeficiency lentiviral infections in natural and nonnatural hosts. Blood 118:847–854

Brenchley JM, Hill BJ, Ambrozak DR, Price DA, Guenaga FJ, Casazza JP, Kuruppu J, Yazdani J, Migueles SA, Connors M et al (2004a) T-cell subsets that harbor human immunodeficiency virus (HIV) in vivo: implications for HIV pathogenesis. J Virol 78(3):1160–1168

Brenchley JM, Schacker TW, Ruff LE, Price DA, Taylor JH, Beilman GJ, Nguyen PL, Khoruts A, Larson M, Haase AT et al (2004b) CD4+ T cell depletion during all stages of HIV disease occurs predominantly in the gastrointestinal tract. J Exp Med 200(6):749–759

Brenchley JM, Price DA, Schacker TW, Asher TE, Silvestri G, Rao S, Kazzaz Z, Bornstein E, Lambotte O, Altmann D et al (2006) Microbial translocation is a cause of systemic immune activation in chronic HIV infection. Nat Med 12(12):1365–1371

Brenchley JM, Paiardini M, Knox KS, Asher AI, Cervasi B, Asher TE, Scheinberg P, Price DA, Hage CA, Kholi LM et al (2008) Differential Th17 CD4 T-cell depletion in pathogenic and nonpathogenic lentiviral infections. Blood 112(7):2826–2835

Brown CR, Czapiga M, Kabat J, Dang Q, Ourmanov I, Nishimura Y, Martin MA, Hirsch VM (2007) Unique pathology in simian immunodeficiency virus-infected rapid progressor macaques is consistent with a pathogenesis distinct from that of classical AIDS. J Virol 81(11):5594–5606

Campbell-Yesufu OT, Gandhi RT (2011) Update on human immunodeficiency virus (HIV)-2 infection. Clin Infect Dis 52(6):780–787

CDC (1981a) Follow-up on Kaposi's Sarcoma and pneumocystis pneumonia. MMWR Morb Mortal Wkly Rep 30:409–410

CDC (1981b) Kaposi's sarcoma and pneumocycstis pneumonia among homosexual men—New York City and California. MMWR Morb Mortal Wkly Rep 30:305–308

CDC (1981c) Pneumocycstis pneumonia—Los Angeles. MMWR Morb Mort Wkly Rep 30:1–3

CDC (1982a) Epidemiologic aspects of the current outbreak of Kaposi's sarcoma and opportunistic infections. N Engl J Med 306(4):248–252

CDC (1982b) Epidemiologic notes and reports update on Kaposi's sarcoma and opportunistic infections in previously healthy persons—United States. MMWR Morb Mortal Wkly Rep 31:300–301

CDC (1982c) Opportunistic infections and Kaposi's sarcoma among Haitians in the United States. MMWR Morb Mortal Wkly Rep 31:353–354

Cecchinato V, Trindade CJ, Laurence A, Heraud JM, Brenchley JM, Ferrari MG, Zaffiri L, Tryniszewska E, Tsai WP, Vaccari M et al (2008) Altered balance between Th17 and Th1 cells at mucosal sites predicts AIDS progression in simian immunodeficiency virus-infected macaques. Mucosal Immunol 1(4):279–288

Chen Z, Kwon D, Jin Z, Monard S, Telfer P, Jones MS, Lu CY, Aguilar RF, Ho DD, Marx PA (1998) Natural infection of a homozygous delta24 CCR5 red-capped mangabey with an R2b-tropic simian immunodeficiency virus. J Exp Med 188(11):2057–2065

Chen J, Powell D, Hu WS (2006) High frequency of genetic recombination is a common feature of primate lentivirus replication. J Virol 80(19):9651–9658
Chin J (2007) The AIDS pandemic : the collision of epidemiology with political correctness. Radcliffe, Oxford, Seattle, p 230, xiv
Chitnis A, Rawls D, Moore J (2000) Origin of HIV type 1 in colonial French Equatorial Africa? AIDS Res Hum Retroviruses 16(1):5–8
Clavel F, Guetard D, Brun-Vezinet F, Chamaret S, Rey MA, Santos-Ferreira MO, Laurent AG, Dauguet C, Katlama C, Rouzioux C et al (1986) Isolation of a new human retrovirus from West African patients with AIDS. Science 233(4761):343–346
Collaborative Group on AIDS Incubation and HIV Survival including the CASCADE EU Concerted Action (2000) Time from HIV-1 seroconversion to AIDS and death before widespread use of highly-active antiretroviral therapy: a collaborative re-analysis. Lancet 355(9210):1131–1137
Committee on Long-Term Care of Chimpanzees IfLAR, Commission on Life Sciences, National Research Council (1997) Chimpanzees in research: strategies for their ethical care, management, and use. National Academy Press, Washington, DC
Courgnaud V, Salemi M, Pourrut X, Mpoudi-Ngole E, Abela B, Auzel P, Bibollet-Ruche F, Hahn B, Vandamme AM, Delaporte E et al (2002) Characterization of a novel simian immunodeficiency virus with a vpu gene from greater spot-nosed monkeys (Cercopithecus nictitans) provides new insights into simian/human immunodeficiency virus phylogeny. J Virol 76(16):8298–8309
Courgnaud V, Abela B, Pourrut X, Mpoudi-Ngole E, Loul S, Delaporte E, Peeters M (2003) Identification of a new simian immunodeficiency virus lineage with a vpu gene present among different cercopithecus monkeys (C. mona, C. cephus, and C. nictitans) from Cameroon. J Virol 77(23):12523–12534
Daniel MD, King NW, Letvin NL, Hunt RD, Sehgal PK, Desrosiers RC (1984) A new type D retrovirus isolated from macaques with an immunodeficiency syndrome. Science 223(4636): 602–605
Dazza MC, Ekwalanga M, Nende M, Shamamba KB, Bitshi P, Paraskevis D, Saragosti S (2005) Characterization of a novel vpu-harboring simian immunodeficiency virus from a Dent's Mona monkey (Cercopithecus mona denti). J Virol 79(13):8560–8571
de Sousa JD, Muller V, Lemey P, Vandamme AM (2010) High GUD incidence in the early 20 century created a particularly permissive time window for the origin and initial spread of epidemic HIV strains. PLoS One 5(4):e9936
Dean M, Carrington M, Winkler C, Huttley GA, Smith MW, Allikmets R, Goedert JJ, Buchbinder SP, Vittinghoff E, Gomperts E et al (1996) Genetic restriction of HIV-1 infection and progression to AIDS by a deletion allele of the CKR5 structural gene. Hemophilia Growth and Development Study, Multicenter AIDS Cohort Study, Multicenter Hemophilia Cohort Study, San Francisco City Cohort, ALIVE Study. Science 273(5283):1856–1862
Deeks SG, Kitchen CM, Liu L, Guo H, Gascon R, Narvaez AB, Hunt P, Martin JN, Kahn JO, Levy J et al (2004) Immune activation set point during early HIV infection predicts subsequent CD4+ T-cell changes independent of viral load. Blood 104(4):942–947
Diaz A, Alos L, Leon A, Mozos A, Caballero M, Martinez A, Plana M, Gallart T, Gil C, Leal M et al (2010) Factors associated with collagen deposition in lymphoid tissue in long-term treated HIV-infected patients. AIDS 24(13):2029–2039
Doisne JM, Urrutia A, Lacabaratz-Porret C, Goujard C, Meyer L, Chaix ML, Sinet M, Venet A (2004) CD8+ T cells specific for EBV, cytomegalovirus, and influenza virus are activated during primary HIV infection. J Immunol 173(4):2410–2418
Edwards BH, Bansal A, Sabbaj S, Bakari J, Mulligan MJ, Goepfert PA (2002) Magnitude of functional CD8+ T-cell responses to the gag protein of human immunodeficiency virus type 1 correlates inversely with viral load in plasma. J Virol 76(5):2298–2305
Ellis BR, Munene E, Elliott D, Robinson J, Otsyula MG, Michael SF (2004) Seroprevalence of simian immunodeficiency virus in wild and captive born Sykes' monkeys (Cercopithecus mitis) in Kenya. Retrovirology 1:34

Estes JD, Li Q, Reynolds MR, Wietgrefe S, Duan L, Schacker T, Picker LJ, Watkins DI, Lifson JD, Reilly C et al (2006) Premature induction of an immunosuppressive regulatory T cell response during acute simian immunodeficiency virus infection. J Infect Dis 193(5):703–712

Estes JD, Wietgrefe S, Schacker T, Southern P, Beilman G, Reilly C, Milush JM, Lifson JD, Sodora DL, Carlis JV (2007) Simian immunodeficiency virus-induced lymphatic tissue fibrosis is mediated by transforming growth factor beta 1-positive regulatory T cells and begins in early infection. J Infect Dis 195(4):551–561

Estes J, Baker JV, Brenchley JM, Khoruts A, Barthold JL, Bantle A, Reilly CS, Beilman GJ, George ME, Douek DC et al (2008) Collagen deposition limits immune reconstitution in the gut. J Infect Dis 198(4):456–464

Etienne L, Nerrienet E, LeBreton M, Bibila GT, Foupouapouognigni Y, Rousset D, Nana A, Djoko CF, Tamoufe U, Aghokeng AF et al (2011) Characterization of a new simian immunodeficiency virus strain in a naturally infected Pan troglodytes troglodytes chimpanzee with AIDS related symptoms. Retrovirology 8:4

Fabre PH, Rodrigues A, Douzery EJ (2009) Patterns of macroevolution among Primates inferred from a supermatrix of mitochondrial and nuclear DNA. Mol Phylogenet Evol 53(3):808–825

Favre D, Lederer S, Kanwar B, Ma ZM, Proll S, Kasakow Z, Mold J, Swainson L, Barbour JD, Baskin CR et al (2009) Critical loss of the balance between Th17 and T regulatory cell populations in pathogenic SIV infection. PLoS Pathog 5(2):e1000295

Fleming DT, Wasserheit JN (1999) From epidemiological synergy to public health policy and practice: the contribution of other sexually transmitted diseases to sexual transmission of HIV infection. Sex Transm Infect 75(1):3–17

Ganusov VV, De Boer RJ (2007) Do most lymphocytes in humans really reside in the gut? Trends Immunol 28(12):514–518

Gea-Banacloche JC, Migueles SA, Martino L, Shupert WL, McNeil AC, Sabbaghian MS, Ehler L, Prussin C, Stevens R, Lambert L et al (2000) Maintenance of large numbers of virus-specific CD8+ T cells in HIV-infected progressors and long-term nonprogressors. J Immunol 165(2):1082–1092

Gibbs JS, Lackner AA, Lang SM, Simon MA, Sehgal PK, Daniel MD, Desrosiers RC (1995) Progression to AIDS in the absence of a gene for vpr or vpx. J Virol 69(4):2378–2383

Gifford RJ, Katzourakis A, Tristem M, Pybus OG, Winters M, Shafer RW (2008) A transitional endogenous lentivirus from the genome of a basal primate and implications for lentivirus evolution. Proc Natl Acad Sci USA 105(51):20362–20367

Gilbert MT, Rambaut A, Wlasiuk G, Spira TJ, Pitchenik AE, Worobey M (2007) The emergence of HIV/AIDS in the Americas and beyond. Proc Natl Acad Sci USA 104(47):18566–18570

Gilbert C, Maxfield DG, Goodman SM, Feschotte C (2009) Parallel germline infiltration of a lentivirus in two Malagasy lemurs. PLoS Genet 5(3):e1000425

Gordon SN, Klatt NR, Bosinger SE, Brenchley JM, Milush JM, Engram JC, Dunham RM, Paiardini M, Klucking S, Danesh A et al (2007) Severe depletion of mucosal CD4+ T cells in AIDS-free simian immunodeficiency virus-infected sooty mangabeys. J Immunol 179(5):3026–3034

Gottlieb MS, Schroff R, Schanker HM, Weisman JD, Fan PT, Wolf RA, Saxon A (1981) Pneumocystis carinii pneumonia and mucosal candidiasis in previously healthy homosexual men: evidence of a new acquired cellular immunodeficiency. N Engl J Med 305(24): 1425–1431

Gougeon ML, Lecoeur H, Boudet F, Ledru E, Marzabal S, Boullier S, Roue R, Nagata S, Heeney J (1997) Lack of chronic immune activation in HIV-infected chimpanzees correlates with the resistance of T cells to Fas/Apo-1 (CD95)-induced apoptosis and preservation of a T helper 1 phenotype. J Immunol 158(6):2964–2976

Guadalupe M, Reay E, Sankaran S, Prindiville T, Flamm J, McNeil A, Dandekar S (2003) Severe CD4+ T-cell depletion in gut lymphoid tissue during primary human immunodeficiency virus type 1 infection and substantial delay in restoration following highly active antiretroviral therapy. J Virol 77(21):11708–11717

Gueye A, Diop OM, Ploquin MJ, Kornfeld C, Faye A, Cumont MC, Hurtrel B, Barre-Sinoussi F, Muller-Trutwin MC (2004) Viral load in tissues during the early and chronic phase of non-pathogenic SIVagm infection. J Med Primatol 33(2):83–97

Gupta RK, Mlcochova P, Pelchen-Matthews A, Petit SJ, Mattiuzzo G, Pillay D, Takeuchi Y, Marsh M, Towers GJ (2009) Simian immunodeficiency virus envelope glycoprotein counteracts tetherin/BST-2/CD317 by intracellular sequestration. Proc Natl Acad Sci USA 106(49): 20889–20894

Gurtler LG, Hauser PH, Eberle J, von Brunn A, Knapp S, Zekeng L, Tsague JM, Kaptue L (1994) A new subtype of human immunodeficiency virus type 1 (MVP-5180) from Cameroon. J Virol 68(3):1581–1585

Guyader M, Emerman M, Sonigo P, Clavel F, Montagnier L, Alizon M (1987) Genome organization and transactivation of the human immunodeficiency virus type 2. Nature 326(6114): 662–669

Hardy GA, Sieg SF, Rodriguez B, Jiang W, Asaad R, Lederman MM, Harding CV (2009) Desensitization to type I interferon in HIV-1 infection correlates with markers of immune activation and disease progression. Blood 113(22):5497–5505

Harris LD, Tabb B, Sodora DL, Paiardini M, Klatt NR, Douek DC, Silvestri G, Muller-Trutwin M, Vasile-Pandrea I, Apetrei C et al (2010) Down-regulation of robust acute type I interferon responses distinguishes nonpathogenic simian immunodeficiency virus (SIV) infection of natural hosts from pathogenic SIV infection of rhesus macaques. J Virol 84(15):7886–7891

Hatziioannou T, Ambrose Z, Chung NP, Piatak M Jr, Yuan F, Trubey CM, Coalter V, Kiser R, Schneider D, Smedley J et al (2009) A macaque model of HIV-1 infection. Proc Natl Acad Sci USA 106(11):4425–4429

Hazenberg MD, Otto SA, Cohen Stuart JW, Verschuren MC, Borleffs JC, Boucher CA, Coutinho RA, Lange JM, Rinke de Wit TF, Tsegaye A et al (2000) Increased cell division but not thymic dysfunction rapidly affects the T-cell receptor excision circle content of the naive T cell population in HIV-1 infection. Nat Med 6(9):1036–1042

Hazenberg MD, Otto SA, van Benthem BH, Roos MT, Coutinho RA, Lange JM, Hamann D, Prins M, Miedema F (2003) Persistent immune activation in HIV-1 infection is associated with progression to AIDS. AIDS 17(13):1881–1888

Heeney JL (1995) AIDS: a disease of impaired Th-cell renewal? Immunol Today 16(11): 515–520

Heeney JL, Rutjens E, Verschoor EJ, Niphuis H, ten Haaft P, Rouse S, McClure H, Balla-Jhagjhoorsingh S, Bogers W, Salas M et al (2006) Transmission of simian immunodeficiency virus SIVcpz and the evolution of infection in the presence and absence of concurrent human immunodeficiency virus type 1 infection in chimpanzees. J Virol 80(14):7208–7218

Hirsch VM, Dapolito G, Johnson PR, Elkins WR, London WT, Montali RJ, Goldstein S, Brown C (1995) Induction of AIDS by simian immunodeficiency virus from an African green monkey: species-specific variation in pathogenicity correlates with the extent of in vivo replication. J Virol 69(2):955–967

Hirsch VM, Sharkey ME, Brown CR, Brichacek B, Goldstein S, Wakefield J, Byrum R, Elkins WR, Hahn BH, Lifson JD et al (1998) Vpx is required for dissemination and pathogenesis of SIV(SM) PBj: evidence of macrophage-dependent viral amplification. Nat Med 4(12): 1401–1408

Holmgren B, da Silva Z, Vastrup P, Larsen O, Andersson S, Ravn H, Aaby P (2007) Mortality associated with HIV-1, HIV-2, and HTLV-I single and dual infections in a middle-aged and older population in Guinea-Bissau. Retrovirology 4:85

Hout DR, Gomez ML, Pacyniak E, Gomez LM, Inbody SH, Mulcahy ER, Culley N, Pinson DM, Powers MF, Wong SW et al (2005) Scrambling of the amino acids within the transmembrane domain of Vpu results in a simian-human immunodeficiency virus (SHIVTM) that is less pathogenic for pig-tailed macaques. Virology 339(1):56–69

Huet T, Cheynier R, Meyerhans A, Roelants G, Wain-Hobson S (1990) Genetic organization of a chimpanzee lentivirus related to HIV-1. Nature 345(6273):356–359

Jacquelin B, Mayau V, Targat B, Liovat AS, Kunkel D, Petitjean G, Dillies MA, Roques P, Butor C, Silvestri G et al (2009) Nonpathogenic SIV infection of African green monkeys induces a strong but rapidly controlled type I IFN response. J Clin Invest 119(12):3544–3555

Jaffar S, Grant AD, Whitworth J, Smith PG, Whittle H (2004) The natural history of HIV-1 and HIV-2 infections in adults in Africa: a literature review. Bull World Health Organ 82(6):462–469

Jin MJ, Hui H, Robertson DL, Muller MC, Barre-Sinoussi F, Hirsch VM, Allan JS, Shaw GM, Sharp PM, Hahn BH (1994) Mosaic genome structure of simian immunodeficiency virus from west African green monkeys. EMBO J 13(12):2935–2947

Juompan LY, Hutchinson K, Montefiori DC, Nidtha S, Villinger F, Novembre FJ (2008) Analysis of the immune responses in chimpanzees infected with HIV type 1 isolates. AIDS Res Hum Retroviruses 24(4):573–586

Katzourakis A, Tristem M, Pybus OG, Gifford RJ (2007) Discovery and analysis of the first endogenous lentivirus. Proc Natl Acad Sci USA 104(15):6261–6265

Kaufmann GR, Cunningham P, Kelleher AD, Zaunders J, Carr A, Vizzard J, Law M, Cooper DA (1998) Patterns of viral dynamics during primary human immunodeficiency virus type 1 infection. The Sydney Primary HIV Infection Study Group. J Infect Dis 178(6):1812–1815

Keckesova Z, Ylinen LM, Towers GJ, Gifford RJ, Katzourakis A (2009) Identification of a RELIK orthologue in the European hare (Lepus europaeus) reveals a minimum age of 12 million years for the lagomorph lentiviruses. Virology 384(1):7–11

Keele BF, Van Heuverswyn F, Li Y, Bailes E, Takehisa J, Santiago ML, Bibollet-Ruche F, Chen Y, Wain LV, Liegeois F et al (2006) Chimpanzee reservoirs of pandemic and nonpandemic HIV-1. Science 313(5786):523–526

Keele BF, Jones JH, Terio KA, Estes JD, Rudicell RS, Wilson ML, Li Y, Learn GH, Beasley TM, Schumacher-Stankey J et al (2009) Increased mortality and AIDS-like immunopathology in wild chimpanzees infected with SIVcpz. Nature 460(7254):515–519

Kibaya RS, Bautista CT, Sawe FK, Shaffer DN, Sateren WB, Scott PT, Michael NL, Robb ML, Birx DL, de Souza MS (2008) Reference ranges for the clinical laboratory derived from a rural population in Kericho, Kenya. PLoS One 3(10):e3327

Kirchhoff F (2009) Is the high virulence of HIV-1 an unfortunate coincidence of primate lentiviral evolution? Nat Rev Microbiol 7(6):467–476

Kirchhoff F (2010) Immune evasion and counteraction of restriction factors by HIV-1 and other primate lentiviruses. Cell Host Microbe 8(1):55–67

Klimkait T, Strebel K, Hoggan MD, Martin MA, Orenstein JM (1990) The human immunodeficiency virus type 1-specific protein vpu is required for efficient virus maturation and release. J Virol 64(2):621–629

Kornfeld C, Ploquin MJ, Pandrea I, Faye A, Onanga R, Apetrei C, Poaty-Mavoungou V, Rouquet P, Estaquier J, Mortara L et al (2005) Antiinflammatory profiles during primary SIV infection in African green monkeys are associated with protection against AIDS. J Clin Invest 115(4):1082–1091

Kumarasamy N, Venkatesh KK, Devaleenol B, Poongulali S, Yephthomi T, Pradeep A, Saghayam S, Flanigan T, Mayer KH, Solomon S (2009) Factors associated with mortality among HIV-infected patients in the era of highly active antiretroviral therapy in southern India. Int J Infect Dis 14(2):e127–131

Laguette N, Sobhian B, Casartelli N, Ringeard M, Chable-Bessia C, Segeral E, Yatim A et al (2011) SAMHD1 is the dendritic- and myeloid-cell-specific HIV-1 restriction factor counteracted by Vpx. Nature 474(7353):654–657

Le Tortorec A, Neil SJ (2009) Antagonism to and intracellular sequestration of human tetherin by the human immunodeficiency virus type 2 envelope glycoprotein. J Virol 83(22):11966–11978

Lederer S, Favre D, Walters KA, Proll S, Kanwar B, Kasakow Z, Baskin CR, Palermo R, McCune JM, Katze MG (2009) Transcriptional profiling in pathogenic and non-pathogenic SIV infections reveals significant distinctions in kinetics and tissue compartmentalization. PLoS Pathog 5(2):e1000296

Leendertz SA, Locatelli S, Boesch C, Kucherer C, Formenty P, Liegeois F, Ayouba A, Peeters M, Leendertz FH (2011) No evidence for transmission of SIVwrc from western red colobus

monkeys (Piliocolobus badius badius) to wild West African chimpanzees (Pan troglodytes verus) despite high exposure through hunting. BMC Microbiol 11(1):24
Li Q, Smith AJ, Schacker TW, Carlis JV, Duan L, Reilly CS, Haase AT (2009) Microarray analysis of lymphatic tissue reveals stage-specific, gene expression signatures in HIV-1 infection. J Immunol 183(3):1975–1982
Liang SC, Tan XY, Luxenberg DP, Karim R, Dunussi-Joannopoulos K, Collins M, Fouser LA (2006) Interleukin (IL)-22 and IL-17 are coexpressed by Th17 cells and cooperatively enhance expression of antimicrobial peptides. J Exp Med 203(10):2271–2279
Liegeois F, Boue V, Mouacha F, Butel C, Mve-Ondo B, Pourrut X, Leroy E, Peeters M, Rouet F (2012) New STLV-3 strains and a divergent SIVmus strain identified in non-human primate bushmeat in Gabon. Retrovirology 9(1):28
Ling B, Apetrei C, Pandrea I, Veazey RS, Lackner AA, Gormus B, Marx PA (2004) Classic AIDS in a sooty mangabey after an 18-year natural infection. J Virol 78(16):8902–8908
Ling B, Veazey RS, Hart M, Lackner AA, Kuroda M, Pahar B, Marx PA (2007) Early restoration of mucosal CD4 memory CCR5 T cells in the gut of SIV-infected rhesus predicts long term non-progression. AIDS 21(18):2377–2385
Locatelli S, Liegeois F, Lafay B, Roeder AD, Bruford MW, Formenty P, Noe R, Delaporte E, Peeters M (2008) Prevalence and genetic diversity of simian immunodeficiency virus infection in wild-living red colobus monkeys (Piliocolobus badius badius) from the Tai forest, Cote d'Ivoire SIVwrc in wild-living western red colobus monkeys. Infect Genet Evol 8(1):1–14
Lozano Reina JM, Favre D, Kasakow Z, Mayau V, Nugeyre MT, Ka T, Faye A, Miller CJ, Scott-Algara D, McCune JM et al (2009) Gag p27-specific B- and T-cell responses in Simian immunodeficiency virus SIVagm-infected African green monkeys. J Virol 83(6):2770–2777
Mandl JN, Barry AP, Vanderford TH, Kozyr N, Chavan R, Klucking S, Barrat FJ, Coffman RL, Staprans SI, Feinberg MB (2008) Divergent TLR7 and TLR9 signaling and type I interferon production distinguish pathogenic and nonpathogenic AIDS virus infections. Nat Med 14(10):1077–1087
Martinez-Steele E, Awasana AA, Corrah T, Sabally S, van der Sande M, Jaye A, Togun T, Sarge-Njie R, McConkey SJ, Whittle H et al (2007) Is HIV-2- induced AIDS different from HIV-1-associated AIDS? Data from a West African clinic. AIDS 21(3):317–324
Marx PA, Alcabes PG, Drucker E (2001) Serial human passage of simian immunodeficiency virus by unsterile injections and the emergence of epidemic human immunodeficiency virus in Africa. Philos Trans R Soc Lond B Biol Sci 356(1410):911–920
Mattapallil JJ, Douek DC, Hill B, Nishimura Y, Martin M, Roederer M (2005) Massive infection and loss of memory CD4+ T cells in multiple tissues during acute SIV infection. Nature 434(7037):1093–1097
May RM, Gupta S, McLean AR (2001) Infectious disease dynamics: What characterizes a successful invader? Philos Trans R Soc Lond B Biol Sci 356(1410):901–910
McClure HM, Anderson DC, Fultz PN, Ansari AA, Lockwood E, Brodie A (1989) Spectrum of disease in macaque monkeys chronically infected with SIV/SMM. Vet Immunol Immunopathol 21(1):13–24
Mehandru S, Poles MA, Tenner-Racz K, Horowitz A, Hurley A, Hogan C, Boden D, Racz P, Markowitz M (2004) Primary HIV-1 infection is associated with preferential depletion of CD4+ T lymphocytes from effector sites in the gastrointestinal tract. J Exp Med 200(6):761–770
Melkus MW, Estes JD, Padgett-Thomas A, Gatlin J, Denton PW, Othieno FA, Wege AK, Haase AT, Garcia JV (2006) Humanized mice mount specific adaptive and innate immune responses to EBV and TSST-1. Nat Med 12(11):1316–1322
Mellors JW, Rinaldo CR Jr, Gupta P, White RM, Todd JA, Kingsley LA (1996) Prognosis in HIV-1 infection predicted by the quantity of virus in plasma. Science 272(5265):1167–1170
Meythaler M, Martinot A, Wang Z, Pryputniewicz S, Kasheta M, Ling B, Marx PA, O'Neil S, Kaur A (2009) Differential CD4+ T-lymphocyte apoptosis and bystander T-cell activation in rhesus macaques and sooty mangabeys during acute simian immunodeficiency virus infection. J Virol 83(2):572–583

Milush JM, Mir KD, Sundaravaradan V, Gordon SN, Engram J, Cano CA, Reeves JD, Anton E, O'Neill E, Butler E et al (2011) Lack of clinical AIDS in SIV-infected sooty mangabeys with significant CD4+ T cell loss is associated with double-negative T cells. J Clin Invest 121(3):1102–1110

Molez JF (1998) The historical question of acquired immunodeficiency syndrome in the 1960s in the Congo River basin area in relation to cryptococcal meningitis. Am J Trop Med Hyg 58(3):273–276

Moll M, Andersson SK, Smed-Sorensen A, Sandberg JK (2010) Inhibition of lipid antigen presentation in dendritic cells by HIV-1 Vpu interference with CD1d recycling from endosomal compartments. Blood 116(11):1876–1884

Morgan D, Mahe C, Mayanja B, Okongo JM, Lubega R, Whitworth JA (2002) HIV-1 infection in rural Africa: is there a difference in median time to AIDS and survival compared with that in industrialized countries? AIDS 16(4):597–603

Muller MC, Saksena NK, Nerrienet E, Chappey C, Herve VM, Durand JP, Legal-Campodonico P, Lang MC, Digoutte JP, Georges AJ et al (1993) Simian immunodeficiency viruses from central and western Africa: evidence for a new species-specific lentivirus in tantalus monkeys. J Virol 67(3):1227–1235

Murphey-Corb M, Martin LN, Rangan SR, Baskin GB, Gormus BJ, Wolf RH, Andes WA, West M, Montelaro RC (1986) Isolation of an HTLV-III-related retrovirus from macaques with simian AIDS and its possible origin in asymptomatic mangabeys. Nature 321(6068):435–437

Neel C, Etienne L, Li Y, Takehisa J, Rudicell RS, Bass IN, Moudindo J, Mebenga A, Esteban A, Van Heuverswyn F et al (2010) Molecular epidemiology of simian immunodeficiency virus infection in wild-living gorillas. J Virol 84(3):1464–1476

Neil SJ, Zang T, Bieniasz PD (2008) Tetherin inhibits retrovirus release and is antagonized by HIV-1 Vpu. Nature 451(7177):425–430

Nowroozalizadeh S, Mansson F, da Silva Z, Repits J, Dabo B, Pereira C, Biague A, Albert J, Nielsen J, Aaby P et al (2010) Microbial translocation correlates with the severity of both HIV-1 and HIV-2 infections. J Infect Dis 201(8):1150–1154

Ogg GS, Jin X, Bonhoeffer S, Dunbar PR, Nowak MA, Monard S, Segal JP, Cao Y, Rowland-Jones SL, Cerundolo V et al (1998) Quantitation of HIV-1-specific cytotoxic T lymphocytes and plasma load of viral RNA. Science 279(5359):2103–2106

Ostrowski MA, Chun TW, Justement SJ, Motola I, Spinelli MA, Adelsberger J, Ehler LA, Mizell SB, Hallahan CW, Fauci AS (1999) Both memory and CD45RA+/CD62L+ naive CD4(+) T cells are infected in human immunodeficiency virus type 1-infected individuals. J Virol 73(8):6430–6435

Paiardini M, Cervasi B, Reyes-Aviles E, Micci L, Ortiz AM, Chahroudi A, Vinton C, Gordon SN, Bosinger SE, Francella N et al (2011) Low levels of SIV infection in sooty mangabey central memory CD4(+) T cells are associated with limited CCR5 expression. Nat Med 17(7):830–836

Pandrea I, Apetrei C, Gordon S, Barbercheck J, Dufour J, Bohm R, Sumpter B, Roques P, Marx PA, Hirsch VM et al (2007a) Paucity of CD4+CCR5+ T cells is a typical feature of natural SIV hosts. Blood 109(3):1069–1076

Pandrea IV, Gautam R, Ribeiro RM, Brenchley JM, Butler IF, Pattison M, Rasmussen T, Marx PA, Silvestri G, Lackner AA et al (2007b) Acute loss of intestinal CD4+ T cells is not predictive of simian immunodeficiency virus virulence. J Immunol 179(5):3035–3046

Phair J, Munoz A, Detels R, Kaslow R, Rinaldo C, Saah A (1990) The risk of Pneumocystis carinii pneumonia among men infected with human immunodeficiency virus type 1. Multicenter AIDS Cohort Study Group. N Engl J Med 322(3):161–165

Phillips A, Lee CA, Elford J, Janossy G, Bofill M, Timms A, Kernoff PB (1989) Prediction of progression to AIDS by analysis of CD4 lymphocyte counts in a haemophilic cohort. AIDS 3(11):737–741

Piot P, Quinn TC, Taelman H, Feinsod FM, Minlangu KB, Wobin O, Mbendi N, Mazebo P, Ndangi K, Stevens W et al (1984) Acquired immunodeficiency syndrome in a heterosexual population in Zaire. Lancet 2(8394):65–69

Plantier JC, Leoz M, Dickerson JE, De Oliveira F, Cordonnier F, Lemee V, Damond F, Robertson DL, Simon F (2009) A new human immunodeficiency virus derived from gorillas. Nat Med 15(8):871–872
Poulsen AG, Aaby P, Larsen O, Jensen H, Naucler A, Lisse IM, Christiansen CB, Dias F, Melbye M (1997) 9-year HIV-2-associated mortality in an urban community in Bissau, West Africa. Lancet 349(9056):911–914
Quinn TC, Mann JM, Curran JW, Piot P (1986) AIDS in Africa: an epidemiologic paradigm. Science 234(4779):955–963
Redd AD, Dabitao D, Bream JH, Charvat B, Laeyendecker O, Kiwanuka N, Lutalo T, Kigozi G, Tobian AA, Gamiel J et al (2009) Microbial translocation, the innate cytokine response, and HIV-1 disease progression in Africa. Proc Natl Acad Sci USA 106(16):6718–6723
Ricard D, Wilkins A, N'Gum PT, Hayes R, Morgan G, Da Silva AP, Whittle H (1994) The effects of HIV-2 infection in a rural area of Guinea-Bissau. AIDS 8(7):977–982
Riddick NE, Hermann EA, Loftin LM, Elliott ST, Wey WC, Cervasi B, Taaffe J, Engram JC, Li B, Else JG et al (2010) A novel CCR5 mutation common in sooty mangabeys reveals SIVsmm infection of CCR5-null natural hosts and efficient alternative coreceptor use in vivo. PLoS Pathog 6(8):e1001064
Rieder P, Joos B, von Wyl V, Kuster H, Grube C, Leemann C, Boni J, Yerly S, Klimkait T, Burgisser P et al (2010) HIV-1 transmission after cessation of early antiretroviral therapy among men having sex with men. AIDS 24(8):1177–1183
Rodriguez B, Sethi AK, Cheruvu VK, Mackay W, Bosch RJ, Kitahata M, Boswell SL, Mathews WC, Bangsberg DR, Martin J et al (2006) Predictive value of plasma HIV RNA level on rate of CD4 T-cell decline in untreated HIV infection. JAMA 296(12):1498–1506
Rutjens E, Vermeulen J, Verstrepen B, Hofman S, Prins JM, Srivastava I, Heeney JL, Koopman G (2008) Chimpanzee CD4+ T cells are relatively insensitive to HIV-1 envelope-mediated inhibition of CD154 up-regulation. Eur J Immunol 38(4):1164–1172
Rutjens E, Mazza S, Biassoni R, Koopman G, Ugolotti E, Fogli M, Dubbes R, Costa P, Mingari MC, Greenwood EJ et al (2010) CD8+ NK cells are predominant in chimpanzees, characterized by high NCR expression and cytokine production, and preserved in chronic HIV-1 infection. Eur J Immunol 40(5):1440–1450
Saez-Cirion A, Lacabaratz C, Lambotte O, Versmisse P, Urrutia A, Boufassa F, Barre-Sinoussi F, Delfraissy JF, Sinet M, Pancino G et al (2007) HIV controllers exhibit potent CD8 T cell capacity to suppress HIV infection ex vivo and peculiar cytotoxic T lymphocyte activation phenotype. Proc Natl Acad Sci USA 104(16):6776–6781
Schacker TW, Nguyen PL, Beilman GJ, Wolinsky S, Larson M, Reilly C, Haase AT (2002) Collagen deposition in HIV-1 infected lymphatic tissues and T cell homeostasis. J Clin Invest 110(8):1133–1139
Schellens IM, Borghans JA, Jansen CA, De Cuyper IM, Geskus RB, van Baarle D, Miedema F (2008) Abundance of early functional HIV-specific CD8+ T cells does not predict AIDS-free survival time. PLoS One 3(7):e2745
Schindler M, Munch J, Kutsch O, Li H, Santiago ML, Bibollet-Ruche F, Muller-Trutwin MC, Novembre FJ, Peeters M, Courgnaud V et al (2006) Nef-mediated suppression of T cell activation was lost in a lentiviral lineage that gave rise to HIV-1. Cell 125(6):1055–1067
Schmokel J, Sauter D, Schindler M, Leendertz FH, Bailes E, Dazza MC, Saragosti S, Bibollet-Ruche F, Peeters M, Hahn BH et al (2010) The presence of a vpu gene and the lack of Nef-mediated downmodulation of T cell receptor-CD3 are not always linked in primate lentiviruses. J Virol 85(2):742–752
Sedaghat AR, German J, Teslovich TM, Cofrancesco J Jr, Jie CC, Talbot CC Jr, Siliciano RF (2008) Chronic CD4+ T-cell activation and depletion in human immunodeficiency virus type 1 infection: type I interferon-mediated disruption of T-cell dynamics. J Virol 82(4):1870–1883
Seiler BM, Dick EJ Jr, Guardado-Mendoza R, VandeBerg JL, Williams JT, Mubiru JN, Hubbard GB (2009) Spontaneous heart disease in the adult chimpanzee (Pan troglodytes). J Med Primatol 38(1):51–58

Shah AH, Sowrirajan B, Davis ZB, Ward JP, Campbell EM, Planelles V, Barker E (2010) Degranulation of natural killer cells following interaction with HIV-1-infected cells is hindered by downmodulation of NTB-A by Vpu. Cell Host Microbe 8(5):397–409

Sharova N, Wu Y, Zhu X, Stranska R, Kaushik R, Sharkey M, Stevenson M (2008) Primate lentiviral Vpx commandeers DDB1 to counteract a macrophage restriction. PLoS Pathog 4(5): e1000057

Sharp PM, Hahn BH (2011) Origins of HIV and the AIDS pandemic. Cold Spring Harb Perspect Med 1(1):a006841

Sharp PM, Bailes E, Gao F, Beer BE, Hirsch VM, Hahn BH (2000) Origins and evolution of AIDS viruses: estimating the time-scale. Biochem Soc Trans 28(2):275–282

Sharp PM, Shaw GM, Hahn BH (2005) Simian immunodeficiency virus infection of chimpanzees. J Virol 79(7):3891–3902

Sieg SF, Rodriguez B, Asaad R, Jiang W, Bazdar DA, Lederman MM (2005) Peripheral S-phase T cells in HIV disease have a central memory phenotype and rarely have evidence of recent T cell receptor engagement. J Infect Dis 192(1):62–70

Silvestri G, Sodora DL, Koup RA, Paiardini M, O'Neil SP, McClure HM, Staprans SI, Feinberg MB (2003) Nonpathogenic SIV infection of sooty mangabeys is characterized by limited bystander immunopathology despite chronic high-level viremia. Immunity 18(3):441–452

Silvestri G, Fedanov A, Germon S, Kozyr N, Kaiser WJ, Garber DA, McClure H, Feinberg MB, Staprans SI (2005) Divergent host responses during primary simian immunodeficiency virus SIVsm infection of natural sooty mangabey and nonnatural rhesus macaque hosts. J Virol 79(7):4043–4054

Singh DK, McCormick C, Pacyniak E, Lawrence K, Dalton SB, Pinson DM, Sun F, Berman NE, Calvert M, Gunderson RS et al (2001) A simian human immunodeficiency virus with a nonfunctional Vpu (deltavpuSHIV(KU-1bMC33)) isolated from a macaque with neuroAIDS has selected for mutations in env and nef that contributed to its pathogenic phenotype. Virology 282(1):123–140

Singh DK, Griffin DM, Pacyniak E, Jackson M, Werle MJ, Wisdom B, Sun F, Hout DR, Pinson DM, Gunderson RS et al (2003) The presence of the casein kinase II phosphorylation sites of Vpu enhances the CD4(+) T cell loss caused by the simian-human immunodeficiency virus SHIV(KU-lbMC33) in pig-tailed macaques. Virology 313(2):435–451

Souquiere S, Bibollet-Ruche F, Robertson DL, Makuwa M, Apetrei C, Onanga R, Kornfeld C, Plantier JC, Gao F, Abernethy K et al (2001) Wild Mandrillus sphinx are carriers of two types of lentivirus. J Virol 75(15):7086–7096

Stacey AR, Norris PJ, Qin L, Haygreen EA, Taylor E, Heitman J, Lebedeva M, DeCamp A, Li D, Grove D et al (2009) Induction of a striking systemic cytokine cascade prior to peak viremia in acute human immunodeficiency virus type 1 infection, in contrast to more modest and delayed responses in acute hepatitis B and C virus infections. J Virol 83(8):3719–3733

Stephens EB, McCormick C, Pacyniak E, Griffin D, Pinson DM, Sun F, Nothnick W, Wong SW, Gunderson R, Berman NE et al (2002) Deletion of the vpu sequences prior to the env in a simian-human immunodeficiency virus results in enhanced Env precursor synthesis but is less pathogenic for pig-tailed macaques. Virology 293(2):252–261

Strebel K, Klimkait T, Martin MA (1988) A novel gene of HIV-1, vpu, and its 16-kilodalton product. Science 241(4870):1221–1223

Stremlau M, Owens CM, Perron MJ, Kiessling M, Autissier P, Sodroski J (2004) The cytoplasmic body component TRIM5alpha restricts HIV-1 infection in Old World monkeys. Nature 427(6977):848–853

Switzer WM, Qari SH, Wolfe ND, Burke DS, Folks TM, Heneine W (2006) Ancient origin and molecular features of the novel human T-lymphotropic virus type 3 revealed by complete genome analysis. J Virol 80(15):7427–7438

Taaffe J, Chahroudi A, Engram J, Sumpter B, Meeker T, Ratcliffe S, Paiardini M, Else J, Silvestri G (2010) A five-year longitudinal analysis of sooty mangabeys naturally infected with simian immunodeficiency virus reveals a slow but progressive decline in CD4+ T-cell count whose magnitude is not predicted by viral load or immune activation. J Virol 84(11):5476–5484

Takemura T, Hayami M (2004) Phylogenetic analysis of SIV derived from mandrill and drill. Front Biosci 9:513–520
Traina-Dorge V, Blanchard J, Martin L, Murphey-Corb M (1992) Immunodeficiency and lymphoproliferative disease in an African green monkey dually infected with SIV and STLV-I. AIDS Res Hum Retroviruses 8(1):97–100
Tugume SB, Piwowar EM, Lutalo T, Mugyenyi PN, Grant RM, Mangeni FW, Pattishall K, Katongole-Mbidde E (1995) Hematological reference ranges among healthy Ugandans. Clin Diagn Lab Immunol 2(2):233–235
Vallari A, Bodelle P, Ngansop C, Makamche F, Ndembi N, Mbanya D, Kaptue L, Gurtler LG, McArthur CP, Devare SG et al (2010a) Four new HIV-1 group N isolates from Cameroon: prevalence continues to be low. AIDS Res Hum Retroviruses 26(1):109–115
Vallari A, Holzmayer V, Harris B, Yamaguchi J, Ngansop C, Makamche F, Mbanya D, Kaptue L, Ndembi N, Gurtler L et al (2010b) Confirmation of putative HIV-1 group P in Cameroon. J Virol 85(3):1403–1407
van der Loeff MF, Larke N, Kaye S, Berry N, Ariyoshi K, Alabi A, van Tienen C, Leligdowicz A, Sarge-Njie R, da Silva Z et al (2010) Undetectable plasma viral load predicts normal survival in HIV-2-infected people in a West African village. Retrovirology 7:46
Van Dooren S, Salemi M, Vandamme AM (2001) Dating the origin of the African human T-cell lymphotropic virus type-i (HTLV-I) subtypes. Mol Biol Evol 18(4):661–671
Van Heuverswyn F, Li Y, Neel C, Bailes E, Keele BF, Liu W, Loul S, Butel C, Liegeois F, Bienvenue Y et al (2006) Human immunodeficiency viruses: SIV infection in wild gorillas. Nature 444(7116):164
Veazey RS, DeMaria M, Chalifoux LV, Shvetz DE, Pauley DR, Knight HL, Rosenzweig M, Johnson RP, Desrosiers RC, Lackner AA (1998) Gastrointestinal tract as a major site of CD4+ T cell depletion and viral replication in SIV infection. Science 280(5362):427–431
Vergne L, Bourgeois A, Mpoudi-Ngole E, Mougnutou R, Mbuagbaw J, Liegeois F, Laurent C, Butel C, Zekeng L, Delaporte E et al (2003) Biological and genetic characteristics of HIV infections in Cameroon reveals dual group M and O infections and a correlation between SI-inducing phenotype of the predominant CRF02_AG variant and disease stage. Virology 310(2):254–266
Vidal N, Peeters M, Mulanga-Kabeya C, Nzilambi N, Robertson D, Ilunga W, Sema H, Tshimanga K, Bongo B, Delaporte E (2000) Unprecedented degree of human immunodeficiency virus type 1 (HIV-1) group M genetic diversity in the Democratic Republic of Congo suggests that the HIV-1 pandemic originated in Central Africa. J Virol 74(22):10498–10507
Wertheim JO, Worobey M (2009) Dating the age of the SIV lineages that gave rise to HIV-1 and HIV-2. PLoS Comput Biol 5(5):e1000377
Wilkinson DA, Operskalski EA, Busch MP, Mosley JW, Koup RA (1998) A 32-bp deletion within the CCR5 locus protects against transmission of parenterally acquired human immunodeficiency virus but does not affect progression to AIDS-defining illness. J Infect Dis 178(4): 1163–1166
Willey RL, Maldarelli F, Martin MA, Strebel K (1992) Human immunodeficiency virus type 1 Vpu protein induces rapid degradation of CD4. J Virol 66(12):7193–7200
Worobey M, Gemmel M, Teuwen DE, Haselkorn T, Kunstman K, Bunce M, Muyembe JJ, Kabongo JM, Kalengayi RM, Van Marck E et al (2008) Direct evidence of extensive diversity of HIV-1 in Kinshasa by 1960. Nature 455(7213):661–664
Worobey M, Telfer P, Souquiere S, Hunter M, Coleman CA, Metzger MJ, Reed P, Makuwa M, Hearn G, Honarvar S et al (2010) Island biogeography reveals the deep history of SIV. Science 329(5998):1487
Zeng M, Smith AJ, Wietgrefe SW, Southern PJ, Schacker TW, Reilly CS, Estes JD, Burton GF, Silvestri G, Lifson JD et al (2011) Cumulative mechanisms of lymphoid tissue fibrosis and T cell depletion in HIV-1 and SIV infections. J Clin Invest 121(3):998–1008
Zhang F, Wilson SJ, Landford WC, Virgen B, Gregory D, Johnson MC, Munch J, Kirchhoff F, Bieniasz PD, Hatziioannou T (2009) Nef proteins from simian immunodeficiency viruses are tetherin antagonists. Cell Host Microbe 6(1):54–67

Part III
Primates, Pathogens and Health

Microbial Exposures and Other Early Childhood Influences on the Subsequent Function of the Immune System

Graham A.W. Rook

Introduction

The developed countries have undergone massive increases in the prevalence of a wide range of chronic inflammatory disorders including allergies, autoimmune diseases and inflammatory bowel disease. Rigorous meta-analyses that check the diagnostic criteria used have confirmed that these massive increases are real (discussed detail in a later section) (Eder et al. 2006; Elliott et al. 2005). The increases were seen first in Northern countries and so correlated with economic development and standards of hygiene. In Europe this phenomenon can be traced back to the nineteenth century. In 1873 Charles Harrison Blackley, working in England, noted that hay fever was associated with exposure to pollen, but he also remarked "farmers rarely experience the condition" (Blackley 1873). Indeed hay fever began to be regarded as a mark of wealth, education and sophistication. In the 1880 s Morell Mackenzie, a British physician, went so far as to state "As, therefore, summer sneezing goes hand-in-hand with culture, we may, perhaps infer that the higher we rise in the intellectual scale, the more is the tendency developed" (Mackenzie 1887). Recent studies have rediscovered and definitively confirmed the protective effect of the farming environment (Riedler et al. 2001; von Ehrenstein et al. 2000; von Mutius and Vercelli 2010) and shown that contact with animals such as dogs is also protective (Ownby et al. 2002). In addition to these observations on allergic disorders, a link between hygiene and an autoimmune disease was explicitly suggested in 1966, when it was reported that the prevalence of multiple sclerosis (MS) showed a positive correlation with sanitation in Israel (Leibowitz et al. 1966). However, it was not until 1989 that the term "hygiene hypothesis" was coined following the observation that in young adults, a history of hay fever was inversely related to the number of

G.A.W. Rook (✉)
Department of Infection, University College London, London, UK
e-mail: g.rook@ucl.ac.uk

J.F. Brinkworth and K. Pechenkina (eds.), *Primates, Pathogens, and Evolution*, Developments in Primatology: Progress and Prospects,
DOI 10.1007/978-1-4614-7181-3_11, © Springer Science+Business Media New York 2013

siblings (especially older male siblings) in their family when they were 11 years old (Strachan 1989). Then Matricardi and colleagues found that army recruits with evidence of infections attributable to faecal–oral transmission were less likely to have allergic manifestations (Matricardi et al. 1998). Such data were considered consistent with a protective influence of postnatal infection that might be lost in the presence of modern hygiene (Matricardi et al. 1998; Strachan 1989; Strachan et al. 1996). A few years later it was pointed out that type 1 diabetes (T1D; caused by autoimmune destruction of the insulin-secreting β-cells in the pancreas) is increasing at the same rate and in the same countries (mostly rich and developed) as the allergic disorders (Stene and Nafstad 2001). Similarly, a parallel rise in inflammatory bowel diseases (IBD; Crohn's disease and ulcerative colitis) had clearly started at the beginning of the twentieth century, rising from rare and sporadic in 1900 to 400–500/100,000 by the 1990s in rich northern developed countries (the epidemiological basis for these assertions is expanded below) (Elliott et al. 2005).

How can we make sense of all this? There are two obvious problems. First, allergic disorders, autoimmune diseases and inflammatory bowel diseases involve different immune mechanisms. Allergic disorders are mediated largely by T helper 2 cells (Th2) secreting IL-4, IL-5 and IL-13, while the autoimmune disorders that are increasing (multiple sclerosis (MS) and T1D) involve T helper 1 (Th1) and T helper 17 (Th17 cells) secreting IFN-γ, IL-17 and other mediators. The various forms of IBD (Crohn's disease, ulcerative colitis and celiac disease) involve a variety of effector cell types. Is there an "umbrella" concept associated with hygiene or economic development that can explain increases in pathologies that involve such diverse effector cell types?

The second problem is the diversity of the epidemiology. What is the connection between economic development, latitude, the farming environment, pet ownership and dirty older siblings? Indeed, is there any possibility of finding a mechanism with such broad explanatory power?

Since about 1998 it has been apparent that the first problem can be solved by the view that the increase in the prevalence of these chronic inflammatory diseases is due at least in part to defects in maturation of the immune system, leading notably to defective immunoregulation (Rook and Stanford 1998). Thus, one cause of these increases lies in a broad imbalance between immunoregulatory mechanisms and effector mechanisms (whether Th1, Th17 or Th2). A failure of immunoregulatory mechanisms can indeed lead to simultaneous increases in diverse types of pathology. We know this because genetic defects of Foxp3, a transcription factor that plays a crucial role in the development and function of regulatory lymphocytes, leads to the X-linked autoimmunity–allergic dysregulation syndrome (XLAAD) that includes aspects of allergy, autoimmunity and enteropathy (Wildin et al. 2002). So although the recent increases in chronic inflammatory disorders are too rapid to be due to genetic changes, there could be a lifestyle-associated factor that has a broad detrimental effect on regulation of the immune system.

More recently the second problem (i.e. the factor that is common to economic deprivation, farms, siblings, dogs and poor hygiene) has also been resolved. As will be explained in detail below, we now know that the correct functioning of the immune system is heavily dependent on the microbial input that it receives in utero, in the neonatal period and also throughout life (Ege et al. 2008; Rook 2010). Epidemiological and experimental studies have progressively identified organisms that have appropriate immunoregulatory properties, are depleted from modern environments and can downregulate chronic inflammatory disorders in experimental models. Some of these are now entering clinical trials in humans. In the context of this book, the interesting thing about the identified organisms is that not only do they resolve the farm/dog/sibling dilemma, but they also take us back at least as far as early hominins. The organisms identified are viruses, bacteria and helminths, often harmless, that coevolved with humans but are depleted from the modern urban environment. These can be divided into three overlapping groups: (1) organisms that form part of the coevolved human microbiota that are altered by modern diets, living conditions and antibiotics; (2) infections that will have been present in early humans, usually harmless, transmitted by the faecal–oral route very early in life, that have been depleted since the second epidemiological transition; and (3) harmless environmental organisms in mud, untreated water and fermenting vegetable material that are eliminated by the modern city lifestyle. I refer to these organisms collectively as "Old Friends", to emphasise our long association with them and our dependency on their presence (Rook 2010). Below, I discuss the evolutionary significance and mechanisms of the immunoregulatory effects of these organisms.

As so often in medicine, a Darwinian approach has huge explanatory power. But it is emphasised that this chapter is drawn from the medical literature and discusses the effects on the immune system of recent human cultural and technological developments. I leave to others the interesting task of asking whether there were other similar immunological turning points at earlier stages in hominoid evolution.

This chapter starts with a general account of the role of the immunoregulatory "Old Friends", in which immunology has been kept to a minimum. Those who want more detail of the immunology and molecular mechanisms will find that in the later section entitled "Mechanisms".

Evolved Dependence and the Environment of Evolutionary Adaptedness

The concept of "evolved dependence" provides essential background to this discussion. This concept refers to situations where an organism has become adapted to the presence of a partner and can no longer perform well without that partner (de Mazancourt et al. 2005). It was originally used to describe endosymbiosis. A classical example was seen in the laboratory environment when an amoeba (*Amoeba discoides*) became infected with a bacterium (Jeon 1972). Initially this infection severely compromised the growth of the amoebae. However, after 5 years the relationship between the two species had changed, and neither organism could survive without the other. This indicates genetic changes leading to dependence. For instance, an enzyme that is encoded in the genome of both species might be dropped from the genome of one of them. Access to that gene is now "entrusted" to the other species. This idea is at first surprising to most readers, but it is in fact rather commonplace. For instance, most mammals can synthesise vitamin C, but large primates and guinea pigs have lost the relevant pathways. In effect, man and guinea pig are now in a state of evolved dependence on fruit and vegetables. Of course the same is true for many other genes involved in the synthesis of vitamins and other essential nutrients that we have to consume after other organisms have created them for us (Resta 2009).

The role of this evolved dependence is most clearly seen in germ-free animals, propagated by caesarean section under sterile conditions and maintained in a sterile environment. The immune systems of germ-free animals fail to develop correctly and are functionally distorted. There is lack of cellularity and, above all, lack of effective immunoregulation (discussed in detail later). In 2005 Mazmanian and colleagues showed that a single polysaccharide from an intestinal commensal, *Bacteroides fragilis*, could partly correct these developmental abnormalities (Mazmanian et al. 2005). More recently they have shown, using three different models of intestinal inflammation, that the same polysaccharide, given by mouth, can turn on crucial anti-inflammatory, immunoregulatory pathways (Mazmanian et al. 2008). In the discussion of the latter paper they state:

> We propose that the mammalian genome does not encode for all functions required for immunological development but rather that mammals depend on critical interactions with their microbiome (the collective genomes of the microbiota) for health.

To put it even more simply, some genes needed for the development and regulation of the mammalian immune system might have been "entrusted" to microorganisms: a clear example of "evolved dependence". It is obvious that these organisms have to be those with which mammals have coevolved for a very long time and that were always present. They cannot be organisms that merely cause sporadic infections or high death rates. The latter can modify the human genome by elimination of susceptible genotypes, but they cannot be entrusted with the role of supplying genes and functions that we need.

Environment of Evolutionary Adaptedness

It may be useful to think about the state of evolved dependence seen in the mammalian immune system in the context of the Environment of Evolutionary Adaptedness (EEA). The term EEA was first used in 1969 by John Bowlby, who was concerned that those aspects of human behaviour that are genetically determined might be adapted to the hunter-gatherer existence rather than to modern city life (Bowlby 1971 first published by Hogarth press in 1969). The basis for this was the view that since the start of agriculture and pastoralism about 10,000 years ago, much human adaptation to new environments has been cultural and technological rather than genetic. For example, we have not adapted genetically to living in cold places: we have learnt to make fur coats and electric heaters. Humans easily detect problems within the physical environment and invent appropriate technological adaptations. But humans are not equipped with a sense that tells us when the environment is inappropriate for our immune systems. Only since the development of modern biology have we begun to understand that we have an immune system and that we need to think in a Darwinian and anthropological way about the nature and timing of the inputs that it receives.

Epidemiological Transitions

It is possible to predict the identity of the organisms on which our immune systems have evolved dependence (i.e. that were part of the immune system's EEA) by considering the history of man's changing microbial exposures. In 1971 Omran coined the term "epidemiological transition" to describe the major watersheds in human development that led to massive changes in mortality (discussed in Armelagos et al. 2005). Palaeolithic populations carried organisms inherited from primate ancestors ("heirloom" species), including many viruses, as shown in Fig. 1 (Armelagos et al. 2005; Van Blerkom 2003). In addition they would have been exposed to zoonoses that they picked up as they scavenged carrion (Armelagos et al. 2005; Despres et al. 1992; Hoberg 2006). Phylogenetic trees for these organisms confirm the extremely ancient association with humans, probably for a million years or more (Despres et al. 1992; Hoberg 2006). Finally, Palaeolithic populations will have consumed several milligrams of harmless environmental saprophytes every day, since these are ubiquitous in soil and untreated water (Delmont et al. 2011). We have called these "pseudocommensals" because of their inevitable continuous presence until chlorination and purification of water in the modern era. The organisms that have been found to be important for the hygiene hypothesis belong within these three categories.

About 10,000 years ago, the shift to agriculture and husbandry created the first (Neolithic) epidemiological transition (Fig. 1) (Armelagos et al. 2005). This will have had little effect on exposure to the "pseudocommensals" or to the heirloom species. However, the more sedentary lifestyle increased faecal–oral transmission and combined with husbandry caused prolonged contact with animals. The latter led to adaptation to humans of a number of animal viruses shown in Fig. 1 (Van Blerkom 2003). However, the viruses acquired during the Neolithic such as influenza (B and C), smallpox, mumps and measles cannot have become endemic until populations were large enough. This required communities of several hundreds of thousands, which did not occur until the appearance of cities 2000–3000 years ago (Armelagos et al. 2005). Since this represents only 100–150 generations from the present, extremely strong selection pressure would have been required for evolved dependence to appear, and this seems unlikely. Moreover, most humans did not live in such large groups, and these viruses were, for example, absent from pre-Columbian American populations. In any case, one would not expect "evolved

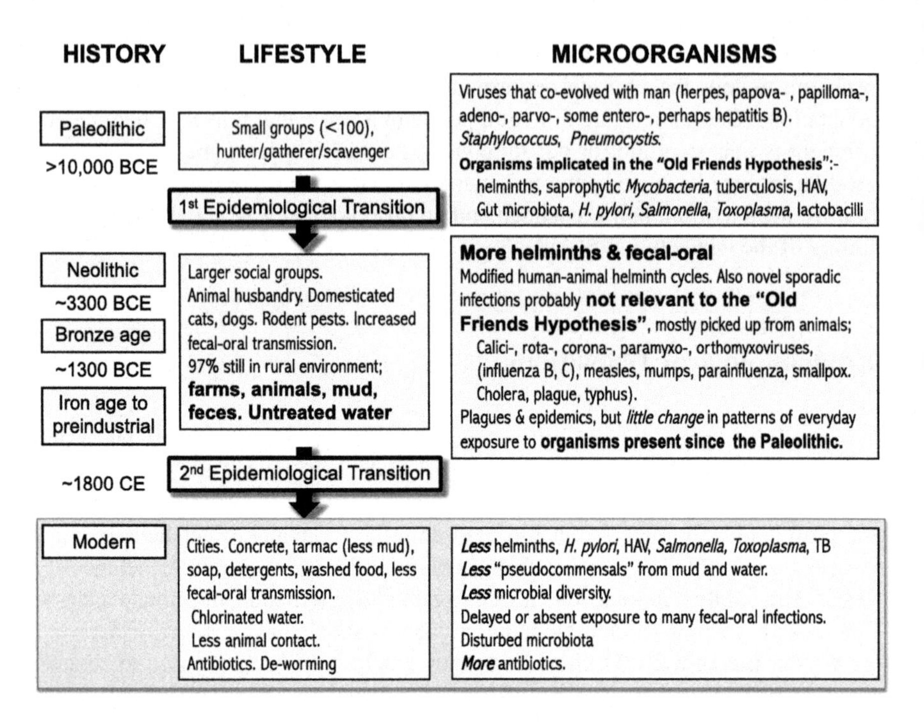

Fig. 1 Aspects of man's microbiological history that are most relevant to the hygiene hypothesis. Epidemiological data, laboratory and animal models and preliminary clinical trials investigating the hygiene hypothesis implicate organisms that are thought to have accompanied mammalian and human evolution. This relationship was long enough for the establishment of evolved dependence on these organisms, that must be tolerated, and so have developed roles in the initiation of regulatory pathways. Organisms that evolved during the Neolithic are less likely to be relevant in this context, and the first epidemiological transition did not reduce human contact with organisms associated with animals, faeces and mud. On the other hand, the second epidemiological transition has led to gene–environment misfit, as the "Old Friends" from the Palaeolithic were progressively removed from the modern environment

dependence" on contact with organisms that were both dangerous and sporadic, and as explained later, these "recently" acquired viruses tend to *cause or exacerbate* chronic inflammatory disorders, rather than prevent them.

In short, there were dramatic changes to the human microbial environment after the first epidemiological transition, but this *did not* result in loss of exposure to the organisms that had coevolved with humans and are now implicated by epidemiology, experimental models and clinical trials in the hygiene hypothesis, as detailed below. Until the modern era more than 97 % of the population still lived in rural environments, close to mud, animals and faeces, which were the sources of these organisms. The situation did not change until the mid-nineteenth century. Since then some populations have undergone a second epidemiological transition in which concrete, tarmac, diminished contact with animals, public health measures and, more recently, antibiotics have resulted in diminished (or *delayed*) exposure to the "Old Friends" that were present in earlier eras. The dramatic effects of the second epidemiological transition are revealed by epidemiological studies of the allergic disorders. The prevalence of symptoms of asthma, allergic rhinoconjunctivitis and atopic eczema varied between 20-fold and 60-fold in different regions. For asthma symptoms, the highest 12-month prevalences were in the UK, Australia, New Zealand and the Republic of Ireland. The lowest prevalences were found in Eastern European countries, Indonesia, Greece, China, Taiwan, Uzbekistan, India and Ethiopia (ISAAC Steering Committee 1998).

How Real Are the Modern Increases in Chronic Inflammatory Disorders?

It must be emphasised that these progressive increases in the incidences of chronic inflammatory diseases that start at the second epidemiological transition are real and not merely due to changing awareness and diagnostic methods in different countries.

The allergies are mainly due to unregulated and inappropriate immune responses to trivial allergens in the air or food: a failure of immunoregulatory mechanisms (Larche 2007; Strickland and Holt 2011). There are several types of study that confirm the recent increases in allergic disorders. First, we can perform studies in the same developed countries at intervals of a decade or more, using standardised diagnostic methods. The very large increases in allergic asthma during the twentieth century were confirmed in a massive meta-analysis that included more than 40 studies, most of them performed during the last 2 decades (Eder et al. 2006). Second, we can look at urban–rural differences in developing countries. Allergies increase first in hygienic urban settings (Robinson et al. 2011). A third approach is to look at societies that are undergoing the second epidemiological transition and document changing disease patterns. A striking example is Karelia, an area populated by a genetically homogeneous group, but partitioned between ultra-modern Finland and developmentally retarded Russia (Seiskari et al. 2007b). Allergic sensitisation was lower in the Russian Karelian children, while their antibodies to microorganisms and the abundance and diversity of microorganisms in their homes were much higher (Pakarinen et al. 2008; Seiskari et al. 2007b). A fourth approach applicable to the allergies is to document the effect of prolonged treatment of helminths so that, at least as far as these organisms are concerned, the population begins to resemble modern urban humans. Deworming Vietnamese schoolchildren for 12 months (Flohr et al. 2010) and still more prolonged treatment of children in Venezuela or Gabon all led to increased allergen sensitisation and skin prick test responses (Lynch et al. 1993; van den Biggelaar et al. 2004).

In autoimmune diseases the immune system's regulatory police force (such as regulatory T lymphocytes (Treg)) is failing to stop the immune system from attacking the host's own tissues (Long and Buckner 2011). Here the Karelians are again interesting. The autoimmune inflammatory disorder, type 1 diabetes (where the immune system destroys the insulin-secreting β-cells in the pancreas), is six times more prevalent in Finnish Karelians than in Russian Karelians despite the fact that the genotypes associated with susceptibility are at the same frequency in the two populations (Kondrashova et al. 2005). Meanwhile striking and repeatedly confirmed large global variations in the incidence of multiple sclerosis (Rosati 2001) (an autoimmune inflammatory disorder where the immune system attacks the brain) are found to correlate inversely with the prevalence of the helminth *Trichuris trichiura*, which is an excellent surrogate marker of economic underdevelopment and of the likelihood of exposure to other developing country infections (Fleming and Cook 2006).

Inflammatory bowel diseases (IBD) are at least partly due to inappropriate, unregulated immune responses to bowel contents (Boden and Snapper 2008). The early medical literature suggests that IBD had been extremely rare before the 1900s (Kirsner 1995). Then during the early twentieth century IBD was uncommon, sporadic and usually seen in individuals from the upper classes, urban areas and northern latitudes (Kirsner 1995). Later in the twentieth century, the incidence increased steadily, as proven in repeat studies in individual countries (England, USA, Scotland, Wales, Denmark, Sweden, Israel), using reproducible methods (reviewed and referenced in Elliott et al. 2005). Meanwhile the greater incidence of IBD in hygienic urban settings continues to be obvious, and the incidence increases (using standardised diagnostic criteria) as hygiene and modernisation progress (Klement et al. 2008).

The Critical Organisms and Their Immunological Role

These considerations allow prediction of the organisms involved in the "Old Friends" hypothesis—that is to say, organisms that triggered immunoregulatory mechanisms in the past but are now lacking from the modern environment. From a Darwinian perspective we would expect the relevant organisms to have been present, inevitably and continuously, from relatively early in the evolution of the immune system ("Old Friends"). One would also anticipate a reliable mode of transmission such as the faecal–oral route, often accompanied by the ability to establish carrier states that facilitate such transmission. The following section identifies the organisms implicated as "Old Friends" that are necessary for the maturation of the regulatory pathways of the immune system (Figs. 1 and 2). Note that the mechanisms, and the evidence that the immunoregulation-enhancing properties of these organisms can be demonstrated in experimental models, are discussed later in a separate section entitled "*Mechanisms*".

Gut Microbiota

Proof, using modern methods, that the microbiota of city-dwelling European (EU) children differs dramatically from that of rural Africans was obtained recently (De Filippo et al. 2010). The faecal microbiota of children from Burkina Faso (BF) had

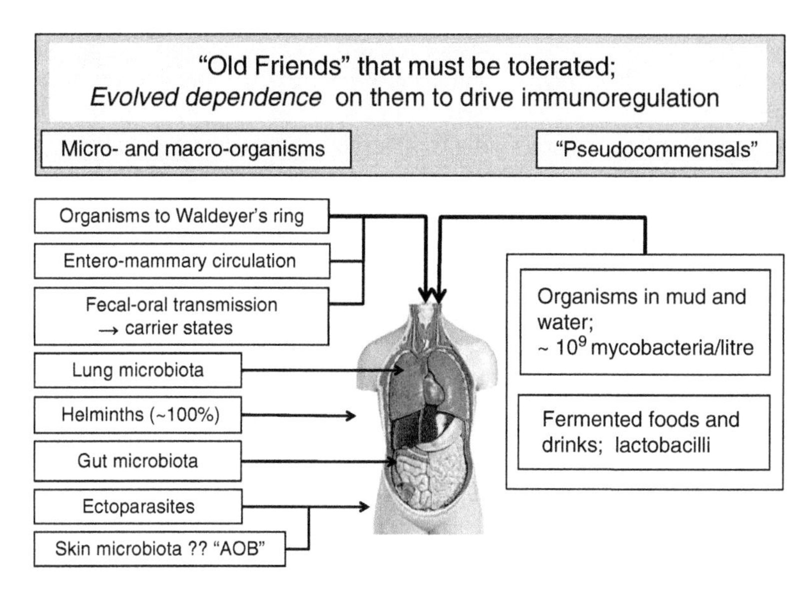

Fig. 2 Locations of some of the organisms that can be regarded as "Old Friends" and that play a role in setting up immunoregulatory circuits. *AOB* ammonia-oxidising bacteria

more Bacteroidetes and less Firmicutes and Enterobacteriaceae. Only the BF children had organisms (*Prevotella* and *Xylanibacter*) known to express enzymes for hydrolysis of cellulose and xylan which may have contributed to the fact that BF children had more short-chain fatty acids (SCFA) than did EU children (De Filippo et al. 2010). SCFA have a protective anti-inflammatory role in the gut (dos Santos et al. 2010; Maslowski et al. 2009). There was also significantly greater microbial richness and biodiversity in BF samples than in EU samples (De Filippo et al. 2010). This is important because a decrease in the abundance and biodiversity of Firmicutes has been observed repeatedly in Crohn's disease patients (Sokol et al. 2008). Interestingly, *Faecalibacterium prausnitzii,* an anti-inflammatory commensal bacterium that also contributes to SCFA generation, was present in the microbiota of the BF children (De Filippo et al. 2010) but is often lacking from European CD patients (Sokol et al. 2008).

The virome of the microbiota is now also being studied. Most of the viruses are bacteriophages, but how they vary or affect the immunoregulatory effects of the bacterial microbiota is not yet known (Reyes et al. 2010).

Organisms Introduced by Pets

Prenatal exposure to household pets influences fetal immunoglobulin E production (Aichbhaumik et al. 2008), and neonatal exposure to dogs reduces subsequent risk of allergic sensitisation (Ownby et al. 2002). It is now known that owning a cat or dog strikingly increases the abundance and diversity of bacterial taxa in house dust (Fujimura et al. 2010). The role of prenatal exposure is supported by the observation that pregnant women (not sensitised to dog) who had a dog or cat in the home had higher circulating regulatory T cell (Treg) levels compared with women who did not have pets (Wegienka et al. 2009). Treg readily suppress allergic responses (discussed in the mechanisms section below).

Chronic Infections Transmitted by the Faecal–Oral Route

Organisms highlighted in recent studies of the hygiene hypothesis that are transmitted by the faecal–oral route include *H. pylori, Salmonella*, hepatitis A virus (HAV), enteroviruses and *Toxoplasma gondii* (Fig. 1 and Table 1) (Matricardi et al. 2000; Pelosi et al. 2005; Seiskari et al. 2007a; Umetsu et al. 2005).

A number of studies have demonstrated an association between infection with the hepatitis A virus (HAV) and protection against the development of asthma (Matricardi et al. 2000; Umetsu et al. 2005). This infection was probably universal and coevolved with humans (Van Blerkom 2003). It was transmitted to the neonate by faecal–oral contamination and harmless if transmitted in this way (it is of course dangerous in the modern world when infection occurs in an adult). Its incidence rapidly declined during the twentieth century. HAV directly infects lymphocytes via the receptor TIM-1 (T-cell immunoglobulin and mucin domain containing 1). This is thought to alter differential survival of T cell subsets and so bias the immune response away from allergy-mediating Th2 and towards immunoregulation (Umetsu et al. 2005).

Viruses and Asthma: Induction *Rather than Protection*

Apart from HAV, mentioned in the previous paragraph, most studies implicate childhood viruses as *triggers* of asthma rather than as protective organisms. The viruses most frequently implicated as inducers of allergic disorders are human rhinovirus (HRV) (Jackson 2010) and respiratory syncytial virus (RSV) (reviewed inYoo et al. 2007). A prospective study of 95,310 children from birth through early childhood showed that the risk of developing asthma was greatest for children born approximately 4 months before the winter virus peak (Wu et al. 2008). This finding implied a role for early exposure to winter viruses in the triggering of later asthma (Wu et al. 2008). Animal models provide further support. It is possible to provoke a chronic inflammatory Th2-biased airway disease in mice using Sendai virus. This is a mouse parainfluenza-type I virus that is similar to RSV. In this mouse model acute lung disease appears 3 weeks after infection when IL-13-producing CD4+ Th2 cells are recruited to the airways. Seven weeks after infection there is a chronic phase associated with continuing production of IL-13 and Th2-mediated inflammation (Holtzman et al. 2009).

Table 1 Organisms inducing immunoregulation, identified by epidemiology and/or testing in experimental models of chronic inflammatory disease

Organism or location	Disease or model or effect	References
Faecal–oral transmission		
Helicobacter pylori	Allergies (epidemiology)	Matricardi et al. (2000)
Salmonella	Allergies (epidemiology)	Pelosi et al. (2005)
Toxoplasma	Allergies (epidemiology)	Matricardi et al. (2002), Matricardi et al. (2000)
Viruses		
Enteroviruses	Allergies (epidemiology)	Seiskari et al. (2007b)
Hepatitis A virus	Asthma (epidemiology)	Matricardi et al. (2002), Umetsu and DeKruyff (2010)
Viruses protective if infected early but trigger disease if late (Filippi and von Herrath 2008)		
Coxsackievirus B	Type 1 diabetes (T1D) (mouse model)	Serreze et al. (2000)
Rotavirus	T1D (mouse model)	Harrison et al. (2008)
Helminths		
Many species	Allergies (human studies)	Flohr et al. (2010), Huang et al. (2002), Scrivener et al. (2001), van den Biggelaar et al. (2004), Yazdanbakhsh and Wahyuni (2005)
Assorted natural infection	Multiple sclerosis (MS) (human studies)	Correale et al. (2008), Correale and Farez (2011)
Trichuris trichiura	MS (correlation)	Fleming and Cook (2006)
Enterobius vermicularis	T1D (correlation)	Gale (2002)
Various species	Inflammatory bowel disease (IBD) (epidemiology)	Koloski et al. (2008), Weinstock and Elliott (2009)
Animal models treated with helminths		
Heligmosomoides polygyrus	Allergy, T1D, colitis	Reviewed in Osada and Kanazawa (2010)
Schistosoma mansoni	Allergy, T1D, EAE, colitis, arthritis	Reviewed in Osada and Kanazawa (2010)
Strongyloides stercoralis	Allergy	Reviewed in Osada and Kanazawa (2010)
Fasciola hepatica	EAE	Reviewed in Osada and Kanazawa (2010)
Trichinella spiralis	T1D, EAE	Reviewed in Osada and Kanazawa (2010)
Hymenolepis diminuta	Colitis, arthritis	Reviewed in Osada and Kanazawa (2010)
Human clinical trials with helminths		
Trichuris suis	MS	Fleming et al. (2011)
Trichuris suis	IBD (Crohn's disease, ulcerative colitis)	Summers et al. (2005a, b)
Necator americanus	Asthma	Feary et al. (2010)

(continued)

Table 1 (continued)

Organism or location	Disease or model or effect	References
Gut microbiota		
Segmented filamentous bacteria	Th17 cells (mice)	Gaboriau-Routhiau et al. (2009), Ivanov et al. (2009), Wu et al. (2010)
Clostridia species	Treg in lamina propria (mice)	Geuking et al. (2011)
Bacillus fragilis	IL-10 and Treg (mice)	Round et al. (2011)
Faecalibacterium prausnitzii	Crohn's disease (human and animal studies)	Sokol et al. (2008)
Other microbiota		
Skin microbiota; ammonia-oxidising bacteria	Nitrite, nitric oxide	Whitlock and Feelisch (2009)
Lung microbiota	Asthma	Huang et al. (2011)
Oral and periodontal microbiota	IBD	Singhal et al. (2011)
Gut organisms transported to breast milk	? immunoregulation	Donnet-Hughes et al. (2010)
Ectoparasites		
Various	Response to TLR agonists in vitro	Friberg et al. (2010)
Environmental saprophyte		
Mycobacterium vaccae	Allergy (mouse, dog)	Ricklin-Gutzwiller et al. (2007), Zuany-Amorim et al. (2002)

KEY: *T1D* type 1 diabetes, *IBD* inflammatory bowel disease, *EAE* experimental autoimmune encephalomyelitis, *TLR* toll-like receptor

Viruses and Autoimmunity: A Question of Timing?

Enteroviruses such as Coxsackievirus B (CVB) and other faecal–oral viruses such as rotavirus have been implicated as triggers of type 1 diabetes (Filippi and von Herrath 2008). Weaker evidence implicates mumps virus, cytomegalovirus and rubella virus (Filippi and von Herrath 2008). However, timing is crucial, and viruses such as Coxsackieviruses (Serreze et al. 2000) or rotaviruses (Harrison et al. 2008) or lymphocytic choriomeningitis virus (LCM) that provoke autoimmunity when mice are infected at weaning or later can be protective when given very soon after birth (Filippi and von Herrath 2008; Harrison et al. 2008; Serreze et al. 2000). These findings suggest another twist to the hygiene hypothesis. Modern hygiene may cause *delayed faecal–oral transmission*, so that the virus infection occurs later than was normal during early human evolution. Consequently the immune system is at an inappropriate stage of maturation, and levels of antibody obtained transplacentally are lower.

Helminths

It is estimated that in 1947, about 36 % of the population of Europe carried helminths such as *Enterobius vermicularis, Trichuris trichiura* and *Ascaris lumbricoides* (Stoll 1947). Now even pinworm (*E. vermicularis*) has become a rarity in Europe (Gale 2002). A number of studies have reported inverse correlations between indicators of helminth burden and allergic sensitisation to environmental allergens (Araujo et al. 2000; Cooper et al. 2003; Hagel et al. 1993; Lynch et al. 1983; Nyan et al. 2001). More importantly, the risk of wheeze was reduced in individuals with hookworm (*Necator americanus*) infection in Ethiopia (Scrivener et al. 2001), and *Enterobius* infestation was negatively correlated with asthma and rhinitis in primary school children in Taiwan (Huang et al. 2002). Similarly it was suggested that infection with *Schistosoma mansoni* was associated with milder forms of asthma (Medeiros et al. 2003).

If helminths protect from chronic inflammatory disorders, the protection should be lost when the helminths are eliminated by anti-helminthic treatments. In fact this effect constitutes strong evidence for the protective role. Short periods of treatment (<12 months) did not change the prevalence of atopy or clinical signs of allergy (Cooper et al. 2006). By contrast, deworming Vietnamese schoolchildren for 12 months (Flohr et al. 2010) and still more prolonged treatment of children in Venezuela or Gabon all led to increased allergen sensitisation and skin prick test responses (Lynch et al. 1993; van den Biggelaar et al. 2004). These studies did not reveal simultaneous increases in clinical allergies such as eczema, wheeze or rhinitis, but this would probably require still longer periods of treatment and follow-up (Flohr et al. 2010).

The inverse relationship between the multiple sclerosis (MS) and the prevalence of helminth infections was mentioned earlier (Fleming and Cook 2006; Rosati 2001). Correale and colleagues have shown that patients with MS who become infected with helminths have a strikingly diminished rate of disease progression (Correale and Farez 2007). Similarly, a recent study has also shown exacerbation of MS after treatment of helminth infestation in patients suffering from this disorder (Correale and Farez 2011). A Phase 1 clinical trial using ingestion of living eggs of *Trichuris suis* (the pig whipworm) to treat MS has shown that this parasite is well tolerated, and favourable trends were observed in exploratory magnetic resonance imaging (MRI) and immunological assessments (Fleming et al. 2011).

As far as inflammatory bowel disease (IBD; Crohn's disease and ulcerative colitis) is concerned, the epidemiological data are less strong than for allergic disorders, because IBD is less common. Nevertheless, analyses of the available data conclude that exposure to helminths is one of the environmental factors most convincingly associated with a low risk of IBD (Koloski et al. 2008; Weinstock and Elliott 2009). Two well-documented anecdotes are also informative. A 12-year-old girl with occasional gastrointestinal symptoms was found to be heavily infected with *Enterobius vermicularis*. Curing the *Enterobius* infection led to severe active ulcerative colitis (Buning et al. 2008). Interestingly, a patient suffering from severe ulcerative colitis treated himself with the human whipworm (*Trichuris trichiura*) and underwent a clear remission (Broadhurst et al. 2010). Similarly, clinical trials with *Trichuris suis* have given significant results (Summers et al. 2005a, b).

Microbiota of the Skin

The role of recent changes to the microbial microbiota of the skin, lung and breast has received almost no attention. Before the invention of modern soaps and detergents, the skin was probably colonised by ammonia-oxidising bacteria (AOB) (Fig. 2). These are ubiquitous in soil, but they are exquisitely sensitive to alkylbenzene sulfonate detergents (Whitlock and Feelisch 2009). AOB can convert the high concentrations of urea and ammonia found in human sweat into nitrite and NO which are efficiently absorbed via the skin, so this source of nitrite might have been biologically significant for immunoregulation in which NO is known to play a major role (Whitlock and Feelisch 2009).

Microbiota of the Lung

The lung is not sterile. It is estimated that there are about 2,000 bacterial genomes per square centimetre of the surface of the bronchial tree, though there is a risk of contamination from the bronchoscope used to gather the samples. Pathogenic Proteobacteria, particularly *Haemophilus* spp., were more abundant in asthmatic airways (Hilty et al. 2010). Interestingly bacterial concentrations and diversity were significantly higher among asthmatic patients, as determined by assaying 16S ribosomal RNA amplicons and the relative abundance of members of several bacterial families correlated with bronchial hyperresponsiveness (Huang et al. 2011). Nothing is yet known of differences between lung microbiota of hunter-gatherers and modern urban man, but they must be striking (Fig. 2). This will be a difficult area to study because truly sterile samples of material from healthy individuals probably cannot be obtained.

The Role of Milk

Breast milk is not sterile (Donnet-Hughes et al. 2010), and there appears to be an "entero-mammary" circulation. Human peripheral blood mononuclear cells and breast milk cells contain bacteria during lactation, due to increased translocation from the gut to Peyer's patches and thence into circulating dendritic cells. Bacteria can also be seen in the glandular tissue of healthy breast and in mononuclear cells in breast milk, which when "sterile" contains small numbers of cultivable organisms

(<10^3). The mononuclear cells seem to be partially matured dendritic cells that might promote tolerogenic responses in the neonate (Donnet-Hughes et al. 2010). It remains to be seen whether this is an important part of the colonisation and education of the neonatal gut and immune system, but if it is, it must be severely disrupted in many modern humans (Fig. 2).

Not only does human milk contain bacteria as described above, but it also contains complex oligosaccharides that favour growth and establishment of selected strains such as *Bifidobacterium longum subsp. infantis* in the infant gut (Zivkovic et al. 2010). Thus, the decline in breastfeeding due to pressure to use formula milk substitutes must be altering infant microbiota.

Environmental "Pseudocommensals"

Mud and untreated water contain milligrams of various saprophytic bacterial species per litre. Therefore, milligram quantities were inevitably consumed by everyone until modern chlorinated water supplies were developed. We have designated these "pseudocommensals" because they were consumed regularly and inevitably throughout mammalian evolution (Fig. 2). These too turn out to have immunoregulatory roles and are currently entering further clinical trials (Ricklin-Gutzwiller et al. 2007; Zuany-Amorim et al. 2002). Some non-colonising lactobacillus strains present in fermenting vegetable matter must have been encountered daily, and many *Lactobacillus* strains are immunoregulatory (Smits et al. 2005) and are entering clinical trials as probiotics (Sheil et al. 2007).

Ectoparasites

In a study of wild rodents, it emerged that various aspects of the innate immune response were profoundly affected by the load of ectoparasites (i.e. fleas, lice, etc.) (Friberg et al. 2010). It is perhaps not surprising that repeated "injections" of pharmacologically active materials, many of them designed to modulate the host immune response, can provoke lasting systemic effects. Such exposures must be greatly diminished in modern society, but further work is needed to determine the importance of this factor (Friberg et al. 2010).

Malaria and Other Organisms Not Yet Implicated in the Old Friends Hypothesis

Malaria clearly needs to be studied in this context, especially in view of its importance in human evolution. *Plasmodium vivax* induces Treg population expansion in humans (Bueno et al. 2010). Similarly, placental malaria, whether active at delivery or resolved, leads to an expanded population in cord blood of malaria-specific FOXP3(+) Treg and a larger expansion of non-malaria-reactive Treg revealed by stimulation in vitro (Flanagan et al. 2010). In a mouse model *Plasmodium chabaudi* infection will attenuate the course of experimental autoimmune encephalomyelitis (EAE), an experimentally induced autoimmune disease which is widely regarded as a model of human MS (Farias et al. 2011).

Endemic malaria also causes selection of mutations within the immune system that facilitate control of the parasite (Sotgiu et al. 2007). In the Sardinian population this has included selection of proinflammatory variants of genes encoding tumour necrosis factor (TNF) and the human lymphocyte antigens (HLA) that help to protect against malaria. Unfortunately, in the absence of malaria, these variants are risk factors for MS. It has recently been suggested that the disappearance of malaria might be relevant to increasing levels of MS noted in Sardinia immediately after elimination of the parasite (Sotgiu et al. 2007).

Other protozoa have also received little or no attention in this context. For example, what about the very ancient *Entamoeba*, *Giardia* and *Trichomonas* all of which have lost their mitochondria and so become obligate parasites with a close association with humans? This area needs to be explored.

Evidence from Animal Models and Clinical Trials

If the depletion of the organisms listed in the previous sections is contributing to the increases in disorders of immunoregulation that have occurred since the second epidemiological transition, it should be possible to demonstrate that administration of these organisms can prevent and/or treat the same conditions in animal models and in human clinical trials. A few examples have been cited in context above, and a detailed analysis would be beyond the scope of this review, but this issue is expanded briefly below.

Animal Models

There are numerous experimental models in which exposure to microorganisms that were ubiquitous during mammalian evolutionary history but are currently "missing" from the environment in rich countries (or from animal units with Specific Pathogen-Free facilities) will treat allergy (Ricklin-Gutzwiller et al. 2007; Zuany-Amorim et al. 2002), autoimmunity (Farias et al. 2011; Osada and Kanazawa 2010; Zaccone et al. 2003) or intestinal inflammation (Elliott et al. 1999, 2003). More examples are listed in Table 1. The dominant mechanism revealed in these models is the induction of immunoregulatory, anti-inflammatory cells, particularly Treg.

Clinical Trials

Clinical trials using hookworm (*Necator americanus*) (Blount et al. 2009) or *Trichuris suis* have been completed (Summers et al. 2005a, b). The focus has been on allergies and inflammatory bowel disease (Blount et al. 2009; Summers et al. 2005a, b) with significant benefit in IBD. In Brisbane a trial of hookworm has been completed in celiac disease (http://clinicaltrials.gov/show/nct00671138). Hookworm did not have much impact, if any, on clinicopathological measures of celiac disease in the context of a robust gluten challenge. However, it does seem that hookworm dampens the gluten-specific Th1 and Th17 responses (John Croese, personal communication). Trials in multiple sclerosis are also in progress. *Trichuris suis* eggs by mouth were well tolerated, and favourable trends were observed in exploratory MRI and immunological assessments during a Phase 1 study (Fleming et al. 2011). Further efficacy studies are, therefore, in progress.

There is clearly also enormous scope for the modulation of gut microbiota with probiotics or prebiotics, and this area is likely to develop rapidly. So far the most convincing results have come from certain gastroenterological conditions, while in other clinical areas results have been variable (Round and Mazmanian 2009; Yan and Polk 2010).

Mechanisms

How and why do the Old Friends modulate our immune systems? Numerous mechanisms exist, but there are two related underlying evolutionary principles that probably apply to all the "Old Friends". First, the encounters with a broad range of microbial molecules that trigger the innate immune system via pattern recognition receptors (PRR) such as Toll-like receptors (TLR), and also experience of diverse microbial antigens, may be a necessary *maturation* stimulus for the immune system. Second, the "Old Friends" persist as commensals (or "pseudocommensals"), carrier states or chronic subclinical infections. Therefore, the host–"Old Friend" relationship evolved so that rather than provoking needless damaging aggressive immune responses, an anti-inflammatory equilibrium is established. This translates into the priming of immunoregulation. Maturation and immunoregulation are considered separately below, though they are clearly overlapping concepts.

Maturation of the Immune System In Utero and in the Neonate

It is suggested that the neonatal immune system defaults to a Th2 bias, in the absence of stimuli that drive maturation (Pfefferle et al. 2010). Since this is the arm of the

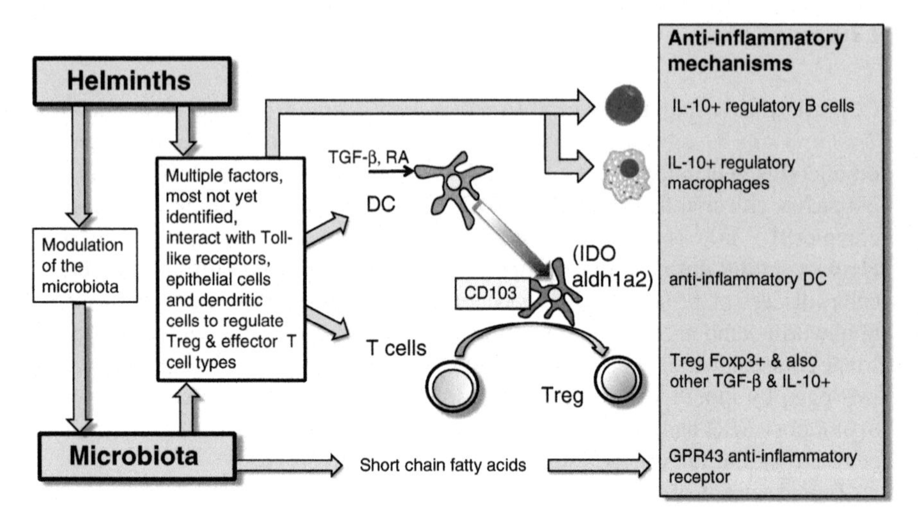

Fig. 3 Some mechanisms involved in the immunoregulatory properties of the "Old Friends" and microbiota. A key pathway is the modification of DC so that they tend to drive Treg. Such DC also process self-antigens, allergens, etc. and so drive crucial specific regulatory cell populations. The intestinal helminths also exert indirect effects by modulating the microbiota. The individual pathways (some taken from experimental systems in the mouse, as illustrations) are referenced in the main text: *RA* retinoic acid, *DC* dendritic cell, *Aldh1a2* retinaldehyde dehydrogenase 2, *GPR43* G-protein-coupled receptor 43, *RegIIIγ* regenerating islet-derived 3γ, *CD103* an integrin and marker of intestinal regulatory DC, *IDO* indoleamine 2,3, dioxygenase

immune response that drives allergic disorders, this point is of some interest. It now seems that children exposed to high quantities and high diversity of microbial materials are protected from subsequent allergic disorders (Ege et al. 2011). Interestingly this exposure can take place before birth. Exposing pregnant women to barns and farm animals during gestation reduced the risk of allergic disease in their offspring (Ege et al. 2008; von Mutius and Radon 2008). Similarly, prenatal exposure to household pets reduces fetal immunoglobulin E production (Aichbhaumik et al. 2008). Maturation following microbial exposures seems to push the system towards Th1, with increased production of the Th1 signature cytokine, IFN-γ (Pfefferle et al. 2010). Interestingly, this protective effect of *prenatal* exposure to microbial components has been reproduced in a mouse allergy model and strongly suggests that epigenetic factors are involved (Conrad et al. 2009). It is also possible that accelerated maturation and Th1 bias induced by microbial exposure indirectly protects from allergy by facilitating rapid removal of the viruses that are implicated in allergic disorders (RSV and HRV discussed earlier) (Jackson 2010; Yoo et al. 2007).

These points all raise the interesting issue that it might take several generations for the full consequences of loss of exposure to "Old Friends" to be manifested. This greatly complicates the epidemiology.

However, maturation and Th1 bias is not the whole story and might in fact be less important than priming of immunoregulation. For example, farm exposures during pregnancy increased the number and function of cord blood Treg cells, in addition to reducing the extent of Th2 cytokine secretion after in vitro stimulation (Schaub et al. 2009). Immunoregulation is discussed in the next sections.

Initiation of Immunoregulation

The Old Friends persist as commensals, carrier states or chronic subclinical infections and so coevolved a role as inducers of immunoregulation. This avoids pointless and damaging inflammatory attacks on organisms that are essential to our health (such as gut microbiota) or impossible to remove (such as established helminth infections). For instance, a futile effort to destroy *Brugia malayi* microfilariae results in lymphatic blockage and elephantiasis (Babu et al. 2006). Thus, many helminths induce expansion of Treg populations. For instance, the percentage of circulating Treg was positively related to the level of infection with *Schistosoma haematobium* in Zimbabwean children aged 8–13 years (Nausch et al. 2011). Some mechanisms are summarised in Fig. 3.

A frequent mechanism is modulation of dendritic cells (DC) such that these drive Treg rather than Th1, Th17 or Th2 effector cells (referenced in Rook 2009). These modified DC can be regarded as DCreg. Then the constitutive presence of the "Old Friends" causes continuous background activation of the DCreg and of Treg specific for the Old Friends themselves, resulting in background bystander suppression of inflammation. Meanwhile these DCreg inevitably sample self, gut contents and allergens and so induce Treg specific for the illicit target antigens of the three groups of chronic inflammatory disorder. Release of the anti-inflammatory cytokines, IL-10 and TGF-β is often involved in the anti-inflammatory effects of these cells (Zuany-Amorim et al. 2002).

Helminths and Immunoregulation

A striking example of this in human autoimmunity is a recent experiment of nature. Patients in Argentina suffering from multiple sclerosis were followed up for 4.6 years. It was found that those who developed parasite infections (which were not treated) had significantly fewer exacerbations than those who did not (Correale and Farez 2007). Moreover, they also developed regulatory lymphocytes that specifically responded to myelin basic protein by releasing IL-10 and TGF-β. In other words, the presence of the parasite appeared to drive the development of regulatory cells that recognised the auto-antigen and inhibited the autoimmune disease process. The parasites acted as "Treg adjuvants".

Little is yet known about the precise molecular signals involved in immunoregulation by helminths. For some helminths it seems that complex oligosaccharides are important, especially those that mimic human oligosaccharides. Dendritic cells (DC) express many different C-type lectin receptors on their membranes, which vary within distinct DC subsets. For example, DC-SIGN (full name Dendritic Cell-Specific Intercellular adhesion molecule-3-Grabbing Non-integrin), macrophage galactose-type C-type lectin (MGL, CD301) and the mannose receptor (MR) are all expressed by DCs and shown to interact with host-like glycans of helminths in ways that are thought to provoke immunoregulatory changes that assist parasite persistence (van Die and Cummings 2010). *Heligmosomoides polygyrus* (a helminth that has shown immunoregulatory effects in several mouse models) secretes a molecule that binds the TGF-βRII and causes FoxP3-negative T cells to become functional Foxp3+ Treg (Grainger et al. 2010). But this helminth also modulates DC function in an anti-inflammatory way (Hang et al. 2010) and may exert indirect immunomodulatory effects via induced changes in the bacterial microbiota (Walk et al. 2010).

Gut Microbiota and Immunoregulation

This topic was extensively reviewed recently, and Fig. 3 lists some of the known mechanisms (Ehlers and Rook 2011; Round and Mazmanian 2009). In the 1980s it was revealed that defined alterations to the microbiota could reproducibly either increase or decrease susceptibility to autoimmune arthritis (Kohashi et al. 1985). Similar findings have been published recently using experimental autoimmune encephalomyelitis (EAE), a mouse model of multiple sclerosis (Lee et al. 2010). Modulation of the bowel microbiota could alter susceptibility to EAE by mechanisms that involved the ability of intestinal DC to prime Th1, Th17 or Treg responses (Lee et al. 2010). Some bacteria have also been found to enhance numbers and activity of Treg (Atarashi et al. 2011) or even to secrete single molecules that lead directly to expansion of Treg populations (Round et al. 2011). Short-chain fatty acids (SCFA) produced by many gut bacteria also have an anti-inflammatory role in the gut. SCFAs bind the G-protein-coupled receptor 43 (GPR43, also known as FFAR2) and exert an anti-inflammatory effect that proved relevant in models of colitis, arthritis and asthma (Maslowski et al. 2009). Other mechanisms include induction of regulatory B cells (Correale et al. 2008) and regulatory macrophages (Schnoeller et al. 2008), modulation of Treg/Th17 balance (Ivanov et al. 2008) and indirect effects via epithelial cell products that cause DC to drive Foxp3+ cells with gut-homing properties (Iliev et al. 2009), and induced secretion of REGIIIγ, a C-type lectin with bactericidal effects on Gram-positive bacteria (Cash et al. 2006).

Interestingly there is evidence that very clean Specific Pathogen-Free (SPF) mice (i.e. animals with a "normal" microbiota, in theory lacking pathogens) have abnormally functioning Treg that can fail to secrete IL-10 and can switch function to an aggressive cell type (Erdman et al. 2010). Humans in rich Western cities are not SPF, but some modern babies must be getting close to the SPF state, with less diverse commensals and microbiota and little exposure to pathogens.

Viruses and Immunoregulation

As outlined earlier, there is strong epidemiological evidence that neonatal infection with HAV can protect against allergic disorders. The cellular receptor for HAV is TIM-1 (T-cell immunoglobulin domain and mucin domain). TIM-1 is an important atopy susceptibility gene. Furthermore, recent studies indicate that TIM-1 is a receptor for phosphatidylserine, a marker of apoptotic cells (Umetsu and DeKruyff 2010). It is not yet clear how this translates into effects on immunoregulatory pathways, but there seems to be an effect on the Th2/Treg balance.

It was stated above that viruses such as Coxsackieviruses (Serreze et al. 2000) or rotaviruses (Harrison et al. 2008) or LCM that provoke autoimmunity when given

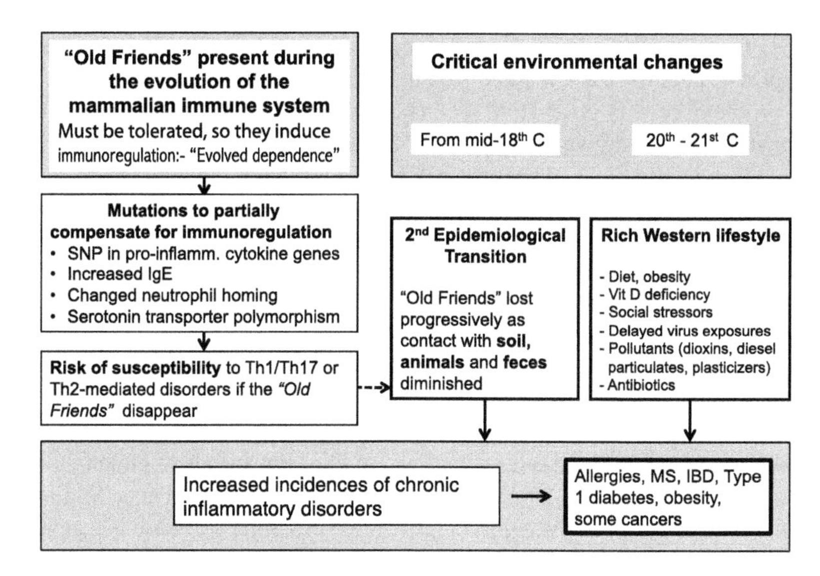

Fig. 4 Interaction of genetics and loss of the "Old Friends". The Old Friends had to be tolerated and so coevolved roles as triggers of immunoregulatory pathways. In areas with very high loads of these and other organisms, particularly helminths, compensatory genetic variants accumulated, to partially restore inflammatory responses. In the absence of the Old Friends, not only is immunoregulation inadequately primed, but also these genetic variants cause excessive inflammation and become risk factors for chronic inflammatory disorders. Thus, genetic variants that were advantageous and did not cause disease in the past start to do so in the absence of the Old Friends (referenced in main text). Several aspects of modern life are potentially interacting with the lack of "Old Friends" at the level of immunoregulation. Obesity is associated with altered gut microbiota and excessive release of proinflammatory cytokines. Lack of vitamin D exacerbates immunodysregulation, as does the triggering of Th17 cells by dioxins (changes in the microbiota also impact on Th17 development). Diesel particulates drive Th2 cells. Social stressors also drive inflammation, and delayed exposure to childhood viruses may cause them trigger allergy and autoimmunity

late (for instance, at weaning) can be protective when given very early (Filippi and von Herrath 2008; Harrison et al. 2008; Serreze et al. 2000). The mechanism again involves immunoregulation. Such viruses, at least in a mouse model, can activate invariant natural killer T cells that induce TGF-β-producing plasmacytoid DCs (pDC) in the pancreatic lymph nodes (Diana et al. 2011). These regulatory pDC then drive development of CD4+CD25+ Tregs that synergise with upregulated programmed cell death-1 ligand 1 (PD-L1) to shut off the autoimmune response (Filippi et al. 2009).

Compensatory Genetic Variants

In parts of the world where there was a heavy load of organisms causing immunoregulation, there has been selection for single nucleotide polymorphisms (SNP) or other variants that partially compensate for the immunoregulation. This is seen for several proinflammatory cytokines (Fumagalli et al. 2009) and IgE (Barnes et al.

2005). There is also an increased frequency of a truncated form of the serotonin transporter that also has a marked proinflammatory effect (Fredericks et al. 2010). The problem here is clear (Fig. 4). As soon as the immunoregulation-inducing organisms are withdrawn by the modern lifestyle, these genetic variants lead to excessive inflammation and become risk factors for chronic inflammatory disorders (Barnes et al. 2005; Fredericks et al. 2010; Fumagalli et al. 2009). The selection in Sardinia of TNF and HLA variants that protect from malaria, but predispose to MS in the absence of malaria, is another example of this that was quoted earlier (Sotgiu et al. 2007). These effects constitute a second layer of evolved dependence on the continuing presence of the "Old Friends" (Fig. 4).

This is important because work that identifies proximate "causes" for diseases that were rare or nonexistent before the second epidemiological transition may merely be unravelling a problem that would be irrelevant if the microbial status could be returned to that seen in the Palaeolithic. For instance, gluten-associated enteropathies might be an "artefact" of poorly immunoregulated guts. Similarly the recent claim to have discovered that the "cause" of Crohn's disease is a genetically determined defect in the homing of neutrophils is difficult to reconcile with the fact that 100 years ago the disease barely existed. It is the recent environmental changes that have caused this phenotype to become a risk factor (Smith et al. 2009) (Fig. 4).

Interactions with Other Changes in Modern Lifestyles

It would be foolish to assume that decreased exposure to microbial "Old Friends" is the only reason for the increasing frequency of chronic inflammatory disorders in developed countries. Other aspects of modern life that deviate from the EEA of our primate ancestors must contribute and are likely to interact with and amplify the immunoregulatory deficit resulting from the altered microbial environment (Fig. 4).

Diminished Nonmicrobial Immunoregulatory Exposures

Modern children may also be deprived of relevant *nonmicrobial* immunoregulatory exposures. For instance, the powerful protection from allergic disorders provided by early exposure to cowsheds (Riedler et al. 2001) might be in part due to exposure to immunomodulatory arabinogalactans derived from grass (Peters et al. 2010), rather than to microorganisms. Exposure to such grass dust oligosaccharides will have been inevitable during those phases of human evolution that took place on grasslands. Similarly the protective effect of unpasteurised farm milk direct from the cow is not understood (von Mutius 2010), but is probably not entirely microbial.

Obesity and Diet

The modern human diet encourages obesity, and adipose tissue releases proinflammatory mediators that will exacerbate the effects of any immunoregulatory defects (Collins and Bercik 2009). Moreover, obesity is associated with phylum-level changes in the microbiota and reduced bacterial diversity (Turnbaugh et al. 2009), which will have immunoregulatory consequences (Round and Mazmanian 2009). Psychological stress also modulates gut microbiota and gut permeability, while both obesity and stress result in greater release of proinflammatory cytokines (Collins and Bercik 2009). This all leads to a vicious circle, because the tendency to develop obesity is modulated by the nature of the microbiota. For example, low levels of Bifidobacteria and high levels of *Staphylococcus aureus* in infant microbiota may predict the development of obesity later in life (Kalliomaki et al. 2008).

Vitamin D

Humans need sunlight to drive formation of vitamin D3, which is rarely present at adequate levels in the diet. Vitamin D is involved in driving regulatory cells such as Treg (Xystrakis et al. 2006). In rich developed countries deficiency of vitamin D is increasingly common, partly because of fears of melanoma. This deficiency is implicated in the increases in chronic inflammatory disorders, some cancers and allergic disorders (Brehm et al. 2010; Herr et al. 2011; Honeyman and Harrison 2009). Moreover, vitamin D may also protect from allergic disorders indirectly by enhancing immunity to respiratory viruses (Sabetta et al. 2010), some of which are implicated in the causation of allergic disorders (Jackson 2010; Yoo et al. 2007).

This area has become deeply controversial following a recent report from the Institute of Medicine in the USA (Ross et al. 2011) that recommended much lower levels of vitamin D than most workers in the field would want to see (Heaney and Holick 2011). The recommended levels are certainly likely to be lower than those that would have been found in scantily clad hunter-gatherer humans.

Pollution

Several modern environmental pollutants might increase the incidence of chronic inflammatory disorders. Dioxins, which drive Th17 cells via the aryl hydrocarbon receptor (Veldhoen et al. 2008), will also encourage inflammatory responses of the type seen in autoimmune disease. Diesel particulates contribute to Th2 responses and allergic sensitisation (Riedl and Diaz-Sanchez 2005). There must be many other novel molecules in our environments that have significant though poorly documented proinflammatory effects.

Broader Clinical Implications

The realisation that the hygiene or "Old Friends" hypothesis is largely a matter of immunoregulation leads to the possibility that several other groups of disease might also be increasing as a consequence of diminished exposure to organisms with which we coevolved (Rook 2010). Two examples are particularly worthy of mention. First, chronically raised levels of proinflammatory cytokines cause symptoms of depression and are often found in depressed patients (Raison et al. 2010). Second, chronic inflammation is oncogenic, and once tumours have developed, ongoing inflammation releases growth factors and angiogenic factors that encourage tumour growth. The epidemiology of some human cancers that are increasing in frequency is similar to that of the allergic disorders and of T1D, and it is likely that failed immunoregulation is playing a role (Rook and Dalgleish 2011). Thus, the consequences of a lifestyle that no longer resembles the EEA of the primate or even that of early hominins may be very broad and still not fully revealed.

Conclusions

This chapter has attempted to describe the microbiological history of humankind from the hunter-gatherer past to the modern city lifestyle. Clearly this is impossible, and the result is a superficial overview. Nevertheless, some simple and important principles emerge. The immune system evolved three overlapping types of interaction with the microbial world. First it had to combat infection with pathogens. Second it had to "manage" and stabilise complex interactions with organisms that are part of our physiology and so essential for health (microbiota and pseudocommensals). Third it had to evolve ways to coexist with chronic infections that could not be eliminated (many helminths). The last two types of interaction involve immunoregulation and tolerance, and the immune system is not surprisingly in a state of evolved dependence on these organisms to drive maturation and set up the correct regulatory pathways. It is interesting that these microbial roles were not upset until the second epidemiological transition, when organisms that had coexisted with mammals from the very distant past—largely from faeces, mud and other animals—became depleted. Thus, we now see massive increases in chronic inflammatory disorders in rich countries, and the same increases are beginning to appear in developing countries. There is of course a limitation to this approach. We can measure rather accurately the changing patterns of disease and of microbial exposures during the twentieth and early twenty-first centuries, and we can also observe what happens in contemporary hunter-gatherer and subsistence farmer populations as they undergo the second epidemiological transition, lose their "Old Friends" and start to develop chronic inflammatory disorders. But extrapolation from these contemporary observations to our distant past is more uncertain. Nevertheless, phylogenetic trees are increasingly confirming the antiquity of our association with the "Old Friends", and archaeology is strengthening the view that contemporary hunter-gatherers and subsistence farmers do indeed reflect life as it was in Palaeolithic and Neolithic times. The exciting conclusion is that it should be possible to reverse the increases in chronic inflammatory disorders that plague developing countries by manipulating the microbial environments and by devising drugs based on their immunoregulatory components.

References

Aichbhaumik N, Zoratti EM, Strickler R, Wegienka G, Ownby DR, Havstad S, Johnson CC (2008) Prenatal exposure to household pets influences fetal immunoglobulin E production. Clin Exp Allergy 38(11):1787–1794

Araujo MI, Lopes AA, Medeiros M, Cruz AA, Sousa-Atta L, Sole D, Carvalho EM (2000) Inverse association between skin response to aeroallergens and Schistosoma mansoni infection. Int Arch Allergy Immunol 123(2):145–148

Armelagos GJ, Brown PJ, Turner B (2005) Evolutionary, historical and political economic perspectives on health and disease. Soc Sci Med 61(4):755–765

Atarashi K, Tanoue T, Shima T, Imaoka A, Kuwahara T, Momose Y, Cheng G, Yamasaki S, Saito T, Ohba Y et al (2011) Induction of colonic regulatory T cells by indigenous Clostridium species. Science 331:337–341

Babu S, Blauvelt CP, Kumaraswami V, Nutman TB (2006) Regulatory networks induced by live parasites impair both Th1 and Th2 pathways in patent lymphatic filariasis: implications for parasite persistence. J Immunol 176(5):3248–3256

Barnes KC, Grant AV, Gao P (2005) A review of the genetic epidemiology of resistance to parasitic disease and atopic asthma: common variants for common phenotypes? Curr Opin Allergy Clin Immunol 5(5):379–385

Blackley CH (1873) Experimental researches on the causes and nature of Catarrhus Aestivus (Hay-fever and Hay-asthma). Baillière Tindall and Cox, London

Blount D, Hooi D, Feary J, Venn A, Telford G, Brown A, Britton J, Pritchard D (2009) Immunological profiles of subjects recruited for a randomized, placebo controlled clinical trial of hookworm infection. Am J Trop Med Hyg 81:911–916

Boden EK, Snapper SB (2008) Regulatory T cells in inflammatory bowel disease. Curr Opin Gastroenterol 24(6):733–741

Bowlby J (1971) Attachment and loss, Volume 1: Attachment. Penguin, Harmondsworth, Middlesex, England, p 478 (first published by Hogarth press in 1969)

Brehm JM, Schuemann B, Fuhlbrigge AL, Hollis BW, Strunk RC, Zeiger RS, Weiss ST, Litonjua AA (2010) Serum vitamin D levels and severe asthma exacerbations in the Childhood Asthma Management Program study. J Allergy Clin Immunol 126(1):52–58, e55

Broadhurst MJ, Leung JM, Kashyap V, McCune JM, Mahadevan U, McKerrow JH, Loke P (2010) IL-22+ CD4+ T cells are associated with therapeutic trichuris trichiura infection in an ulcerative colitis patient. Sci Transl Med 2(60):60ra88

Bueno LL, Morais CG, Araujo FF, Gomes JA, Correa-Oliveira R, Soares IS, Lacerda MV, Fujiwara RT, Braga EM (2010) *Plasmodium vivax*: induction of CD4+CD25+FoxP3+ regulatory T

cells during infection are directly associated with level of circulating parasites. PLoS One 5(3):e9623
Buning J, Homann N, von Smolinski D, Borcherding F, Noack F, Stolte M, Kohl M, Lehnert H, Ludwig D (2008) Helminths as governors of inflammatory bowel disease. Gut 57(8):1182–1183
Cash HL, Whitham CV, Behrendt CL, Hooper LV (2006) Symbiotic bacteria direct expression of an intestinal bactericidal lectin. Science 313(5790):1126–1130
Collins SM, Bercik P (2009) The relationship between intestinal microbiota and the central nervous system in normal gastrointestinal function and disease. Gastroenterology 136(6):2003–2014
Conrad ML, Ferstl R, Teich R, Brand S, Blumer N, Yildirim AO, Patrascan CC, Hanuszkiewicz A, Akira S, Wagner H et al (2009) Maternal TLR signaling is required for prenatal asthma protection by the nonpathogenic microbe Acinetobacter lwoffii F78. J Exp Med 206(13):2869–2877
Cooper PJ, Chico ME, Rodrigues LC, Ordonez M, Strachan D, Griffin GE, Nutman TB (2003) Reduced risk of atopy among school-age children infected with geohelminth parasites in a rural area of the tropics. J Allergy Clin Immunol 111(5):995–1000
Cooper PJ, Chico ME, Vaca MG, Moncayo AL, Bland JM, Mafla E, Sanchez F, Rodrigues LC, Strachan DP, Griffin GE (2006) Effect of albendazole treatments on the prevalence of atopy in children living in communities endemic for geohelminth parasites: a cluster-randomised trial. Lancet 367(9522):1598–1603
Correale J, Farez M (2007) Association between parasite infection and immune responses in multiple sclerosis. Ann Neurol 61(2):97–108
Correale J, Farez MF (2011) The impact of parasite infections on the course of multiple sclerosis. J Neuroimmunol 233(1–2):6–11
Correale J, Farez M, Razzitte G (2008) Helminth infections associated with multiple sclerosis induce regulatory B cells. Ann Neurol 64(2):187–199
De Filippo C, Cavalieri D, Di Paola M, Ramazzotti M, Poullet JB, Massart S, Collini S, Pieraccini G, Lionetti P (2010) Impact of diet in shaping gut microbiota revealed by a comparative study in children from Europe and rural Africa. Proc Natl Acad Sci USA 107(33):14691–14696
de Mazancourt C, Loreau M, Dieckmann U (2005) Understanding mutualism when there is adaptation to the partner. J Ecol 93:305–314
Delmont TO, Robe P, Cecillon S, Clark IM, Constancias F, Simonet P, Hirsch PR, Vogel TM (2011) Accessing the soil metagenome for studies of microbial diversity. Appl Environ Microbiol 77(4):1315–1324
Despres L, Imbert-Establet D, Combes C, Bonhomme F (1992) Molecular evidence linking hominid evolution to recent radiation of schistosomes (Platyhelminthes: Trematoda). Mol Phylogenet Evol 1(4):295–304
Diana J, Vedran Brezar V, Beaudoin L, Dalod M, Mellor A, Tafuri A, von Herrath M, Boitard C, Mallone R, Lehuen A (2011) Viral infection prevents diabetes by inducing regulatory T cells through NKT cell–plasmacytoid dendritic cell interplay. J Exp Med 208:729–745
Donnet-Hughes A, Perez PF, Dore J, Leclerc M, Levenez F, Benyacoub J, Serrant P, Segura-Roggero I, Schiffrin EJ (2010) Potential role of the intestinal microbiota of the mother in neonatal immune education. Proc Nutr Soc 69(3):407–415
dos Santos VM, Muller M, de Vos WM (2010) Systems biology of the gut: the interplay of food, microbiota and host at the mucosal interface. Curr Opin Biotechnol 21:1–12
Eder W, Ege MJ, von Mutius E (2006) The asthma epidemic. N Engl J Med 355(21):2226–2235
Ege MJ, Herzum I, Buchele G, Krauss-Etschmann S, Lauener RP, Roponen M, Hyvarinen A, Vuitton DA, Riedler J, Brunekreef B et al (2008) Prenatal exposure to a farm environment modifies atopic sensitization at birth. J Allergy Clin Immunol 122(2):407–412, 412 e401–404
Ege MJ, Mayer M, Normand AC, Genuneit J, Cookson WO, Braun-Fahrlander C, Heederik D, Piarroux R, von Mutius E (2011) Exposure to environmental microorganisms and childhood asthma. N Engl J Med 364(8):701–709
Ehlers S, Rook GAW (2011) The role of bacterial and parasitic infections in chronic inflammatory disorders and autoimmunity. In: Kaufmann SHE, Rouse BT, Sacks DL (eds) Immunology of infectious diseases. American Society for Microbiology, Washington, DC, pp 521–535

Elliott DE, Crawford C, Lie J, Blum A, Metwali A, Qadir K, Weinstock JV (1999) Exposure to helminthic parasites protects mice from intestinal inflammation. Gastroenterology 16:A706
Elliott DE, Li J, Blum A, Metwali A, Qadir K, Urban JF Jr, Weinstock JV (2003) Exposure to schistosome eggs protects mice from TNBS-induced colitis. Am J Physiol 284(3):G385–G391
Elliott DE, Summers RW, Weinstock JV (2005) Helminths and the modulation of mucosal inflammation. Curr Opin Gastroenterol 21(1):51–58
Erdman SE, Rao VP, Olipitz W, Taylor CL, Jackson EA, Levkovich T, Lee CW, Horwitz BH, Fox JG, Ge Z et al (2010) Unifying roles for regulatory T cells and inflammation in cancer. Int J Cancer 126(7):1651–1665
Farias AS, Talaisys RL, Blanco YC, Lopes SC, Longhini AL, Pradella F, Santos LM, Costa FT (2011) Regulatory T cell induction during *Plasmodium chabaudi* infection modifies the clinical course of experimental autoimmune encephalomyelitis. PLoS One 6(3):e17849
Feary JR, Venn AJ, Mortimer K, Brown AP, Hooi D, Falcone FH, Pritchard DI, Britton JR (2010) Experimental hookworm infection: a randomized placebo-controlled trial in asthma. Clin Exp Allergy 40(2):299–306
Filippi CM, von Herrath MG (2008) Viral trigger for type 1 diabetes: pros and cons. Diabetes 57(11):2863–2871
Filippi CM, Estes EA, Oldham JE, von Herrath MG (2009) Immunoregulatory mechanisms triggered by viral infections protect from type 1 diabetes in mice. J Clin Invest 119(6):1515–1523
Flanagan KL, Halliday A, Burl S, Landgraf K, Jagne YJ, Noho-Konteh F, Townend J, Miles DJ, van der Sande M, Whittle H et al (2010) The effect of placental malaria infection on cord blood and maternal immunoregulatory responses at birth. Eur J Immunol 40(4):1062–1072
Fleming JO, Cook TD (2006) Multiple sclerosis and the hygiene hypothesis. Neurology 67(11):2085–2086
Fleming J, Isaak A, Lee J, Luzzio C, Carrithers M, Cook T, Field A, Boland J, Fabry Z (2011) Probiotic helminth administration in relapsing-remitting multiple sclerosis: a phase 1 study. Mult Scler 17(6):743–754
Flohr C, Tuyen LN, Quinnell RJ, Lewis S, Minh TT, Campbell J, Simmons C, Telford G, Brown A, Hien TT et al (2010) Reduced helminth burden increases allergen skin sensitization but not clinical allergy: a randomized, double-blind, placebo-controlled trial in Vietnam. Clin Exp Allergy 40(1):131–142
Fredericks CA, Drabant EM, Edge MD, Tillie JM, Hallmayer J, Ramel W, Kuo JR, Mackey S, Gross JJ, Dhabhar FS (2010) Healthy young women with serotonin transporter SS polymorphism show a pro-inflammatory bias under resting and stress conditions. Brain Behav Immun 24:350–357
Friberg IM, Bradley JE, Jackson JA (2010) Macroparasites, innate immunity and immunoregulation: developing natural models. Trends Parasitol 26:540–549
Fujimura KE, Johnson CC, Ownby DR, Cox MJ, Brodie EL, Havstad SL, Zoratti EM, Woodcroft KJ, Bobbitt KR, Wegienka G et al (2010) Man's best friend? The effect of pet ownership on house dust microbial communities. J Allergy Clin Immunol 126(2):410–412, 412 e411–413
Fumagalli M, Pozzoli U, Cagliani R, Comi GP, Riva S, Clerici M, Bresolin N, Sironi M (2009) Parasites represent a major selective force for interleukin genes and shape the genetic predisposition to autoimmune conditions. J Exp Med 206(6):1395–1408
Gaboriau-Routhiau V, Rakotobe S, Lecuyer E, Mulder I, Lan A, Bridonneau C, Rochet V, Pisi A, De Paepe M, Brandi G et al (2009) The key role of segmented filamentous bacteria in the coordinated maturation of gut helper T cell responses. Immunity 31(4):677–689
Gale EA (2002) A missing link in the hygiene hypothesis? Diabetologia 45(4):588–594
Geuking MB, Cahenzli J, Lawson MA, Ng DC, Slack E, Hapfelmeier S, McCoy KD, Macpherson AJ (2011) Intestinal bacterial colonization induces mutualistic regulatory T cell responses. Immunity 34(5):794–806
Grainger JR, Smith KA, Hewitson JP, McSorley HJ, Harcus Y, Filbey KJ, Finney CAM, Greenwood EJD, Knox DP, Wilson MS et al (2010) Helminth secretions induce de novo T cell Foxp3 expression and regulatory function through the TGF-beta pathway. J Exp Med 207(11):2331–2341

Hagel I, Lynch NR, Perez M, Di Prisco MC, Lopez R, Rojas E (1993) Modulation of the allergic reactivity of slum children by helminthic infection. Parasite Immunol 15(6):311–315

Hang L, Setiawan T, Blum AM, Urban J, Stoyanoff K, Arihiro S, Reinecker HC, Weinstock JV (2010) Heligmosomoides polygyrus infection can inhibit colitis through direct interaction with innate immunity. J Immunol 185(6):3184–3189

Harrison LC, Honeyman MC, Morahan G, Wentworth JM, Elkassaby S, Colman PG, Fourlanos S (2008) Type 1 diabetes: lessons for other autoimmune diseases? J Autoimmun 31(3):306–310

Heaney RP, Holick MF (2011) Why the IOM recommendations for vitamin D are deficient. J Bone Miner Res 26(3):455–457

Herr C, Greulich T, Koczulla RA, Meyer S, Zakharkina T, Branscheidt M, Eschmann R, Bals R (2011) The role of vitamin D in pulmonary disease: COPD, asthma, infection, and cancer. Respir Res 12:31

Hilty M, Burke C, Pedro H, Cardenas P, Bush A, Bossley C, Davies J, Ervine A, Poulter L, Pachter L et al (2010) Disordered microbial communities in asthmatic airways. PLoS One 5(1):e8578

Hoberg EP (2006) Phylogeny of Taenia: species definitions and origins of human parasites. Parasitol Int 55(Suppl):S23–S30

Holtzman MJ, Byers DE, Benoit LA, Battaile JT, You Y, Agapov E, Park C, Grayson MH, Kim EY, Patel AC (2009) Immune pathways for translating viral infection into chronic airway disease. Adv Immunol 102:245–276

Honeyman MC, Harrison LC (2009) Alternative and additional mechanisms to the hygiene hypothesis. In: Rook GAW (ed) The hygiene hypothesis and Darwinian medicine. Birkhäuser, Basel, pp 279–298

Huang SL, Tsai PF, Yeh YF (2002) Negative association of Enterobius infestation with asthma and rhinitis in primary school children in Taipei. Clin Exp Allergy 32(7):1029–1032

Huang YJ, Nelson CE, Brodie EL, Desantis TZ, Baek MS, Liu J, Woyke T, Allgaier M, Bristow J, Wiener-Kronish JP et al (2011) Airway microbiota and bronchial hyperresponsiveness in patients with suboptimally controlled asthma. J Allergy Clin Immunol 127(2):372–381, e371–373

Iliev ID, Mileti E, Matteoli G, Chieppa M, Rescigno M (2009) Intestinal epithelial cells promote colitis-protective regulatory T-cell differentiation through dendritic cell conditioning. Mucosal Immunol 2(4):340–350

Ivanov II, Frutos Rde L, Manel N, Yoshinaga K, Rifkin DB, Sartor RB, Finlay BB, Littman DR (2008) Specific microbiota direct the differentiation of IL-17-producing T-helper cells in the mucosa of the small intestine. Cell Host Microbe 4(4):337–349

Ivanov II, Atarashi K, Manel N, Brodie EL, Shima T, Karaoz U, Wei D, Goldfarb KC, Santee CA, Lynch SV et al (2009) Induction of intestinal Th17 cells by segmented filamentous bacteria. Cell 139(3):485–498

Jackson DJ (2010) The role of rhinovirus infections in the development of early childhood asthma. Curr Opin Allergy Clin Immunol 10(2):133–138

Jeon KW (1972) Development of cellular dependence on infective organisms: micrurgical studies in amoebas. Science 176:1122–1123

Kalliomaki M, Collado MC, Salminen S, Isolauri E (2008) Early differences in fecal microbiota composition in children may predict overweight. Am J Clin Nutr 87(3):534–538

Kirsner JB (1995) The historical basis of idiopathic inflammatory bowel diseases. Inflamm Bowel Dis 1:2–26

Klement E, Lysy J, Hoshen M, Avitan M, Goldin E, Israeli E (2008) Childhood hygiene is associated with the risk for inflammatory bowel disease: a population-based study. Am J Gastroenterol 103(7):1775–1782

Kohashi O, Kohashi Y, Takahashi T, Ozawa A, Shigematsu N (1985) Reverse effect of gram-positive bacteria vs. gram-negative bacteria on adjuvant-induced arthritis in germfree rats. Microbiol Immunol 29:487–497

Koloski NA, Bret L, Radford-Smith G (2008) Hygiene hypothesis in inflammatory bowel disease: a critical review of the literature. World J Gastroenterol 14(2):165–173

Kondrashova A, Reunanen A, Romanov A, Karvonen A, Viskari H, Vesikari T, Ilonen J, Knip M, Hyoty H (2005) A six-fold gradient in the incidence of type 1 diabetes at the eastern border of Finland. Ann Med 37(1):67–72

Larche M (2007) Regulatory T cells in allergy and asthma. Chest 132(3):1007–1014

Lee YK, Menezes JS, Umesaki Y, Mazmanian SK (2010) Proinflammatory T-cell responses to gut microbiota promote experimental autoimmune encephalomyelitis. Proc Natl Acad Sci USA. doi:10.1073/pnas.1000082107

Leibowitz U, Antonovsky A, Medalie JM, Smith HA, Halpern L, Alter M (1966) Epidemiological study of multiple sclerosis in Israel. II. Multiple sclerosis and level of sanitation. J Neurol Neurosurg Psychiatry 29(1):60–68

Long SA, Buckner JH (2011) CD4+FOXP3+ T regulatory cells in human autoimmunity: more than a numbers game. J Immunol 187(5):2061–2066

Lynch NR, Lopez R, Isturiz G, Tenias-Salazar E (1983) Allergic reactivity and helminthic infection in Amerindians of the Amazon Basin. Int Arch Allergy Appl Immunol 72(4):369–372

Lynch NR, Hagel I, Perez M, Di Prisco MC, Lopez R, Alvarez N (1993) Effect of anthelmintic treatment on the allergic reactivity of children in a tropical slum. J Allergy Clin Immunol 92(3):404–411

Mackenzie M (1887) Hay fever and paroxysmal sneezing: their etiology and treatment. Churchill, London

Maslowski KM, Vieira AT, Ng A, Kranich J, Sierro F, Yu D, Schilter HC, Rolph MS, Mackay F, Artis D et al (2009) Regulation of inflammatory responses by gut microbiota and chemoattractant receptor GPR43. Nature 461(7268):1282–1286

Matricardi PM, Franzinelli F, Franco A, Caprio G, Murru F, Cioffi D, Ferrigno L, Palermo A, Ciccarelli N, Rosmini F (1998) Sibship size, birth order, and atopy in 11,371 Italian young men. J Allergy Clin Immunol 101:439–444

Matricardi PM, Rosmini F, Riondino S, Fortini M, Ferrigno L, Rapicetta M, Bonini S (2000) Exposure to foodborne and orofecal microbes versus airborne viruses in relation to atopy and allergic asthma; epidemiological study. Br Med J 320:412–417

Matricardi PM, Rosmini F, Panetta V, Ferrigno L, Bonini S (2002) Hay fever and asthma in relation to markers of infection in the United States. J Allergy Clin Immunol 110(3):381–387

Mazmanian SK, Liu CH, Tzianabos AO, Kasper DL (2005) An immunomodulatory molecule of symbiotic bacteria directs maturation of the host immune system. Cell 122(1):107–118

Mazmanian SK, Round JL, Kasper DL (2008) A microbial symbiosis factor prevents intestinal inflammatory disease. Nature 453(7195):620–625

Medeiros M Jr, Figueiredo JP, Almeida MC, Matos MA, Araujo MI, Cruz AA, Atta AM, Rego MA, de Jesus AR, Taketomi EA et al (2003) Schistosoma mansoni infection is associated with a reduced course of asthma. J Allergy Clin Immunol 111(5):947–951

Nausch N, Midzi N, Mduluza T, Maizels RM, Mutapi F (2011) Regulatory and activated T cells in human Schistosoma haematobium infections. PLoS One 6(2):e16860

Nyan OA, Walraven GE, Banya WA, Milligan P, Van Der Sande M, Ceesay SM, Del Prete G, McAdam KP (2001) Atopy, intestinal helminth infection and total serum IgE in rural and urban adult Gambian communities. Clin Exp Allergy 31(11):1672–1678

Osada Y, Kanazawa T (2010) Parasitic helminths: new weapons against immunological disorders. J Biomed Biotechnol 2010:743–758

Ownby DR, Johnson CC, Peterson EL (2002) Exposure to dogs and cats in the first year of life and risk of allergic sensitization at 6 to 7 years of age. JAMA 288(8):963–972

Pakarinen J, Hyvarinen A, Salkinoja-Salonen M, Laitinen S, Nevalainen A, Makela MJ, Haahtela T, von Hertzen L (2008) Predominance of Gram-positive bacteria in house dust in the low-allergy risk Russian Karelia. Environ Microbiol 10(12):3317–3325

Pelosi U, Porcedda G, Tiddia F, Tripodi S, Tozzi AE, Panetta V, Pintor C, Matricardi PM (2005) The inverse association of salmonellosis in infancy with allergic rhinoconjunctivitis and asthma at school-age: a longitudinal study. Allergy 60(5):626–630

Peters M, Kauth M, Scherner O, Gehlhar K, Steffen I, Wentker P, von Mutius E, Holst O, Bufe A (2010) Arabinogalactan isolated from cowshed dust extract protects mice from allergic airway inflammation and sensitization. J Allergy Clin Immunol 126(3):648–656, e641–644

Pfefferle PI, Buchele G, Blumer N, Roponen M, Ege MJ, Krauss-Etschmann S, Genuneit J, Hyvarinen A, Hirvonen MR, Lauener R et al (2010) Cord blood cytokines are modulated by maternal farming activities and consumption of farm dairy products during pregnancy: the PASTURE Study. J Allergy Clin Immunol 125(1):108–115 e101–103

Raison CL, Lowry CA, Rook GAW (2010) Inflammation, sanitation and consternation: loss of contact with co-evolved, tolerogenic micro-organisms and the pathophysiology and treatment of major depression. Arch Gen Psychiatry 67(12):1211–1224

Resta SC (2009) Effects of probiotics and commensals on intestinal epithelial physiology: implications for nutrient handling. J Physiol 587(Pt 17):4169–4174

Reyes A, Haynes M, Hanson N, Angly FE, Heath AC, Rohwer F, Gordon JI (2010) Viruses in the faecal microbiota of monozygotic twins and their mothers. Nature 466(7304):334–338

Ricklin-Gutzwiller ME, Reist M, Peel JE, Seewald W, Brunet LR, Roosje PJ (2007) Intradermal injection of heat-killed Mycobacterium vaccae in dogs with atopic dermatitis: a multicentre pilot study. Vet Dermatol 18(2):87–93

Riedl M, Diaz-Sanchez D (2005) Biology of diesel exhaust effects on respiratory function. J Allergy Clin Immunol 115(2):221–228, quiz 229

Riedler J, Braun-Fahrlander C, Eder W, Schreuer M, Waser M, Maisch S, Carr D, Schierl R, Nowak D, von Mutius E (2001) Exposure to farming in early life and development of asthma and allergy: a cross-sectional survey. Lancet 358(9288):1129–1133

Robinson CL, Baumann LM, Romero K, Combe JM, Gomez A, Gilman RH, Cabrera L, Gonzalvez G, Hansel NN, Wise RA et al (2011) Effect of urbanisation on asthma, allergy and airways inflammation in a developing country setting. Thorax 66(12):1051–1057

Rook GAW (2009) The broader implications of the hygiene hypothesis. Immunology 126:3–11

Rook GAW (2010) 99th Dahlem conference on infection, inflammation and chronic inflammatory disorders: Darwinian medicine and the 'hygiene' or 'old friends' hypothesis. Clin Exp Immunol 160(1):70–79

Rook GAW, Dalgleish A (2011) Infection, immunoregulation and cancer. Immunol Rev 240:141–159

Rook GAW, Stanford JL (1998) Give us this day our daily germs. Immunol Today 19:113–116

Rosati G (2001) The prevalence of multiple sclerosis in the world: an update. Neurol Sci 22(2):117–139

Ross AC, Manson JE, Abrams SA, Aloia JF, Brannon PM, Clinton SK, Durazo-Arvizu RA, Gallagher JC, Gallo RL, Jones G et al (2011) The 2011 report on dietary reference intakes for calcium and vitamin D from the Institute of Medicine: what clinicians need to know. J Clin Endocrinol Metab 96(1):53–58

Round JL, Mazmanian SK (2009) The gut microbiota shapes intestinal immune responses during health and disease. Nat Rev Immunol 9(5):313–323

Round JL, Lee SM, Li J, Tran G, Jabri B, Chatila TA, Mazmanian SK (2011) The Toll-like receptor 2 pathway establishes colonization by a commensal of the human microbiota. Science 332(6032):974–977

Sabetta JR, DePetrillo P, Cipriani RJ, Smardin J, Burns LA, Landry ML (2010) Serum 25-hydroxyvitamin d and the incidence of acute viral respiratory tract infections in healthy adults. PLoS One 5(6):e11088

Schaub B, Liu J, Hoppler S, Schleich I, Huehn J, Olek S, Wieczorek G, Illi S, von Mutius E (2009) Maternal farm exposure modulates neonatal immune mechanisms through regulatory T cells. J Allergy Clin Immunol 123(4):774–782.e775

Schnoeller C, Rausch S, Pillai S, Avagyan A, Wittig BM, Loddenkemper C, Hamann A, Hamelmann E, Lucius R, Hartmann S (2008) A helminth immunomodulator reduces allergic and inflammatory responses by induction of IL-10-producing macrophages. J Immunol 180(6):4265–4272

Scrivener S, Yemaneberhan H, Zebenigus M, Tilahun D, Girma S, Ali S, McElroy P, Custovic A, Woodcock A, Pritchard D et al (2001) Independent effects of intestinal parasite infection and domestic allergen exposure on risk of wheeze in Ethiopia: a nested case–control study. Lancet 358(9292):1493–1499

Seiskari T, Kondrashova A, Viskari H, Kaila M, Haapala AM, Aittoniemi J, Virta M, Hurme M, Uibo R, Knip M et al (2007) Allergic sensitization and microbial load—a comparison between Finland and Russian Karelia. Clin Exp Immunol 148(1):47–52

Serreze DV, Ottendorfer EW, Ellis TM, Gauntt CJ, Atkinson MA (2000) Acceleration of type 1 diabetes by a coxsackievirus infection requires a preexisting critical mass of autoreactive T-cells in pancreatic islets. Diabetes 49(5):708–711

Sheil B, Shanahan F, O'Mahony L (2007) Probiotic effects on inflammatory bowel disease. J Nutr 137(3 Suppl 2):819S–824S

Singhal S, Dian D, Keshavarzian A, Fogg L, Fields JZ, Farhadi A (2011) The role of oral hygiene in inflammatory bowel disease. Dig Dis Sci 56(1):170–175

Smith AM, Rahman FZ, Hayee B, Graham SJ, Marks DJ, Sewell GW, Palmer CD, Wilde J, Foxwell BM, Gloger IS et al (2009) Disordered macrophage cytokine secretion underlies impaired acute inflammation and bacterial clearance in Crohn's disease. J Exp Med 206(9):1883–1897

Smits HH, Engering A, van der Kleij D, de Jong EC, Schipper K, van Capel TM, Zaat BA, Yazdanbakhsh M, Wierenga EA, van Kooyk Y et al (2005) Selective probiotic bacteria induce IL-10-producing regulatory T cells in vitro by modulating dendritic cell function through dendritic cell-specific intercellular adhesion molecule 3-grabbing nonintegrin. J Allergy Clin Immunol 115(6):1260–1267

Sokol H, Pigneur B, Watterlot L, Lakhdari O, Bermudez-Humaran LG, Gratadoux JJ, Blugeon S, Bridonneau C, Furet JP, Corthier G et al (2008) Faecalibacterium prausnitzii is an anti-inflammatory commensal bacterium identified by gut microbiota analysis of Crohn disease patients. Proc Natl Acad Sci USA 105(43):16731–16736

Sotgiu S, Sannella AR, Conti B, Arru G, Fois ML, Sanna A, Severini C, Morale MC, Marchetti B, Rosati G et al (2007) Multiple sclerosis and anti-*Plasmodium falciparum* innate immune response. J Neuroimmunol 185(1–2):201–207

Steering Committee ISAAC (1998) Worldwide variation in prevalence of symptoms of asthma, allergic rhinoconjunctivitis, and atopic eczema: ISAAC. The International Study of Asthma and Allergies in Childhood (ISAAC) Steering Committee. Lancet 351(9111):1225–1232

Stene LC, Nafstad P (2001) Relation between occurrence of type 1 diabetes and asthma. Lancet 357:607

Stoll NR (1947) This wormy world. J Parasitol 33(1):1–18

Strachan DP (1989) Hay fever, hygiene, and household size. Br Med J 299(6710):1259–1260

Strachan DP, Taylor EM, Carpenter RG (1996) Family structure, neonatal infection, and hay fever in adolescence. Arch Dis Child 74:422–426

Strickland DH, Holt PG (2011) T regulatory cells in childhood asthma. Trends Immunol 32(9):420–427

Summers RW, Elliott DE, Urban JF Jr, Thompson R, Weinstock JV (2005a) Trichuris suis therapy in Crohn's disease. Gut 54(1):87–90

Summers RW, Elliott DE, Urban JF Jr, Thompson RA, Weinstock JV (2005b) Trichuris suis therapy for active ulcerative colitis: a randomized controlled trial. Gastroenterology 128(4):825–832

Turnbaugh PJ, Hamady M, Yatsunenko T, Cantarel BL, Duncan A, Ley RE, Sogin ML, Jones WJ, Roe BA, Affourtit JP et al (2009) A core gut microbiome in obese and lean twins. Nature 457(7228):480–484

Umetsu DT, DeKruyff RH (2010) Microbes, apoptosis and TIM-1 in the development of asthma. Clin Exp Immunol 160(1):125–129

Umetsu DT, McIntire JJ, DeKruyff RH (2005) TIM-1, hepatitis A virus and the hygiene theory of atopy: association of TIM-1 with atopy. J Pediatr Gastroenterol Nutr 40(Suppl 1):S43

Van Blerkom LM (2003) Role of viruses in human evolution. Am J Phys Anthropol Suppl 37:14–46

van den Biggelaar AH, Rodrigues LC, van Ree R, van der Zee JS, Hoeksma-Kruize YC, Souverijn JH, Missinou MA, Borrmann S, Kremsner PG, Yazdanbakhsh M (2004) Long-term treatment of intestinal helminths increases mite skin-test reactivity in Gabonese schoolchildren. J Infect Dis 189(5):892–900 van Die I, Cummings RD (2010) Glycan gimmickry by parasitic helminths: a strategy for modulating the host immune response? Glycobiology 20(1):2–12

Veldhoen M, Hirota K, Westendorf AM, Buer J, Dumoutier L, Renauld JC, Stockinger B (2008) The aryl hydrocarbon receptor links TH17-cell-mediated autoimmunity to environmental toxins. Nature 453(7191):106–109

von Ehrenstein OS, von Mutius E, Illi S, Baumann L, Bohm O, von Kries R (2000) Reduced risk of hay fever and asthma among children of farmers. Clin Exp Allergy 30:187–193

von Mutius E (2010) 99th Dahlem conference on infection, inflammation and chronic inflammatory disorders: farm lifestyles and the hygiene hypothesis. Clin Exp Immunol 160(1):130–135

von Mutius E, Radon K (2008) Living on a farm: impact on asthma induction and clinical course. Immunol Allergy Clin North Am 28(3):631–647

von Mutius E, Vercelli D (2010) Farm living: effects on childhood asthma and allergy. Nat Rev Immunol 10(12):861–868

Walk ST, Blum AM, Ewing SA, Weinstock JV, Young VB (2010) Alteration of the murine gut microbiota during infection with the parasitic helminth Heligmosomoides polygyrus. Inflamm Bowel Dis 16(11):1841–1849

Wegienka G, Havstad S, Zoratti EM, Woodcroft KJ, Bobbitt KR, Ownby DR, Johnson CC (2009) Regulatory T cells in prenatal blood samples: variability with pet exposure and sensitization. J Reprod Immunol 81(1):74–81

Weinstock JV, Elliott DE (2009) Helminths and the IBD hygiene hypothesis. Inflamm Bowel Dis 15(1):128–133

Whitlock DR, Feelisch M (2009) Soil bacteria, nitrite, and the skin. In: Rook GAW (ed) The hygiene hypothesis and Darwinian medicine. Basel, Birkhäuser, pp 103–116

Wildin RS, Smyk-Pearson S, Filipovich AH (2002) Clinical and molecular features of the immunodysregulation, polyendocrinopathy, enteropathy, X linked (IPEX) syndrome. J Med Genet 39(8):537–545

Wu P, Dupont WD, Griffin MR, Carroll KN, Mitchel EF, Gebretsadik T, Hartert TV (2008) Evidence of a causal role of winter virus infection during infancy in early childhood asthma. Am J Respir Crit Care Med 178(11):1123–1129

Wu HJ, Ivanov II, Darce J, Hattori K, Shima T, Umesaki Y, Littman DR, Benoist C, Mathis D (2010) Gut-residing segmented filamentous bacteria drive autoimmune arthritis via T helper 17 cells. Immunity 32(6):815–827

Xystrakis E, Kusumakar S, Boswell S, Peek E, Urry Z, Richards DF, Adikibi T, Pridgeon C, Dallman M, Loke TK et al (2006) Reversing the defective induction of IL-10-secreting regulatory T cells in glucocorticoid-resistant asthma patients. J Clin Invest 116(1):146–155

Yan F, Polk DB (2010) Probiotics: progress toward novel therapies for intestinal diseases. Curr Opin Gastroenterol 26(2):95–101

Yazdanbakhsh M, Wahyuni S (2005) The role of helminth infections in protection from atopic disorders. Curr Opin Allergy Clin Immunol 5(5):386–391

Yoo J, Tcheurekdjian H, Lynch SV, Cabana M, Boushey HA (2007) Microbial manipulation of immune function for asthma prevention: inferences from clinical trials. Proc Am Thorac Soc 4(3):277–282

Zaccone P, Fehervari Z, Jones FM, Sidobre S, Kronenberg M, Dunne DW, Cooke A (2003) Schistosoma mansoni antigens modulate the activity of the innate immune response and prevent onset of type 1 diabetes. Eur J Immunol 33(5):1439–1449

Zivkovic AM, German JB, Lebrilla CB, Mills DA (2010) Microbes and health Sackler colloquium: human milk glycobiome and its impact on the infant gastrointestinal microbiota. Proc Natl Acad Sci USA 108:4653–4658. doi:10.1073/pnas.1000083107, www.pnas.org/cgi/doi/

Zuany-Amorim C, Sawicka E, Manlius C, Le Moine A, Brunet LR, Kemeny DM, Bowen G, Rook G, Walker C (2002) Suppression of airway eosinophilia by killed Mycobacterium vaccae-induced allergen-specific regulatory T-cells. Nat Med 8:625–629

Make New Friends and Keep the Old? Parasite Coinfection and Comorbidity in *Homo sapiens*

Melanie Martin, Aaron D. Blackwell, Michael Gurven, and Hillard Kaplan

Introduction

Across species, the fitness costs of parasitic infection have been a major force shaping host adaptations to avoid infection (Hart 2009; Schmid-Hempel 2003; Sheldon and Verhulst 1996), diminish the cost of infection (Minchella 1985; Råberg et al. 2009), and even advertise resistance to infection to possible mates (Hamilton and Zuk 1982; Moller 1990). These adaptations, in turn, have shaped selection on parasite transmission and virulence, leading to coevolved host–parasite systems. Host–parasite interactions are further shaped by local environments and proximate host factors that influence transmission risk and infectious outcomes, including age, sex, and nutritional and immune status (Anderson and May 1981; Anderson 1991; Quinnell et al. 1995; Schad and Anderson 1985; Woolhouse 1992).

Host–parasite interactions are often observed and modeled as hosts interacting with a single parasite species. Yet as is increasingly observed in animal populations, including humans, coinfection with two or more species (alternately termed "multiple-species" or "polyparasitic" infection) may be the rule in nature (Howard et al. 2001; Booth et al. 1998; Bordes and Morand 2009; Pullan and Brooker 2008). Coinfecting species may include any number of "typical parasites" (e.g., helminths, flukes, tapeworms) and/or pathogens (e.g., bacteria, viruses, protozoa), each with different associated exposure risks, infectious sites, reproductive strategies, virulence,

M. Martin (✉) • A.D. Blackwell • M. Gurven
Department of Anthropology, University of California Santa Barbara,
Santa Barbara, CA 93106-3210, USA
e-mail: melaniemartin@umail.ucsb.edu

H. Kaplan
University of New Mexico, Department of Anthropology, MSC01-1040,
Anthropology 1, 87131, Albuquerque, NM, USA

Tsimane Health and Life History Project

J.F. Brinkworth and K. Pechenkina (eds.), *Primates, Pathogens, and Evolution*,
Developments in Primatology: Progress and Prospects,
DOI 10.1007/978-1-4614-7181-3_12, © Springer Science+Business Media New York 2013

and associated immune responses (Alizon 2008; May et al. 2009; Rigaud et al. 2010; de Roode et al. 2005; Van Baalen and Sabelis 1995). Of particular interest is the role of immune responses in mediating coinfection risk, as an immune response generated by one species may either increase or decrease a host's susceptibility to infection with another (Christensen et. al. 1987; Cox 2001; Supali et al. 2010). At the same time, infecting species may competitively inhibit establishment or replication by other species (Lim and Heyneman 1972; Fredensborg and Poulin 2005). As such, coinfection may increase, decrease, or have no effect on host fitness, depending on the individual species involved (Fellous and Koella 2009).

For humans—whose habitats span from hot, humid jungles to dry deserts, frozen tundras, and sterile office buildings—infection risk may be especially varied. Differences in parasitic and pathogenic exposure appear to have been a major force shaping genetic variation across human populations (Fumagalli et al. 2011). Given its ubiquity in nature and among nonindustrialized populations (Howard et al. 2001), multiple-species infection was likely equally common and varied among human and hominin ancestral populations. Along with more transient infections, hominin ancestors would have harbored multiple symbiotic organisms common to other mammals: commensal bacteria, pseudo-commensals, ectoparasites, and helminths (Armelagos and Harper 2005; Rook 2008). As proposed by the "hygiene hypothesis," continuous exposure to these organisms during mammalian evolution may have favored the evolution of immunoregulatory systems that required their antigenic input to develop appropriately (Jackson et al. 2008). In modern industrialized, hygienic environments, infection with many of these "old friends" (particularly helminths) is exceedingly rare, and consequently, disorders of immunoregulation (i.e., allergy, asthma, chronic inflammatory conditions) have become increasingly common (Rook 2008).

However, these old friends are also not without costs. First, helminth-induced immunoregulation, which downregulates proinflammatory responses, may decrease resistance to other parasites and more virulent pathogens, resulting in increased infection intensity or exacerbated immunopathology (Graham et al. 2005; Pullan and Brooker 2008). Second, exposure to helminth coinfection may increase investment in immune function (Bordes and Morand 2009), which may divert energy away from other fitness-enhancing allocations, such as growth and reproduction (Sheldon and Verhulst 1996; Adamo 2001; Uller et al. 2006; Blackwell et al. 2010; Muehlenbein et al. 2010). Given the risk of helminth coinfection in ancestral environments, potentially divergent immune responses, and the costs of increased investment in immune function, several questions arise. How does helminth coinfection risk and associated morbidity vary across environments and with different interacting species? How costly are multiple-species infections involving helminths in humans? What multiple-species infections would have been typical for ancestral populations? Finally, how have recent environmental changes altered helminth coinfection risk in modern populations, and what are the consequences for human health?

In this chapter, we review known aspects of immune responses to helminths and other parasitic and pathogenic threats and consider how coinfections involving helminths may affect human immune function and health, both past and present. We first review host, environmental, and parasitic characteristics that influence the likelihood of helminth coinfection. We then examine helminth-protozoa coinfection and helminth-associated morbidity among the Tsimane of lowland Bolivia. The Tsimane are a subsistence-scale, forager-horticulturalist population afflicted with a high burden of both parasites and pathogens. We examine coinfection involving helminths and protozoa and interactions between helminths and the risk of other infections and inflammatory conditions. Although many aspects of the Tsimane environment are unlikely to match those of ancestral hunter-gatherer populations, the Tsimane disease ecology is likely more representative of ancestral conditions than that of a contemporary industrialized or transitioning population. Our intent is to provide an example of the complex interactions between infecting species that would likely have been present through much of human history.

Multiple-Species Infections in Humans

Across human populations, associations between coinfecting species and infectious outcomes vary widely (Hagel et al. 2011; Howard et al. 2001; Walson and John-Stewart 2007). This variation may result from (1) individual host and parasite factors influencing transmission and infection risk and (2) direct and indirect interactions between coinfecting species (Cox 2001; Karvonen et al. 2009; Lello and Hussell 2008). First, factors that mediate the risk of single-species transmission influence the likelihood of multiple-species infection. Host factors influencing susceptibility to infection include age, sex, socioeconomic status, physical condition, nutritional status, sanitation and hygiene, work and labor demands, access to medical care, prior exposure or immunization to infecting species, water supply, and interactions with infected or reservoir hosts (Esrey et al. 1991; Haswell-Elkins et al. 1987; Sayasone et al. 2011). Local ecological features (e.g., seasonality, soil, streams, ponds, etc.) and the life cycles, growth requirements, and density and distribution of parasites in a given environment further influence transmission risk (Anderson 1991; Hall and Holland 2000; Holland 2009). Parasites with similar transmission routes (e.g., soil transmitted or waterborne) and microclimatic requirements for growth and replication are more likely to coinfect hosts (Ellis et al. 2007; Fleming et al. 2006; Haswell-Elkins et al. 1987; Supali et al. 2010).

Once transmitted, infecting species must also establish and replicate. Typical parasites such as helminths do not replicate inside hosts; eggs are excreted and larval life stages occur in soil or animal vectors. Consequently, infection intensity and pathogenicity depend on the infectious dose of initial and secondary infections (Anderson and May 1979; May and Anderson 1979; Lafferty and Kuris 2002). In contrast, pathogens replicate asexually in hosts and effects on hosts are independent of the initial infectious dose (Lafferty and Kuris 2002).

For coinfecting species, the sequential order of establishment, infectious dose, and density of established parasites influence infectious outcomes of secondarily invading species (Fellous and Koella 2009). Coinfecting species may inhibit the establishment of new parasites through direct competition or competitive inhibition (e.g., monopolizing host resources), ultimately reducing infection intensity or pathogenesis of one or more species (Lafferty et al. 1994; Fredensborg and Poulin 2005). Direct competition is more likely when species inhabit similar locales in the host (e.g., skin, lung, gut, blood, or lymphatic system) (Karvonen et al. 2009). Finally, individual species can indirectly influence establishment and clearance of coinfecting species across locales through antagonistic or synergistic immune responses.

Immune Responses to Parasites and Pathogens

Mammalian, and indeed all vertebrate, immune systems have evolved to counter invasions from diverse parasites and pathogens. Responses may be optimized to clear infection and/or minimize damage from infection, depending on the infectious agent's own evolved strategy to evade or exploit host immune pathways (Allen and Maizels 2011). The immune system makes strong phenotypic commitments in response to infection, which may lead to biased immune responses that in turn influence coinfection outcomes (Bradley and Jackson 2008). Each of the several different types of immune defense, therefore, has its own costs and benefits, and organisms must allocate resources appropriately to invest in defenses that are useful for local pathogens (Long and Nanthakumar 2004; McDade 2005).

The vertebrate immune system is generally divided into two levels of response: innate and adaptive immunity. Innate immunity is the first line of defense, found in all plants and animals; it recognizes and responds to generic signals of invasion (e.g., unchanging structures on bacteria cell walls) with nonspecific responses including inflammation, induction of acute-phase proteins (e.g., C-reactive protein), activation of the complement system (a cascade of proteins that assist antibodies and phagocytic cells in pathogen clearance), and activation and recruitment of white blood cells, or leukocytes, to target and clear infected host cells and extracellular viruses, bacteria, and protozoa.

Adaptive immunity is found only in vertebrates and, compared to innate immunity, is highly specific, highly flexible in its recognition capabilities, and capable of antigen-specific memory. Importantly, helminth diversity—a proxy for coinfection risk—may have selected for increased investment in adaptive immunity during mammalian evolution (Bordes and Morand 2009). Adaptive immunity is activated when particles from invading organisms (antigens) are engulfed and processed by phagocytic cells of the innate immune system. These cells present the antigens to effector cells of the adaptive system (T and B cells), which are then activated and clonally expanded. Activated B cells release antibodies, which bind to antigen and facilitate pathogen clearance. Activated B cells may also develop into memory B cells, which are the basis of acquired immunity.

Activated T cells undergo further differentiation into various subgroups of T cells that direct different immune responses. These subgroups include cytotoxic T cells, helper T cells (T_H), and regulatory T cells (T_{reg} cells). Helper T cells are further differentiated into T_H1, T_H2, and T_H17 cells based on the cytokines they are associated with. In brief, T_H1 cells and associated cytokines such as IFN-γ stimulate inflammatory and cell-killing activity important in clearance of pathogens (e.g., protozoa, trypanosomes, bacteria, viruses). T_H2-associated cytokines (primarily IL-4, IL-5, IL-13) stimulate antibodies including immunoglobulin E (IgE), IgG1, and (in humans) IgG4 production, as well as basophils, eosinophils, and mast cells, which are important in mediating clearance and tissue repair associated with typical parasites (e.g., helminths, flatworms) (Allen and Maizels 2011). T_H1 and T_H2 responses are directly antagonistic, with IL-4 inhibiting IFN-γ production and vice versa (Maizels and Yazdanbakhsh 2003).

Clearance of coinfecting species that provoke similar immune responses may be enhanced through cross-immunity (Lello and Hussell 2008; Supali et al. 2010), which can also diminish the likelihood of future coinfection with other commonly associated species (Karvonen et al. 2009). Conversely, immune responses directed against one species may suppress responses against other species if the coinfecting species invoke antagonistic immune responses. As is discussed below, increasing evidence suggests that helminths may bias immune function in a manner that increases susceptibility to viral and bacterial infections.

Helminth-Induced Immune Responses and Coinfection in Humans

Helminths are a large category of parasite known to have significant effects on host fitness. The term "helminth" refers collectively to wormlike parasites and encompasses two phyla of major human parasites: Platyhelminthes (flatworms), which include tapeworms (e.g., *Taenia* spp. and *Hymenolepis* spp.) and flukes (e.g., *Schistosoma mansoni* and *Schistosoma japonicum*), and Nematoda (roundworms), which include ascarids (e.g., *Ascaris lumbricoides*), filarial worms (e.g., *Wuchereria bancrofti*), pinworm (e.g., *Enterobius vermicularis*), whipworm (*Trichuris trichiura*), threadworm (*Strongyloides stercoralis*), and hookworm (referring to *Ancylostoma duodenale* and *Necator americanus*, which are often undifferentiated in microscopic identification). Helminths—which are long lived, grow to sexual maturity in hosts, but do not replicate in hosts—evoke relatively gentle immune responses in mammals, quite distinct from the strong inflammatory responses evoked by transient microbial pathogens that present imminent threats to host fitness (Jackson et al. 2008; Allen and Maizels 2011).

Helminths shift T cell populations towards a T_H2 immune response, with corresponding decreases in T_H1 and proinflammatory responses (Cooper et al. 2000; Fallon and Mangan 2007; Fox et al. 2000; Hewitson et al. 2009; Maizels and Yazdanbakhsh 2003; Yazdanbakhsh et al. 2002). T_H2 cells activated by helminths in mucosal tissues induce production of IL-13 and IL-4 cytokines that drive mucosal and muscular responses to dislodge the parasites. In non-mucosal tissues, T_H2-induced pathways and innate immune cells such as eosinophils, basophils, and mast cells help drive parasite killing. IgE secreted from B cells is important in protecting hosts from extraintestinal and encysted stages of helminths and may facilitate antibody-induced larval killing following concomitant or secondary infections (Allen and Maizels 2011).

Many helminths (as well as commensal bacteria) also induce T_{reg} activity in order to enhance their own survival in the host (Maizels et al. 2009). T_{reg} cells release cytokines that suppress T_H1 and T_H2 responses in order to minimize immunopathology and epithelial damage caused by immune activation (Rook 2008; Round and Mazmanian 2009). T_{reg} activity may also promote production of IgG4 over IgE and reduce expulsion of worms from the host (Mingomataj et al. 2006). Enhanced and spontaneous production of T_{reg} cytokines has also been observed in children with chronic *A. lumbricoides* or *T. trichiura* infection and *A. lumbricoides/T. trichiura* coinfection, suggesting endemic exposure to multiple helminths promotes stronger immunoregulation (Turner et al. 2008; Figueiredo et al. 2010).

As such, the prototypical T_H2/T_{reg} response induced by helminths may be better characterized as a "tolerance" response that contains the extent of helminth infection while limiting damage to the host (Jackson et al. 2008). In this case, the interests of host and parasite may align. Chronic exposure to helminths, which present low or intermediate threats to host fitness compared to pathogens, would favor a continual tolerance response over successive, highly inflammatory responses that would be energetically costly and highly immunopathogenic to hosts. A tolerance response would also be favored by helminths, as the T_{reg} response enhances their own long-term survival in the host, while T_H2 responses may limit establishment and competition by secondary invaders, allowing established parasites to monopolize host resources (Jackson et al. 2008).

The tolerance response suggests that mammals share a deep coevolutionary legacy with helminths. The IgE antibody, which is integral to antihelminth responses, is a derived innovation in the mammalian lineage (Jackson et al. 2008). However, more recent coevolution is also apparent: in humans, helminths appear to have played a major role in genetic population divergence since the appearance of anatomically modern humans within the last 200,000 years (Fumagalli et al. 2010). More recent evidence of helminth infection in human history comes from mummified remains dating to approximately 30,000 BP in the Old World (*A. lumbricoides*) and 7,837 BP in the New World (*E. vermicularis*) (Gonçalves et al. 2003), as well as historical writings from classical Egyptian and Greek physicians (Cox 2002).

Today, however, helminth exposure and associated immune phenotypes are varied across human populations. Hygiene, medicine, and socioeconomic development

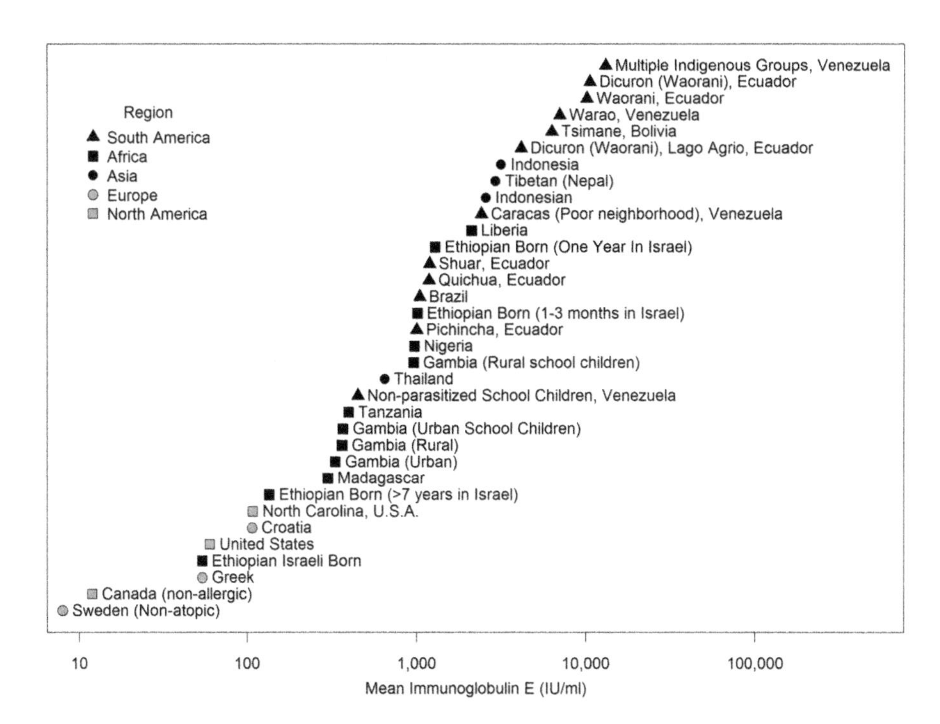

Fig. 1 Variation in human IgE by population. Values are geometric means for each published population (Adapted from Blackwell et al. (2010), which contains the complete references for each population)

have drastically reduced early helminth exposure in many industrialized populations. Average IgE levels in North America and Europe are as much as 200 times lower than those observed in subsistence-scale populations (Fig. 1). The highest IgE levels are found among lowland indigenous groups in Ecuador (Buckley et al. 1985; Kaplan et al. 1980; Kron et al. 2000) and Venezuela (Hagel et al. 2006; Lynch et al. 1983), which have reported geometric mean IgE in excess of 10,000 IU/ml. In contrast, geometric mean IgE in the USA is 52 IU/ml (Blackwell et al. 2011). Although genetic factors have been shown to influence IgE levels (Weidinger et al. 2008) and IgE levels show relatively high heritability when parents and offspring experience similar environments (Grant et al. 2008), differences between populations appear to be influenced largely by exposure to helminths (Cooper et al. 2008) and other parasites and pathogens, including malaria (Perlmann et al. 1994, 1999). Immigrants who move from areas with endemic helminth infections to those with low endemicity show an eventual drop in IgE levels, although it may take a decade or more for levels to fall significantly (Iancovici Kidon et al. 2005; Kalyoncu and Stålenheim 1992).

While the low IgE levels observed in North America and Europe are consistent with low levels of helminth exposure, individuals in these populations with allergic diseases such as asthma *do* show elevated IgE levels (e.g., Bergmann et al. 1995; Holford-Strevens et al. 1984; Lindberg and Arroyave 1986). Variations on the "hygiene" and "old friends" hypotheses posit that a major factor in the rise of allergic, autoimmune, and inflammatory disorders in industrialized populations today is a mismatch between a human immune system that coevolved with "old friends" (e.g., helminths and commensals) and a modern hygienic environment in which these "old friends" are largely absent (Rook 2008). Without early and regular antigenic input from helminths, immune system development is altered, resulting in "inappropriate" T_H2 immune responses to harmless environmental antigens. When induced by helminths, those same immune responses may depress allergic reactions (Maizels 2005; Wilson and Maizels 2004), to the extent that prescribed low-dose infections may be effective clinical treatments (Blount et al. 2009; Feary et al. 2010). The downregulation of inflammatory pathways induced by low-grade helminth infections may also protect against diabetes (Maizels et al. 2009), obesity (Wu et al. 2011), and immunopathology associated with opportunistic bacterial infections (Anthony et al. 2008).

At the same time, chronic helminth infection can be costly to hosts. Presently, the most common human helminth infections involve intestinal nematodes, especially the soil-transmitted helminths (STH) *A. lumbricoides, T. trichiura,* and hookworm. Adult STH reside and replicate in the host's gut and pass eggs through host feces. *A. lumbricoides* and *T. trichiura* eggs are ingested by hosts, whereas hookworm eggs penetrate the skin, generally the bottom of the feet (Hotez et al. 2008; Jackson et al. 2008). STH are widespread but are most prevalent in tropic and subtropic regions (Chan et al. 1994; Silva et al. 2003). In many nonindustrialized nations, STH infections are endemic and—despite increased efforts to improve sanitary conditions, access to health care, and implement large-scale control programs—remain a significant cause of morbidity, particularly among children (Bethony et al. 2006). Complications ensuing from STH infections in children and adults (e.g., anemia, growth faltering, reduced work output, and impaired cognitive ability), compounded with poor nutrition and poverty, likely contribute to poor economic growth in these areas (Guyatt 2000).

Coinfections involving multiple STH, or at least one STH and another parasite or pathogen, are exceedingly common in nonindustrialized populations (Hotez et al. 2008). Multiple STH infections are often associated with increased infection intensity, egg output, and morbidity (Booth et al. 1998; Brooker et al. 2000; Ellis et al. 2007; Pullan and Brooker 2008). It is also increasingly documented that helminth infection and helminth-typical immune biasing may diminish immune responses to vaccines, viruses, and bacteria, resulting in increased susceptibility to other infectious diseases (e.g., HIV/AIDS (Bentwich et al. 1995); BCG, typhoid, measles, and polio vaccines (Labeaud et al. 2009); tuberculosis (Lienhardt et al. 2002)). In sum, helminths may interact with human immune function and other host factors in a myriad of complex ways that have varying implications for human health. The risks and effects of helminth coinfection therefore, while of clear clinical and epidemiological significance, are also of relevance to researchers working across evolutionary and ecological fields.

Helminth Coinfection and Morbidity in the Tsimane of Bolivia

Human immune pathways may have been selected to counter the disease ecologies of our predecessors, which included constant exposure to multiple parasites and pathogens. Unfortunately, much of our understanding of human immune function and health has derived from populations living under evolutionarily novel conditions in which common parasites and pathogens are largely absent. Wider surveying of the patterns of helminth coinfection and comorbidity across a range of environments are needed to better understand the varying consequences of endemic helminth exposure—and lack thereof—in both nonindustrialized and industrialized populations.

In this section we present and review data on helminth coinfection and comorbidity patterns in an Amazonian, small-scale subsistence population, the Tsimane of lowland Bolivia. We focus specifically on the risk of coinfection involving helminths and *Giardia lamblia* (aka *Giardia intestinalis, Giardia duodenalis*), a common Tsimane intestinal protozoan. We then examine the links between helminth infection and other medical diagnoses. This research provides an example of the interactions that may occur when multiple species and conditions afflict a single population (Table 1).

Overview of the Tsimane

The Tsimane are a subsistence-level Amerindian population (pop. ~10,000) scattered across approximately 120 villages along the Maniquí River and surrounding forest areas. Most Tsimane have minimal access to medical care, market foods, or wage labor opportunities and subsist primarily on locally cultivated plantains, rice, manioc, and corn, hunted game, and wild fish (Gurven et al. 2007). Tsimane live in large family clusters in open-air huts with thatched-palm roofs. Few villages have wells or other clean water sources; water is generally obtained from nearby rivers and streams and rarely boiled. As of yet, no village has electricity or sewage. The Tsimane do not maintain outhouses but urinate and defecate privately in surrounding foliage. Domestic animals (dogs, cats, pigs, and chickens) are owned by individual families but are rarely penned and roam freely around villages and familial spaces. Despite economic impoverishment, the Tsimane are food secure. Nearly 70 % of the average adult diet is comprised of locally cultivated rice, plantain, manioc, and maize, with the remaining 30 % of the diet comprised of hunted game, river fish, and cultivated or foraged fruits and nuts. There is little wasting indicative of protein malnutrition in children (Foster et al. 2005), and the prevalence of underweight (body mass index < 18.5) among reproductive aged females is <2 %.

Table 1 Characteristics of common Tsimane intestinal parasites

Parasite/type	% infected Latin Am/Caribbean	Transmission route	Infection site	Immune response	Age peak	Associated morbidity
Hookworm (helminth)[a]	9 %	Skin penetration Oral	Small intestine	T_H2/T_{reg}	Adulthood	Intestinal blood loss; iron-deficiency anemia; protein malnutrition
Ascaris lumbricoides (helminth)	15 %	Fecal–oral	Small intestine	T_H2/T_{reg}	5–10 year	Lactose intolerance; vitamin A deficiency; intestinal obstruction; hepatopancreatic ascariasis
Trichuris trichiura (helminth)	18 %	Fecal–oral	Cecum Colon	T_H2/T_{reg}	5–10 year	Colitis; Trichuris dysentery syndrome; rectal prolapse; impaired nutrition
Giardia lamblia (protozoan)	Unknown	Contaminated water Fecal–oral	Small intestine	T_H1/T_H2	Weaning infants, children	Diarrhea, flatulence, vomiting, intestinal mucosal damage; fat, sugar, and vitamin malabsorption; lactose, vitamin A deficiency

References: Bethony et al. (2006), Brooker et al. (2004), Hotez et al. (2008), Ortega and Adam (1997), Wolfe (1992)
[a]Hookworm=*Ancylostoma duodenale/Necator americanus*

Methods of Tsimane Data Collection

Since 2002, the Tsimane have been participants in the ongoing Tsimane Health and Life History Project (THLHP). THLHP researchers have worked extensively in Tsimane villages, collecting demographic, anthropological, and biomedical data while also providing primary medical care. The data presented in this chapter was collected from participants seen by a mobile team of THLHP physicians, who traveled annually through Tsimane villages from 2007 to 2010.

Patients seen by THLHP physicians were given routine physical exams (patient history, symptom investigation, blood pressure and temperature, height and weight). Physicians administered vitamins, antibiotics, and antihelmintics as warranted, following on-site analysis of participant blood and fecal samples. Ethnographic and epidemiological information on the Tsimane, methods for age estimation, subject sampling, biomarker collection, and physician diagnostics have been described elsewhere (Gurven et al. 2007, 2008, 2009).

Results of parasitic infection presented and reviewed in this chapter were obtained through community sampling and patient diagnostics conducted from 2004 to 2010. Fecal samples collected by THLHP researchers were analyzed using two methods. From 2004 to 2008, fecal samples were analyzed for the presence of helminth eggs, larvae, and protozoa by direct identification on wet mounts. Beginning in 2007, fecal samples were also preserved in 10 % formalin solution following direct identification and later quantitatively analyzed using a modified Percoll (Amersham Pharmacia) technique (Eberl et al. 2002). Methods of fecal sample collection and parasite identification using both methods have been described in greater detail elsewhere (Vasunilashorn et al. 2010; Blackwell et al. 2011). Data presented here were aggregated from the two methods, with individuals coded as either infected or not infected if helminths were detected by either method ($n=3{,}628$).

Helminth and Protozoan Infections Among the Tsimane

Tsimane exposure to multiple gastrointestinal parasites and pathogens is endemic and lifelong. From fecal samples collected from 2004 to 2010, we estimate that 77 % of Tsimane are infected with at least one intestinal parasite (Table 2). The most common infections are hookworm, *G. lamblia*, and *A. lumbricoides*, infecting 51 %, 37 %, and 15 % of Tsimane, respectively. Females are more likely to be infected with *A. lumbricoides* than males (OR = 1.33, $\chi^2=9.30$, $p=0.002$) and less likely to be infected with *S. stercoralis* (OR = 0.70, $\chi^2=4.57$, $p=0.03$), while other infections do not vary significantly by sex. *T. trichiura* and *S. stercoralis* are relatively uncommon, infecting only 3.6 % and 3.7 % of subjects. Hookworm infections are less common in children 10 and younger than in Tsimane over age 10 (37 % vs. 55 %) and show a steady increase with age (Fig. 2). Nearly 1/3 of the Tsimane

Table 2 Prevalence of Tsimane single- and multiple-species infections

	All ages (*n*=3,628)		≤10 (*n*=893)		≥10 (*n*=2,735)	
	N	%	*N*	%	*N*	%
Infection						
Hookworm	1,842	50.8	328	36.7	1,514	55.4
G. lamblia	1,336	36.8	308	34.5	1,028	37.6
A. lumbricoides	544	15.0	132	14.8	412	15.1
T. trichiura	129	3.6	16	1.8	113	4.1
S. stercoralis	135	3.7	18	2.0	117	4.3
Any infection	**2,801**	**77.2**	**581**	**65.1**	**2,220**	**81.2**
Hookworm only	956	26.4	176	19.7	780	28.5
G. lamblia only	680	18.7	186	20.8	494	18.1
A. lumbricoides only	133	3.7	32	3.6	101	3.7
T. trichiura only	14	0.4	2	0.2	12	0.4
S. stercoralis only	12	0.3	2	0.2	10	0.4
Total single-species infections	**1,795**	**49.5**	**398**	**44.5**	**1,397**	**51.1**
Hookworm and *G. lamblia*	422	11.6	61	6.8	361	13.2
Hookworm and *A. lumbricoides*	194	5.3	47	5.3	147	5.4
Hookworm and *S. stercoralis*	67	1.8	5	0.6	62	2.3
A. lumbricoides and *G. lamblia*	65	1.8	17	1.9	48	1.8
Hookworm and *T. trichiura*	29	0.8	3	0.3	26	1.0
T. trichiura and *G. lamblia*	14	0.4	2	0.2	12	0.4
S. stercoralis and *G. lamblia*	12	0.3	7	0.8	5	0.2
A. lumbricoides and *T. trichiura*	12	0.3	1	0.1	11	0.4
A. lumbricoides and *S. stercoralis*	6	0.2	1	0.1	5	0.2
Total 2-species infections	**821**	**22.5**	**144**	**16.1**	**677**	**24.8**
Hookworm, *A. lumbricoides, G. lamblia*	79	2.2	26	2.9	53	1.9
Hookworm, *S. stercoralis, G. lamblia*	25	0.7	2	0.2	23	0.8
Hookworm, *A. lumbricoides, T. trichiura*	21	0.6	2	0.2	19	0.7
Hookworm, *T. trichiura, G. lamblia,*	20	0.6	3	0.3	17	0.6
Hookworm, *A. lumbricoides, S. stercoralis*	15	0.4	2	0.2	13	0.5
Hookworm, *S. stercoralis, T. trichiura*	5	0.1	0	0.0	5	0.2
A. lumbricoides, T. trichiura, G. lamblia	8	0.2	2	0.2	6	0.2
A. lumbricoides, S. stercoralis, G. lamblia	3	0.1	1	0.1	2	0.1
Total 3-species infections	**176**	**4.9**	**38**	**4.1**	**138**	**5.0**
Hookworm, *A lumbricoides, T. trichiura, G. lamblia*	4	0.1	1	0.1	3	0.1
Hookworm, *A. lumbricoides, S. stercoralis, G. lamblia*	3	0.1	0	0.0	3	0.1
Hookworm, *S. stercoralis, T. trichiura, G. lamblia*	1	0.0	0	0.0	1	0.0
Hookworm, *A. lumbricoides, S. stercoralis, T. trichiura*	1	0.0	0	0.0	1	0.0
Total 4-species infections	**9**	**0.2**	**1**	**0.1**	**8**	**0.3**
Total 2+ species infections	**1,006**	**27.6**	**183**	**20.2**	**823**	**30.1**

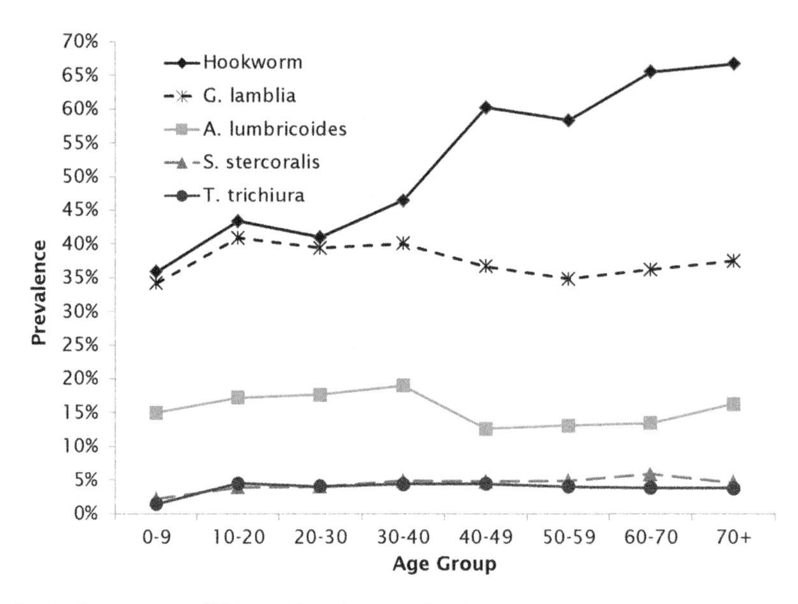

Fig. 2 Total prevalence of Tsimane intestinal parasites by age

population harbors a multiple-species infection, though prevalence rates of different coinfectious combinations are varied (Table 2). Hookworm is present in 85 % and *G. lamblia* present in 65 % of all coinfections, while 55 % of coinfections involve both hookworm and *G. lamblia.*

Helminth Coinfection and Infection Intensity Risk Among the Tsimane

Half of all multiple-species infections observed involved infection with at least two helminth species. As has been shown elsewhere (e.g., Howard et al. 2001), we found that individual helminth infection increases the risk of coinfection with other helminths. The strongest positive association we observed was between *A. lumbricoides* and *T. trichiura,* with *T. trichiura*-infected subjects nearly six times as likely to be coinfected with *A. lumbricoides* (Table 3). Each helminth infection was associated with higher odds of infection with another helminth. *A. lumbricoides* (OR = 1.40, $p<0.001$), *T. trichiura* (OR = 1.47, $p=0.05$), and *S. stercoralis* (OR = 3.28, $p<0.001$) were all predictive of hookworm infection, while hookworm was associated with *A. lumbricoides* (OR = 2.32, $p<0.001$), and *A. lumbricoides* was associated with *T. trichiura* (OR = 2.54, $p<0.001$). Howard et al. (2001), in a survey of 60 international studies of helminth coinfection, found that in ~70 % of cases, the risk of coinfection with *A. lumbricoides* and *T. trichiura* was significantly higher than would be expected by independent transmission. In the same study, increased risks

Table 3 Odds ratios for infection with one parasite given infection with another

	Dependent				
Independent	*G. lamblia*	Hookworm	*A. lumbricoides*	*T. trichiura*	*S. stercoralis*
G. lamblia		0.54***	0.64*	0.89^{ns}	0.51^{t}
Hookworm	0.54***		2.32***	1.14^{ns}	1.77^{ns}
A. lumbricoides	0.72**	1.40***		2.54**	0.97^{ns}
T. trichiura	1.13^{ns}	1.47*	2.03^{t}		1.11^{ns}
S. stercoralis	0.83^{ns}	3.28***	1.08^{ns}	1.10^{ns}	
Age (decades)	1.04*	2.22***	0.87*	1.05^{ns}	1.06^{ns}
Sex (male)	1.03^{ns}	1.10^{ns}	0.44**	0.77^{ns}	0.91^{ns}

Odds ratios were calculated in binomial logistic mixed models with all independent variables in one model, controlling for age, sex, and repeat observations

Parameter significance: $^{t}p \leq 0.10$; $^{*}p \leq 0.05$; $^{**}p \leq 0.01$; $^{***}p \leq 0.001$; ^{ns}nonsignificant

for hookworm-A. *lumbricoides* and hookworm-*T. trichiura* coinfections were also widely documented (Howard et al. 2001).

Significantly higher infection intensity associated with helminth coinfection has been widely reported (Brooker et al. 2000; Howard et al. 2002); we observed a similar relationship in the Tsimane. Among infected Tsimane patients, hookworm egg count was significantly higher in the presence of *S. stercoralis* ($\beta = 170.7$ eggs/g, $p < 0.001$). *A. lumbricoides* intensity was significantly increased by hookworm coinfection ($\beta = 416.2$ eggs/g, $p = 0.004$). *S. stercoralis* intensity was higher with *A. lumbricoides* ($\beta = 397.4$ eggs/g, $p < 0.001$), but not if an individual was infected by both *A. lumbricoides* and hookworm ($\beta = -417.4$, $p < 0.001$). There were no significant effects of coinfection on *T. trichiura* intensity. As a caution, the relationship between coinfection and infection intensity is somewhat difficult to parse. In many studies, high-intensity infection with one species is associated with increased coinfection risk, but it is unclear if high-intensity infections predispose hosts to coinfection or vice versa (Raso et al. 2004; Fleming et al. 2006). Host factors such as age or nutritional status may also predispose hosts to coinfection and higher infection intensity (Pullan and Brooker 2008).

Given the high prevalence of helminth infection among the Tsimane (77 %), the prevalence of helminth coinfection (14 %) in the Tsimane is much lower than rates reported for many other tropical, underdeveloped populations (e.g., 58 % Tanzania, Booth et al. 1998; 49 % Kenya, Brooker et al. 2000; 68 % Rwanda, Mupfasoni et al. 2009). The lower helminth coinfection rates in the Tsimane may be due to the relatively lower prevalence of helminths other than hookworm. As compared to the referred African populations, the Tsimane may also be less negatively impacted by recent environmental and socioeconomic changes (e.g., food insecurity, malnutrition, increased population density, environmental degradation) that may increase host susceptibility to coinfection and/or exposure to multiple parasites—including more virulent parasites and pathogens such as *Schistosoma* and *Plasmodium* spp. It is also worth nothing that even in the absence of malnutrition and more virulent coinfecting species, coinfection in the Tsimane was associated with increased infection intensity.

Table 4 Odds of infection based on receipt of antihelmintic or antiprotozoal drugs at the previous medical visit

	Any helminth	Hookworm	*A. lumbricoides*	*G. lamblia*
(Intercept)	0.41***	0.33***	0.05***	0.77**
Received antihelmintic	1.15	1.12	0.41***	1.29*
Received antiprotozoal	1.46**	1.61***	0.83	0.93
Age (years)	1.02***	1.02***	1.00	1.00
Sex (male)	1.06	1.13ˈ	0.72*	1.05
Dist. to San Borja (10 km)	1.14***	1.09***	1.19***	0.92***

Parameter values are odds ratios estimated in separate generalized logistic mixed model for each parasite or pathogen.
Parameter significance: ˈ$p \leq 0.10$; *$p \leq 0.05$; **$p \leq 0.01$; ***$p \leq 0.001$

Previously, we have also shown that early and chronic elevated IgE—characteristic of endemic helminth exposure—is also associated with growth deficits (Blackwell et al. 2010). Thus, endemic exposure to multiple helminths may be a significant cause of morbidity in the Tsimane. Future research with the Tsimane will investigate host factors that increase susceptibility to helminth coinfection and infection intensity and will evaluate if certain helminth coinfections and higher infection intensity are associated with increased morbidity.

Evidence of Hookworm and Giardia lamblia *Antagonism Among the Tsimane*

In contrast to the higher odds of helminth coinfection, we have found that the risk of helminth-giardia coinfection among the Tsimane is significantly less common than would be predicted by independent transmission. *G. lamblia* infection was associated with significantly lower odds of infection with both hookworm (OR = 0.54, $p < 0.001$) and *A. lumbricoides* (OR = 0.64, $p < 0.001$), while hookworm and *A. lumbricoides* were conversely associated with lower odds of *G. lamblia* infection (OR = 0.54, $p < 0.001$; OR = 0.72, $p < 0.001$). To put the size of this effect into perspective, 31 % of those infected with any helminth were also infected with *G. lamblia*, compared to 45 % of those without helminth infection. Of those with *G. lamblia* infection, 49 % were infected with at least one helminth, compared to 64 % of those without *G. lamblia*.

Given the apparent antagonism between helminth and *G. lamblia* infection, we examined how treatment for helminths affected later risk of infection with *G. lamblia*, and vice versa (Table 4). Receiving an antihelmintic had no effect on the odds of being infected with any helminth (OR = 1.15, $p = 0.28$) or with hookworm (OR = 1.12, $p = 0.34$) but did reduce the odds of *A. lumbricoides* infection (OR = 0.41, $p < 0.001$). However, receipt of antihelmintics was also associated with increased odds for *G. lamblia* infection (OR = 1.29, $p = 0.03$). Antiprotozoal agents had no effect on the odds of *G. lamblia* infection 1 year later (OR = 0.93, $p = 0.55$) but were associated with increased odds of hookworm infection at the subsequent visit ($OR=1.60$, $p<0.001$).

Our results are consistent with those reported by Rousham (1994), who observed an increase in *G. lamblia* prevalence following mebendazole treatment for *A. lumbricoides* and *T. trichiura* infection. The apparent antagonism may reflect competitive inhibition or cross-immunity. In a murine model, *G. lamblia*, which reside on microvilli in the small intestine, were inhibited by *Trichinella spiralis* when these helminths inhabited the small intestine but not at later stages when they moved to muscular tissue, suggesting a physical rather than immune interaction between the two species (Chunge et al. 1992). In our study, only *A. lumbricoides* and hookworm, both of which inhabit the small intestine, were negatively associated with *G. lamblia*, whereas *T. trichiura* (located further down in the large intestine) was not, supporting these earlier observations.

Studies have also shown that *G. lamblia* clearance and protective immunity are mediated by mixed T_H1 and T_H2 cytokine production (characterized by both INF-γ and IL-4), as well as T_H2 antibody responses (IgA, IgG, IgE) (Abdul-Wahid and Faubert 2008; Jiménez et al. 2009; Matowicka-Karna et al. 2009). Therefore, it is possible that helminth-induced T_H2 activity may provide some cross-immunity against *G. lamblia*. However, Hagel et al. (2011) found a higher prevalence of *G. lamblia* in association with *A. lumbricoides* infection—but only at moderate intensity and in conjunction with increased T_{reg} cytokine activity—suggesting helminth-giardia coinfection risk may only be increased with helminth-induced T_{reg} activity. Future studies with the Tsimane will examine the range of immune parameters alternately associated with helminth and protozoan infection and may shed further light on the role of immune responses in helminth-giardia antagonism.

Helminth Infection Is Associated with Altered Odds for Respiratory and Inflammatory Diagnoses

It has been suggested that the immunomodulatory properties of helminths may protect against allergies, autoimmune, and inflammatory disorders. However, helminth-induced immune biasing may increase susceptibility to other infectious diseases. To examine potential interactions between helminths, *G. lamblia*, and other medical conditions, we grouped THLHP patient disease diagnoses into broad categories representing the most common types of complaint. Excluding diagnoses of helminthiasis and giardiasis, these included gastrointestinal problems (43 % of 3,391 patient examinations), muscle or back pain (34 %), upper respiratory illnesses (28 %), urinary tract infections (13 % cases), fungal infections (8 %), arthritis (6 %), skin infections (3 % cases), and traumatic burns or injuries (2 %). For analysis, we divided the sample by age into children ≤ 16 years of age and adults over age 16 since many diagnoses were not equally prevalent in children and adults (Table 5). Controlling for age, sex, and village location, helminth infections were associated with greater odds of upper respiratory infection in children ($OR=1.33$, $p=0.04$).

Table 5 Association between current helminth and giardia infection and likelihood of medical diagnosis during medical visit

			Odds ratios				
Sample	Medical diagnosis (cases)	Cases (%)	Helminth infected	Giardia infected	Age (years)	Sex (male)	Dist. to town (per 10 km)
Children ≤ 16 years	Gastrointestinal problems	454 (42 %)	1.09	0.88	0.97[t]	1.01	0.91**
n = 894	Fungal infections	70 (6 %)	1.17	0.88	1.00	1.23	1.06
obs = 1,086	Upper respiratory infections	474 (44 %)	1.33*	0.99	0.93***	0.73*	1.11***
	Urinary tract infections	19 (2 %)	0.58	0.75	1.24	0.46	0.62
	Skin infections	57 (5 %)	0.84	0.37	0.79[t]	0.46	1.13
	Trauma	17 (2 %)	0.54	0.68	1.13	0.51	0.65
	Muscle or back pain	10 (1 %)	0.50	0.28	1.54	1.45	1.33
Adults > 16	Gastrointestinal problems	989 (43 %)	1.31**	1.01	1.00	1.02	0.97
n = 1,439	Fungal infections	186 (8 %)	1.00	0.49*	0.99	1.09	1.02
obs = 2,305	Upper respiratory infections	474 (21 %)	1.09	0.84	0.99***	0.97	0.99
	Urinary tract infections	423 (18 %)	1.03	0.96	1.00	0.35***	0.88***
	Skin infections	35 (2 %)	0.20	0.53	0.99	1.01	1.07
	Trauma	55 (2 %)	1.75	1.07	0.95	1.35	0.81
	Arthritis[a]	198 (9 %)	0.68*	1.13	1.06***	0.48***	1.13**
	Muscle or back pain	1,143 (50 %)	0.72***	1.32**	1.00	2.01***	1.07***

Parameter values are odds ratios estimated in separate generalized logistic mixed model for each medical diagnosis

Parameter significance: [t]$p \leq 0.10$; *$p \leq 0.05$; **$p \leq 0.01$; ***$p \leq 0.001$

[a]There were no children with arthritis

In adults, helminths were associated with greater odds of non-giardia gastrointestinal problems (OR = 1.31, $p < 0.01$) and with reduced odds of both arthritis (OR = 0.68, $p = 0.04$) and muscle or back pain (OR = 0.72, $p < 0.001$). *G. lamblia* was associated with reduced odds of adult fungal infection (OR = 0.49, $p = 0.02$) and greater odds of adult muscle or back pain (OR = 1.32, $p < 0.01$). Neither was significantly associated with trauma, skin infection, or urinary tract infection.

The observation of increased odds of respiratory infection in children but reduced odds of arthritis and muscle or back pain in older adults is consistent with predictions based on the immunomodulatory effects of helminths and elaborated on in the hygiene hypothesis. In the absence of helminth and commensal-induced immunoregulation, inflammation may be excessive, leading to autoimmune disorders such as arthritis. Previously we have shown that IgE, a marker of past and current helminth infection, is associated with lower inflammatory markers, such as CRP (Blackwell et al. 2010), and lower total cholesterol (Vasunilashorn et al. 2010). However, as observed in the Tsimane, helminth infection may have varying effects on comorbidity during different life stages, with different clinical implications. Helminths may be protective in preventing inflammatory disorders during adulthood, but their immunoregulatory effects may increase susceptibility to infectious disease at younger ages.

Limitations and Future Directions

There are several limitations of our analysis that limit wider extrapolation of our results and interpretations. First, infectious status among the Tsimane is diagnosed on the basis of a single fecal sample only, which may underestimate the true prevalence of both single- and multiple-species infections in this population. Second, THLHP community sampling does not permit us to conclusively discriminate between acute, chronic, or resolving infections, which may influence our interpretations of parasite associations in hosts. We also as yet do not have data on the possible risk of coinfection involving helminths and more virulent diseases known to afflict the Tsimane, including leishmaniasis, dengue fever, tuberculosis, leptospirosis, and other common viral and bacterial infections. While we have shown that endemic helminth infection in the Tsimane may be protective against a common intestinal protozoan, helminth-induced T_H2/T_{reg} biasing may increase susceptibility to coinfection with these more virulent pathogens.

Finally, although we have shown that the prevalence of helminth coinfection among the Tsimane is lower than that of several African populations, Tsimane coinfection rates may have increased in recent decades and may continue to increase in coming years. The Tsimane have only been permanently settled in their current territory since the mid-twentieth century (Gurven et al. 2007; Huanca 2006). Increased sedentism, population growth and density, interactions with domesticates, fecal contamination of local water sources, and emerging social threats (e.g., prostitution and alcohol abuse) have likely increased the rate of exposure to existing and novel parasites and other infectious diseases. Meanwhile, hygienic conditions, access to medical care, vaccine coverage, and antibiotic usage—though improving—remain poor. Novel patterns of coinfection, and their effects on Tsimane health and immune function, may yet be emerging. Moreover, while the Tsimane environment may be more similar to a recent ancestral environment than that of a contemporary industrialized population, it is by no means identical to the disease ecology of our hominin ancestors.

Future research with the Tsimane will examine the role of additional host and environmental factors on coinfection susceptibility, such as regional and seasonal risks, and additional variation in immune and morbidity markers associated with different coinfectious combinations and infection intensities. In particular, more sensitive surveys of pathogen prevalence and helminth-pathogen risk are needed in this population. Such research may help to identify current risk factors associated with a wider range of multiple-species infections and may help predict how emerging infectious threats are likely to interact with current endemic intestinal parasites and associated immune responses.

Conclusions

In humans, the prevalence and distribution of a range of multiple-species infections has been increasingly well documented in epidemiological research but remains understudied from an evolutionary or ecological perspective. Researchers interested in human-pathogen coevolution should consider the additional challenges posed by parasitic coinfection. There are many factors that influence parasite coinfection risk and coinfectious outcomes in hosts. In this chapter, we focused on coinfections involving STH, which are the most widespread human intestinal parasite and induce characteristic immunoregulatory responses suspected to influence a range of coinfectious outcomes.

Chronic helminth infection, which may have been the norm during human ancestry, is associated with varying risks and benefits in modern human populations. While the prevalence of multiple helminth infection among the Tsimane is lower than that reported for several African populations, infection with a single helminth species in the Tsimane was associated with helminth coinfection risk, and multiple helminth infections were associated with increased infection intensity.

Our results suggest that helminth infection during childhood increases risk of respiratory illness, which may indicate increased susceptibility to bacterial and viral infection due to helminth-induced immune biasing. These results have implications for vaccination programs and the spread of epidemic diseases among the Tsimane and require further investigation. Other work has suggested that early and chronic elevated IgE, characteristic of endemic helminth exposure, is also associated with growth deficits (Blackwell et al. 2010). Therefore, helminth infection may still pose a substantial threat to health and well-being in the Tsimane and other nonindustrialized populations.

At the same time, helminth-giardia coinfection in the Tsimane occurs less frequently than would be predicted by independent transmission. The lower risk of helminth-giardia coinfection may be mediated by direct competition or controlled by strong T_H2 mechanisms invoked by chronic helminth challenge. Future research is needed to elucidate the mechanisms of helminth-giardia antagonism, particularly given the clinical implications of increased infection risk following antihelmintic or antiprotozoal administration. The results reviewed here also suggest that helminth infection and elevated IgE in adulthood is associated with lower incidence of inflammatory-associated morbidity (Blackwell et al. 2010; Vasunilashorn et al. 2010). These findings are all consistent with the proposal that some inflammation-linked "diseases of modernity," such as obesity and heart disease, may be due to the absence of "old friends" with which our immune systems coevolved.

We intend this discussion of the costs, benefits, and altered risks of coinfection associated with helminths to illustrate the importance of considering multiple-species infections and the role of infectious communities in affecting the evolution of immune responses in hosts, the virulence of pathogens, and implications for health and treatment. In sum, we have presented evidence of a complex trade-off in the risks and benefits of helminth infection in humans. These trade-offs are likely to vary by life stage, environment, and host factors influencing coinfection and morbidity risk, which are constantly changing across the human landscape. These factors may influence differences in patterns of coinfection prevalence and associated morbidity observed today and must be considered in models of ancestral parasite and pathogen coinfection risk. Identifying and understanding both the ancestral and emerging risks of coinfection pose an important challenge for researchers working across varied human populations, which will be best met by an integrated cultural, epidemiological, ecological, and evolutionary approach.

References

Abdul-Wahid A, Faubert G (2008) Characterization of the local immune response to cyst antigens during the acute and elimination phases of primary murine giardiasis. Int J Parasitol 38:691–703

Adamo S (2001) Changes in lifetime immunocompetence in male and female Gryllus texensis (formerly G. integer): trade-offs between immunity and reproduction. Anim Behav 62:417–425

Alizon S (2008) Decreased overall virulence in coinfected hosts leads to the persistence of virulent parasites. Am Nat 172:E67–E79

Allen JE, Maizels RM (2011) Diversity and dialogue in immunity to helminths. Nat Rev Immunol 11:375–388

Anderson RM (1991) Populations and infectious diseases: ecology or epidemiology? J Anim Ecol 60:1–50

Anderson R, May R (1979) Population biology of infectious diseases: Part I. Nature 280: 361–367

Anderson RM, May RM (1981) The population dynamics of micro-parasites and their invertebrate hosts. Phil Trans R Soc B 291:451–524

Anthony RM, Rutitzky LI, Urban JFU Jr, Stadecker MJ, Gause WC (2008) Protective immune mechanisms in helminth infection. Nat Rev Immunol 7:975–987

Armelagos GJ, Harper KN (2005) Genomics at the origins of agriculture: Part II. Evol Anthropol 14:109–121

Bentwich Z, Kalinkovich A, Weisman Z (1995) Immune activation is a dominant factor in the pathogenesis of African AIDS. Immunol Today 16:187–191

Bergmann RL, Schulz J, Gunther S, Dudenhausen JW, Bergmann KE, Bauer CP, Dorsch W, Schmidt E, Luck W, Lau S (1995) Determinants of cord-blood IgE concentrations in 6401 German neonates. Allergy 50:65–71

Bethony J, Brooker S, Albonico M, Geiger SM, Loukas A, Diemert D, Hotez PJ (2006) Soil-transmitted helminth infections: ascariasis, trichuriasis, and hookworm. Lancet 367:1521–1532

Blackwell AD, Snodgrass JJ, Madimenos FC, Sugiyama LS (2010) Life history, immune function, and intestinal helminths: trade-offs among immunoglobulin E, C-reactive protein, and growth in an Amazonian population. Am J Hum Biol 22:836–848

Blackwell AD, Gurven MD, Sugiyama LS, Madimenos FC, Liebert MA, Martin MA, Kaplan HS, Snodgrass JJ (2011) Evidence for a peak shift in a humoral response to helminths: age profiles of IgE in the Shuar of Ecuador, the Tsimane of Bolivia, and the U.S. NHANES. PLoS Negl Trop Dis 5:12

Blount D, Hooi D, Feary J, Venn A, Telford G, Brown A, Britton J, Pritchard D (2009) Immunologic profiles of persons recruited for a randomized, placebo-controlled clinical trial of hookworm infection. Am J Trop Med Hyg 81:911–916

Booth M, Bundy DA, Albonico M, Chwaya HM, Alawi KS, Savioli L (1998) Associations among multiple geohelminth species infections in schoolchildren from Pemba Island. Parasitology 116:85–93

Bordes F, Morand S (2009) Coevolution between multiple helminth infestations and basal immune investment in mammals: cumulative effects of polyparasitism? Parasitol Res 106:33–37

Bradley JE, Jackson JA (2008) Measuring immune system variation to help understand host-pathogen community dynamics. Parasitology 135:807–823

Brooker S, Miguel EA, Moulin S, Luoba AI, Bundy DA, Kremer M (2000) Epidemiology of single and multiple species of helminth infections among school children in Busia District, Kenya. East Afr Med J 77:157–161

Brooker S, Bethony J, Hotez P (2004) Human hookworm infection in the 21st century. Adv Parasitol 58:197–288

Buckley CE, Larrick JW, Kaplan JE (1985) Population differences in cutaneous methacholine reactivity and circulating IgE concentrations. J Allergy Clin Immunol 76:847

Chan MS, Medley GF, Jamison D, Bundy DA (1994) The evaluation of potential global morbidity attributable to intestinal nematode infections. Parasitology 109:373–387

Christensen NO, Nansen P, Fagbemi BO, Monrad J (1987) Heterologous antagonistic and synergistic interactions between helminths and between helminths and protozoans in concurrent experimental infection of mammalian hosts. Parasitol Res 73:387–410

Chunge RN, Nagelkerke N, Karumba PN, Kaleli N, Wamwea M, Mutiso N, Andala EO, Gachoya J, Kiarie R, Kinoti SN (1992) Longitudinal study of young children in Kenya: intestinal parasitic infection with special reference to *Giardia lamblia,* its prevalence, incidence and duration, and its association with diarrhoea and with other parasites. Acta Trop 50:39–49

Cooper PJ, Chico ME, Sandoval C, Espinel I, Guevara A, Kennedy MW, Urban JF Jr, Griffin GE, Nutman TB (2000) Human infection with *Ascaris lumbricoides* is associated with a polarized cytokine response. J Infect Dis 182:1207–1213

Cooper PJ, Alexander N, Moncayo A-L, Benitez SM, Chico ME, Vaca MG, Griffin GE (2008) Environmental determinants of total IgE among school children living in the rural tropics: importance of geohelminth infections and effect of anthelmintic treatment. BMC Immunol 9:33

Cox FE (2001) Concomitant infections, parasites and immune responses. Parasitology 122:S23–S38

Cox FEG (2002) History of human parasitology. Clin Microbiol Rev 15:595–612

de Roode JC, Helinski MEH, Anwar MA, Read AF (2005) Dynamics of multiple infection and within-host competition in genetically diverse malaria infections. Am Nat 166:531–542
Eberl M, Hagan P, Ljubojevic S, Thomas AW, Wilson RA (2002) A novel and sensitive method to monitor helminth infections by faecal sampling. Acta Trop 83:183–187
Ellis MK, Raso G, Li Y-S, Rong Z, Chen H-G, McManus DP (2007) Familial aggregation of human susceptibility to co- and multiple helminth infections in a population from the Poyang Lake region, China. Int J Parasitol 37:1153–1161
Esrey SA, Potash JB, Roberts L, Shiff C (1991) Effects of improved water supply and sanitation on ascariasis, diarrhoea, dracunculiasis, hookworm infection, schistosomiasis, and trachoma. Bull World Health Organ 69:609–621
Fallon PG, Mangan NE (2007) Suppression of TH2-type allergic reactions by helminth infection. Nat Rev Immunol 7:220–230
Feary JR, Venn AJ, Mortimer K, Brown AP, Hooi D, Falcone FH, Pritchard DI, Britton JR (2010) Experimental hookworm infection: a randomized placebo-controlled trial in asthma. Clin Exp Allergy 40:299–306
Fellous S, Koella JC (2009) Infectious dose affects the outcome of the within-host competition between parasites. Am Nat 173:E177–E184
Figueiredo CA, Barreto ML, Rodrigues LC, Cooper PJ, Silva NB, Amorim LD, Alcantara-Neves NM (2010) Chronic intestinal helminth infections are associated with immune hyporesponsiveness and induction of a regulatory network. Infect Immun 78:3160–3167
Fleming FM, Brooker S, Geiger SM, Caldas IR, Correa-oliveira R, Hotez PJ, Bethony JM (2006) Synergistic associations between hookworm and other helminth species in a rural community in Brazil. Trop Med Int Health 11:56–64
Foster Z, Byron E, Reyes-García V, Huanca T, Vadez V, Apaza L, Pérez E, Tanner S, Gutierrez Y, Sandstrom B, Yakhedts A, Osborn C, Gody RA, Leonard WR (2005) Physical growth and nutritional status of Tsimane' Amerindian children of lowland Bolivia. Am J Phys Anthropol 126:343–351
Fox JG, Beck P, Dangler CA, Whary MT, Wang TC, Shi HN, Nagler-Anderson C (2000) Concurrent enteric helminth infection modulates inflammation and gastric immune responses and reduces helicobacter-induced gastric atrophy. Nat Med 6:536–542
Fredensborg BL, Poulin R (2005) Larval helminths in intermediate hosts: does competition early in life determine the fitness of adult parasites? Int J Parasitol 35:1061–1070
Fumagalli M, Pozzoli U, Cagliani R, Comi GP, Bresolin N, Clerici M, Sironi M (2010) The landscape of human genes involved in the immune response to parasitic worms. BMC Evol Biol 10:264
Fumagalli M, Sironi M, Pozzoli U, Ferrer-Admettla A, Pattini L, Nielsen R (2011) Signatures of environmental genetic adaptation pinpoint pathogens as the main selective pressure through human evolution. PLoS Genet 7:e1002355
Gonçalves MLC, Araújo A, Ferreira LF (2003) Human intestinal parasites in the past: new findings and a review. Mem Inst Oswaldo Cruz 98(Suppl I):103–118
Graham AL, Lamb TJ, Read AF, Allen JE (2005) Malaria-filaria coinfection in mice makes malarial disease more severe unless filarial infection achieves patency. J Infect Dis 191:410–421
Grant AV, Araujo MI, Ponte EV, Oliveira RR, Cruz AA, Barnes KC, Beaty TH (2008) High heritability but uncertain mode of inheritance for total serum IgE level and Schistosoma mansoni infection intensity in a schistosomiasis-endemic Brazilian population. J Infect Dis 198:1227–1236
Gurven M, Kaplan H, Supa AZ (2007) Mortality experience of Tsimane Amerindians of Bolivia: regional variation and temporal trends. Am J Hum Biol 19:376–398
Gurven M, Kaplan H, Winking J, Finch C, Crimmins EM (2008) Aging and inflammation in two epidemiological worlds. J Gerontol A Biol Sci Med Sci 63:196–199
Gurven M, Kaplan H, Winking J, Rodriguez DE, Vasunilashorn S, Kim JK, Finch C, Crimmins E (2009) Inflammation and infection do not promote arterial aging and cardiovascular disease risk factors among lean horticulturalists. PLoS One 4:e6590
Guyatt H (2000) Do intestinal nematodes affect productivity in adulthood? Parasitol Today 16:153–158

Hagel I, Cabrera M, Sánchez P, Rodríguez P, Lattouf JJ (2006) Role of the low affinity IgE receptor (CD23) on the IgE response against *Ascaris lumbricoides* in Warao Amerindian children from Venezuela. Invest Clin 47:241–251

Hagel I, Cabrera M, Puccio F, Santaella C, Buvat E, Infante B, Zabala M, Cordero R, Di Prisco MC (2011) Co-infection with *Ascaris lumbricoides* modulates protective immune responses against *Giardia duodenalis* in school Venezuelan rural children. Acta Trop 117:189–195

Hall A, Holland C (2000) Geographical variation in *Ascaris lumbricoides* fecundity and its implications for helminth control. Parasitology 16:540–544

Hamilton WD, Zuk M (1982) Heritable true fitness and bright birds: a role for parasites ? Science 218:384–387

Hart BL (2009) Adaptations to parasites: an ethological approach. J Parasitol 78:256–265

Haswell-Elkins MR, Elkins DB, Anderson RM (1987) Evidence for predisposition in humans to infection with Ascaris, hookworm, Enterobius and Trichuris in a South Indian fishing community. Parasitology 95:323–337

Hewitson JP, Grainger JR, Maizels RM (2009) Helminth immunoregulation: the role of parasite secreted proteins in modulating host immunity. Mol Biochem Parasitol 167:1–11

Holford-Strevens V, Warren P, Wong C, Manfreda J (1984) Serum total immunoglobulin E levels in Canadian adults. J Allergy Clin Immunol 73:516–522

Holland CV (2009) Predisposition to ascariasis: patterns, mechanisms and implications. Parasitology 136:1537–1547

Hotez PJ, Brindley PJ, Bethony JM, King CH, Pearce EJ, Jacobson J (2008) Helminth infections: the great neglected tropical diseases. J Clin Invest 118:1311–1321

Howard SC, Donnelly CA, Chan MS (2001) Methods for estimation of associations between multiple species parasite infections. Parasitology 122:233–251

Howard SC, Donnelly CA, Kabatereine NB, Ratard RC, Brooker S (2002) Spatial and intensity-dependent variations in associations between multiple species helminth infections. Acta Trop 83:141–149

Huanca T (2006) Oral tradition, landscape, and identity in tropical forest. Wa-Gui, La Paz

Iancovici Kidon M, Stein M, Geller-Bernstein C, Weisman Z, Steinberg S, Greenberg Z, Handzel ZT, Bentwich Z (2005) Serum immunoglobulin E levels in Israeli-Ethiopian children: environment and genetics. Isr Med Assoc J 7:799–802

Jackson JA, Frieber IM, Little S, Bradley JE (2008) Review series on helminths, immune modulation and the hygiene hypothesis: immunity against helminths and immunological phenomena in modern human populations: coevolutionary legacies? Immunology 126:18–27

Jiménez JC, Pinon A, Dive D, Capron M, Dei-Cas E, Convit J (2009) Antibody response in children infected with *Giardia intestinalis* before and after treatment with Secnidazole. Am J Trop Med Hyg 80:11–15

Kalyoncu AF, Stålenheim G (1992) Serum IgE levels and allergic spectra in immigrants to Sweden. Allergy 47:277–280

Kaplan JE, Larrick JW, Yost JA (1980) Hyperimmunoglobulinemia E in the Waorani, an isolated Amerindian population. Am J Trop Med Hyg 29:1012–1017

Karvonen A, Seppälä O, Tellervo VE (2009) Host immunization shapes interspecific associations in trematode parasites. J Anim Ecol 78:945–952

Kron MA, Ammunariz M, Pandey J, Guzman JR (2000) Hyperimmunoglobulinemia E in the absence of atopy and filarial infection: the Huaorani of Ecuador. Allergy Asthma Proc 21:335–341

Labeaud AD, Malhotra I, King MJ, King CL, King CH (2009) Do antenatal parasite infections devalue childhood vaccination? PLoS Negl Trop Dis 3:1–6

Lafferty KD, Kuris AM (2002) Trophic strategies, animal diversity and body size. Trends Ecol Evol 17:507–513

Lafferty KD, Sammond DT, Kuris AM (1994) Analysis of larval trematode communities. Ecology 75:2275–2285

Lello J, Hussell T (2008) Functional group/guild modelling of inter-specific pathogen interactions: a potential tool for predicting the consequences of co-infection. Parasitology 135:825–839

Lienhardt C, Azzurri A, Amedei A, Fielding K, Sillah J, Sow OY, Bah B, Benagiano M, Diallo A, Manetti R, Maneh K, Gustafson P, Bennett S, D'Elios MM, McAdam K, Del Prete G (2002) Active tuberculosis in Africa is associated with reduced Th1 and increased Th2 activity in vivo. Eur J Immunol 32:1605–1613

Lim HK, Heyneman D (1972) Intramolluscan inter-trematode antagonism: a review of factors influencing the host-parasite system and its possible role in biological control. Adv Parasitol 10:191–268

Lindberg R, Arroyave C (1986) Levels of IgE in serum from normal children and allergic children as measured by an enzyme immunoassay. J Allergy Clin Immunol 78:614–618

Long KZ, Nanthakumar N (2004) Energetic and nutritional regulation of the adaptive immune response and trade-offs in ecological immunology. Am J Hum Biol 16:499–507

Lynch NR, Lopez R, Isturiz G, Tenias-Salazar E (1983) Allergic reactivity and helminthic infection in Amerindians of the Amazon basin. Int Arch Allergy Appl Immunol 72:369–372

Maizels RM (2005) Infections and allergy - helminths, hygiene and host immune regulation. Curr Opin Immunol 17:656–661

Maizels RM, Yazdanbakhsh M (2003) Immune regulation by helminth parasites: cellular and molecular mechanisms. Nat Rev Immunol 3:733–744

Maizels RM, Pearce EJ, Artis D, Yazdanbakhsh M, Wynn TA (2009) Regulation of pathogenesis and immunity in helminth infections. J Exp Med 206:2059–2066

Matowicka-Karna J, Dymicka-Piekarska V, Kemona H (2009) IFN-gamma, IL-5, IL-6 and IgE in patients infected with *Giardia intestinalis*. Folia Histochem Cytobiol 47:93–97

May RM, Anderson RM (1979) Population biology of infectious diseases: Part II. Nature 280:455–461

May RM, Nowak MA, Nowak A (2009) Coinfection and the evolution of parasite virulence. Proc Biol Sci 261:209–215

McDade TW (2005) Life history, maintenance, and the early origins of immune function. Am J Hum Biol 17:81–94

Minchella DJ (1985) Host life-history variation in response to parasitism. Parasitology 90:205–216

Mingomataj EC, Xhixha F, Gjata E (2006) Helminths can protect themselves against rejection inhibiting hostile respiratory allergy symptoms. Allergy 61:400–406

Moller AP (1990) Parasites and sexual selection: current status of the Hamilton and Zuk hypothesis. J Evol Biol 3:319–328

Muehlenbein MP, Hirschtick JL, Bonner JZ, Swartz AM (2010) Toward quantifying the usage costs of human immunity: altered metabolic rates and hormone levels during acute immune activation in men. Am J Hum Biol 22:546–556

Mupfasoni D, Karibushi B, Koukounari A, Ruberanziza E, Kaberkuka T, Kramer MH, Mukabayire O, Kabera M, Nizeyimana V, Deville MA, Ruxin J, Webster JP, Fenwick A (2009) Polyparasite helminth infections and their association to anemia and undernutrition in Northern Rwanda. PLoS Negl Trop Dis 3:e517

Ortega YR, Adam R (1997) Giardia: overview and update. Clin Infect Dis 25:545–549

Perlmann H, Helmby H, Hagstedt M, Carlson J, Larsson PH, Troye-Blomberg M, Perlmann P (1994) IgE elevation and IgE anti-malarial antibodies in *Plasmodium falciparum* malaria: association of high IgE levels with cerebral malaria. Clin Exp Immunol 97:284–292

Perlmann P, Perlmann H, ElGhazali G, Blomberg MT (1999) IgE and tumor necrosis factor in malaria infection. Immunol Lett 65:29–33

Pullan R, Brooker S (2008) The health impact of polyparasitism in humans: are we underestimating the burden of parasitic diseases? Parasitology 135:783–794

Quinnell RJ, Grafen A, Woolhouse MEJ (1995) Changes in parasite aggregation with age: a discrete infection model. Parasitology 111:635–644

Råberg L, Graham AL, Read AF (2009) Decomposing health: tolerance and resistance to parasites in animals. Philos Trans R Soc Lond B Biol Sci 364:37–49

Raso G, Luginbühl A, Adjoua CA, Tian-Bi NT, Silué KD, Matthys B, Vounatsou P, Wang Y, Dumas ME, Holmes E, Singer BH, Tanner M, N'Goran E, Utzinger J (2004) Multiple parasite infec-

tions and their relationship to self-reported morbidity in a community of rural Côte d'Ivoire. Int J Epidemiol 33:1092–1102
Rigaud T, Perrot-Minnot M-J, Brown MJF (2010) Parasite and host assemblages: embracing the reality will improve our knowledge of parasite transmission and virulence. Proc Biol Sci 277:3693–3702
Rook GAW (2008) Review series on helminths, immune modulation and the hygiene hypothesis: the broader implications of the hygiene hypothesis. Immunology 126:3–11
Round JL, Mazmanian SK (2009) The gut microbiota shapes intestinal immune responses during health and disease. Nat Rev Immunol 9:313–324
Rousham EK (1994) An increase in *Giardia duodenalis* infection among children receiving periodic antihelmintic treatment in Bangladesh. J Trop Pediatr 40:329–333
Sayasone S, Mak TK, Vanmany M, Rasphone O, Vounatsou P, Utzinger J, Akkhavong K, Odermatt P (2011) Helminth and intestinal protozoa infections, multiparasitism and risk factors in Champasack Province, Lao People's Democratic Republic. PLoS Negl Trop Dis 5:e1037
Schad GA, Anderson RM (1985) Predisposition to hookworm infection in humans. Science 228:1537–1540
Schmid-Hempel P (2003) Variation in immune defence as a question of evolutionary ecology. Proc Biol Sci 270:357–366
Sheldon BC, Verhulst S (1996) Ecological immunology: costly parasite defenses and trade-offs in evolutionary ecology. Trends Ecol Evol 11:317–321
Silva NRD, Brooker S, Hotez PJ, Montresor A, Engels D, Savioli L (2003) Soil-transmitted helminth infections: updating the global picture. Trends Parasitol 19:547–551
Supali T, Verweij JJ, Wiria AE, Djuardi Y, Hamid F, Kaisar MMM, Wammes LJ, van Lieshout L, Luty AJF, Sartono E, Yazdanbakhsh M (2010) Polyparasitism and its impact on the immune system. Int J Parasitol 40:1171–1176
Turner JD, Jackson JA, Faulkner H, Behnke J, Else KJ, Kamgno J, Boussinesq M, Bradley JE (2008) Intensity of intestinal infection with multiple worm species is related to regulatory cytokine output and immune hyporesponsiveness. J Infect Dis 197:1204–1212
Uller T, Isaksson C, Olsson M (2006) Immune challenge reduces reproductive output and growth in a lizard. Funct Ecol 20:873–879
Van Baalen M, Sabelis MW (1995) The dynamics of multiple infection and the evolution of virulence. Am Nat 146:881–910
Vasunilashorn S, Crimmins EM, Kim JK, Winking J, Gurven M, Kaplan H, Finch CE (2010) Blood lipids, infection, and inflammatory markers in the Tsimane of Bolivia. Am J Hum Biol 22:731–740
Walson JL, John-Stewart G (2007) Treatment of helminth co-infection in individuals with HIV-1: a systematic review of the literature. PLoS Negl Trop Dis 1:e102
Weidinger S, Gieger C, Rodriguez E, Baurecht H, Mempel M, Klopp N, Gohlke H, Wagenpfeil S, Ollert M, Ring J, Behrendt H, Heinrich J, Novak N, Bieber T, Kramer U, Berdel D, von Berg A, Bauer CP, Herbath O, Koletzko S, Prokisch H, Mehta D, Meitinger T, Depner M, von Mutuis E, Liang L, Moffatt M, Cookson W, Kabesch M, Wichmann HE, Illig T (2008) Genome-wide scan on total serum IgE levels identifies *FCER1A* as novel susceptibility locus. PLoS Genet 4:e1000166
Wilson MS, Maizels RM (2004) Regulation of allergy and autoimmunity in helminth infection. Clin Rev Allergy Immunol 26:35–50
Wolfe M (1992) Giardiasis. Clin Microbiol Rev 5:93–100
Woolhouse ME (1992) A theoretical framework for the immunoepidemiology of helminth infection. Parasite Immunol 14:563–578
Wu D, Molofsky AB, Liang H-E, Ricardo-Gonzalez RR, Jouihan HA, Bando JK, Chawla A, Locksley RM (2011) Eosinophils sustain adipose alternatively activated macrophages associated with glucose homeostasis. Science 332:243–247
Yazdanbakhsh M, Kremsner PG, van Ree R (2002) Allergy, parasites, and the hygiene hypothesis. Science 296:490–494

Primates, Pathogens, and Evolution: A Context for Understanding Emerging Disease

Kristin N. Harper, Molly K. Zuckerman, Bethany L. Turner, and George J. Armelagos

Introduction

Nonhuman primates (NHPs) are an important source of human infectious disease, likely due to our close phylogenetic relationship (Wolfe et al. 2007). Major human diseases that appear to have originated in NHPs include malaria (Liu et al. 2010), AIDS (Chen et al. 1997; Gao et al. 1999), and perhaps even hepatitis B infection (Chen et al. 1997). In addition, NHPs serve as a reservoir for infections such as yellow fever, monkeypox, and Ebola; indeed, it has been estimated that while primates constitute only 0.5 % of all vertebrate species, they have contributed approximately 20 % of major infectious diseases among humans (Wolfe et al. 2007). Conversely, human pathogens[1] can have devastating effects on NHPs, especially the great apes, all of which are listed as endangered species and some of which, like gorillas, are critically endangered. Infectious diseases have had substantial negative impacts on

[1]Here we use the term "pathogen" broadly, to include both microparasites (viruses, bacteria, and fungi) and macroparasites (such as worms).

K.N. Harper (✉)
Department of Environmental Health Sciences, Columbia University, Black Building, Rm 1618, 650 W 168th Street, New York, NY 10032, USA
e-mail: kh2383@columbia.edu

M.K. Zuckerman
Department of Anthropology and Middle Eastern Cultures, Mississippi State University, Mississippi State, MS 39762, USA
e-mail: mzuckerman@anthro.msstate.edu

B.L. Turner
Department of Anthropology, Georgia State University, Atlanta, GA 30303, USA
e-mail: antblt@langate.gsu.edu

G.J. Armelagos
Department of Anthropology, Emory University, Atlanta, GA 30322, USA
e-mail: antga@learnlink.emory.edu

J.F. Brinkworth and K. Pechenkina (eds.), *Primates, Pathogens, and Evolution*, Developments in Primatology: Progress and Prospects,
DOI 10.1007/978-1-4614-7181-3_13, © Springer Science+Business Media New York 2013

wild great ape populations, exacerbating existing threats posed by habitat loss, human encroachment, and hunting (Boesch and Boesch-Achermann 2000; Leendertz et al. 2004; Wolfe et al. 1998). Whether occurring as epidemics, small outbreaks, or single deaths, infectious diseases can negatively affect the viability of small or isolated NHP populations, due to the characteristically low reproductive rate of many NHP species, especially great apes (Boesch and Boesch-Achermann 2000; Ferber 2000; Goodall 1970, 1983; Nishida 1990; Wallis 2000; Wolfe et al. 1998).

Understanding how pathogens move between human and NHP populations and sometimes become established in a novel host requires knowledge of a pathogen's evolutionary history in its natural host, as well as the potential for transmission in both its natural and novel hosts. However, establishing these evolutionary and epidemiological relationships has been complicated by methodological missteps and conceptual pitfalls. Molecular methods allow us to construct a better picture of the past disease-scape of primates, and the past few years have witnessed a veritable explosion of information regarding major pathogen transmission events between humans and NHPs. These data have helped to elucidate pathogens' histories, though much remains to be done. In this chapter, we discuss some of the theoretical issues and methodological advances involved in reconstructing the evolutionary relationships between pathogenic organisms, humans, and NHPs. Using examples of major human diseases and their causative agents, specifically malaria (*Plasmodium* spp.) and HIV (human immunodeficiency virus), we discuss the implications for understanding emerging infectious diseases in both humans and our closest relatives. First, we explore the differentiation of the human disease-scape, attempting to reconstruct which pathogens diverged with their human hosts over evolutionary time. Next, we outline which human pathogens are believed to have resulted from cross-species transmission from NHPs. Finally, we discuss examples of cross-species parasite transmission from humans to NHPs and consider their ramifications upon conservation biology.

Differentiation of the Human Disease-Scape

We begin this chapter by exploring the differentiation of the human disease-scape from the pathogen profiles of our closest NHP relatives. We frame this discussion in terms of "heirloom" pathogens and "souvenir" species—the former representing those that we inherited from our most recent common ancestor with chimpanzees and the latter those that entered human populations via a host species switching event.

Theoretical Considerations

Studies of human disease ecology suggest that infectious diseases affecting human populations can generally be grouped into two broad categories: those with a long-standing, millennial relationship with humans and the latter those that have been more recently acquired. Sprent (1969a, b) was the first to make this distinction, identifying two distinct classes of microbes that afflicted hunter-gatherers in the Paleolithic: "heirloom species" and "souvenir species." Heirloom species are pathogens that originated in our anthropoid ancestors and continued to infect hominins and eventually modern humans. In contrast to heirloom species, souvenir species are newer evolutionary acquisitions that are typically "picked up" via exposure to zoonotic reservoirs or vectors (Kilks 1990). These diseases are generally zoonoses, i.e., infectious diseases that can be transmitted from animals to humans. Zoonoses can be contracted through a number of routes: from sympatric reservoir host species through cross transfer, through insect or animal bites, or through the preparation and consumption of contaminated animal flesh. It appears that many of these souvenir species are capable of temporary host switches without significant levels of adaptation, as long as the novel host is sufficiently similar to the natural host in terms of resources available to the pathogen (Kellog 1896, in Brooks and Ferrao 2005, p. 1292). Thus, these microbes can remain specialized in their natural hosts while infecting multiple hosts, including humans, across a wider ecological niche. In some cases, however, a souvenir species enters the human population and stays there permanently, adapting to its novel host and becoming established; indeed, as will be discussed, this category of souvenir species includes some of the major pathogens that affect human societies.

Given the complexity of pathogen host specificity and adaptation, it should come as no surprise that reconstructing the history of human pathogens is a difficult endeavor. Often, in studying the origins of human pathogens, it is necessary to gather samples from our close NHP relatives to assess the presence or absence of a pathogen in a given species as well as to determine its genetic relationship to human variants. As the examples ahead will show, this process frequently represents a limiting factor in our search for disease origins due to the difficulty inherent in collecting biological samples from wild NHPs, many of which live in remote areas of the world as well as being endangered and enjoying special protections. In order to illustrate the effect that host switching and limited sampling of NHPs can have on our understanding of a pathogen's history in humans, we present the cautionary tale of malignant malaria. By beginning our discussion of human pathogen origins with this example, we hope to demonstrate the careful attention to sampling necessary when considering the evolutionary history of even intensively studied pathogens such as *Plasmodium falciparum*, never mind the myriad less-studied ones discussed in this volume (e.g., *Treponema, Toxoplasma*, *Pediculus*, and *Pthirus* lice).

Malaria: From Heirloom to Established Souvenir Species

Malignant malaria, which is caused by *P. falciparum*, is the most dangerous form of the disease, with the highest rates of complications and mortality. This species of *Plasmodium* alone is responsible for an estimated 515 million episodes of illness (Snow et al. 2005) and nearly one million deaths annually (WHO 2010). The pathogen has exerted tremendous selective pressure upon the human genome in recent history (Hamblin et al. 2002; Sabeti et al. 2002; Tishkoff et al. 2001), but fundamental questions persist about the *Plasmodium*'s history in humans.

Early molecular studies indicated that the protozoan's closest relative was *P. reichenowi* (Escalante and Ayala 1994), a parasite species found in chimpanzees. This species was, for many years, represented by only a single strain isolated from a chimp captured in the Democratic Republic of Congo several decades ago (Collins et al. 1986). For this reason, many researchers believed that humans and chimpanzees both harbored their own distinct malaria strains; humans were infected by *P. falciparum*, while chimps carried *P. reichenowi*. This was consistent with a scenario in which each host species possessed its own heirloom *Plasmodium* species.

More intensive sampling, enabled by advances in noninvasive techniques, has recently generated a novel twist on the origins of this pathogen by suggesting that *P. falciparum* should instead be considered a souvenir species in humans. By examining blood as well as noninvasive samples collected from many wild chimpanzees, researchers identified many new *Plasmodium* isolates among chimpanzee populations (Prugnolle et al. 2010; Rich et al. 2009). Sequence analysis showed that the global genetic diversity of human *P. falciparum* was very low relative to the diversity of chimpanzee *P. reichenowi* strains, indicating that the human species had diverged more recently than the chimpanzee species. This finding was inconsistent with *P. falciparum* being an heirloom pathogen in humans. Moreover, a more comprehensive study of fecal samples from nearly 3,000 wild great apes, including chimpanzees, gorillas, and bonobos from throughout Central Africa, prompted consideration of an alternative scenario regarding *P. falciparum's* original host (Liu et al. 2010). This study demonstrated that the gorilla branch of the *Plasmodium* phylogeny included the strains most closely related genetically to human *falciparum* strains. Thus, it would appear that all circulating *P. falciparum* strains might be the result of a single, very successful cross-species transmission event from gorillas to humans.

Given the propensity of *Plasmodium* to switch primate hosts (Garamszegi 2009) and the numerous twists and turns in the history of malaria thus far, it has been noted that only further in-depth sampling of wild animals will confirm that no still-closer relative is going undetected and that the current story is the correct one (Prugnolle et al. 2011). Even with regard to a well-studied disease such as malaria, our understanding may well change with future developments, which hinge on sample availability. As Wolfe et al. (2007) note, while resolving the debate surrounding the origin of malaria will not necessarily assist with global eradication of the disease, it may contribute to our broader understanding of the dynamics of disease emergence. With the potential importance of such knowledge in mind, as well as the complexity involved in determining a pathogen's history, we now turn to a discussion of which human infections appear to be due to heirloom pathogens and which appear to be caused by souvenir species instead.

Heirloom Pathogens and Human Paleolithic Ecology

During the Paleolithic, small human population sizes and, to a lesser extent, low population density would have limited the diversity of possible heirloom species affecting human and hominin populations by preventing sustained transmission of many viruses and bacteria (Dunn et al. 2010, p. 2590). However, some species of parasites and bacteria appear to have thrived in this setting. Characteristically, heirloom pathogens are organisms that are able to persist in small, dispersed populations; generate incomplete or short-lived immunity; or have a chronic course, enabling prolonged transmission to new hosts. Several potential heirloom species have been identified, including macroparasites such as head and body lice (*Pediculus humanus*) (Reed et al. 2004) and pinworms (*Enterobius vermicularis*) (Hugot et al. 1999) as well as Staphylococci (Cockburn 1967, 1971; Sprent 1962, 1969a). Other possible heirloom pathogens include the causative agents of yaws (*Treponema pallidum* subsp. *pertenue*) (Harper et al. 2008) and typhoid (*Salmonella typhi*) (Roumagnac et al. 2006).

The identification of extant pathogen species that belong to heirloom lineages has been controversial, however, as the example of *P. falciparum*, above, illustrates. Novel genetic data have assisted in this process, helping to clarify which species most likely belong in the heirloom category by providing information on pathogen divergence times and host histories. In addition, studies of host specificity in pathogens that infect wild NHP populations suggest that some major classes of parasites, such as helminths, are more likely to be heirlooms due to their species specificity than are protozoa and viruses, which are typically able to infect a wider swathe of hosts (Pedersen et al. 2005).

Unique patterns of human behavior have no doubt guided the adaptation of heirloom pathogens. Weiss and Wrangham (1999), for example, note that while chimpanzees and humans share 95–99 % of their DNA (Olson and Varki 2003), we only share approximately 50 % of our pathogens. More specifically, pathogens and parasites affecting wild NHP species primarily include helminths, viruses, and protozoa, while pathogens affecting humans are dominated by fungi and bacteria (Pedersen et al. 2005) (Fig. 1). The reasons for these differences are poorly understood but likely reflect divergent characteristics of host ecology and behavior.

The evolution of human herpesviruses may offer one example of an heirloom pathogen molded over time by uniquely human behaviors. Phylogenetic studies indicate that all eight members of the human herpesvirus family (Herpesviridae) likely derive from an ancestral viral genome which infected the last common ancestor of hominins and great apes (Gentry et al. 1988). Herpes simplex virus (HSV) spreads from sites of initial infection in skin or mucosal surfaces to neuronal cell bodies in order to establish latent infection, forming a long-term relationship with its host. This long latency would have allowed HSV to persist in small, low-density Paleolithic populations. There are two types of herpes simplex: HSV-1 primarily produces oral herpes infection, while HSV-2 primarily produces genital infection. HSV-2 appears to be the only type of herpesvirus among primates that is

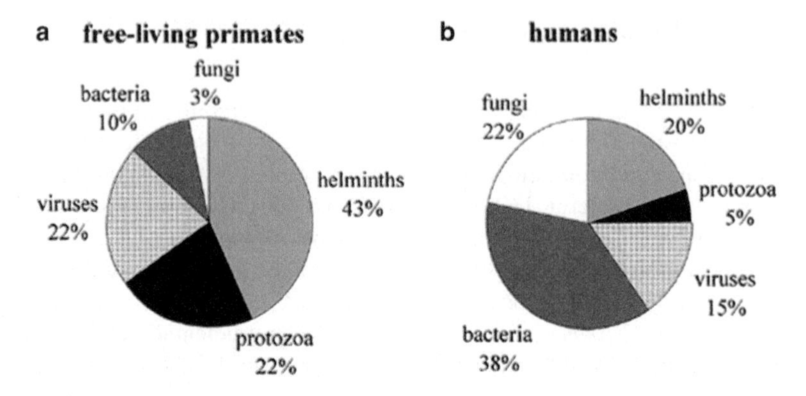

Fig. 1 Comparison of the taxonomic distribution of parasites in (**a**) free-living nonhuman primates and (**b**) humans. This data was based on a survey of 369 nonhuman primates and 1,415 humans (Figure reproduced from Pedersen et al. (2005))

transmitted primarily via sexual contact, and some have suggested that it may owe its existence to uniquely human sexual practices. The divergence of HSV-1 and HSV-2 dates back approximately 8–10 million years (McGeouch et al. 1995) and presumably reflects the development of species-specific tropisms for the epithelium of the oropharynx and the urogenital tract, respectively. For HSV-1 and HSV-2 to take on these distinct tropisms, oral and genital sites had to become microbiologically isolated from each other, while oral–oral and genital–genital contact between the hosts had to be maintained. McGeouch et al. (1995) have suggested that the evolution of continual sexual attractiveness of hominin females throughout the entire menstrual cycle, with an expected attendant increase in the frequency of sexual intercourse, and the adoption of close face-to-face mating among hominins, which may have facilitated the practice of kissing, provided the necessary conditions for the evolutionary divergence of HSV-1 and HSV-2. This hypothesis, of course, awaits rigorous testing, and more generally, the issue of how the behaviors of particular primate species provide niches conducive to sexual transmission remains a subject for further study.

Even absent the unique behavioral practices that characterize humans, however, millions of years of evolution in different hosts ensures divergence among the microbes that inhabit the bodies of humans and NHPs. For instance, the gut microbial communities of the great apes, including humans, have evolved independently with their hosts, diverging over the years in a manner consistent with host speciation patterns (Ochman et al. 2010). Common shared ancestors are hypothesized to exist between gut microflora species of humans and chimpanzees (Ushida et al. 2010) and also strepsirrhines (Bo et al. 2010). This suggests that some of the symbiotic microbes in human and NHP guts could be heirlooms. Understanding the intersecting roles of physiological similarity, behavioral divergence, environmental context, and microbial colonization is therefore crucial when reconstructing the evolutionary histories of heirloom microbes, both beneficial and pathogenic, in humans and NHPs.

Souvenir Species and Changing Human Ecology

Hominins, particularly the genus *Homo*, became increasingly generalized as they evolved new strategies for inhabiting and manipulating new environments. Modern humans, in particular, have increased the scope of their exposure to pathogens via the environmental modification and plant and animal domestication associated with the adoption and intensification of agropastoralism. Numerous souvenir pathogens have come to infect human populations, some of which have their origins among NHPs. Many of these primarily remain residents of their reservoir animal hosts. For example, though limited human-to-human transfer may occur in pathogens such as SARS or Ebola, they must be considered what Weiss (2009) designates "temporary exhibits" in humans. There are cases, though, in which a souvenir species adapts to its novel host so well that it becomes established and flourishes within human populations. One example, already discussed, is *P. falciparum*. Another major example comes in the form of HIV/AIDS, which we discuss here.

HIV: A Souvenir Species with a Complicated Past

Similar to the example of malaria discussed above, investigation into the NHP origins of HIV/SIV has demonstrated that continued research on even the most well-studied disease agents can yield important and surprising insights into their origins and evolution. HIV represents the best-known example of an NHP pathogen transmitted to and then sustained within humans, and it remains one of the most serious pandemics in history. The UNAIDS *Report on the Global AIDS Epidemic 2010* estimates that 33.3 million people are infected with HIV worldwide, among them 2.6 million children. It is estimated that some 25 million people worldwide have died from HIV-/AIDS-related diseases (UNAIDS 2010). The majority of individuals with HIV live in sub-Saharan Africa (UNAIDS 2010). Adding to the disease burden in sub-Saharan Africa and other regions with high rates of HIV is the fact that this infection disproportionately affects young adults (Patton et al. 2009), leading to high morbidity and mortality among those individuals who would otherwise be among the most economically active in their societies. A result of this pattern of infection is a demographic crisis in which young children and older adults carry the burden of looking after themselves, each other, and those who are infected. The fact that a disproportionate burden of infection is found in countries with high rates of political and economic inequality compounds this crisis even further (Fox 2012).

There are two types of HIV: HIV-1 and HIV-2. Scholars have understood for some time that both types evolved from the simian immunodeficiency viruses (SIVs), with HIV-1 deriving from the SIV variant of chimps (SIVcpz) and HIV-2 from the SIV of sooty mangabeys (SIVsmm) (Gao et al. 1999; Hahn et al. 2000; Hirsch et al. 1995; Peeters et al. 2002; Weiss and Wrangham 1999). SIV infection is quite common in NHPs over 40 species-specific SIV variants have been documented in

African monkeys (Sharp and Hahn 2010). Moreover, cross-species transmission of SIV has been postulated from African green monkeys to both patas monkeys (Bibollet-Ruche et al. 1996) and yellow baboons (Jin et al. 1994). In a survey of 788 wild-caught NHPs from Cameroon, serological evidence of SIV infection was present in 13 of the 16 primate species tested, with about 20 % of total samples testing seropositive (Peeters et al. 2002). SIV has also been demonstrated in sooty mangabey bushmeat samples from rural Sierra Leone, underscoring how the initial transfer of HIV-2 may have occurred (Apetrei et al. 2005). Therefore, NHPs represent a substantial potential reservoir of continued SIV transmission to humans.

Our understanding of where human HIV strains originated has become more nuanced with time. For example, it has long been understood that the HIV-1 lineage, as a whole, originated from chimpanzees; however, one group within the lineage may have a slightly different history than the others. HIV-1 group O (named for its "outlier" status) accounts for a relatively small proportion of HIV cases and is rarely found outside of Cameroon. Intensive sampling of wild chimpanzees and gorillas has demonstrated that HIV-1 group O appears to be most closely related to an SIV strain circulating in gorillas (SIVgor) (Van Heuverswyn et al. 2006). Thus, it is possible either that humans initially contracted the group O virus from gorillas or that chimpanzees independently transmitted the virus to both gorillas and humans.

How often does SIV take hold in humans, establishing sustained infection in its new host and spreading to other people? Determining the answer to this question requires extensive and sensitive surveillance of a sort that has yet to be conducted, but it appears that SIV infection in humans in close contact with NHPs is relatively common. In Cameroon, one man was found to be infected with a virus serologically related to SIVmnd, a version of the virus found in mandrills (Souquière et al. 2001), demonstrating that NHPs other than the great apes and sooty mangabeys are capable of transmitting SIV to humans. In another instance, a Cameroonian woman who reported no contact with great apes or bushmeat was found to harbor a novel HIV virus closely related to SIVgor (Plantier et al. 2009). The virus's high replication rate in this patient and the ease of its isolation in culture suggest that this novel variant had adapted to human cells and may have spread from human to human at some low level. Finally, 23 individuals out of a sample of 2,436 people at high risk for exposure through poaching and bushmeat consumption tested seropositive for SIV, though no active infections were demonstrated in this group (Djoko et al. 2012). Thus, it appears that SIV transmission to humans may not be a rare event, and a low level of human-to-human transmission may even occur at times.

What is the fate of individuals infected with SIV? Strains of SIV that infect mangabeys and macaques appear to be very closely related to HIV-2, which is less virulent than HIV-1. Accidental infection of two different laboratory workers with these SIV strains did not result in AIDS-like symptoms, despite the presence of SIV and HIV-2 antibodies (Khabbaz et al. 1994). This suggests that SIV infections in humans may not necessarily be harmful, which could prove lifesaving for individuals in Central Africa who have tested seropositive. However, Peeters et al. (2002)

have noted that recombination between SIVs and circulating HIVs may pose a threat to human health if it results in novel strains with an increased ability to exploit our species.

In conclusion, though the sequence of events that allowed for permanent establishment and spread of HIV-1 and HIV-2 in humans remains uncertain (Pepin 2011), molecular and epidemiological studies are providing a richer context for understanding these souvenir pathogens. As discussed, it has been known for some time that at least two host switches, one from chimpanzees and one from sooty mangabeys, resulted in HIV-1 and HIV-2, respectively. Recent research, though, has demonstrated that a third host switch from gorillas may have resulted in one rare HIV subtype (HIV-1 group O) (Van Heuverswyn et al. 2006). Additionally, molecular epidemiological studies have demonstrated that cross-species exposure to SIV is ongoing (Djoko et al. 2012) and may at times even result in low-level transmission among humans (Plantier et al. 2009). Thus, the potential for novel forms of HIV to take root in humans appears to be present, though most SIV transmission events seem to fizzle out quickly.

Major Factors Facilitating Adaptation of NHP Pathogens to Humans

Malaria and HIV, both discussed extensively in this chapter, are examples of pathogens that originally came from NHPs but found a permanent home in humans. There are multiple reasons why a given NHP host species may or may not become an established source of infection for humans. Phylogenetic relatedness and physical and environmental proximity are two primary—and tightly interwoven—factors which are likely to affect this dynamic. In an assessment of the animal origins of 25 major human infectious human diseases, Wolfe et al. (2007) partially attributed the finding that the majority of tropical infectious diseases arose in the Old World rather than the New World to the greater genetic distance separating New World monkeys and humans. It represents roughly twice the genetic distance between Old World monkeys and humans and many times that between humans and Old World apes. At the same time, physical proximity and shared habitats are likely to play a substantial role. For example, Wolfe et al. (2007) partially attributed the high number of souvenir infections arising in the Old World to the greater evolutionary time available for transfers between primates and humans there (c. 7–8 million years) as compared to the New World (c. < 14,000 years).

On a regional level, some populations have maintained fairly frequent exposure to NHPs. In Africa, South America, and Asia, in particular, communities living in close proximity to NHPs often become involved in activities associated with a high risk of exposure to NHP pathogens. Bushmeat handling and consumption provides one of the most effective means for the spread of pathogens from NHPs to humans

Fig. 2 Nonhuman primate bushmeat confiscated in the Lomako Forest of the Equateur province in the Democratic Republic of Congo. (**a**) Fresh remains of the lower half of a black mangabey (*Lophocebus aterrimus*) found at the Iyema study site in 2010. It lies atop a pile of leaves that were used to wrap it for easier carrying. (**b**) A pile of smoked bushmeat confiscated in 2007 by the members of the Institut Congolais pour la Conservation de la Nature, who work as park guards. The pile includes bonobos, monkeys, antelope, crocodiles, and red river hogs. (**c**) Close-up of a skewered and smoked monkey from the pile, most likely a black mangabey (Photo Credits: Amy Cobden)

(Fig. 2). Almost 100 % of villagers in rural forested areas of Cameroon have reported eating NHPs, with over 70 % involved in hunting and 30 % active in butchering (Wolfe et al. 2004a). Both activities involve repeated contact with potentially infective body fluids and tissues. They also generate opportunities for pathogen transmission to other individuals and communities linked by the bushmeat trade. A market survey of two cities in Equatorial Guinea recorded 4,222 primate carcasses on sale over 424 days (Fa et al. 1995), illustrating the great importance of NHP hunting to local economies.

The prevalence of human infections derived from NHPs suggests that these pathogens are able to exploit between-species interactions effectively. For example, Wolfe et al. (2005) found that bushmeat hunters in rural Cameroon are infected with a wide variety of human T-lymphotropic viruses (HTLVs), which are linked to leukemia, lymphoma, and HTLV-associated myelopathy, as well as multiple simian T-lymphotropic virus (STLV)-1-like viruses. The high diversity of these viruses in Cameroon indicates ongoing cross-species transmission from NHPs to humans. Similarly, at least three independent examples of NHP-to-human transmission of simian foamy virus (SFV), which to date has not been linked to any signs of disease in humans, have been confirmed in hunters (Wolfe et al. 2004b). Each event derived from distinct NHP lineages—De Brazza's guenons, mandrills, and gorillas—indicating that many species can potentially infect humans. Bites and scratches by NHPs, in particular, appear to be a very efficient means of transmitting viruses. In one study, over 35 % of hunters reporting such wounds were seropositive for SFV infection, and almost 2 % of serum samples belonging to adults from Cameroon were found to be seropositive as well, underscoring the high prevalence of cross-species transmission opportunities (Calattini et al. 2007).

While the factors leading to introduction of NHP pathogens into human populations are increasingly well characterized, the conditions that facilitate their establishment in our species are poorly understood. For example, though studies of Cameroonian hunters suggest that exposure to STLVs via NHP blood contributes to a greater diversity of HTLVs than is found in other populations (Wolfe et al. 2005), only HTLV-1 and HTLV-2 appear to have established themselves worldwide. Similarly, reports of the presence of dual HIV-1 and SFV infections in a commercial sex worker from the Democratic Republic of Congo and in a blood donor from Cameroon suggest the potential for SFV to be transmitted from human to human via sex or blood, as is HIV (Switzer et al. 2008). Such human-to-human transmission events appear to occur relatively rarely in the case of SFV, however. At present, our ability to predict which souvenir pathogens will "take off," flourish, and become established within human populations is poor. Nonetheless, continued attention to the movement of pathogens across the NHP-human interface may contribute to our knowledge of this process.

It is worrisome that the trend of increased contact between NHPs and humans seems to be intensifying. Population pressure, expanded ecotourism and conservation programs, and ever more powerful technological advances are enabling humans to encroach further and further into NHP habitats (Auzel and Hardin 2001). For instance, the villages surrounding logging concessions in Equatorial Africa have grown rapidly, often having increased from a few hundred individuals to several thousand. Political instability and forced migration, from Liberia into Sierra Leone, for example, have also played a role in the dramatic redistribution of human populations into areas of increased contact with NHPs (Hodges and Heistermann 2003). This new proximity, paired with the desirability of NHPs as prey, has increased opportunities for cross-species transmission events. In addition, people in these areas can utilize new and improved roads to transport bushmeat from remote villages to major cities, greatly increasing the number of humans who come into contact with NHP carcasses. In truth, contact with wild NHPs now spans continents, reaching consumers who have never set foot in wild NHP habitat (Ellicott 2011); examination of bushmeat samples confiscated at US airports has revealed NHP tissue infected with SFV and herpesviruses (Smith et al. 2012). It is expected that these trends will intensify in the future, in the absence of decisive community-level and governmental actions to regulate ecotourism and constrain environmental destruction and the bushmeat trade.

Examples of Major Diseases Transmitted from Humans to Nonhuman Primates

Naturally, the increasing proximity between humans and NHPs also leads to greater opportunity for human pathogens to infect both captive and wild NHPs. In general, as the level of interaction between humans and NHPs increases, so does the risk of transmission of diseases such as measles and tuberculosis (Wolfe et al. 1998). Not surprisingly, there are many well-documented examples of captive NHPs becoming infected with human pathogens, with "immunologically naïve" great apes proving especially susceptible. For instance, there are frequent reports of tuberculosis infections of human origin among captive NHPs (Montali et al. 2001). Poliovirus can also infect chimpanzees and gorillas, as well as more distantly related anthropoids, like Colobus monkeys (Brack 1987; Suleman et al. 1984). As such, accidental exposure to infected laboratory workers has led to poliovirus infections of chimpanzees and gorillas since the 1940s (Ruch 1959). In another example, *Arcobacter butzleri*, which is a member of the same bacterial family as *Campylobacter* (Campylobacteraceae) and is associated with chronic diarrhea in humans, was implicated in a spate of cases of chronic diarrhea among captive primate populations at a research center (Andersen et al. 1993).

There is accumulating evidence for similar episodes of disease transmission in the wild (Adams et al. 1999; Homsy 1999; Wallis 2000; Wolfe et al. 1998; Woodford et al. 2002). There are a number of "likely" instances of human-to-NHP transmission. In perhaps the most infamous instance, in 1966, six chimpanzees at Gombe Stream National Park in Tanzania died from a polio-like virus, and six others were paralyzed for life, shortly after a polio epidemic swept through neighboring human settlements (Wallis 2000). Unfortunately, as no biological samples were collected from the animals, it was impossible to verify if the epidemic was due to a poliovirus introduced by local human populations or researchers (Wolfe et al. 1998). Confirmed examples of human to wild NHP transmission are relatively rare, due to the difficulty inherent in collecting samples but include *Cryptosporidium* infections in mountain gorillas (Nizeyi et al. 2002) as well as the cases discussed below.

Respiratory diseases, in particular, have been recognized as a major source of morbidity and mortality among free-living NHPs. This class of diseases is widely regarded as the most important cause of morbidity and mortality among wild great apes habituated to the presence of humans, whether due to research, tourism, or human communities living in close proximity (Goodall 1986; Hanamura et al. 2007; Homsy 1999; Nishida 1990; Woodford et al. 2002). For example, a serological survey demonstrated that 100 % of macaques at a temple in Katmandu were seropositive for antibodies to the measles virus (Jones-Engel et al. 2006). Of more potential significance for NHP conservation efforts, about half of long-term chimpanzee research populations have shown major population declines that are likely a consequence of respiratory disease (Hill et al. 2001). For instance, Köndgen et al. (2008) have documented transmission of two common strains of human paramyxoviruses—human respiratory syncytial virus (hRSV) and human metapneumovirus (hMPV)—from humans to chimpanzees at a research station in the Taï Forest in Côte d'Ivoire. These two viruses are common causes of respiratory disease in humans. They are the leading causes of lower respiratory disease in children and, in developing countries, are a major source of infant mortality (Boivin et al. 2003; Weber et al. 1998). Transmission of the viruses from humans to the chimps of the Taï Forest resulted in five discrete epidemics during the study period, each accompanied by high morbidity, with an average of 92.2 % of individuals showing clinical symptoms. In three of the epidemics, 18–34 % of chimpanzees in the study population succumbed to the infection. Viral strains sampled from the deceased chimpanzees were found to be closely related to strains circulating in contemporaneous, worldwide human epidemics, indicating a link between the epidemic and the continuous flow of outside ecotourists and researchers at the station (Köndgen et al. 2008).

Unfortunately, such epidemics are not unique to the Taï Forest. hMPV was also identified in association with acute and fatal respiratory illness outbreaks in the chimpanzees of Mahale Mountains National Park, in Tanzania (Kaur et al. 2008). Additionally, hMPV was documented in association with two mountain gorilla deaths in Rwanda (Palacios et al. 2010). Thus, molecular epidemiology has confirmed that human pathogens are responsible for many of the "mysterious" ailments currently driving population declines in NHPs. Not surprisingly, fatal outbreaks of respiratory disease at Mahale Mountains National Park have coincided with peak tourist season (Kaur et al. 2008). Similarly, wild chimpanzees in Kibale National Park, Uganda, have been found to harbor *E. coli* strains genetically similar to those carried by the humans they come into proximity with via research or tourism. This underscores the fluid transmission of microbes from humans to NHPs made possible by high rates of ecotourism and conservation-oriented research (Goldberg et al. 2007).

As noted, while it is clear that human-derived pathogens have been responsible for swift epidemics and dramatic declines in NHP populations, often the identity of the agent responsible for a given epidemic remains merely suspected or wholly unknown. For example, outbreaks of gastrointestinal illness (Goodall 1983) and respiratory disease (Ferber 2000) suspected to originate in human populations have been recorded among chimpanzees in Tanzania from the 1960s onwards. Similarly, suspected cases of measles among gorillas were documented in Rwanda in 1988. Finally, suspected but unconfirmed cases of scabies among gorillas have been reported in multiple regions (Kalema-Zikusoka et al. 2002). Our inability to determine the etiological agent responsible for most NHP diseases is due in large part to the difficulty of acquiring samples for diagnostic testing from wild NHPs. Systematic screening for the pathogens involved in NHP fatalities is performed infrequently, even though such investigations can reveal both the causal agents and, given the right molecular data, their transmission dynamics (Leendertz et al. 2006).

There is no evidence—yet—of sustained transmission of human pathogens among NHPs. It is probable that many human pathogens that have adapted to large populations with constantly replenished pools of susceptible hosts (i.e., crowd diseases) "burn out" after rapidly infecting small groups of NHPs. The possibility of sustained transmission of human pathogens in NHPs is certainly possible, however, especially for infections with long latent periods. Kaur et al. (2008) note that persistent infection with hMPV in the absence of respiratory symptoms has been demonstrated in humans; if the same is true in NHPs, then the potential of individual animals to carry the disease from one group to another via emigration could have devastating consequences. Implementing rigorous, systematic monitoring of infectious disease outbreaks as part of modern conservation practice is being strongly encouraged (e.g., Leendertz et al. 2006) and gradually implemented (e.g., the Great Ape Health Monitoring Unit (http://www.eva.mpg.de/primat/GAHMU/ index.htm) and the Mountain Gorilla Veterinary Project (http://www.gorilla doctors.org/)). It seems reasonable that given increased surveillance of NHP infections, examples of human microbes capable of sustained transmission among NHPs will be identified.

Conclusion

The surveillance of humans living in close proximity to NHPs has revealed tantalizing clues about the process of disease transmission from NHPs to humans. Perhaps intensive study of circulating strains of HIV/SIV, HTLV/STLV, and SFV will increase our knowledge of the features which characterize cross-species transmission events that result in subsequent sustained transmission between humans. In addition, the explosion of knowledge surrounding pathogens of NHP origin, such as *P. falciparum* and HIV, in the last few years underscores the need to delve into the disease-scape of closely related NHPs to better understand our own infections. For instance, it was not until 2010 that researchers demonstrated that wild chimps in the Taï Forest appeared to be naturally infected with five different *Plasmodium* species (Kaiser et al. 2010). Casting a wider net for pathogens may lead to similar advances in our understanding of the other pathogens that make up the NHP disease-scape.

It is likely that the rapid development of sophisticated, noninvasive means of NHP sampling will provide insight into the existence and/or prevalence of various pathogens. These noninvasive methods include approaches similar to those developed in primatology to study endocrinology (e.g., Deschner et al. 2003; Hodges and Heistermann 2003), characterize NHP genetics (e.g., Boesch et al. 2006; Bradley et al. 2004; Vigilant et al. 2001), and perform urine assessment (e.g., Knott 1996; Krief et al. 2005). For instance, fecal samples have already been used to assay SIV and malaria in NHPs. In addition, a recent retrospective study of the epidemics in the Taï Forest suggests that it is possible to monitor infections caused by paramyxoviruses, such as hMPV and respiratory syncytial virus, using fecal samples (Köndgen et al. 2010). A pioneering study using aDNA techniques to study curated early-twentieth-century NHP skeletal material even suggests that we can explore the history of viruses such as STLV using museum specimens (Calvignac et al. 2008).

Such noninvasive approaches may also help address the role of human pathogens in NHP demographic declines. There are many unresolved questions. Do human pathogens associated with epidemics in NHPs tend to rapidly infect small groups before burning out? Or are some human-derived pathogens circulating continuously and even adapting to their new hosts? In some cases, whether or not the pathogens sweeping through NHP populations derive from humans is not clear. For example, *S. pneumoniae* was recently found to be responsible for clusters of sudden death in the chimpanzees of the Taï Forest (Chi et al. 2007). However, comparison to sequences obtained from people living nearby suggested that the pathogen might not be of human origin. Chimpanzees in different areas, but in frequent contact with one another, were also found to harbor distinct *S. pneumoniae* clones. If not from humans, from what reservoir did this pathogen arise? Similarly, adenoviruses were isolated from two chimpanzees with signs of acute respiratory disease in Mahale Mountains National Park (Tong et al. 2010). Sequence analysis identified two distinct viruses, but their origin was unclear; they could have been acquired from humans, acquired from another species, or circulating among chimpanzees for some time. Further investigation may shed light on the processes underlying such outbreaks.

Continued study of the relationship between humans, pathogens, and NHPs confers several substantial benefits. First, knowledge in this area is fundamental when assessing the impact of pathogen exchange between species (Wolfe et al. 2007), including research on the footprint of natural selection imposed by various infectious diseases upon the human genome (e.g., Hamblin et al. 2002; Sabeti et al. 2002; Tishkoff et al. 2001). Second, understanding more about how NHP pathogens are introduced into human populations and then spread has practical implications. According to Wolfe et al. (2007), benefits include a better understanding of disease emergence and the potential for novel laboratory models helpful in studying public health threats. Applications might include the development of indicators useful in monitoring pathogen transmission between NHPs and high-risk individuals, such as hunters and wildlife veterinarians; predicting which NHP pathogens might represent a future threat; and detecting and even controlling local human outbreaks before they become epidemics (Wolfe et al. 2007). Third, some scholars have argued that wild NHPs can serve as "sentinel species" for predicting disease outbreaks among humans (Leendertz et al. 2006; Rouquet et al. 2005).

Finally, studying the relationship between pathogens, NHPs, and humans has important conservation implications. Köndgen et al. (2008) and others have argued that the close proximity between NHPs and humans, which is critical to both research and ecotourism programs, represents a serious threat to the existence of wild primate populations. Obviously, this represents a dilemma, as both of these activities have clear benefits for conservation efforts, whether via suppressing poaching, generating income for local communities, or creating additional knowledge about primate biology and behavior. Do the benefits associated with conservation efforts outweigh the health costs for apes and other NHPs wrought by increased contact between NHPs and humans? Research efforts should be directed towards reducing deleterious health outcomes, and an improved understanding of the evolutionary trajectory and dynamics of pathogen exchange between humans and NHPs stands to make a substantial contribution to this effort. For instance, research on disease transmission from humans to NHPs can be used to perfect targeted strategies for preventing infection, including close monitoring of the health and behaviors of human observers and workers in conservation, scientific, and veterinary contexts (Homsy 1999; Nizeyi et al. 2002; Woodford et al. 2002).

Findings from the studies discussed above have already been used to generate specific recommendations for reducing the negative effects of tourists, local communities, and researchers upon NHPs, especially endangered great ape communities (see Ryan and Walsh 2011). These guidelines include a variety of strategies for limiting disease spillover into NHPs via the use of facemasks, minimum approach distances, limited-duration visits, and strict hygiene protocols. They also encompass the education of and collaboration with local stakeholders to determine optimal rates of tourism for preventing disease transmission while maximizing tourism revenues (maximum sustainable yield concept); vaccinating NHPs and treating infections when they arise; prohibiting human access to restricted areas in order to minimize both direct and indirect contact, such as through human feces, between local humans and NHPs; and establishing health programs for local communities and staff involved in habituating NHPs for tourism and research (Ryan and Walsh 2011). An active area of research focuses on how these disease-mitigating measures can be carried out with the full participation of local communities throughout Africa and other regions home to endangered NHP populations, in order to make these endeavors sustainable, practical, and desirable for the people involved (Ryan and Walsh 2011).

In terms of more research-intensive interventions, observations stemming from invasive and noninvasive tests on chimpanzees involved in the Taï Forest epidemics have been used to generate demographic, clinical, and diagnostic monitoring systems which could potentially enable humans to intervene quickly in future NHP epidemics there. Researchers involved in this ongoing project have strongly encouraged other investigators to implement similar systems at NHP research centers and parks (see Ryan and Walsh 2011). Such efforts would not only protect NHP populations but also objectively document the negative effects of research or ecotourism on NHPs (Köndgen et al. 2008).

In summary, the rapid and extensive destruction of forest ecosystems and changing patterns of contact between humans and NHPs, both stemming from the increase in size and changing distribution of human populations, have changed the disease-scape of all species involved. Someday, as detection and surveillance improves, we may be able to perform comprehensive analyses of the different disease-scapes of humans and NHP species. When this happens, we will be able to explicitly test hypotheses such as whether host genetic similarity correlates neatly with the proportion of pathogens shared between two given species. Moreover, in the future we may learn more about how different primate species react to identical pathogens, which will yield important information on how immune responses and transmission dynamics differ within and between populations. The relationships between NHPs, humans, and pathogens are fluid. Targeted research may help us prevent the worst possible consequences of these constantly shifting associations by allowing us to learn some general lessons about the processes underlying pathogen host switches.

References

Adams H, Sleeman J, New J (1999) A medical survey of tourists visiting Kibale National Park, Uganda, to determine the potential risk of disease transmission to chimpanzees (*Pan troglodytes*) from ecotourism. In: Baer C (ed) Proceedings of the American Association of Zoo Veterinarians. American Association of Zoo Veterinarians, Media, Philadelphia, PA

Andersen K, Kiehlbauch JA, Anderson DC, McClure HM, Wachsmuth IK (1993) *Arcobacter* (Campylobacter) *butzleri*-associated diarrheal illness in a nonhuman primate population. Infect Immun 61(5):2220–2223

Apetrei C, Metzger M, Richardson D, Ling B, Telfer P, Reed P, Robertson D, Marx P (2005) Detection and partial characterization of simian immunodeficiency virus SIVsm strains from bush meat samples from rural Sierra Leone. J Virol 79(4):2631–2636

Auzel P, Hardin R (2001) Colonial history, concessionary politics, and collaborative management of Equatorial African rain forests. In: Bakarr M, Fonseca GD, Mittermeier R, Rylands A, Painemilla K (eds) Hunting and Bushmeat utilization in the African Rain Forest: Perspectives Toward a Blueprint for Conservation Action. Conservation International, Washington

Bibollet-Ruche F, Galat-Luong A, Cuny G, Sarni-Manchado P, Galat G, Durand J-P, Pourrut X, Veas F (1996) Simian immunodeficiency virus infection in a patas monkey (*Erythrocebus patas*): evidence for cross-species transmission from African green monkeys (*Cercopithecus aethiops sabaeus*) in the wild. J Gen Virol 77(773–781)

Bo X, Zun-xi H, Xiao-yan W, Run-chi G, Xiang-hua T, Yue-lin M, Yun-Juan Y, Hui S, Li-da Z (2010) Phylogenetic analysis of the fecal flora of the wild pygmy loris. Am J Primatol 72:699–706

Boesch C, Boesch-Achermann H (2000) The chimpanzees of the Taï Forest: behavioural ecology and evolution. Oxford University Press, Oxford/New York

Boesch C, Kohou G, Nene H, Vigilant L (2006) Male competition and paternity in wild chimpanzees of Taï Forest. Am J Phys Anthropol 130:103–115

Boivin G, De Serres G, Cote S, Gilca R, Abed Y, Rochette L, Bergeron M, Dery P (2003) Human metapneumovirus infections in hospitalized children. Emerg Infect Dis 9:634–640

Brack M (1987) Agents transmissible from simians to man. Springer, Berlin

Bradley B, Doran-Sheehy D, Lukas D, Boesch C, Vigilant L (2004) Dispersed male networks in western Gorillas. Curr Biol 14:510–513

Brooks D, Ferrao A (2005) The historical biogeography of co-evolution: emerging infectious diseases are evolutionary accidents waiting to happen. J Biogeogr 32:1291–1299

Calattini S, Betsem E, Froment A, Mauclère P, Tortevoye P, Schmitt C, Njouom R, Saib A, Gessain A (2007) Simian foamy virus transmission from apes to humans, rural Cameroon. Emerg Infect Dis 13(9):1314–1320

Calvignac S, Terme J, Hensley S, Jalinot P, Greenwood A, Hänni C (2008) Ancient DNA identification of early 20th century simian T-cell leukemia virus type 1. Mol Biol Evol 25(6):1093–1098

Chen Z, Luckay A, Sodora DL, Telfer P, Reed P, Gettie A, Kanu JM, Sadek RF, Yee J, Ho D et al (1997) Human immunodeficiency virus type 2 (HIV-2) seroprevalence and characterization of a distinct HIV-2 genetic subtype from the natural range of simian immunodeficiency virus-infected sooty mangabeys. J Virol 71(5):3953–3960

Chi F, Leider M, Leendertz F, Bergmann C, Boesch C, Schenk S, Pauli G, Ellerbrok H, Hakenbeck R (2007) New *Streptococcus pneumoniae* clones in deceased wild chimpanzees. J Bacteriol 189(16):6085–6088

Cockburn T (1967) Infections of the order primates. In: Cockburn T (ed) Infectious diseases: their evolution and eradication. CC Thomas, Springfield, IL

Cockburn T (1971) Infectious disease in ancient populations. Curr Anthropol 12(1):45–62

Collins W, Skinner J, Pappaioanou M, Broderson J, Mehaffey P (1986) The sporogonic cycle of *Plasmodium reichenowi*. J Parasitol 72(2):292–298

Deschner T, Heistermann M, Hodges K, Boesch C (2003) Timing and probability of ovulation in relation to sex skin swelling in wild West African chimpanzees, *Pan troglodytes verus*. Anim Behav 66:551–560

Djoko C, Wolfe N, Aghokeng A, Lebreton M, Liegeois F, Tamoufe U, Schneider B, Ortiz N, Mbacham W, Carr J et al (2012) Failure to detect simian immunodeficiency virus infection in a large Cameroonian cohort with high non-human primate exposure. Ecohealth 9(1):17–23

Dunn R, Davies T, Harris N, Gavin M (2010) Global drivers of human pathogen richness and prevalence. Proc R Soc B 277:2587–2595

Ellicott C (2011 March 11) Meat from chimpanzees 'is on sale in Britain' in lucrative black market. Daily Mail, London

Escalante A, Ayala F (1994) Phylogeny of the malarial genus *Plasmodium*, derived from rRNA gene sequences. Proc Natl Acad Sci USA 91:11373–11377

Fa J, Juste J, Val JD, Castroviejo J (1995) Impact of market hunting on mammal species in equatorial Guinea. Conserv Biol 9(5):1107–1115

Ferber D (2000) Primatology. Human diseases threaten great apes. Science 289:1277–1278

Fox A (2012) The HIV-poverty thesis re-examines: poverty, wealth or inequality as a social determinant of HIV infection in Sub-Saharan Africa? J Biosoc Sci 25:1–22

Gao F, Bailes E, Robertson D, Chen Y, Rodenburg C, Michael S, Cummins L, Arthur L, Peeters M, Shaw GM et al (1999) Origin of HIV-1 in the chimpanzee *Pan troglodytes troglodytes*. Nature 397:436–441

Garamszegi L (2009) Patterns of co-speciation and host switching in primate malaria parasites. Malar J 8:110–124

Gentry G, Lowe M, Alford G, Nevias R (1988) Sequence analysis of herpesviral enzymes suggest an ancient origin for human sexual behavior. Proc Natl Acad Sci 85:2658–2661

Goldberg T, Gillespie T, Rwego I, Wheeler E, Estoff E, Chapman C (2007) Patterns of gastrointestinal bacterial exchange between chimpanzees and humans involved in research and tourism in western Uganda. Biol Conserv 135:511–517

Goodall J (1970) In the shadow of man. Collins, London

Goodall J (1983) Population dynamics during a 15-year period in one community of free-living chimpanzees in the Gombe National Park, Tanzania. Z Tierpsychol 61:1–60

Goodall J (1986) The chimpanzees of Gombe: patterns of behaviour. Harvard University Press, Cambridge, MA

Hahn B, Shaw G, Cock KD, Sharp P (2000) AIDS as a zoonosis: scientific and public health implications. Science 287(5453):607–614

Hamblin M, Thompson E, Rienzo AD (2002) Complex signatures of natural selection at the Duffy blood group locus. Am J Hum Genet 70(2):369–383

Hanamura S, Kiyono M, Lukasik-Braum M, Mlengeya T, Fujimoto M, Nakamura M, Nishida T (2007) Chimpanzee deaths at Mahale caused by a flu-like disease. Primates 49:77–80

Harper K, Ocampo P, Steiner B, George R, Silverman M, Bolotin S, Pillay A, Saunders N, Armelagos G (2008) On the origin of the treponematoses: a phylogenetic approach. PLoS Negl Trop Dis 2(1):e148

Hill K, Boesch C, Goodall J, Pusey A, Williams J, Wrangham R (2001) Mortality rates among wild chimpanzees. J Hum Evol 40:437–450

Hirsch V, Dapolito G, Goeken R, Campbell B (1995) Phylogeny and natural history of the primate lentiviruses, SIV and HIV. Curr Opin Genet Dev 5:798–806

Hodges J, Heistermann M (2003) Field endocrinology: monitoring hormonal changes in free-ranging primates. In: Setchell J, Curtis D (eds) Field and laboratory methods in primatology: a practical guide. Cambridge University Press, Cambridge, pp 282–294

Homsy J (1999) Ape tourism and human diseases: how close should we get? A critical review of rules and regulations governing Park Management and Tourism for the Wild Mountain Gorilla, *Gorilla gorilla beringei*. Consultancy for the International Gorilla Conservation Program, Nairobi (http://www.igcp.org/pdf/homsy_rev.pdf)

Hugot J, Reinhard K, Gardner S, Morand S (1999) Human enterobiasis in evolution: origin, specificity and transmission. Parasite 6(3):201–208

Jin M, Rogers J, Phillips-Conroy J, Allan J, Desrosiers R, Shaw G, Sharp P, Hahn B (1994) Infection of a yellow baboon with simian immunodeficiency virus from African green monkeys: evidence for cross-species transmission in the wild. J Virol 68(12):8454–8460

Jones-Engel L, Engel G, Schillaci M, Lee B, Heidrich J, Chalise M, Kyes R (2006) Considering human-primate transmission of measles virus through the prism of risk analysis. Am J Primatol 68:868–879

Kaiser M, Löwa A, Ulrich M, Ellerbrok H, Goffe A, Blasse A, Zommers Z, Couacy-Hymann E, Babweteera F, Zuberbühler K et al (2010) Wild chimpanzees infected with 5 *Plasmodium* species. Emerg Infect Dis 16(12):1956–1959

Kalema-Zikusoka G, Kock R, Macfie E (2002) Scabies in free-ranging mountain gorillas (Gorilla beringei beringei) in Bwindi Impenetrable National Park, Uganda. Vet Rec 150:12–15

Kaur T, Singh J, Tong S, Humphrey C, Clevenger D, Tan W, Szekely B, Wang Y, Li Y, Muse E et al (2008) Descriptive epidemiology of fatal respiratory outbreaks and detection of a human-related metapneumovirus in wild chimpanzees (*Pan troglodytes*) at Mahale Mountains National Park, Western Tanzania. Am J Primatol 70:755–765

Khabbaz R, Heneine W, George J, Parekh B, Rowe T, Woods T, Switzer W, McClure H, Murphey-Corb M, Folks TM (1994) Infection of a laboratory worker with simian immunodeficiency virus. N Engl J Med 330:172–177

Kilks M (1990) Helminths as heirlooms and souvenirs: a review of new world paleoparasitology. Parasitol Today 6(4):93–100

Knott C (1996) Monitoring health status of wild orangutans through field analysis of urine. Am J Phys Anthropol 22:139–140

Köndgen S, Kühl H, N'Goran P, Walsh P, Schenk S, Ernst N, Biek R, Formenty P, Mätz-Rensing K, Schweiger B et al (2008) Pandemic human viruses cause decline of endangered great apes. Curr Biol 18:260–264

Köndgen S, Schenk S, Pauli G, Boesch C, Leendertz F (2010) Noninvasive monitoring of respiratory viruses in wild chimpanzees. Ecohealth 7:332–341

Krief S, Huffman M, S'evenet T, Guillot J, Bories C, Hladik C, Wrangham R (2005) Noninvasive monitoring of the health of Pan troglodytes schweinfurthii in the Kibale National Park, Uganda. Int J Primatol 26:467–490

Leendertz F, Ellerbrok H, Boesch C, Couacy-Hymann E, Mätz-Rensing K, Hakenbeck R, Bergmann C, Abaza P, Junglen S, Moebius Y et al (2004) Anthrax kills wild chimpanzees in a tropical rainforest. Nature 430:451–452

Leendertz F, Pauli G, Maetz-Rensing K, Boardman W, Nunn C, Ellerbrok H, Jensen S, Junglen S, Boesch C (2006) Pathogens as drivers of population declines: the importance of systematic monitoring in great apes and other threatened mammals. Biol Conserv 131: 325–337

Liu W, Li Y, Learn G, Rudicell R, Robertson J, Keele B, Ndjango J, Sanz C, Morgan D, Locatelli S et al (2010) Origin of the human malaria parasite *Plasmodium falciparum* in gorillas. Nature 467:420–425

McGeouch D, Cook S, Dolan A, Jamieson F, Telford E (1995) Molecular phylogeny and evolutionary timescale for the family of mammalian herpesviruses. J Mol Biol 247:443–458

Montali R, Mikota S, Cheng L (2001) *Mycobacterium tuberculosis* in zoo and wildlife species. Rev Sci Tech 20(1):291–303

Nishida T (1990) The chimpanzees of the Mahale Mountains. Sexual and life history strategies. University of Tokyo Press, Tokyo

Nizeyi J, Sebunya D, Dasilva A, Cranfield M, Pieniazek N, Graczyk T (2002) Cryptosporidiosis in people sharing habitats with free-ranging mountain gorillas (*Gorilla gorilla beringei*), Uganda. Am J Trop Med Hyg 66:442–444

Ochman H, Worobey M, Kuo C, Ndjango J, Peeters M, Hahn B, Hugenholtz P (2010) Evolutionary relationships of wild hominids recapitulated by gut microbial communities. PLoS Biol 8(11):e1000546

Olson M, Varki A (2003) Sequencing the chimpanzee genome: insights into human evolution and disease. Nat Rev 4:20–28

Palacios G, Lowenstine L, Cranfield M, Gilardi K, Spelman L, Lukasik-Braum M, Kinani J, Mudakikwa A, Nyirakaragire E, Bussetti A et al (2010) Human metapneumovirus infection in wild mountain gorillas, Rwanda. Emerg Infect Dis 17(4):711–713

Patton G, Coffey C, Sawyer S, Viner R, Haller D, Bose K, Vos T, Ferguson J, Mathers C (2009) Global patterns of mortality in young people: a systematic analysis of population health data. Lancet 374(9693):881–892

Pedersen A, Alitzer S, Poss M, Cunningham A, Nunn C (2005) Patterns of host specificity and transmission among parasites of wild primates. Int J Parasitol 35:647–657

Peeters M, Courgnaud V, Abela B, Auzel P, Pourrut X, Bibollet-Ruche F, Loul S, Liegeois F, Butel C, Koulagna D et al (2002) Risk to human health from a plethora of simian immunodeficiency viruses in primate bushmeat. Emerg Infect Dis 8(5):451–457

Pepin J (2011) The origins of AIDS. Cambridge University Press, Cambridge

Plantier J, Leoz M, Dickerson J, Oliveira FD, Cordonnier F, Lemée V, Damond F, Robertson D, Simon F (2009) A new human immunodeficiency virus derived from gorillas. Nat Med 15(8):871–872

Prugnolle F, Durand P, Neel C, Ollomo B, Ayala F, Arnathau C, Etienne L, Mpoudi-Ngole E, Nkoghe D, Leroy E et al (2010) African great apes are natural hosts of multiple related malaria species, including *Plasmodium falciparum*. Proc Natl Acad Sci 107(4):1458–1463

Prugnolle F, Durand P, Ollomo B, Duval L, Ariey F, Arnathau C, Gonzalez J, Leroy E, Renaud F (2011) A fresh look at the origin of *Plasmodium falciparum*, the most malignant malaria agent. PLoS Pathog 7(2):e1001283

Reed D, Smith V, Hammond S, Rogers A, Clayton D (2004) Genetic analysis of lice supports direct contact between modern and archaic humans. PLoS Biol 2(11):e340

Rich S, Leendertz F, Xu G, LeBreton M, Djoko C, Aminake M, Takang E, Diffo J, Pike B, Rosenthal B et al (2009) The origin of malignant malaria. Proc Natl Acad Sci 106(35):14902–14907

Roumagnac P, Weill F, Dolecek C, Baker S, Brisse S, Chinh N, Le T, Acosta C, Farrar J, Dougan G et al (2006) Evolutionary history of *Salmonella typhi*. Science 314:1301–1304

Rouquet P, Froment J, Bermejo M, Yaba P, Delicat A, Rollin P, Leroy E (2005) Wild animal mortality monitoring and human Ebola outbreaks, Gabon and Republic of Congo, 2001–2003. Emerg Infect Dis 11:283–290

Ruch T (1959) Diseases of laboratory primates. W.B. Saunders, Philadelphia, PA

Ryan S, Walsh P (2011) Consequences of non-intervention for infectious disease in African great apes. PLoS One 6(12):e29030

Sabeti P, Reich D, Higgins J, Levine H, Richter D, Schaffner S, Gabriel S, Platko J, Patterson N, McDonald G et al (2002) Detecting recent positive selection in the human genome from haplotype structure. Nature 419(6909):832–837

Sharp P, Hahn B (2010) The evolution of HIV-1 and the origin of AIDS. Philos Trans R Soc B 365:2487–2494

Smith K, Anthony S, Switzer W, Epstein J, Seimon T, Jia H, Sanchez M, Huynh T, Galland G, Shapiro S et al (2012) Zoonotic viruses associated with illegally imported wildlife products. PLoS One 7(1):e29505

Snow R, Guerra C, Noor A, Myint H, Hay S (2005) The global distribution of clinical episodes of *Plasmodium falciparum* malaria. Nature 434:214–217

Souquière S, Bibollet-Ruche F, Robertson D, Makuwa M, Apetrei C, Onanga R, Kornfeld C, Plantier J, Gao F, Abernethy K et al (2001) Wild *Mandrillus sphinx* are carriers of two types of lentivirus. J Virol 75(15):7086–7096

Sprent J (1962) Parasitism, immunity and evolution. In: Leeper G (ed) The evolution of living organisms. Melbourne University Press, Melbourne, pp 149–165

Sprent J (1969a) Evolutionary aspects of immunity of zooparasitic infections. In: Jackson G (ed) Immunity to parasitic animals. Appleton, New York, pp 3–64

Sprent J (1969b) Helminth "zoonoses": an analysis. Helminthol Abstract 38:333–351

Suleman M, Johnson B, Tarara R, Sayer P, Ochieng D, Muli J, Mbete E, Tukei P, Ndirangu D, Kago S et al (1984) An outbreak of poliomyelitis caused by poliovirus type I in captive black and white colobus monkeys (*Colobus abyssinicus kikuyuensis*) in Kenya. Trans R Soc Trop Med Hyg 78(5):665–669

Switzer W, Garcia A, Yang C, Wright A, Kalish M, Folks T, Heneine W (2008) Coinfection with HIV-1 and simian foamy virus in West Central Africans. J Infect Dis 197(10):1389–1393

Tishkoff S, Varkonyi R, Cahinhinan N, Abbes S, Argyropoulos G, Destro-Bisol G, Drousiotou A, Dangerfield B, Lefranc G, Loiselet J et al (2001) Haplotype diversity and linkage disequilibrium at human G6PD: recent origin of alleles that confer malarial resistance. Science 293(5529):455–462

Tong S, Singh J, Ruone S, Humphrey C, Yip C, Lau S, Anderson L, Kaur T (2010) Identification of adenoviruses in fecal specimens from wild chimpanzees (*Pan troglodytes schweinfurthii*) in Western Tanzania. Am J Trop Med Hyg 82(5):967–970

UNAIDS (2010) UNAIDS Report on the Global Epidemic 2010. Joint United Nations Programme on HIV/AIDS

Ushida K, Uwatoko Y, Adachi Y, Gaspard Soumah A, Matsuzawa T (2010) Isolation of Bifidobacteria from feces of chimpanzees in the wild. J Gen Appl Microbiol 56(1):57–60

Van Heuverswyn F, Li Y, Neel C, Bailes E, Keele B, Liu W, Loul S, Butel C, Liegeois F, Beinvenue Y et al (2006) SIV infection in wild gorillas. Nature 444:164

Vigilant L, Hofreiter M, Siedel H, Boesch C (2001) Paternity and relatedness in wild chimpanzee communities. Proc Natl Acad Sci 98:12890–12895

Wallis J (2000) Prevention of disease transmission in primate conservation. Ann N Acad Sci 916:691–693

Weber M, Mulholland E, Greenwood B (1998) Respiratory syncytial virus infection in tropical and developing countries. Trop Med Int Health 3:268–280

Weiss R (2009) Apes, lice, and prehistory. J Biol 8(2):20–28

Weiss R, Wrangham R (1999) From pan to pandemic. Nature 397:385–386

WHO (2010) World Malaria Report 2010. Geneva

Wolfe N, Escalante A, Karesh W, Kilbourn A, Spielman A, Lal A (1998) Wild primate populations in emerging infectious disease research: the missing link? Emerg Infect Dis 4(2):149–158

Wolfe N, Prosser A, Carr J, Tamoufe U, Mpoudi-Ngole E, Torimiro J, LeBreton M, McCutchan F, Birx D, Burke D (2004a) Exposure to nonhuman primates in rural Cameroon. Emerg Infect Dis 10(12):2094–2099

Wolfe N, Switzer W, Carr J, Bhullar V, Shanmugam V, Tamoufe U, Prosser A, Torimiro J, Wright A, Mpoudi-Ngole E et al (2004b) Naturally acquired simian retrovirus infections in central African hunters. Lancet 363(9413):932–937

Wolfe N, Heneine W, Carr J, Garcia A, Shanmugam V, Tamoufe U, Torimiro J, Prosser A, LeBreton M, Mpoudi-Ngole E et al (2005) Emergence of unique primate T-lymphotropic viruses among central African bushmeat hunters. Proc Natl Acad Sci 102(22):7994–7999

Wolfe N, Panosian Dunavan C, Diamond J (2007) Origins of major human infectious diseases. Nature 447:279–283

Woodford M, Butynski T, Karesh W (2002) Habituating the great apes: the disease risks. Oryx 36:153–160

Index

J.F. Brinkworth and K. Pechenkina (eds.), *Primates, Pathogens, and Evolution*, Developments in Primatology: Progress and Prospects,
DOI 10.1007/978-1-4614-7181-3,

© Springer Science+Business Media New York 2013

I

J

K

R

Printed by Printforce, the Netherlands